The Neurobiological Basis of Memory

Pamela A. Jackson • Andrea A. Chiba
Robert F. Berman • Michael E. Ragozzino
Editors

The Neurobiological Basis of Memory

A System, Attribute, and Process Analysis

Springer

Editors
Pamela A. Jackson
Department of Psychology
Radford University
Radford
Virginia
USA

Andrea A. Chiba
Department of Cognitive Science
University of California, San Diego
La Jolla
California
USA

Robert F. Berman
Department of Neurological Surgery
and Center for Neuroscience
University of California, Davis
Davis
California
USA

Michael E. Ragozzino
Department of Psychology
University of Illinois at Chicago
Chicago
Illinois
USA

ISBN 978-3-319-15758-0 ISBN 978-3-319-15759-7(eBook)
DOI 10.1007/978-3-319-15759-7

Library of Congress Control Number: 2015942459

Springer Cham Heidelberg New York Dordrecht London

Printed on acid-free paper

Springer International Publishing AG Switzerland is part of Springer Science+Business Media (www.springer.com)

Preface

This book is both a tribute to the pioneering research on the neurobiology of learning and memory carried out by Raymond P. Kesner and a summary of much of the current thinking about the nature and organization of memory systems in the brain. The book was a direct outgrowth from a Festschrift held in Ray Kesner's honor on January 2nd and 3rd, 2013 in Salt Lake City, Utah. The speakers and attendees at that event included numerous colleagues and collaborators, as well as many of the students and postdoctoral researchers who have worked and interacted with Ray over more than 40 years of his career (see photograph of conference participants below). There are 18 chapters, including a summary of the "Attribute Model of Memory" by Ray Kesner, a personal account of his life and career, a chapter with letters and comments about Ray and his work from many of the major researchers in the field of memory and learning, and an epilogue. The book is organized into four major sections. The first section contains chapters focusing on the role of the hippocampus in processing spatial and temporal attributes of memory. The second section moves beyond the hippocampus to consider how neural activity in limbic cortex, prefrontal cortex, and basal ganglia contributes to memory and behavioral flexibility. The third section reviews current research applying basic concepts of the Kesner Attribute Model to understanding neurological disorders, including traumatic brain injury, Huntington's disease, and Fragile X-related disorders. While each chapter reflects the current research of the authors, each also attempts to place their research within the general context of multiple memory systems in the brain and, in particular, the attribute model proposed by Kesner. The fourth section contains personal tributes to the life and scientific work of Ray Kesner.

Ray's interest in the neuroanatomical substrates of memory began while a graduate student with Garth Thomas at the University of Illinois in the mid-1960s, where he studied the role of the midbrain reticular formation in learning. This was followed by postdoctoral training with Robert Doty at the University of Rochester where he found that mild seizures induced by electrical stimulation in the amygdala or hippocampus of cats resulted in amnesia. Ray began his career as an assistant professor in the Department of Psychology at the University of Utah where he remained focused on memory and learning throughout his career. Ray maintained an active, visible, and productive research program for more than 40 years, publishing more than 250 peer-reviewed publications and more than 80 chapters on the neurobiology of memory and learning. He was an early proponent of the idea of multiple memory systems in the brain, has played a major role in the development of these ideas, and has provided much of the scientific discoveries to support these ideas. He was one of the first to incorporate the concepts and principles of cognitive neuroscience into his thinking and experiments. One of Ray's important contributions was the recognition that behavioral tests used to assess human memory could be used to explore the neurobiology of memory in animal models and, conversely, tests of memory developed for animal studies could also be adapted for use in humans. This aspect of Ray's research resulted in the ability to better translate results between animal models and humans, and represents a significant advance for the field. Ray's body of work on memory spans research in animal models and in humans, including patients with hypoxic memory loss. Ray, as reflected in these chapters, was also an excellent mentor and trained many individuals who have gone on to develop independent research careers. These include the four editors of this book, Rob Berman (doctoral student from 1972 to 1977), University of California Davis, Pam Jackson (postdoctoral fellow), Radford University, Mike Ragozzino (postdoctoral fellow), University of Illinois at Chicago, and Andrea Chiba (doctoral student), University of California at San Diego, as well as several of the chapter authors including: Ramona Hopkins (doctoral student), Brigham Young University; Paul Gilbert (doctoral student), San Diego State University; Inah Lee (doctoral student), Seoul National University; Bill DeCoteau (doctoral student), St. Lawrence University; Yoon Cho (postdoctoral fellow), University of Bordeaux; Ryan Hunsaker (undergraduate student), University of Utah; Brock Kirwan (undergraduate student), Brigham Young University; Christy Weeden (doctoral student), NIMH.

Kesner Festschrift attendees matched to photo:

1. Dipan Patel
2. Kasia Bieszczad
3. Timothy Allen
4. Arnold Bakker
5. Ashley Fricks-Gleason
6. Andrea Chiba
7. Norbert J. Fortin
8. Joe L. Martinez, Jr.
9. Craig Stark
10. Brock Kirwan
11. Paul E. Gilbert
12. Jeffrey B. Rosen
13. Michael A. Yassa
14. Mike E. Ragozzino
15. Karen Wilcox
16. Matthew Shapiro
17. Howard Eichenbaum
18. Wayne Pratt
19. Jared Roberts
20. Julie R. Korenberg
21. James L. McGaugh
22. Kristen A. Keefe
23. Raymond P. Kesner
24. Thomas N. Parks
25. Inah Lee
26. Mark E. Stanton
27. Joe Williams
28. M. Ryan Hunsaker
29. William DeCoteau
30. Veronique Bohbot
31. Naomi Goodrich-Hunsaker
32. Jeffrey Long
33. Sheri J. Y. Mizumori
34. Robert F. Berman
35. Jeanine Stefanucci
36. David Cook
37. Brent Cooper
38. Sarah Creem-Regehr
39. Pamela Jackson
40. Emmaline Smith
41. Greg Clark
42. Ramona Hopkins

Acknowledgements

Dear Pam, Andrea, Rob and Mike,

I want to thank you from the bottom of my heart for your hard work and for creating an incredible and unique book with a prologue, eighteen chapters, epilogue, letters, pictures, and a biography. It has been an honor for me and it has increased my admiration for the effort that you put into the development of the book. This book will represent a life-time memory of the years we spent in and outside the lab in trying to understand the neurobiological basis of memory in animals and humans. I also want to thank you for setting up the Festschrift and I want to thank the people who came to present their research in the context of my work and to roast me a bit. Finally, I appreciate the contributions made by all the authors who submitted a chapter.

Sincerely, Ray

The editors would like to acknowledge the support of the Interacting Memory Systems Network, a research network of the Temporal Dynamics of Learning Center[1]. Their generous financial and staff support of the Kesner Festschrift research sessions helped to lay the foundation for this volume. Likewise, the editors acknowledge the generous support from Radford University both for support of the Festschrift and for the valuable time of the senior editor of the volume, Prof. Pamela Jackson. Additionally, the editors would like to thank Prof. Thomas Parks, VP for Research at the University of Utah for his generosity in providing University support for the Festchrift. Finally, the editors would like to thank the Kesner Family (Laya, Benjamin, and Debbie) for their kind support and collection of photos.

[1] Support provided by NSF grant SMA 1041755 to the Temporal Dynamics of Learning Center, an NSF Science of Learning Center.

Contents

Contributors

Phillip M. Baker Department of Psychology, University of Washington, Seattle, WA, USA

Robert F. Berman Department of Neurological Surgery, University of California, Davis, Davis, CA, USA

Andrea A. Chiba Department of Cognitive Science and Program in Neuroscience, Temporal Dynamics of Learning Center, University of California, San Diego, La Jolla, CA, USA

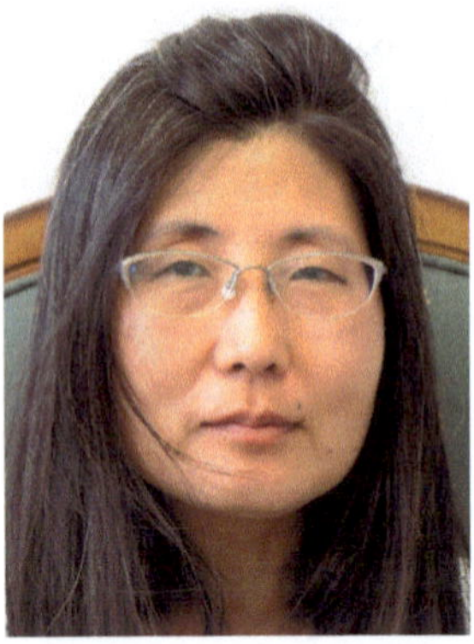

Yoon H. Cho Institut de Neurosciences Cognitives et Intégratives d'Aquitaine, University of Bordeaux, Talence, Cedex, France

Howard Eichenbaum Center for Memory and Brain, Boston University, Boston, MA, USA

Paul E. Gilbert Department of Psychology, San Diego State University, San Diego, CA, USA

Joint Doctoral Program in Clinical Psychology, San Diego State University—University of California San Diego, San Diego, CA, USA

Paul E. Gold Department of Biology, Syracuse University, Syracuse, NY, USA

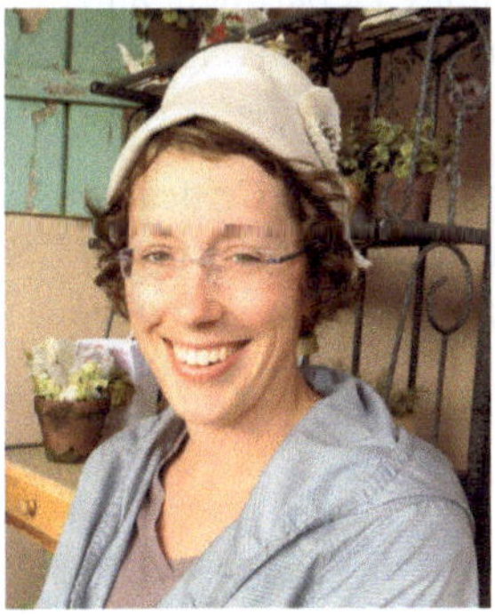

Naomi J. Goodrich-Hunsaker Department of Psychiatry and Behavioral Sciences, MIND Institute, University of California, Davis Medical Center, Sacramento, CA, USA

Department of Psychology, Brigham Young University, Provo, UT, USA

Gene G. Gurkoff Department of Neurological Surgery, University of California, Davis, Davis, CA, USA

Heather M. Holden Joint Doctoral Program in Clinical Psychology, San Diego State University—University of California, San Diego, San Diego, CA, USA

Ramona O. Hopkins Psychology Department and Neuroscience Center, Brigham Young University, Provo, UT, USA

Department of Medicine, Pulmonary and Critical Care Division, Intermountain Medical Center, Murray, UT, USA

Michael R. Hunsaker Center for Integrative Neuroscience and Human Behavior, University of Utah, Salt Lake City, UT, USA

Department of Neurological Surgery, University of California, Davis, Davis, CA, USA

Granite School District, 2500 S. State Street, Salt Lake City, UT, USA

Pamela A. Jackson Department of Psychology, Radford University, Radford, VA, USA

Yannick Jeantet Institut de Neurosciences Cognitives et Intégratives d'Aquitaine, University of Bordeaux, Talence, Cedex, France

Raymond P. Kesner Department of Psychology, University of Utah, Salt Lake City, UT, USA

C. Brock Kirwan Department of Psychology and Neuroscience Center, Brigham Young University, Provo, UT, USA

Robert Komorowski Picower Institute for Learning and Memory, Massachusetts Institute of Technology, Cambridge, MA, USA

Benjamin J. Kraus Center for Memory and Brain, Boston University, Boston, MA, USA

The MathWorks, Inc., Natick, MA, USA

Choong-Hee Lee Department of Brain and Cognitive Sciences, Laboratory for Behavioral Neurophysiology, Seoul National University, Seoul, Republic of Korea

Inah Lee Department of Brain and Cognitive Sciences, Laboratory for Behavioral Neurophysiology, Seoul National University, Seoul, Republic of Korea

Bruce G. Lyeth Department of Neurological Surgery, University of California, Davis, Davis, CA, USA

Christopher J. MacDonald Picower Institute for Learning and Memory, RIKEN-MIT Center for Neural Circuit Genetics, Massachusetts Institute of Technology, Cambridge, MA, USA

Sheri J. Y. Mizumori Department of Psychology, University of Washington, Seattle, WA, USA

Andrea M. Morris Department of Health Policy and Management, University of California, Los Angeles, CA, USA

Michelle I. Nash Department of Psychology and Neuroscience Center, Brigham Young University, Provo, UT, USA

Laleh K. Quinn Department of Cognitive Science, University of California, San Diego, La Jolla, USA

Michael E. Ragozzino Department of Psychology, University of Illinois at Chicago, Chicago, IL, USA

Lara M. Rangel NSF Cognitive Rhythms Collaborative, College of Arts and Sciences Psychology, Boston University, Boston, MA, USA

Jonathan Robitsek Department of Surgery, Jamaica Hospital Medical Center, Jamaica, NY, USA

Edmund T. Rolls Department of Computer Science, University of Warwick, Coventry, UK

Kiarash Shahlaie Department of Neurological Surgery, U.C Davis School of Medicine, Sacramento, CA, USA

David P. Sheppard Department of Psychology, San Diego State University, San Diego, CA, USA

Joint Doctoral Program in Clinical Psychology, San Diego State University—University of California San Diego, San Diego, CA, USA

Department of Psychology, University of Houston, Houston, TX, USA

Christy S. S. Weeden Section on Neuroplasticity, National Institute of Mental Health, National Institutes of Health, Bethesda, MD, USA

Chapter 1
Exploration of the Neurobiological Basis for a Three-System, Multi-attribute Model of Memory

Raymond P. Kesner

The structure and utilization of memory is central to one's knowledge of the past, interpretation of the present, and prediction of the future. Therefore, the understanding of the structural and process components of memory systems at the psychological and neurobiological level is of paramount importance. There have been a number of attempts to divide learning and memory into multiple memory systems. Schacter and Tulving (1994) have suggested that one needs to define memory systems in terms of the kind of information to be represented, the processes associated with the operation of each system, and the neurobiological substrates, including neural structures and mechanisms, that subserve each system. Furthermore, it is likely that within each system there are multiple forms or subsystems associated with each memory system and there are likely to be multiple processes that define the operation of each system. Finally, there are probably multiple neural structures that form the overall substrate of a memory system.

The first model of hippocampal function and the processing of spatial information was described by O'Keefe and Nadel (1978; see Nadel 1994 as well). They developed a memory model with a concentration on space as the critical attribute of specific memories. They further divided the spatial attribute into a locale system, which codes places in the environment into cognitive maps, and a taxon system, which codes motor responses in terms of specific orientations within a spatial environment. In terms of neural mediation of the locale versus taxon system, they propose that the hippocampus is important in mediating only one form of memory, namely spatial, within the locale system and other neural regions as important for subserving the taxon system. With respect to the operation of each system, it was assumed that learning within the locale system is based in part on consolidation processes and is (a) all-or-none, (b) sensitive to interference, (c) involved in separating traces, and (d) flexible, whereas learning in the taxon system is (a) incremental, (b) not sensitive to interference, (c) involved in combining traces, and (d) not flexible.

R. P. Kesner (✉)
Department of Psychology, University of Utah, 380 S 1530 E,
Rm 502, Salt Lake City, UT 84112, USA
e-mail: ray.kesner@psych.utah.edu

P. A. Jackson et al. (eds.), *The Neurobiological Basis of Memory*,
DOI 10.1007/978-3-319-15759-7_1

Even though the hippocampus was assumed to be the mediator to the locale system, the neural circuit subserving spatial information does include a number of neural regions such as the entorhinal cortex, the retrosplenial cortex, the pre-, para-, and postsubiculum, the parietal cortex, and the pre- and infralimbic cortex. Nadel's focus on the hippocampus might be too limiting. The taxon system is large and needs to be differentiated. Furthermore, a genuine neurobiological system analysis requires the identification of neural regions that subserve the response component associated with the taxon system. However, there is no mention of a memory contribution of the prefrontal cortex (PFC) and there is no mention of other brain areas that support memory for other attributes (e.g., amygdala and affect attribute).

A second model of hippocampal function and the processing of spatial information was presented by Olton (1983). He proposed a somewhat different system emphasizing more the importance of process. He suggested that within every learning task there are two types of memories that organize the critical information into two systems, labeled working memory and reference memory (Olton 1983). He suggested that the specific, personal, and temporal context of a situation is coded in working memory. This would translate into memory for events that occur on a specific trial in a task, biasing mnemonic coding toward the processing of incoming data. In contrast, information concerning rules and procedures (general knowledge) of specific situations is coded in reference memory. This would translate into memory for events that happen on all trials in a task, biasing mnemonic coding toward the processing of expectancies based on the organization of the extant memory. The working versus reference memory system emphasizes the role of the hippocampus and interconnected neural systems as the critical substrate of memory for a single process, namely working memory, and the neocortex as the critical neural substrate within reference memory for all forms or attributes of memory. It was assumed that the two memory systems are independent of each other. Different terms have been used to reflect the same distinction including episodic versus semantic memory (Tulving 1983).

The Olton model has some limits in that the emphasis is placed only on the hippocampus and interconnected neural circuits as the neural system subserving working memory for all information. However, it is clear that in the Olton model the hippocampus is limited to working memory for only spatial, temporal, and linguistic information. There is no mention of a memory contribution of the PFC and there is no mention of other brain areas that support memory for other attributes (e.g., caudate and the response attribute). Furthermore, the hippocampus is also involved in processes other than short-term or working memory, such as pattern separation, consolidation, and retrieval of information (Kesner 1996).

A third model and the most popular model of memory was presented by Squire (1994; Squire et al. 2004) and can be characterized as a dual memory system with an emphasis on the hippocampus and medial temporal lobe including perirhinal cortex, parahippocampal cortex, and entorhinal cortex for one component of the model and a composite of other brain structures as the other component. For example, they have suggested that memory can be divided into a medial temporal lobe dependent declarative memory which provides for conscious recollection of facts and events, and a non-hippocampal dependent non-declarative memory which provides for memory without conscious access for skills and habits mediated by the caudate nucleus and interconnected systems. Furthermore, priming is mediated by the neocortex, simple classical conditioning of emotional responses by the amygdala, simple classical conditioning of skeletal musculature by the cerebellum, and nonassociative learning is mediated by reflex pathways. A limitation is that there is no mention of the PFC contribution to memory, in the context of declarative memory different attributes mediated by the amygdala or caudate do not play a role, and the emphasis is primarily on one single process, namely consolidation. Different models have used different terms to reflect the same type of distinction, including a hippocampal dependent explicit memory verses a non-hippocampal dependent implicit memory (Schacter 1987).

A fourth model was presented by Eichenbaum (Cohen and Eichenbaum 1993; Eichenbaum 1994, 2004). They proposed that the declarative memory system is dependent on the hippocampus and provides for a substrate for relational representation of all forms of memory as well as representational flexibility allowing for the retrieval of memories in novel situations. Relational processing is carried out by the hippocampus, but the processing of individual items resides in the perirhinal and parahippocampal cortex. In contrast, a non-declarative system is independent of the hippocampus and is characterized by individual representations and inflexibility in retrieving memories in novel situations. The limitations include that there is no mention of a memory contribution of the PFC as part of the model (see Kesner and Churchwell 2011). Also, there is not enough emphasis on different attributes of memory, and processes such as pattern separation and pattern completion are not incorporated in the model.

Because memory is complex and involves many neural systems in addition to the hippocampus, Kesner (2007) has proposed a three-system (event-based, knowledge-based, and rule-based) multiple attribute-based theoretical model of memory. The model is an extension of models presented above. For example, I have accepted Olton's working–reference memory and Tulving's episodic–semantic dual memory model distinctions and labeled them as event-based memory verses knowledge-based memory, but in addition I have added a third rule-based system subserved by mnemonic processes associated with the PFC. I have also adopted the attribute model described by Underwood (1969) and Spear (1976). They presented a good case that there are many different forms or attributes of memory such as *space, time, response, sensory-perception, reward value (affect), and language*. These attributes are processed by different neural regions and interconnected networks across all three (event-based, knowledge-based, and rule-based) memory systems. This is an enrichment of the previous mentioned memory models that emphasize one or two attributes or do not differentiate among attributes. Finally, each memory system operates in processing mnemonic information based on a unique set of processes that involve more than just consolidation. The selection of some of these processes has been influenced greatly by computational models of specific brain regions (see Rolls and Kesner 2006).

In the three system, multi-attribute model of memory one can characterize each system as composed of the same set of multiple attributes or forms of memory, characterized by a set of process-oriented operating characteristics and mapped onto multiple neural regions and interconnected neural circuits (for more detail see Kesner 1998b, Kesner 2007).

On a psychological level (see Fig. 1.1), the event-based memory system provides for temporary representations of incoming data concerning the present, with an emphasis upon data and events that are usually personal or egocentric and that occur within specific external and internal contexts. The emphasis is upon the processing of new and current information. During initial learning great emphasis is placed on the event-based memory system, which will continue to be of importance even after initial learning in situations where unique or novel trial information needs to be remembered. This system is akin to episodic memory (Tulving 1983) and some aspects of declarative memory (Squire 1994).

The knowledge-based memory system (see Fig. 1.2) provides for more permanent representations of previously stored information in long-term memory and can be thought of as one's general knowledge of the world. The knowledge-based memory system would tend to be of greater importance after a task has been learned given that the situation is invariant and familiar. The organization of these attributes within the knowledge-based memory system can take many forms and are organized as a set of attribute-dependent cognitive maps and their interactions that are unique for each memory. This system is akin to semantic memory (Tulving 1983).

The rule-based memory system (see Fig. 1.3) receives information from the event-based and knowledge-based systems and integrates the information by applying rules and strategies and decisions for subsequent action. In every learning and memory task the subject has to select an appropriate strategy or set of rules to aid in memory consolidation of the task. The processes associated with rule-based memory are most likely mediated by the PFC. In most situations, however, one would expect a contribution of all three systems with a varying proportion of involvement of one relative to the other.

The three memory systems are composed of the same forms, domains, or attributes of memory. Even though there could be many attributes, the most important attributes include *space, time, response, sensory-perception, and reward value (affect)*. In humans a *language* attribute is also added. A spatial (space) attribute within this framework involves memory representations of places or relationships between places. It is exemplified by the ability to encode and remember spatial maps and to localize stimuli in external space. Memory representations of the spatial attribute can be further subdivided into specific spatial features including allocentric spatial distance, egocentric spatial distance, allocentric direction, egocentric direction, and spatial location. A temporal (time) attribute within this framework involves memory representations of the duration of a stimulus, the succession or temporal order of temporally separated events or stimuli, and memory representations of the past. A response attribute within this framework involves memory representations based on feedback from motor responses (often based on proprioceptive and vestibular cues) that occur in specific situations as well as memory representations of stimu-

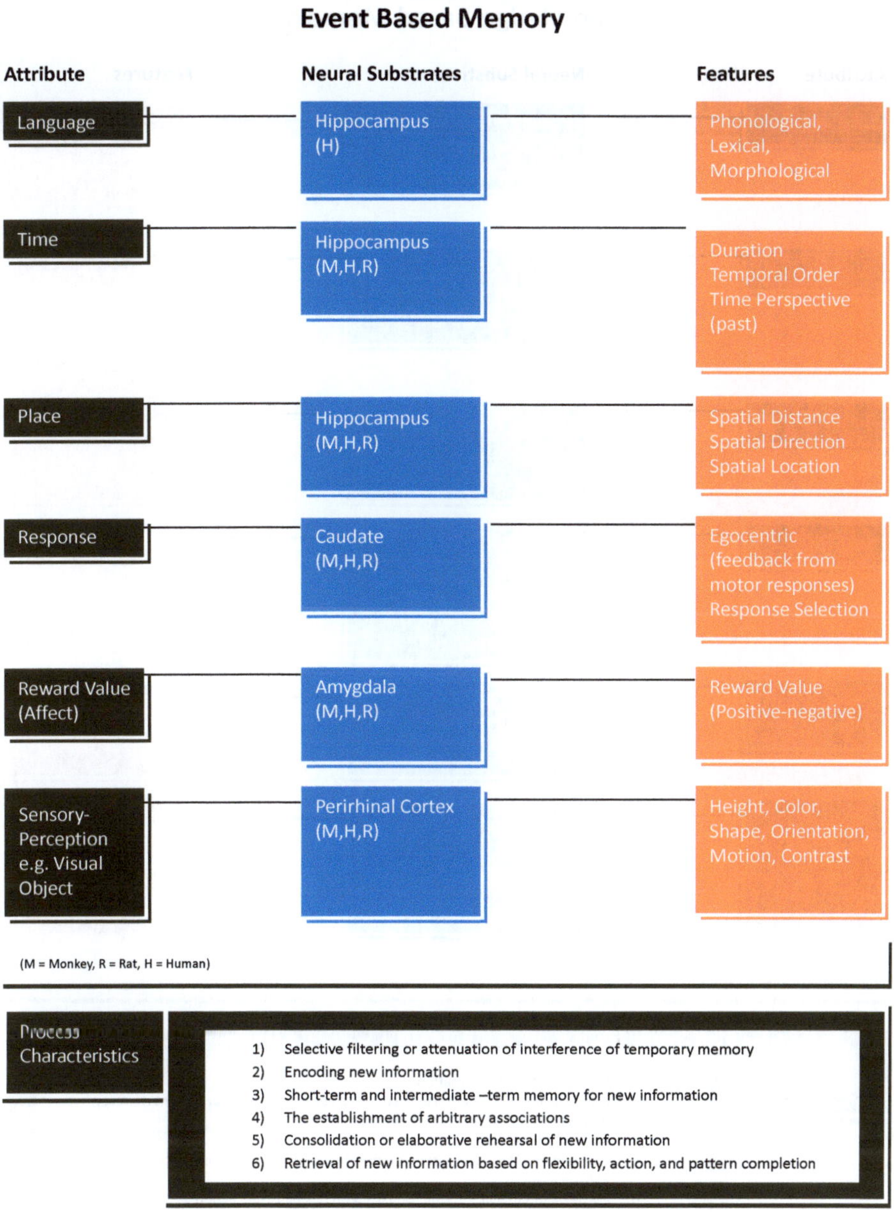

Fig. 1.1 Representation of the neural substrates, features, and process characteristics associated with the event-based memory system for the language, time, place, response, value (affect), and sensory-perception attributes

lus–response associations. A reward value (affect) attribute within this framework involves memory representations of a hedonic continuum of positive and negative values and the associations between stimuli and rewards. A sensory-perceptual attribute within this framework involves memory representations of a set of sensory

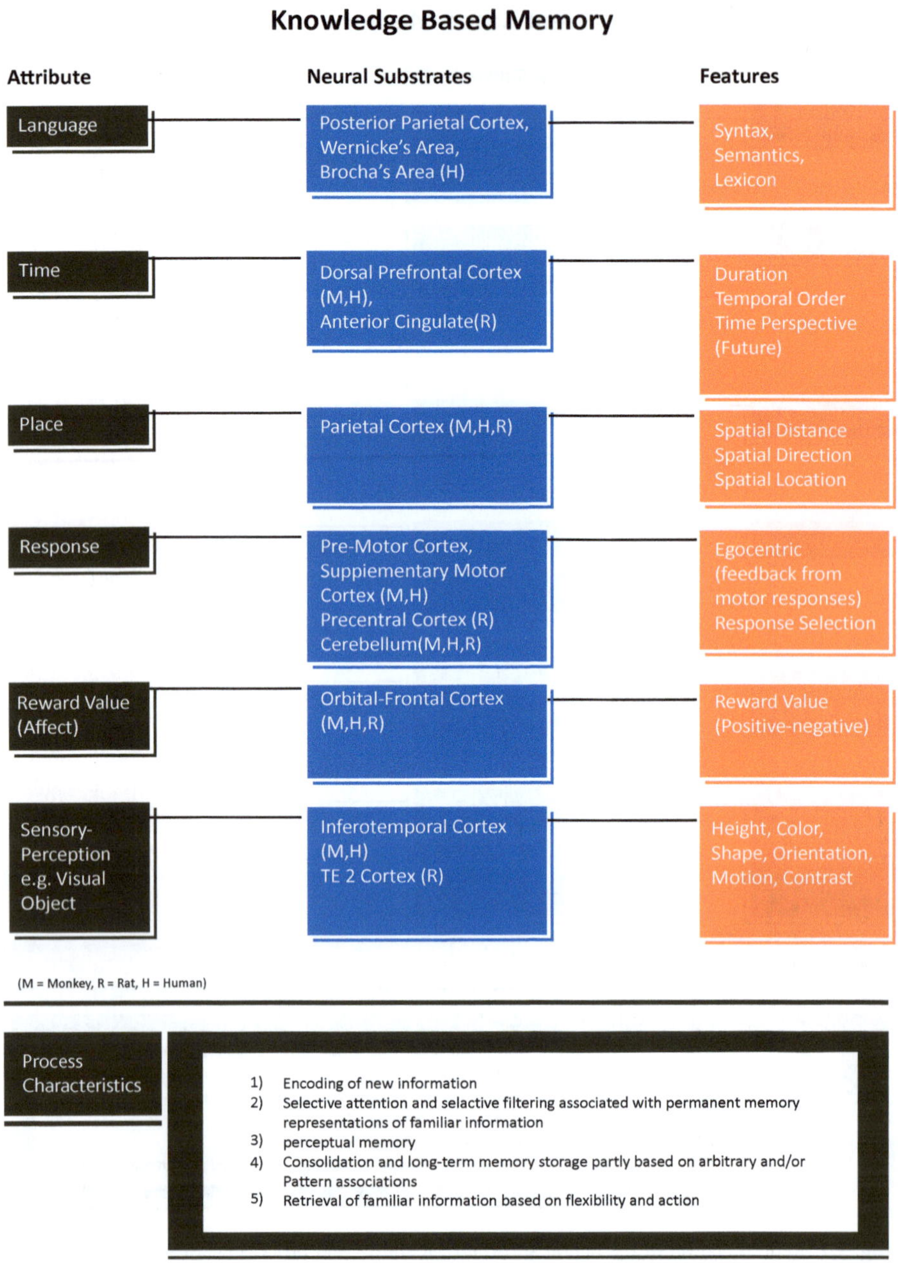

Fig. 1.2 Representation of the neural substrates, features, and process characteristics associated with the knowledge-based memory system for the language, time, place, response, value (affect), and sensory-perception attributes

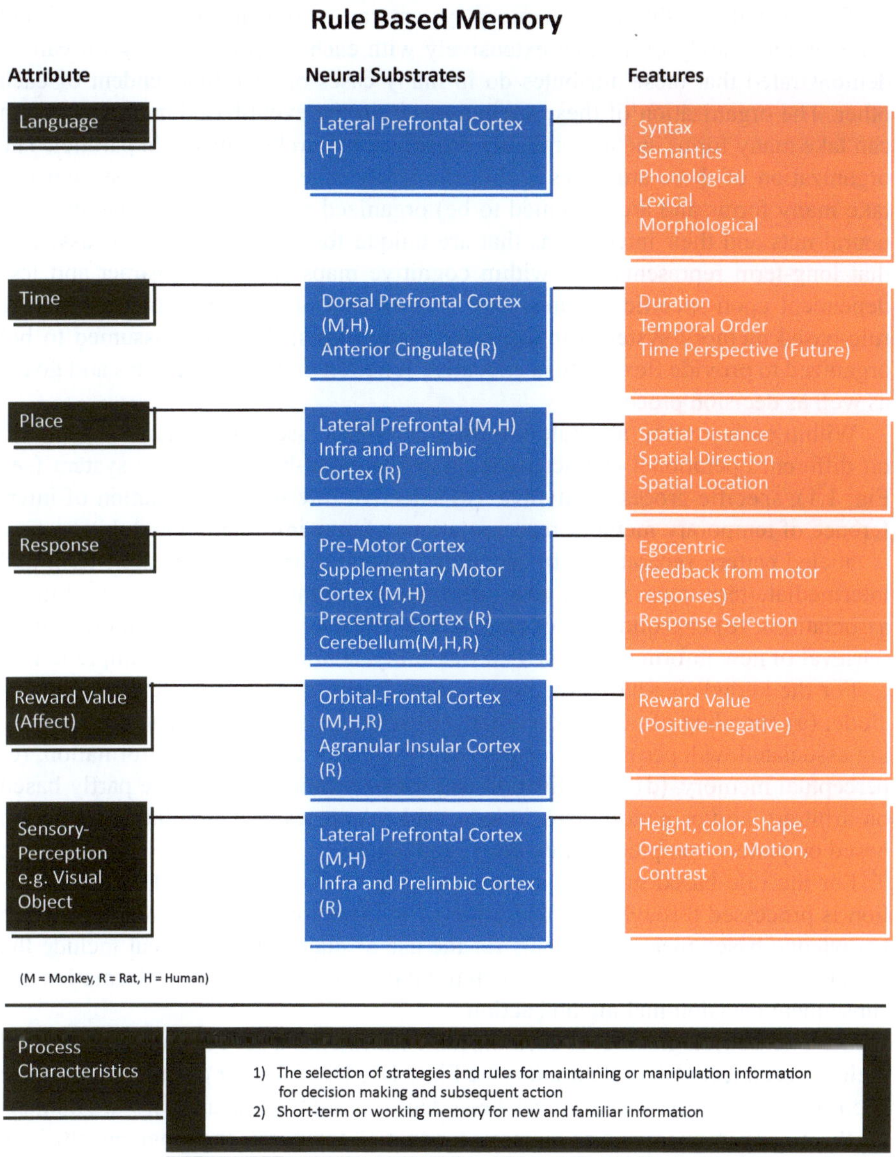

Fig. 1.3 Representation of the neural substrates, features, and process characteristics associated with the rule-based memory system for the language, time, place, response, value (affect), and sensory-perception attributes

stimuli that are organized in the form of cues as part of a specific experience. Each sensory modality (olfaction, auditory, vision, somatosensory, and taste) can be considered part of the sensory-perceptual attribute component of memory. A language attribute within this framework involves memory representations of phonological, lexical, morphological, syntactical, and semantic information.

The attributes within each memory system can be organized in many different ways and are likely to interact extensively with each other even though it can be demonstrated that these attributes do in many cases operate independent of each other. The organization of these attributes within the event-based memory system can take many forms and are probably organized hierarchically and in parallel. The organization of these attributes within the knowledge-based memory system can take many forms and are (assumed to be) organized as a set of cognitive maps or neural nets and their interactions that are unique for each memory. It is assumed that long-term representations within cognitive maps are more abstract and less dependent upon specific features. The organization of these attributes within the rule-based memory system can also take many forms; these are (assumed to be) organized to provide flexibility in executive function in developing rules and goals, as well as decision processes.

Within each system, attribute information is processed in different ways based on different operational characteristics. For the event-based memory system (see Fig. 1.1), specific processes involve: (a) selective filtering or attenuation of interference of temporary memory representations of new information and this process is labeled pattern separation, (b) encoding of new information, (c) short-term and intermediate-term memory for new information, (d) the establishment of arbitrary associations, (e) consolidation or elaborative rehearsal of new information, and (f) retrieval of new information based on flexibility, action, and pattern completion.

For the knowledge-based memory system (see Fig. 1.2), specific processes include: (a) encoding of new information, (b) selective attention and selective filtering associated with permanent memory representations of familiar information, (c) perceptual memory, (d) consolidation and long-term memory storage partly based on arbitrary and/or pattern associations, and (e) retrieval of familiar information based on pattern completion, flexibility, and action.

For the rule-based memory system (see Fig. 1.3), it is assumed that information is processed through the integration of information from the event-based and knowledge-based memory systems for the use of major processes that include the selection of strategies and rules for maintaining or manipulating information for subsequent decision-making and action.

On a neurobiological level each attribute maps onto a set of neural regions and their interconnected neural circuits (see Figs. 1.1, 1.2, 1.3). For example, within the event-based memory system, it has been demonstrated that in animals and humans (a) the hippocampus supports memory for spatial, temporal, and language attribute information, (b) the caudate mediates memory for response attribute information, (c) the amygdala subserves memory for reward value (affect) attribute information, and (d) the perirhinal and extrastriate visual cortex support memory for visual object attribute information as an example of a sensory-perceptual attribute (for more detail see Kesner 1998b, 2007).

Within the knowledge-based memory system, it has been demonstrated that in animals and humans (a) the posterior parietal cortex (PPC) supports memory for spatial attributes, (b) the dorsal and dorsolateral PFC and/or anterior cingulate (AC) support memory for temporal attributes, (c) the premotor, supplementary motor, and cerebellum in monkeys and humans and precentral (PC) cortex and cerebellum in rats support memory for response attributes, (d) the orbital PFC supports memory for reward value (affect) attributes, (e) the inferotemporal cortex in monkeys and humans and TE2 cortex in rats subserves memory for sensory-perceptual attributes, for example, visual objects, and (f) parietal cortex, Broca and Wernicke's areas subserve memory for the language attribute (for more detail see Kesner 1998b, 2007).

Within the rule-based memory system it can be shown that different subdivisions of the PFC support different attributes. For example, (a) the dorsolateral and ventrolateral PFC in humans support spatial, object, and language attributes and the infralimbic and prelimbic (PL) cortex in rats supports spatial and visual object attributes, (b) the premotor and supplementary motor cortex in monkeys and humans and PC cortex in rats support response attributes, (c) the dorsal, dorsolateral, and mid-dorsolateral PFC in monkeys and humans and AC in rats mediate primarily temporal attributes, and (d) the orbital PFC in monkeys and humans and agranular insular cortex in rats support affect attributes (for more detail see Kesner 2000a;2007).

Event-Based Memory System

Given the complexity of memory representations in the brain, how is one to test the neurobiological basis of the attribute model of memory? To test whether different brain regions subserve the processing of different attributes within the event-based memory system, I selected the process of short-term or working memory. The short-term memory task designed to test this consists of a study phase comprising one item (e.g., object, spatial location, motor response, or reward) and then following a delay there is a test phase consisting of two items with one item identical to the study phase and a new item leading to reinforcement for a match or mismatch with the study phase. After the task is learned, lesions of specific neural substrates are used. With this paradigm, it has been shown that there is a triple dissociation among the hippocampus (spatial location), caudate (response), and extra striate visual cortex (visual object; Kesner et al. 1993), a double dissociation between the hippocampus (spatial location) and the amygdala (affect; Gilbert and Kesner 2002a; 2006), a double dissociation between hippocampus (spatial location) and perirhinal cortex (visual object; Gilbert and Kesner 2003a; Kesner 1999), as well as a double dissociation within the hippocampus in terms of spatial (dentate gyrus (DG)) verses temporal (CA1) processing of information (Gilbert et al. 2001). Thus, it appears that within the event-based memory system different neuroanatomical circuits are involved in the processing of different attributes in that they can operate independent of each other.

In subsequent research, I have concentrated on determining the importance of examining multiple processes associated with the event-based memory system, including (1) conjunctive encoding to create a spatial representation, (2) selective filtering or attenuation of interference with encoding of information labeled as pattern separation, especially for spatial location and spatial contextual information, (3) formation of arbitrary associations, (4) retrieval of familiar information based on pattern completion, (5) temporal processing of information including temporal pattern separation, (6) short-term and intermediate-term memory for new information, and (7) promotion of consolidation or elaborative rehearsal of new information. I will concentrate on the different subregions of the hippocampus and will mention other brain areas that subserve the same function for a different attribute given the availability of empirical studies. First, I will examine the role of the DG subregion of the hippocampus in supporting conjunctive encoding to create a spatial representation and selective filtering or attenuation of interference with encoding of information-labeled pattern separation, especially spatial. Second, I will examine the role of the CA3 subregion of the hippocampus in supporting formation of arbitrary associations and retrieval of familiar information based on pattern completion. Third, I will examine the role of the CA1 subregion of the hippocampus in supporting temporal processing of information including temporal pattern separation. I will not discuss intermediate-term memory for new information and promotion of consolidation or elaborative rehearsal of new information because of space limitations.

DG and Conjunctive Encoding

The DG has been shown to receive multiple sensory inputs, including vestibular, olfactory, visual, auditory, and somatosensory, from the perirhinal and lateral entorhinal cortex in conjunction with spatially organized grid cells from the medial entorhinal cortex (Hafting et al. 2005) to represent metric spatial representations. The perforant path input of the DG can be divided into medial and lateral components. The medial component processes spatial information and the lateral component processes nonspatial (e.g., objects, odors) information (Hargreaves et al. 2005; Witter et al. 1989). Based on the idea that the medial perforant path (MPP) input into the DG mediates spatial information via activation of N-methyl-D-aspartate receptor (NMDA) receptors and the lateral perforant path (LPP) input into the DG mediates visual object information via activation of opioid receptors, the following experiment was conducted. Using a paradigm developed by Poucet (1989) rats were tested for detection of a novel spatial change and detection of a novel visual object change while under the influence of direct infusions of AP5 (an NMDA antagonist) or naloxone (a μ-opiate antagonist) into the DG. The results are shown in Fig. 1.4 and indicate that naloxone infusions into the DG disrupted both novelty detection of a spatial location and a visual object, whereas AP5 infusions into the DG disrupted only detection of a novel spatial location, but had no effect on detection of a novel object (Hunsaker et al. 2007). These data suggest that the DG uses conjunctive encoding of visual object and spatial information to provide for a spatial representation that may be based on metric information.

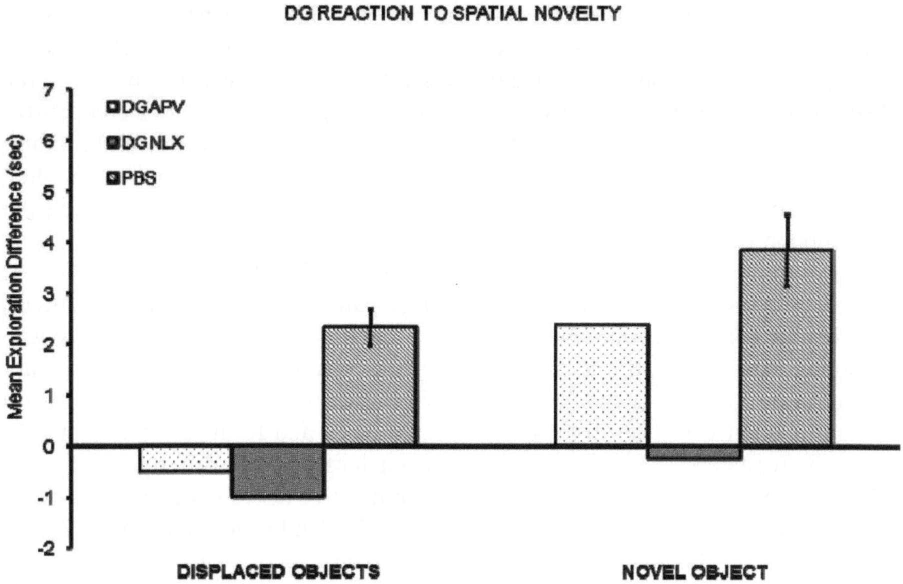

Fig. 1.4 The effects of naloxone (*NLX*), 2-amino-5-phosphonovaleric acid (*APV*), or phosphate buffered saline (*PBS*) infusions within the DG for spatial (*bars* on the *left*) and nonspatial (visual object; *bars* on the *right*) novelty detection

DG and Spatial Pattern Separation

Pattern separation is defined as a process to remove redundancy from similar inputs so that events can be separated from each other and interference can be reduced, and in addition can produce a more orthogonal, sparse, and categorized set of outputs. Computational models have emphasized the importance of the hippocampus in mediating spatial pattern separation, which has been developed extensively by computational models of the subregions of the hippocampus with a special emphasis on the DG. Based on the empirical findings that all sensory inputs are processed by the DG subregion of the hippocampus, it has been suggested that a possible role for the hippocampus might be to provide for sensory markers to demarcate a spatial location, so that the hippocampus can more efficiently mediate spatial information. It is thus possible that one of the main process functions of the hippocampus is to encode and separate spatial locations from each other. This would ensure that new highly processed sensory information is organized within the hippocampus and enhances the possibility of remembering and temporarily storing one place as separate from another place. It is assumed that this is accomplished via pattern separation of spatial information, so that spatial locations can be separated from each other and spatial interference is reduced.

Rolls' (1996) model proposes that pattern separation is facilitated by sparse connections in the mossy fiber system, which connects DG granular cells to CA3 pyramidal neurons. Separation of patterns is accomplished based on the low probability that any two CA3 neurons will receive mossy fiber input synapses from a similar subset of DG cells. Mossy fiber inputs to CA3 from DG are suggested to be essential during learning and may influence which CA3 neurons will fire based on the distributed activity in the DG. Cells of the DG are suggested to act as a competitive learning network with Hebb-like modifiability to reduce redundancy and produce sparse, orthogonal outputs. O'Reilly and McClelland (1994) and Shapiro and Olton (1994) also suggested that the mossy fiber connections between the DG and CA3 may support pattern separation.

To examine the contribution of the DG to spatial pattern separation, Gilbert et al. (2001) tested rats with DG lesions using a paradigm that measured short-term memory for spatial location information as a function of spatial similarity between locations. Specifically, the study was designed to examine the role of the DG subregion in discriminating spatial locations when rats were required to remember a spatial location based on distal environmental cues and to differentiate between the to-be-remembered location and a distractor location with different degrees of similarity or overlap among the distal cues. Rats were tested using a cheeseboard maze apparatus (the cheeseboard is similar to a dry-land water maze with 177 circular, recessed holes on a 119 cm diameter board) on a delayed-match-to-sample for spatial location task. Animals were trained to displace an object that was randomly positioned to cover a baited food well in 1 of 15 locations along a row of food wells. Following a short delay, the animals were required to choose between objects which were identical to the sample phase object: One object was in the same location as the sample phase object and the second object was in a different location along the row of food wells. Rats were rewarded for displacing the object in the same spatial location as the sample phase object (correct choice), but they received no reward for displacing the foil object (incorrect choice). Five spatial separations, from 15 to 105 cm, were used to separate the correct object and the foil object during the choice phase. The results are shown in Fig. 1.5 and indicate that rats with DG lesions were significantly impaired at short spatial separations; however, during the choice phase, performance of DG-lesioned animals increased as a function of greater spatial separation between the correct and foil objects. The performance of rats with DG lesions matched control rats at the largest spatial separation. The graded nature of the impairment and the significant linear improvement in performance as a function of increased separation illustrate a deficit in pattern separation. Based on these results, it was concluded that lesions of the DG decrease the efficiency of spatial pattern separation, which results in impairments on trials with increased spatial proximity and increased spatial similarity among working memory representations. Thus, the DG may function to encode and to separate events in space producing spatial pattern separation. Such a spatial pattern separation ensures that new highly processed sensory information is organized within the hippocampus, which in turn enhances the possibility of encoding and temporarily remembering one spatial location as separate from another.

In further support of the attribute model, it has been shown that lesions of the amygdala, but not hippocampus, disrupt memory-based pattern separation for affect information (Gilbert and Kesner 2002a), lesions of the caudate nucleus, but

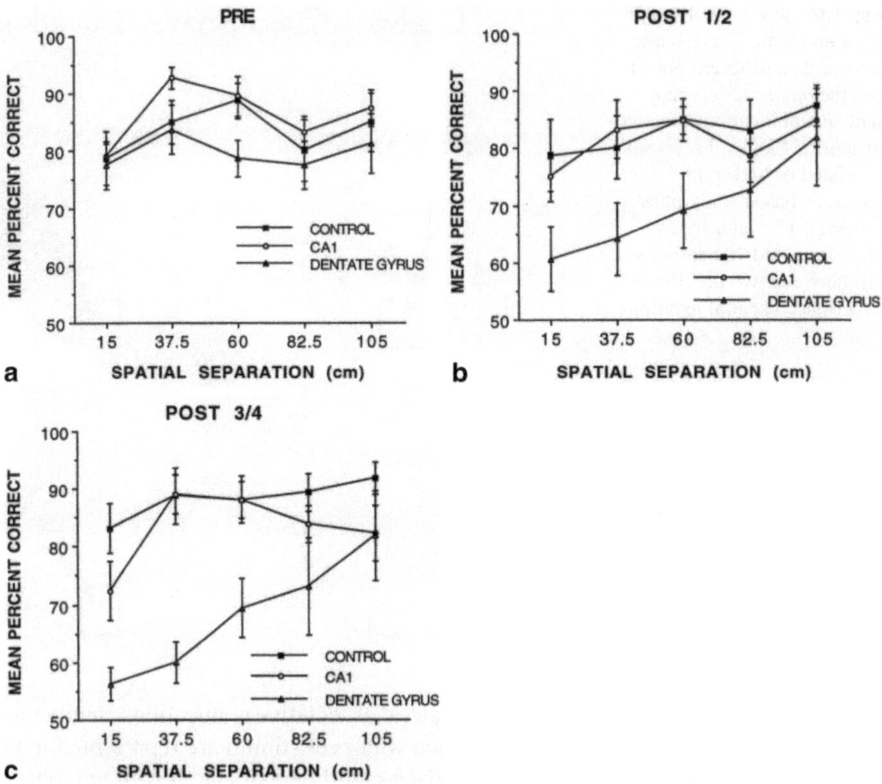

Fig. 1.5 **a** Mean percent correct performance as a function of spatial separation (number of intervening locations) for the control group, CA1 lesion group, and DG lesion group on preoperative trials. **b, c** Mean percent correct performance as a function of spatial separation for the control group, CA1 lesion group, and DG lesion group on two sets 1/2 and 3/4 of 30 postoperative trials. Note the inter-cue distance-dependent impairment in performance in the DG-lesioned group, demonstrating the role of DG in spatial pattern separation

not hippocampus, disrupt memory-based pattern separation for motor responses (Kesner and Gilbert 2006), lesions of the perirhinal cortex, but not hippocampus, disrupt memory-based pattern separation for objects (Gilbert and Kesner 2003a), and ventral DG lesions disrupt memory-based pattern separation for odors (Weeden et al. 2014).

CA3 and Arbitrary Associations

In the standard model (Marr 1971; McNaughton and Morris 1987; Levy 1996; Hasselmo and Wyble 1997; Rolls and Treves 1998; Rolls and Kesner 2006), the CA3 system acts as an auto-association system. This enables arbitrary (especially spatial in animals and likely language for humans as well) associations to be formed within

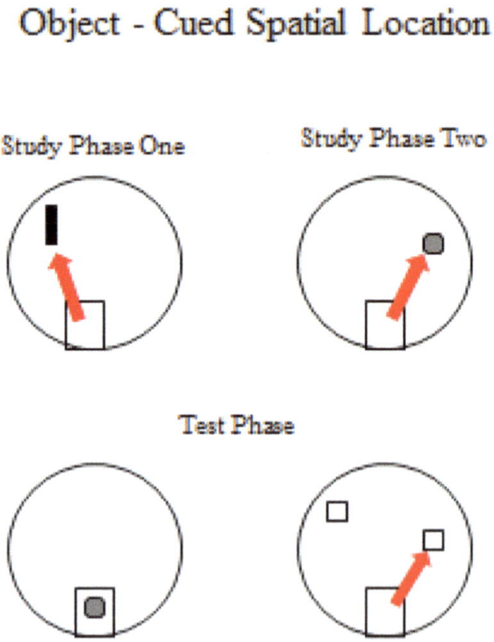

Fig. 1.6 Object-cued spatial location recall. Each shape represents a different object, and the *open squares* represent spatial locations covered by neutral blocks. Each trial consisted of two spatial locations based study phases followed 15 s later by an object cue and 10 s later by a test between two previously experienced spatial locations

the hippocampus. The CA3 recurrent collateral associative connections enable bidirectional associations to be formed between whatever stimuli are represented in the hippocampus, in that, for example, any place could be associated with any object, and in that the object could be recalled with a spatial recall cue, or the place with an object recall cue (Rolls and Treves 1998).

In the Kesner laboratory, a visual object-recall for a spatial location task has been developed based on the Day et al. (2003) experiment. In this task, after training to displace objects for food, rats in the study phase of each trial are placed in the start box (see Fig. 1.6 where each shape represents an object). When the door in front of the start box is opened the rats are allowed to displace one object in one location, and then return to the start box, after which the door is opened again and the rats are allowed to displace a second object in another location. To ensure that each trial was unique, 50 possible objects and 48 locations were used. In the test phase of each trial (see Fig. 1.6 where the open square represents spatial locations covered by a neutral block), the rat is shown one of the previously presented objects (first or second, randomized) in the start box as a cue for which spatial location to choose, and then, after a 10-s delay, the door is opened and the rats must go to the correct location (choosing and displacing one of two identical neutral objects). The rats receive a reward for selecting the correct location that was associated with that specific object cue. A spatial location-cued recall for a visual object task has also been developed (see Fig. 1.7). For the spatial-cued recall for a visual object task, the study phase (See Fig. 1.7 where each shape represents a different object) is the same, but in this case in the test phase (see Fig. 1.7 where the open square represents

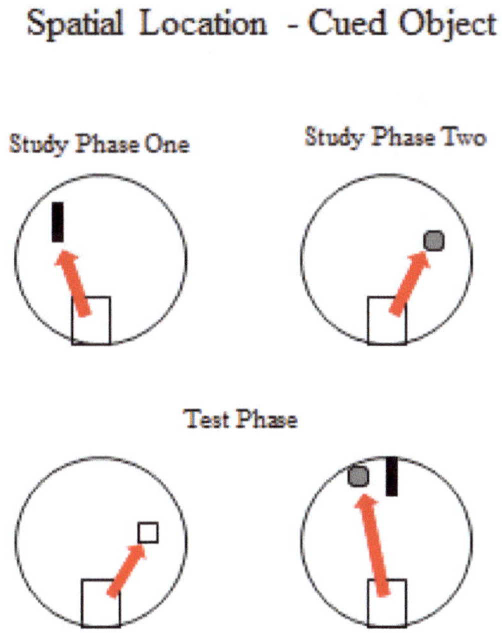

Fig. 1.7 Spatial location-cued object recall. Each shape represents a different object. The *open square* represents a neutral block placed on the correct spatial location as a cue. Each trial consisted of two object-based study phases followed 15 s later by a spatial location cue, and 10 s later by a test between two previously experienced objects

the correct location which is covered by a neutral block given as a cue), when the door is opened the rat is allowed to displace a neutral object in one of the previous locations (first or second, randomized) on the maze as a location cue, return to the start box, and then, after a 10-s delay, the door is opened and the rats must select the correct object (choosing and displacing one of two visual objects placed in different locations than during the study phases). The rats receive a reward for selecting the correct visual object that was associated with the location cue. Rats learn both tasks with 75 % or better accuracy.

Results are shown in Figs. 1.8 and 1.9 and indicate that CA3 lesions produce chance performance on both the object-cued place recall and the place-cued object recall task (Kesner et al. 2008).

The potential implications of such results are that indeed the CA3 supports arbitrary associations as well as episodic memory based on one-trial learning. A control fixed visual conditional to place task with the same delay was not impaired, showing that it is recall after one-trial (or rapid) learning that is impaired. Thus, some hippocampal neurons appear to process spatial recall given an object recall cue. These data are consistent with the prediction of the standard computational model that emphasizes the importance of CA3 in mediating the development of arbitrary associations.

There is anatomical support for CA3 involvement in support of the mediation of associative processes including arbitrary associations. The perforant path from the entorhinal cortex can be divided into a medial and lateral component. It has been suggested that the medial component processes spatial information and that the lateral component processes nonspatial (e.g., object, odor) information (Witter

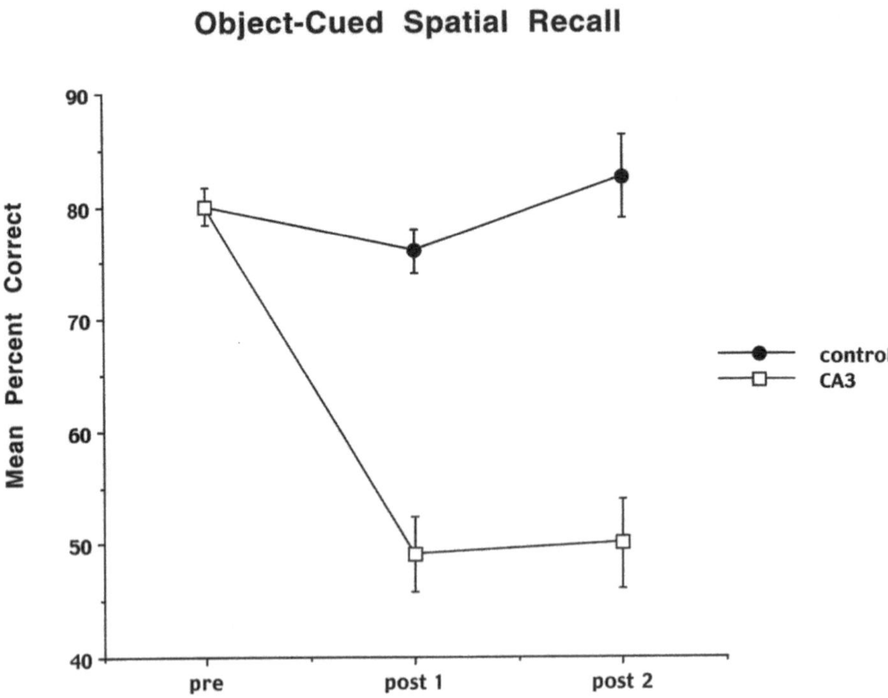

Fig. 1.8 Mean percent correct performance for the control and CA3-lesioned rats on the object-cued spatial location recall task before (pre) and after surgery (30 trials of post 1 and 30 trials of post 2). Note the profound CA3 lesion effect

et al. 1989; Hargreaves et al. 2005). In one study Ferbinteanu et al. (1999) showed that lesions of the MPP disrupted water maze learning, whereas LPP lesions had no effect. In a more recent study based on the idea that the MPP input into the CA3 mediates spatial information via activation of NMDA receptors and the LPP input into the CA3 mediates visual object information via activation of opioid receptors, the following experiment was conducted using the same paradigm described in the dentate and conjunctive encoding section except that direct infusions of AP5 (an NMDA antagonist) or naloxone (a μ-opiate antagonist) into CA3 were administered. The results indicated that naloxone or AP5 infusions into the CA3 disrupted both novelty detection of a spatial location and a visual object (Hunsaker et al. 2007).

CA3 and Pattern Completion

Marr (1971) suggested that hippocampal recurrent collaterals should play a significant role during the retrieval of previously stored information patterns in the face of partial inputs to the hippocampus ("collateral effect" or pattern completion). According to McNaughton and Morris (1987) and Rolls and Treves (1998), an auto-associative network within CA3 should be able to support pattern completion.

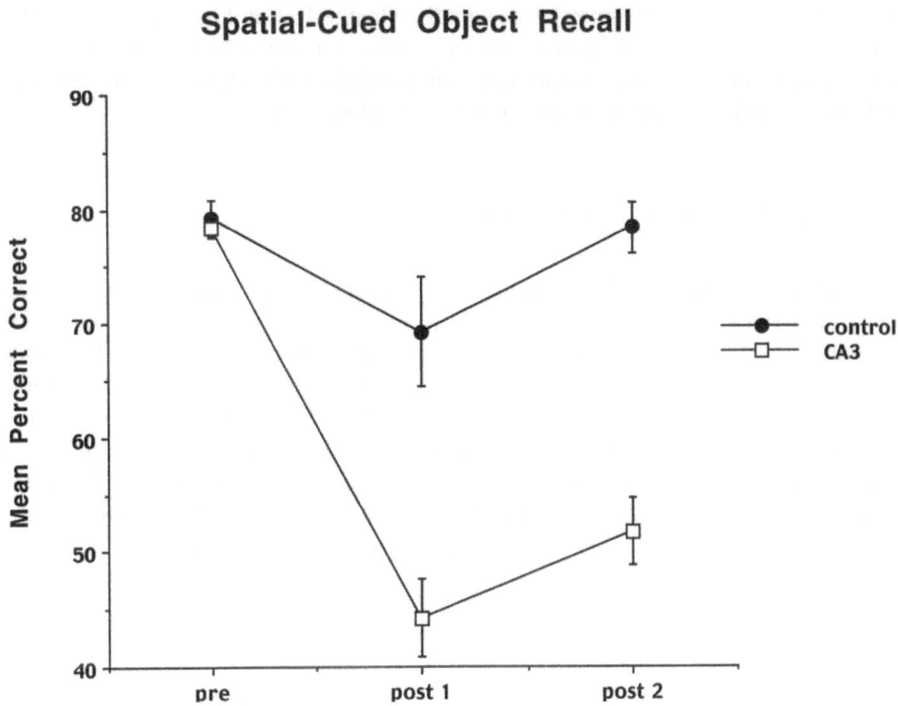

Fig. 1.9 Mean percent correct performance for the control and CA3-lesioned rats on the spatial location-cued object recall task before (pre) and after surgery (30 trials post 1 and 30 trials post 2). Note the profound CA3 lesion effect

Experimental efforts to find evidence of pattern completion within the CA3 region have been successful in recent years. For example, Gold and Kesner (2005) trained rats on a delayed matching-to-sample for a spatial location task to study spatial pattern completion. Animals were tested on the cheeseboard task, which was surrounded by a black curtain with 4 extra-maze cues. In the sample phase of the task, rats were trained to move a small black block covering a food well which could appear in 1 of 5 possible spatial locations that were in front of 4 extra-maze cues (i.e., the rat could see all 4 cues when approaching the spatial location as they were within the 180° visible immediately upon leaving the start box). During the choice phase of the task, rats were required to find the same food well, with the block removed in order to receive a food reward. After reaching stable performance, rats were randomly assigned to receive bilateral intracranial neurotoxic infusions or vehicle control infusions into the CA3 subregion of the hippocampus. Following recovery from surgery, each animal was re-tested on the delayed matching-to-sample task. During the sample phase, the animal was presented with all 4 extra-maze cues; however, the number of available cues (0, 1, 2, 3, or 4 cues) varied during the choice phase. The results are shown in Fig. 1.10 and indicate that control rats performed well on the task regardless of the availability of 1, 2, 3, or 4 cues, suggesting intact spatial pattern completion. Following the CA3 lesion, however, there were impairments in accuracy compared to the controls especially when only 1 or 2 cues were available, suggesting impairment in spatial pattern completion in CA3-lesioned rats (Gold and Kesner 2005). Similar results were observed for naloxone (μ-opioid receptor antagonist) infusions into CA3 (Kesner and Warthen 2010).

CA1 and Temporal Pattern Separation

Estes (1986) summarized data demonstrating that, in human memory, there are fewer errors for distinguishing items (by specifying the order in which they occurred) that are far apart in a sequence than those that are temporally adjacent. This phenomenon is referred to as a temporal distance effect (sometimes referred to as a temporal pattern separation effect (Kesner et al. 2004)). The temporal distance effect is assumed to occur because there is more interference for temporally proximal events than for temporally distant events. Based on these findings, Gilbert et al. (2001) tested memory for the temporal order of items in a one-trial sequence learning paradigm in rodents. In the task, each rat was given one daily trial consisting of a sample

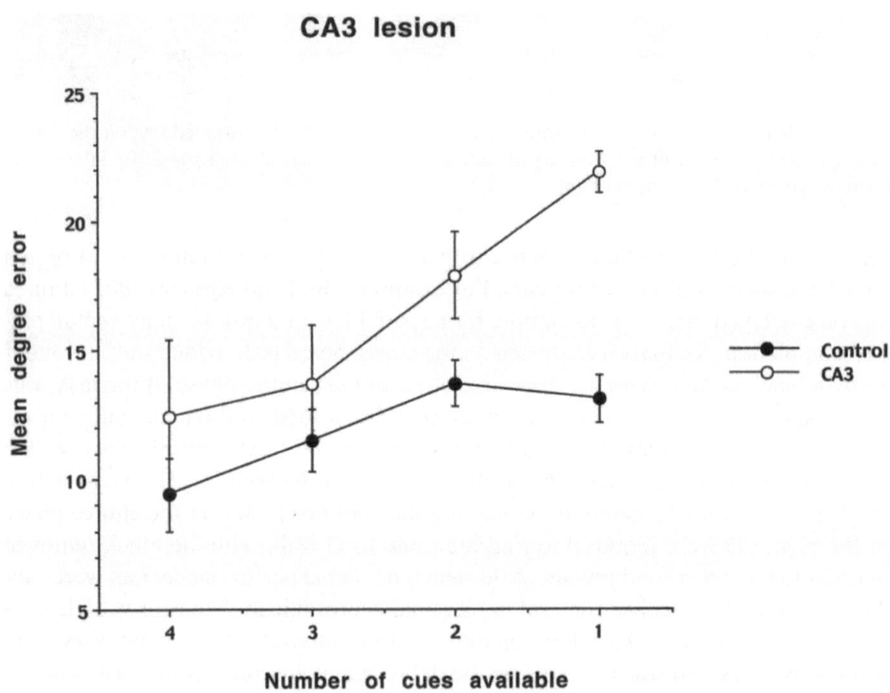

Fig. 1.10 Pattern completion impairment produced by CA3 lesions. The mean (with SEM) degree of error in finding the correct place on the cheeseboard task when rats were tested with 1, 2, 3, or 4 of the extra-maze cues available. A graded impairment in the CA3 lesion group as a function of the number of cues available was found. Prior to surgery the task was learned in the study phase with the 4 cues present. The performance of the control group is also shown

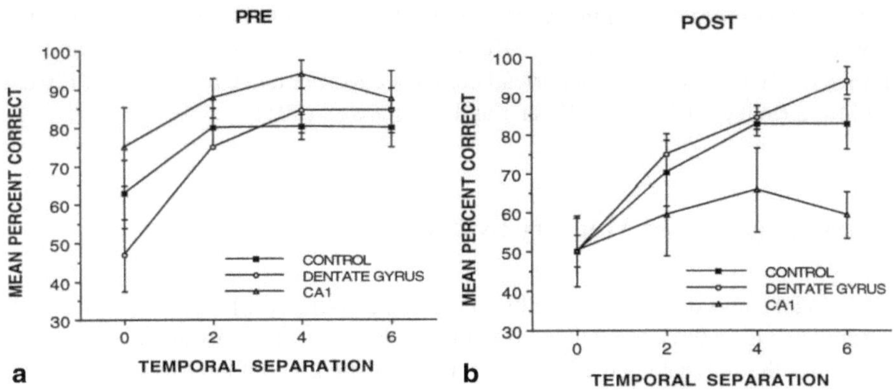

Fig. 1.11 **a** Mean percent correct performance as a function of temporal pattern separation for the control group, DG lesion group, and CA1 lesion group on preoperative trials. **b** Mean percent correct performance as a function of temporal separation for the control group, DG lesion group, and CA1 lesion group on postoperative trials

phase followed by a choice phase. During the sample phase, the animal visited each arm of an 8-arm radial maze once in a randomly predetermined order and was given a reward at the end of each arm. The choice phase began immediately following the presentation of the final arm in the sequence. In the choice phase, two arms were opened simultaneously and the animal was allowed to choose between the arms. To obtain a food reward, the animal had to enter the arm that occurred earlier in the sequence that it had just followed. Temporal separations of 0, 2, 4, and 6 were randomly selected for each choice phase. These values represented the number of arms in the sample phase that intervened between the arms that were used in the test phase. After reaching criterion, rats received CA1, DG, or control lesions. The results are shown in Fig. 1.11 and indicate that control and DG-lesioned rats matched their preoperative performance across all temporal separations. In contrast, rats with CA1 lesions performed at chance across 0, 2, 4, and 6 temporal separations.

The results suggest that the CA1 subregion is involved in memory for spatial location as a function of temporal separation of spatial locations. Thus, lesions of the CA1 decrease efficiency in temporal pattern separation. CA1-lesioned rats cannot separate events across time, perhaps due to an inability to inhibit interference that may be associated with sequentially occurring events. The increase in temporal interference impairs the rat's ability to remember the order of specific events. For additional functions of CA1, see Hunsaker et al. (2008).

In summary, the hippocampus was used to detail the multiple operations that characterize the overall activity of this brain region within the event-based memory system. The processes that were discussed include DG mediation of conjunctive encoding and spatial pattern separation, CA3 mediation of arbitrary associations and pattern completion, and CA1 mediation of temporal pattern separation. It should be noted that there are parallel brain-function relationships between the rodent data and the human data. With the use of similar behavioral paradigms with humans, it can be shown that there is extensive support for the attribute-based theoretical model of memory that is organized into event-, knowledge-, and rule-based memory systems. For review of the hippocampus see Kesner and Hopkins (2006) and Kesner and Goodrich (2010), for a review of parietal cortex see Kesner and Cream-Regehr (2013), and for a review of PFC, see Kesner and Churchwell (2011).

Knowledge-Based Memory System

The model suggests that different brain regions subserve the processing of different attributes within the knowledge-based memory system. To illustrate this, I selected processes that mediate perceptual memory associated within long-term memory including repetition priming and object recognition. The emphasis will be on visual and spatial perceptual processing and object recognition within the knowledge-based system. I will concentrate on temporal cortex (TE2) and make comparisons with the PPC in this section. To study one process associated with the knowledge-based system, a positive priming task was selected. Rats were then trained on tasks that resulted in a positive priming effect as indexed by facilitation of responding following a repetition of a spatial location or a visual object. TE2 lesions produced a deficit in processing positive priming for features of visual objects (a component of the knowledge-based memory system), but the rats performed well in positive priming for spatial location (Kesner, in preparation), whereas PPC lesions produced a deficit in processing positive priming for spatial locations (a component of the knowledge-based memory system), but performed well in positive priming for visual objects (in preparation). Thus, there is a double dissociation between TE2 and PPC for visual object verses spatial location priming. In a somewhat different study, a continuous recognition procedure was used to train rats on a 12-arm radial maze. Each rat was allowed to visit a sequence of 12 arms per day in an order predetermined for that trial. Of the 12 arms visited, either 3 or 4 of the arms were repeated within the running sequence. The arms selected for repetition varied according to lag (0–6), or the number of arms that occurred between the first visit to an arm and its repetition. To gain access to each arm, the animal was required to orient to a cue on the Plexiglas door at the entrance of the arm. Once the animal oriented to the cue, the door was lowered and the latency for the animal to reach the end of the arm was measured. Three groups of rats were trained on the knowledge-based perceptual memory training procedure. The perceptual/implicit memory group received reinforcement at the end of each arm regardless of whether the arm was a novel arm or a repeated arm. This group showed decreased latencies when visiting repeated arms displaying a repetition priming effect. The rats then received PPC, sham-operated, or cortical control lesions. After retesting, the results indicated that relative to the sham-operated and control groups control, the PPC-lesioned rats were impaired in the knowledge-based perceptual memory condition (Chiba et al. 2002).

Using a visual object–place recognition task, TE2-lesioned rats failed to detect a visual object change, whereas PPC-lesioned rats failed to detect a spatial location (Tees 1999) suggesting that the two cortical areas play a distinctive role in perceptual processing of visual verses spatial location information. Similar results were reported by Ho et al. (2011) who showed that rats with TE2 lesions had object recognition problems at 20 min, but not at 5-min delays. Lesions of the rat PPC disrupted retention of a spatial navigation task using either the water maze or dry-land version of the water maze task (DiMattia and Kesner 1988; Kesner et al. 1991; Save and Moghaddam 1996). Furthermore, in a multiple object scene task, PPC lesions disrupted retention of a previously learned discrimination in which rats had to detect a change in the location of the object in a scene, but had no effect in a previously learned discrimination in which the rat had to detect a change in one of the objects (DeCoteau and Kesner 1998). Finally, rats with PPC lesions do not react to a change consisting of removing a stimulus requiring a retrieval-dependent pattern completion process (Save et al. 1992).

Other examples of a role for PPC in storing spatial information into long-term memory include a study by Kesner et al. (1987), who had shown that in an 8-arm maze task PPC lesions placed in rats after training on 4 unbaited and 4 baited arms resulted in a deficit in retrieval from knowledge-based memory, but not from event-based memory. If one assumes that the presentation of unbaited arms reflects the operation of long-term memory and that the presentation of baited arms reflects the operation of event-based memory, then lesions of the PPC only disrupted long-term memory, but not event-based memory.

Finally, there is evidence to suggest that the parietal cortex may be a site for long-term representation of complex spatial information. Cho and Kesner (1996; Cho et al. 1995) have shown that rats with parietal cortex lesions have a nongraded retrograde amnesia for four, but not two previously learned spatial discriminations prior to surgery, suggesting that the deficit cannot be due to a performance or anterograde amnesia problem, but rather appears to be a function of the number or complexity of the spatial information to be stored and to be remembered.

In summary, within the knowledge-based memory system different brain regions process different attributes in support of perceptual processes. Data are presented to support this assertion by demonstrating that the PPC mediates the spatial attribute for spatial perceptual information and spatial recognition, whereas the TE2 cortex mediates the sensory-perceptual attribute for visual object information and visual object recognition.

Rule-Based Memory System

The model assumes that different brain regions subserve the processing of different attributes within the rule-based memory system. I selected a variety of tasks to illustrate this, because processing of mnemonic information is likely to incorporate rules and strategies and is associated with the emphasis on PFC function. Wise et al. (1996) suggested that the subregions of the PFC can be divided on the basis of rules and strategies. Furthermore, they proposed a hierarchy in terms of the complexity of the rules required, which they labeled lower order, higher order, and highest order.

I have proposed that the PFC in the rat can be fractionated in terms of functions associated with a slightly revised rule model that incorporates the rule-based memory system component of the attribute model (Kesner 2000a).

One can organize the subregions of the PFC in the rat according to the schema proposed by Uylings and van Eden (1990). These subregions include the medial PFC which can be subdivided into a dorsal medial region including the PC cortex, the dorsal and ventral AC cortices, and a ventral medial region including the PL and infralimbic as well as medial orbital cortices (PL-IL/MO), the lateral PFC which includes the dorsal and ventral agranular insular and the lateral orbital cortices (AI/LO), and the ventral PFC which includes the ventral orbital and ventrolateral orbital cortices (VLO/VO).

PC Cortex

The PC cortex appears to play an important role in working memory for motor responses requiring temporal processing of information, and paired associate learning. Supporting evidence is based on the findings that lesions of the AC and PC cortices that spare the PL-IL/MO cortex produce a deficit in working memory for motor response information such as working memory for a motor (right-left turn) response (Kesner et al. 1996), acquisition of an egocentric turn response (Kesner et al. 1989), and acquisition of visual-motor associative conditional discriminations (Passingham et al. 1988; Winocur 1991; Winocur and Eskes 1998).

AC Cortex

The AC cortex appears to play an important role in memory requiring temporal processing of information and paired associate learning. These lesions disrupt performance associated with processing of information in complex tasks, such as memory for temporal order of spatial information (Chiba et al. 1994, 1997; Kesner and Holbrook 1987; Kesner 1998a), memory for frequency information (Kesner 1990), use of a prospective code in a spatial 12-arm working memory task (Kesner 1989), and working memory for a list of five spatial locations (Kesner and Holbrook 1987).

The AC and PC cortex lesions, however, do not disrupt acquisition of visual, spatial, or olfactory discrimination (Eichenbaum et al. 1983; Harrison and Mair 1996; Ragozzino et al. 1999a), spatial discrimination reversal, cross-modal switching from visual cue to place or place to visual cue, switching between win-stay and win-shift rules or switching from a delayed nonmatching-to-sample to a delayed matching-to-sample rule (Harrison and Mair 1996; Neave et al. 1994; Ragozzino et al. 1998), spatial location navigation (deBruin et al. 1997; Kesner et al. 1989; King and Corwin 1992), working memory for visual object (Ennaceur et al. 1997; Kesner et al. 1996; Shaw and Aggleton 1993), duration (Jackson et al. 1998), or affect information (Decoteau et al. 1997). There are also no deficits, with a few exceptions, in working memory for spatial information using delayed nonmatching-to-position, delayed spatial alternation or nonmatching-to-sample in a T-maze, 8-arm maze, or continuous spatial recognition memory procedure (Ennaceur et al. 1997; Kesner et al. 1996; Ragozzino et al. 1998). Thus, the data suggest that the AC and PC cortex process rule-dependent working memory for motor response information, conditioned learning with response association as an important component to be learned, and/or higher order cognitive processes, but do not process rule-dependent working memory for visual object, spatial, affect (taste), or time as duration information as well as intramodal or cross-modal shifting of set and acquisition of spatial location navigation.

PL and Infralimbic Plus Medial Orbital Cortex (PL-IL/MO)

The PL-IL/MO cortex appears to play an important role in working memory for visual object and spatial location information as well as rules associated with cross-modal set switching. Supporting evidence is based on the findings that lesions of the PL-IL/MO cortex produced deficits in working memory for spatial information (Delatour and Gisquet-Verrier 1996; Ragozzino et al. 1998; Seamans et al. 1995), working memory for visual object information (Kesner et al. 1996), and cross-modal switching between place and visual cue or visual cue and place as well as motor response and place and place and motor response (Ragozzino et al. 1999a; Ragozzino et al. 1999b). These lesions, however, do not affect the acquisition of spatial, motor response, and visual discriminations, or visual, motor response, and spatial intramodal (reversal) learning (Bussey et al. 1997; Ragozzino et al. 1998; Ragozzino et al. 1999b) or learning of spatial location navigation (Maaswinkel et al. 1996), working memory for affect or motor response (DeCoteau et al. 1997; Ragozzino and Kesner 1998), and no deficit in a visual-response conditional associative task (Bussey et al. 1996). Thus, the data suggest that the PL-IL/MO cortex mediates working memory for spatial and visual object information as well as cross-modal switching involving spatial locations and visual objects as well as spatial locations and motor responses, but is not involved in motor response working memory, visual-response conditional processing or intramodal switching.

Agranular Insular and Lateral Orbital Cortex (AI/LO)

Based on anatomical and behavioral data, the AI/LO cortex appears to play an important role in working memory for affect information usually involving odor and taste. Supporting evidence is based on the findings that lesions of the AI/LO cortex produce deficits in working memory for affect based on taste or odor information (DeCoteau et al. 1997; Otto and Eichenbaum 1992; Ragozzino and Kesner 1999). There is also some evidence that this region plays a role in mediation of cross-modal associations in that many neurons within the AI/LO region fire differentially for a cross-modal association between odors and locations (Lipton et al. 1999). There are also deficits in acquisition and retention of a tactile-odor configuration task (Whishaw et al. 1992). However, there are mild or no significant deficits in odor discrimination or taste preferences (DeCoteau et al. 1997; Eichenbaum et al. 1983; Whishaw et al. 1992), in spatial working memory (Eichenbaum et al. 1983; Ragozzino and Kesner 1998a), and in learning a spatial location navigation task (Corwin et al. 1994). Also, there are no deficits in spatial discrimination or its reversal (Harrison and Mair 1996). Analysis of single cell recording from the agranular insular, lateral orbital, and ventrolateral orbital cortices revealed that there are cells that respond primarily when the animal makes a reliable shift to perform in a go no-go olfactory discrimination task. A few cells reverse their firing selectivity during reversal training, but the exact location of these cells within the agranular insular, lateral orbital, and ventrolateral orbital cortices was not specified (Schoenbaum et al. 1999). Thus, the data suggest that the AI/LO cortex mediates working memory for odor and taste information as well as cross-modal associations with odor and other sensory modalities, but is not involved in spatial processing of information. There is not much data available for the contribution of the ventral orbital and ventrolateral orbital cortices, but lesions in this area in conjunction with lateral orbital cortex contribute to reversal learning (Kim and Ragozzino 2005; McAlonan and Brown 2003).

In summary, based on the Wise et al. 1996 rule model, the PC cortex supports higher order rules for motor responses, the AC cortex supports the highest rules for temporal ordering, paired associate learning, list learning, and planning that include the use of temporal and prospective strategies, the PL-IL/MO cortex supports higher order rules for spatial and visual object information, the AI/LO cortex supports higher order rules for odor and taste information, and the VLO/VO cortex supports lower order rules.

Also there are dissociations based on different attributes characterizing the contribution of (a) the response memory attribute mediated by the PC, but not the AC, PL-IL/MO, or AI/LO cortical regions, (b) the temporal memory attribute mediated by AC, but not PC, PL-IL/MO, or AI/LO cortical regions, (c) the object and spatial memory attributes mediated by the PL-IL/MO region, but not the PC, AC, or AI/LO cortical regions, and (d) the affect memory attribute mediated by AI/LO but not PL-IL/MO cortical region. There is also a clear correspondence between rats and humans in terms of mediation of the abovementioned attributes. For more detail see Kesner (2000a) and Kesner and Churchwell (2011).

Interactions between Event-Based and Rule-Based Memory

Are there interactions between the event-based (e.g., hippocampus) and rule-based memory systems (e.g., IL/PL)? I present two examples based on temporal processing of information. In the first study, Lee and Kesner (2003) examined the dynamic interactions between the PFC and hippocampus by training and testing rats on a delayed nonmatching-to-place task on an 8-arm radial maze. Rats had to remember a single spatial location following short-term delays (i.e., 10 s or 5 min). The results showed that inactivating both regions at the same time resulted in a severe impairment of short-term and intermediate memory for spatial information suggesting that one of the structures needs to function properly for intact processing of short- or intermediate-term spatial memory. Thus, the two regions interact with each other to ensure the processing of spatial information across a dynamic temporal range including both short- and intermediate-term memory. These results provide compelling evidence indicating that a mnemonic time window is a critical factor in dissociating the function of the hippocampal system from that of the medial PFC in a delayed choice task. That is, the dorsal hippocampus and medial PFC appear to process spatial memory in parallel within a short-term range, whereas the dorsal hippocampal function becomes more essential once the critical time window requires spatial memory for a time period exceeding that range. In the second study, rats were also trained on a spatial delayed nonmatch-to-sample working memory task using short- (10 s) and long- (5 min) time delays to evaluate the hypothesis that the intermediate CA1 region of the HPC (iCA1) and PL cortex interact and operate in parallel under different temporal working memory constraints. To assess the functional role of these structures, an inactivation strategy was used in which each subject received bilateral chronic cannula implantation of the iCA1 and PL, allowing one to perform bilateral, contralateral, ipsilateral, and combined bilateral inactivation of structures and structure pairs within each subject. Compared to saline infusions, rats receiving contralateral infusions of muscimol into PL and iCA1 displayed an impairment for the 5-min delay, but not the 10-s delay. In contrast, rats receiving ipsilateral infusions of muscimol into PL and iCA1 displayed no impairment at either delay. These results suggest that there is an interaction in terms of temporal processing of information between the PL and iCA1. However, bilateral infusions of muscimol into both PL and iCA1 resulted in a deficit at both the 5-min and 10-s delay, suggesting that either structure may independently represent spatial information sufficient to successfully complete the task (Churchwell and Kesner 2011). This result is similar to what was reported by Lee and Kesner (2003). The findings of these studies suggest that there are interactions and parallel processing of temporal information between the event and rule based memory systems. From an anatomical point of view, there is a direct one way connection from iCA to PL region (Jay and Witter 1991) and information from PL can reach the hippocampus either via nucleus reuniens or entorhinal cortex (Vertes 2006). This circuit could subserve the functions described above.

Interactions Between Knowledge-Based and Event-Based Memory

Are there interactions and dissociations between different attributes within the knowledge-based (e.g., PPC) and event-based memory systems (e.g., hippocampus)? I present two experiments dissociating knowledge-based perceptual memory verses event-based memory processing of information as well as experiments examining binding of objects and places.

In the first experiment, two spatial continuous recognition training procedures designed to query knowledge-based perceptual memory and event-based episodic memory were employed. A continuous recognition procedure was used to train rats on a 12-arm radial maze. The details of the experimental protocol can be found in the knowledge-based memory section. After training, rats received PPC, hippocampus, or sham-operated and cortical control lesions. After retesting, the results indicated that relative to control and pretraining performance, the PPC-lesioned rats were impaired in the knowledge-based perceptual memory condition, but showed no deficits in the event-based episodic memory condition. In contrast, the hippocampal-lesioned rats were impaired in the event-based episodic memory condition, but showed no deficits in the knowledge-based perceptual memory condition (Chiba et al. 2002).

To have an even better measure of knowledge-based perceptual memory, a new experiment was generated to measure positive as well as negative repetition priming for spatial locations in rats similar to paradigms used with humans. Based on 48 repetition trials, all rats in the positive priming condition ran more quickly to the repeated spatial location. In the negative priming condition, it was assumed that rats not only actively attend to the positive stimulus but also actively inhibit responding to the negative stimulus (Neill and Mathis 1995). Based on 48 repetition trials, all rats in the negative priming condition ran more slowly to the repeated spatial location, because the correct location had resulted in some inhibition on the previous trial. After training, rats received PPC lesions and then were retested. The results indicate that PPC-lesioned rats are impaired for both positive and negative priming (Kesner 2000b). In the positive priming paradigm different rats received lesions of the hippocampus (Kesner 2000b). The results indicate that rats with hippocampal lesions showed normal positive priming. Thus, it appears that the PPC, but not the hippocampus, is directly involved in knowledge-based perceptual memory for spatial location information. The observation that the PPC does not mediate event-based episodic memory is supported by the observations that PPC lesions do not disrupt performance in a 5-choice serial reaction-time task (Muir et al. 1996). The data of both experiments suggest that there is a double dissociation between the two systems indicating that the two systems can operate independent of each other. Thus, a double dissociation appears to exist between PPC and hippocampus for knowledge-based perceptual memory verses event-based episodic memory operations, suggesting that the two neural circuits mediated by the hippocampus and PPC can operate independent of each other. This functional independence would require that spatial information reach the hippocampus and PPC via separate neural pathways. Indeed spatial information that reaches the dorsal lateral thalamus in the rat can be directed to the hippocampus via connections with the pre- and parasubiculum and medial entorhinal cortex and the PPC via direct connections. In the rat there are no direct connections between the PPC and the hippocampus. The parietal cortex and the hippocampus can interact via the entorhinal cortex or the retrosplenial cortex and pre- and parasubiculum (Kohler 1985; Van Groen and Wyss 1990; Witter et al. 1989).

A second possible role for the rodent parietal cortex could be to bind across modalities to maintain the association between landmark and spatial location information. In other words, the parietal cortex may not be involved in memory for a single landmark or a single spatial location, but rather in the processing that assigns a specific landmark to a specific spatial location. To test this hypothesis, rats with small lesions of the parietal cortex were tested in an object/spatial location paired-associate task that required concurrent memory for both object and spatial location information. In addition, memory for landmark only or spatial location only information was also assessed. A deficit in the paired associate task (which requires memory for both landmark and spatial location information), in the absence of deficits in either the landmark only or the spatial location only memory, would support the idea that the PPC is involved in the memory for the combination of landmark and spatial location information. The results indicated that small lesions of the PPC as defined by Reep et al. (1994) and larger PPC lesions disrupted learning of the object–place paired-associate task, but did not disrupt the learning of a spatial or object discrimination (Long and Kesner 1998). Furthermore, lesions of the hippocampus and especially the CA3 subregion of the hippocampus disrupted object–place paired-associate learning (Gilbert and Kesner 2002b, 2003b; Long et al. 1998), although it should be noted that a disruption only occurs when one component of the paired-associate is a spatial location. In a subsequent study unilateral lesions were made to the dorsal hippocampus or posterior PC contralaterally or ipsilaterally. It was hypothesized that if the hippocampus and PC interact, then contralateral-lesioned animals should be markedly impaired compared to ipsilateral lesions. The results indicate that contralateral-lesioned animals were significantly more impaired than animals with ipsilateral lesions during object–place paired-associate learning; however, both groups readily learned single discriminations (i.e., objects or places; Rogers and Kesner 2007). These results suggest that in this case there is an interaction between the PPC and hippocampus.

It appears that both parallel and interactive processing of information characterize the relationships between the PPC (a component of the knowledge-based memory system) and hippocampus (a component of the event-based memory system).

Interactions Between Knowledge-Based Memory and Rule-Based Memory

Are there interactions between different attributes within the knowledge-based memory system (e.g., PPC) and rule-based memory system (e.g., PFC)? I selected egocentric verses allocentric spatial processing to illustrate possible interactions between the knowledge-based and rule-based memory systems. Rats with medial PFC or parietal cortex lesions and sham-operated and non-operated controls were tested for the acquisition of an adjacent arm task where the rats were placed at the end of a randomly selected arm in an 8-arm radial maze and trained to run to the adjacent right or left arm to receive a reinforcement. This task accentuated the importance of egocentric spatial localization. In a second task a cheeseboard spatial navigation task that accentuated the importance of allocentric spatial localization was used.

Results indicated that relative to controls, animals with medial-PFC lesions were impaired on the adjacent arm task but displayed facilitation on the cheeseboard task. In contrast, relative to controls, rats with parietal cortex lesions were impaired on the cheeseboard task but showed no impairment on the adjacent arm task (Kesner et al. 1989; King and Corwin 1992). The data suggest a double dissociation of function between medial PFC and parietal cortex in terms of coding of egocentric versus allocentric spatial information. However, there are data to suggest that in a less structured task such as the water maze, that the PPC can also mediate egocentric spatial information. For example, Save and Poucet (2000) showed that in the Morris water maze PPC-lesioned rats were impaired in finding a hidden platform when 3 salient cues were located in the pool close to the correct location (proximal cues), but they were not impaired when only room cues (distal cues) were available to find the platform. Kolb and Walkey (1987) showed that PPC-lesioned rats were impaired in finding a platform location in a landmark task in which the rats had to associate a visual cue with a site that was spatially discontiguous and where the relevant cue moved relative to the rest of the extra-maze cues. This impairment manifested itself in the adoption of a looping strategy to locate a hidden platform. Foreman et al. (1992) found that the trajectories of rats turning and running between familiar visible targets at opposite ends of an area were less accurate in PPC-lesioned rats than in controls.

It appears that both parallel and potential interactive processing of information characterize the relationships between the PPC (a component of the knowledge-based memory system) and PFC (a component of the rule-based memory system).

Even though the event-based, knowledge-based, and rule-based memory systems are supported by neural substrates and different operating characteristics, the systems can operate independent of each other and there are also important interactions between the three systems. Clearly, for each attribute there is a neural circuit that encompasses all three memory systems in representing specific attribute information. I will present one example depicting the neural substrates and their interconnections associated with the spatial (place) attribute across all three memory systems (see Fig. 1.12). Note that the dorsal lateral thalamus, pre- and parasubiculum, hippocampus, and subiculum represent neural substrates that support the event-based memory system, the entorhinal cortex, parahippocampal gyrus or postrhinal cortex, PPC, and retrosplenial cortex support the knowledge-based memory system, and the lateral PFC or pre- and infralimbic cortex support the rule-based memory system. This circuit provides anatomical support for a possible independence in the operation of the hippocampus as part of the event-based memory system and PPC as part of the knowledge-based memory system in that spatial information that is processed via the dorsal lateral thalamus can activate both the hippocampus and the PPC in parallel. Also, information can reach the lateral PFC or pre- and infralimbic cortex as part of the rule-based memory system via direct connections from the PPC as part of the knowledge-based memory system and hippocampus as part of the event-based memory system. Finally, spatial information can interact with other specific attributes via a series of direct connections including, for example, an interaction with reward value attribute information via hippocampus–amygdala

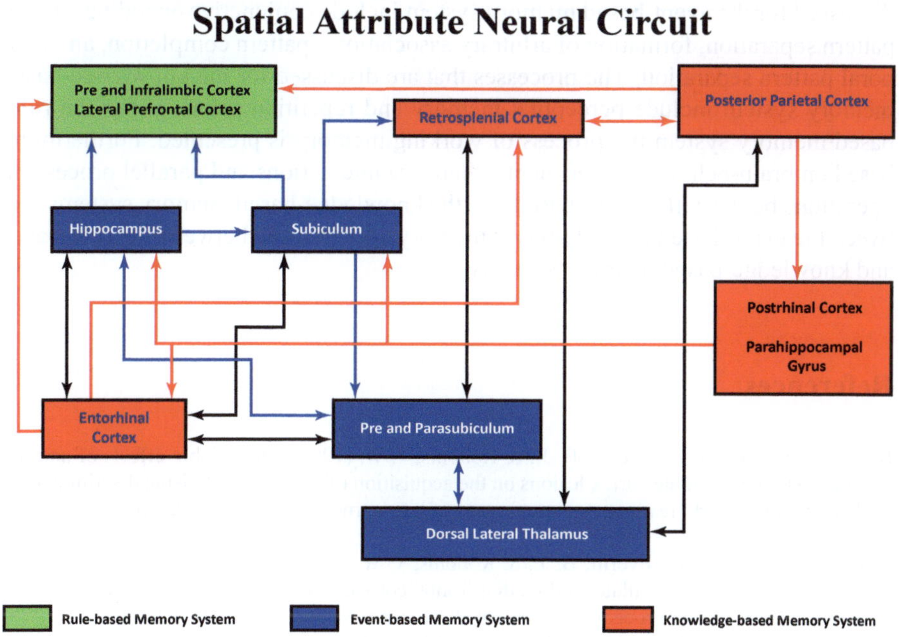

Fig. 1.12 A representation of the spatial attribute neural circuit incorporating neural regions that mediate event-based, knowledge-based, and rule-based memory

connection or lateral PFC–orbital frontal cortex connections and an interaction with response attribute information via hippocampus–caudate or lateral prefrontal–premotor and supplementary motor connections.

Conclusion

In this chapter, I have presented data in support of a neurobiological basis for an attribute model based on different forms or attributes of memory such as *space, time, response, sensory-perception, reward value (affect)* and in humans a *language* attribute is also added. These attributes are processed by different neural regions and interconnected networks across all three (event-based, knowledge-based, and rule-based) memory systems. The model is a major extension of previously mentioned brain-based memory models (Nadel 1994; Olton 1983; Tulving 1983; Squire 1994; Cohen and Eichenbaum 1993). Each memory system operates the processing of mnemonic information based on a unique set of processes. The selection of some of these processes has been influenced greatly by computational models of specific brain regions. For each brain area there are a large number of processes that define the operation of each memory system. The hippocampus is used extensively, but not exclusively, to detail the multiple operations that characterize the overall activity of this brain region within the event-based memory system. The processes that are discussed for the event-based memory system include conjunctive encoding, spatial pattern separation, formation of arbitrary associations, pattern completion, and temporal pattern separation. The processes that are discussed for the knowledge-based memory system include perceptual memory and repetition priming. For the rule-based memory system the process of working memory is presented. Furthermore, based on brain-behavior experiments, there are interactions and parallel processing operations between the event-based and the knowledge-based memory systems, between the event-based and rule-based memory systems, and between the rule-based and knowledge-based memory systems.

References

Bussey, T. J., Muir, J. L., Everitt, B. J., & Robbins, T. W. (1996). Dissociable effects of anterior and posterior cingulate cortex lesions on the acquisition of a conditional visual discrimination: Facilitation of early learning vs. impairment of late learning. *Behavioural Brain Research, 82,* 45–56.

Bussey, T. J., Muir, J. L., Everitt, B. J., & Robbins, T. W. (1997). Triple dissociation of anterior cingulate, posterior cingulate, and medial frontal cortices on visual discrimination tasks using a touchscreen testing procedure for the rat. *Behavioral Neuroscience, 111,* 920–936.

Chiba, A. A., Kesner, R. P., & Reynolds, A. M. (1994). Memory for spatial location as a function of temporal lag in rats: Role of hippocampus and medial prefrontal cortex. *Behavioral and Neural Biology, 61,* 123–131.

Chiba, A. A., Kesner, R. P., & Gibson, C. J. (1997). Memory for temporal order of new and familiar spatial location sequences: Role of the medial prefrontal cortex. *Learning and Memory, 4,* 311–317.

Chiba, A. A., Kesner, R. P., & Jackson, P. (2002). Two forms of spatial memory: A double dissociation between the parietal cortex and the hippocampus in the rat. *Behavioral Neuroscience, 116,* 874–883.

Cho, Y. H., & Kesner, R. P. (1996). Involvement of entorhinal cortex or parietal cortex in long-term spatial discrimination memory in rats: Retrograde amnesia. *Behavioral Neuroscience, 110,* 436–442.

Cho, Y. H., Kesner, R. P., & Brodale, S. (1995). Retrograde and anterograde amnesia for spatial discrimination in rats: Role of hippocampus, entorhinal cortex and parietal cortex. *Psychobiology, 23,* 185–194.

Churchwell, J. C., & Kesner, R. P. (2011). Hippocampal-prefrontal dynamics in spatial working memory: Interactions and independent parallel processing. *Behavioural Brain Research, 225,* 389–395.

Cohen, N. J., & Eichenbaum, H. B. (1993). *Memory, amnesia, and hippocampal function.* Cambridge: MIT Press.

Corwin, J. V., Fussinger, M., Meyer, R. C., King, V. R., & Reep, R. L. (1994). Bilateral destruction of the ventrolateral orbital cortex produces allocentric but not egocentric spatial deficits in rats. *Behavioural Brain Research, 61,* 79–86.

Day, M., Langston, R., & Morris, R. G. (2003). Glutamate-receptor-mediated encoding and retrieval of paired-associate learning. *Nature, 424,* 205–209.

deBruin, J. P. C., Swinkels, W. A. M., & deBrabander, J. M. (1997). Response learning of rats in a Morris water maze: Involvement of the medial prefrontal cortex. *Behavioural Brain Research, 85,* 47–55.

DeCoteau, W. E., & Kesner, R. P. (1998). Effects of hippocampal and parietal cortex lesions on the processing of multiple object scenes. *Behavioral Neuroscience, 112,* 68–82.

DeCoteau, W. E., Kesner, R. P., & Williams, J. M. (1997). Short-term memory for food reward magnitude: The role of the prefrontal cortex. *Behavioural Brain Research, 88,* 239–249.
Delatour, B., & Gisquet-Verrier, P. (1996). Prelimbic cortex specific lesions disrupt delayed variable response tasks in the rat. *Behavioral Neuroscience, 110,* 1282–1298.
DiMattia, B. V., & Kesner, R. P. (1988). Spatial cognitive maps: Differential role of parietal cortex and hippocampal formation. *Behavioral Neuroscience, 102,* 471–480.
Eichenbaum, H. (1994). The hippocampal system and declarative memory in humans and animals: Experimental analysis and historical origins. In D. L. Schacter & E. Tulving (Eds.), *Memory systems 1994* (pp. 39–63). Cambridge: MIT Press.
Eichenbaum, H. (2004). Hippocampus: Cognitive processes and neural representation that underlie declarative memory. *Neuron, 44,* 109–120.
Eichenbaum, H., Clegg, R. A., & Feeley, A. (1983). Reexamination of functional subdivisions of the rodent prefrontal cortex. *Experimental Neurology, 79,* 434–451.
Ennaceur, A., Neave, N., & Aggleton, J. P. (1997). Spontaneous object recognition and object location memory in rats: The effects of lesions in the cingulate cortices, the medial prefrontal cortex, the cingulum bundle and the fornix. *Experimental Brain Research, 113,* 509–519.
Estes, W. K. (1986). Memory for temporal information. In J. A. Michon & J. L. Jackson (Eds.), *Time, mind and behavior* (pp. 151–168). New York: Springer.
Ferbinteanu, J., Holsinger, R. M., & McDonald, R. J. (1999). Lesions of the medial or lateral perforant path have different effects on hippocampal contributions to place learning and fear conditioning to context. *Behavioural Brain Research, 101,* 65–84.
Foreman, N., Save, E., Thinus-Blanc, C., & Buhot, M. C. (1992). Visually guided locomotion, distractibility, and the missing-stimulus effect in hooded rates with unilateral or bilateral lesions of parietal cortex. *Behavioral Neuroscience, 106,* 529–538.
Gilbert, P. E., & Kesner, R. P. (2002a). The amygdala but not the hippocampus is involved in pattern separation based on reward value. *Neurobiology of Learning and Memory, 77,* 338–353.
Gilbert, P. E., & Kesner, R. P. (2002b). Role of the rodent hippocampus in paired-associate learning involving associations between a stimulus and a spatial location. *Behavioral Neuroscience, 116,* 63–71.
Gilbert, P. E., & Kesner, R. P. (2003a). Recognition memory for complex visual discrimination is influenced by stimulus interference in rodents with perirhinal cortex damage. *Learning & Memory, 10,* 525–530.
Gilbert, P. E., & Kesner, R. P. (2003b). Localization of function within the dorsal hippocampus: The role of the CA3 subregion in paired-associate learning. *Behavioral Neuroscience, 117,* 1385–1394.
Gilbert, P. E., & Kesner, R. P. (2006). The role of dorsal CA3 hippocampal subregion in spatial working memory and pattern separation. *Behavioural Brain Research, 169,* 142–149.
Gilbert, P. E., Kesner, R. P., & Lee, I. (2001). Dissociating hippocampal subregions: A double dissociation between dentate gyrus and CA1. *Hippocampus, 11,* 626–636.
Gold, E., & Kesner, R. P. (2005). The role of the CA3 subregion of the dorsal hippocampus in spatial pattern completion in the rat. *Hippocampus, 15,* 808–814.
Hafting, T., Fyhn, M., Molden, S., Moser, M. B., & Moser, E. I. (2005). Microstructure of a spatial map in the entorhinal cortex. *Nature, 436,* 801–806.
Hargreaves, E. L., Rao, G., Lee, I., & Knierim, J. J. (2005). Major dissociation between medial and lateral entorhinal input to dorsal hippocampus. *Science, 308,* 1792–1794.
Harrison, L. D., & Mair, R. G. (1996). A comparison of the effects of frontal cortical and thalamic lesions on measures of spatial learning and memory in the rat. *Behavioural Brain Research, 75,* 195–206.
Hasselmo, M. E., & Wyble, B. P. (1997). Free recall and recognition in a network model of the hippocampus: Simulating effects of scopolamine on human memory function. *Behavioural Brain Research, 89,* 1–34.
Ho, J. W., Narduzzo, K. E., Outram, A., Tinsley, C. J., Henley, J. M., Warburton, E. C., & Brown, M. W. (2011). Contributions of areaTe2 to rat recognition memory. *Learning and Memory, 18,* 493–501.

Hunsaker, M. R., Mooy, G. G., Swift, J. S., & Kesner, R. P. (2007). Dissociations of the medial and lateral perforant path projections into dorsal DG, CA3, and CA1 for spatial and nonspatial (visual object) information processing. *Behavioral Neuroscience, 121,* 742–750.

Hunsaker, M. R., Lee, B., & Kesner, R. P. (2008). Evaluating the temporal context of episodic memory: The role of CA3 and CA1. *Behavioural Brain Research, 188,* 310–315.

Jackson, P. A., Kesner, R. P., & Amann, K. (1998). Memory for duration: Role of hippocampus and medial prefrontal cortex. *Neurobiology of Learning and Memory, 70,* 328–348.

Jay, T. M., & Witter, M. P. (1991). Distribution of hippocampal CA1 and subicular efferents in the prefrontal cortex of the rat studied by means of anterograde transport of Phaseolus vulgaris-leucoagglutinin. *Journal of Comparative Neurology, 313,* 574–586.

Kesner, R. P. (1989). Retrospective and prospective coding of information: Role of the medial prefrontal cortex. *Experimental Brain Research, 74,* 163–167.

Kesner, R. P. (1990). Memory for frequency in rats: Role of the hippocampus and medial prefrontal cortex. *Behavioral and Neural Biology, 53,* 402–410.

Kesner, R. P. (1998a). Neural mediation of memory for time: Role of the hippocampus and medial prefrontal cortex. *Psychonomic Bulletin and Review, 5,* 585–596.

Kesner, R. P. (1998b). Neurobiological views of memory. In J. L. Martinez & R. P. Kesner (Eds.), *The neurobiology of learning and memory* (pp. 361–416). San Diego: Academic.

Kesner, R. P. (1999). Perirhinal cortex and hippocampus mediate parallel processing of object and spatial location information. *Behavioural and Brain Sciences, 22,* 455–479.

Kesner, R. P. (2000a). Subregional analysis of mnemonic functions of the prefrontal cortex in the rat. *Psychobiology, 28,* 219–228.

Kesner, R. P. (2000b). Behavioral analysis of the contribution of the hippocampus and parietal cortex to the processing of information: Interactions and dissociations. *Hippocampus, 10,* 483–490.

Kesner, R. P. (2007). Neurobiological views of memory. In R. P. Kesner & J. L. Martinez (Eds.), *The neurobiology of learning and memory* (2nd ed., pp. 271–304). San Diego: Academic.

Kesner, R. P., & Churchwell, J. C. (2011). An analysis of rat prefrontal cortex in mediating executive function. *Neurobiology of Learning and Memory, 96,* 417–431.

Kesner, R. P., & Creem-Regehr, S. H. (2013). Parietal contributions to spatial cognition. In L. Nadel & D. Waller (Eds.), *Handbook of spatial cognition* (pp. 35–64). Washington: American Psychological Association.

Kesner, R. P., & Gilbert, P. (2006). The role of the medial caudate nucleus, but not the hippocampus, in a matching-to sample task for a motor response. *The European Journal of Neuroscience, 23,* 1888–1894.

Kesner, R. P., & Holbrook, T. (1987). Dissociation of item and order spatial memory in rats following medial prefrontal cortex lesions. *Neuropsychologia, 25,* 653–664.

Kesner, R. P., & Hopkins, R. O. (2006). Mnemonic functions of the hippocampus: A comparison between animals and humans. *Biological Psychology, 73,* 3–18.

Kesner, R. P., & Warthen, D. K. (2010). Implications of CA3 NMDA and opiate receptors for spatial pattern completion in rats. *Hippocampus, 20,* 550–557.

Kesner, R. P., DiMattia, B. V., & Crutcher, K. A. (1987). Evidence for neocortical involvement in reference memory. *Behavioral and Neural Biology, 47,* 40–53.

Kesner, R. P., Farnsworth, G., & DiMattia, B. V. (1989). Double-dissociation of egocentric and allocentric space following medial prefrontal and parietal cortex lesions in the rat. *Behavioral Neuroscience, 103,* 956–961.

Kesner, R. P., Fansworth, G., & Kametani, H. (1991). Role of parietal cortex and hippocampus in representing spatial information. *Cerebral Cortex, 1,* 367–373.

Kesner, R. P., Bolland, B., & Dakis, M. (1993). Memory for spatial locations, motor responses, and objects: Triple dissociations among the hippocampus, caudate nucleus and extrastriate visual cortex. *Experimental Brain Research, 93,* 462–470.

Kesner, R. P., Hunt, M. E., Williams, J. M., & Long, J. M. (1996). Prefrontal cortex and working memory for spatial response, spatial location, and visual object information in the rat. *Cerebral Cortex, 6,* 311–318.

Kesner, R. P., Lee, I., & Gilbert, P. (2004). A behavioral assessment of hippocampal function based on a subregional analysis. *Reviews in the Neuroscience, 15,* 333–351.
Kesner, R. P., Hunsaker, M. R., & Warthen, M. W. (2008). The CA3 subregion of the hippocampus is critical for episodic memory processing by means of relational encoding in rats. *Behavioral Neuroscience, 122,* 1217–1225.
Kim, J., & Ragozzino. M. E. (2005). The involvement of the orbitofrontal cortex in learning under changing task contingencies. *Neurobiology of Learning and Memory, 83,* 125–133.
King, V. R., & Corwin, J. V. (1992). Comparison of hemi-inattention produced by unilateral lesions of the posterior parietal cortex or medial agranular prefrontal cortex in rats: Neglect, extinction, and the role of stimulus distance. *Behavioural Brain Research, 54,* 117–131.
Kohler, C. (1985). Intrinsic projections of the retrohippocampal region in the rat brain. I. The subicular complex. *Journal of Comparative Neurology, 236,* 504–522.
Kolb, B., & Walkey, J. (1987). Behavioural and anatomical studies of the posterior parietal cortex in the rat. *Behavioural Brain Research, 23,* 127–145.
Lee, I., & Kesner, R. P. (2003). Time-dependent relationship between the dorsal hippocampus and the prefrontal cortex in spatial memory. *Journal of Neuroscience, 23,* 1517–1523.
Levy, W. B. (1996). A sequence predicting CA3 is a flexible associator that learns and uses context to solve hippocampal-like tasks. *Hippocampus, 6,* 579–590.
Lipton, P, A., Alvarez, P., & Eichenbaum, H. (1999). Crossmodal associative memory representation in rodent orbitofrontal cortex. *Neuron, 22,* 349–359.
Long, J. M., & Kesner, R. P. (1998). The effects of hippocampal and parietal cortex lesions on memory for egocentric distance and spatial location information in rats. *Behavioral Neuroscience, 112,* 480–495.
Long, J. M., Mellem, J. E., & Kesner, R. P. (1998). The effects of parietal cortex lesions on an object/spatial location paired-associate task in rats. *Psychobiology, 26,* 128–133.
Maaswinkel, H., Gispen, W. H., & Spruijut, B. M. (1996). Effects of an electrolytic lesion of the prelimbic area on anxiety-related and cognitive tasks in the rat. *Behavioural Brain Research, 79,* 51–59.
Marr, D. (1971). Simple memory: A theory for archicortex. *Philosophical Transactions of the Royal Society of London. Series B: Biological Sciences, 262,* 23–81.
McAlonan, K., & Brown, V. J. (2003). Orbital prefrontal cortex mediates reversal learning and not attentional set shifting in the rat. *Behavioural Brain Research, 146,* 97–103.
McNaughton, B. L., & Morris, R. G. M. (1987). Hippocampal synaptic enhancement and information storage within a distributed memory system. *Trends in Neuroscience, 10,* 408–415.
Muir, J. L., Everitt, B. J., & Robbins, T. W. (1996). The cerebral cortex of the rat and visual attentional function: Dissociable effects of mediofrontal, cingulate, anterior dorsolateral, and parietal cortex lesions on a five-choice serial reaction time ask. *Cerebral Cortex, 6,* 470–481.
Nadel, L. (1994). Multiple memory systems: What and why, an update. In D. L. Schacter & E. Tulving (Eds.), *Memory systems 1994* (pp. 39–63). Cambridge: MIT Press.
Neave, N., Lloyd, S., Sahgal, A., & Aggleton, J. P. (1994). Lack of effect of lesions in the anterior cingulate cortex and retrosplenial cortex on certain tests of spatial memory in the rat. *Behavioural Brain Research, 65,* 89–101.
Neill, W. T., & Mathis, K. M. (1995). Transfer-inappropriate processing: Negative priming and related phenomena. *Psychological Learning and Motivation, 38,* 1–44.
O'Keefe, J., & Nadel, L. (1978). *The hippocampus as a cognitive map.* Oxford: Clarendon.
Olton, D. S. (1983). Memory functions and the hippocampus. In W. Seifert (Ed.), *Neurobiology of the hippocampus.* New York: Academic.
O'Reilly, R. C., & McClelland, J. L. (1994). Hippocampal conjunctive encoding, storage, and recall: Avoiding a trade-off. *Hippocampus, 4,* 661–682.
Otto, T., & Eichenbaum, H. (1992). Complementary roles of the orbital prefrontal cortex and the perirhinal-entorhinal cortices in an odor-guided delayed-nonmatching-to-sample task. *Behavioral Neuroscience, 106,* 762–775.
Passingham, R. E., Myers, C., Rawlins, N., Lightfoot, V., Fearn, S., et al. (1988). Premotor cortex in the rat. *Behavioral Neuroscience, 102,* 101–109.

Poucet, B. (1989). Object exploration, habituation, and response to a spatial change in rats following septal or medial frontal cortical damage. *Behavioral Neuroscience, 103,* 1009–1016.
Ragozzino, M. E., & Kesner, R. P. (1998). The effects of muscarinic cholinergic receptor blockade in the rat anterior cingulate and prelimbic/infralimbic cortices on spatial working memory. *Neurobiology of Learning and Memory, 69,* 241–257.
Ragozzino, M. E., & Kesner, R. P. (1999). The role of the agranular insular cortex in working memory for food reward value and allocentric space in rats. *Behavioural Brain Research, 1,* 103–112.
Ragozzino, M. E., Adams, S., & Kesner, R. P. (1998). Differential involvement of the dorsal anterior cingulate and prelimbic-infralimbic areas of the rodent prefrontal cortex in spatial working memory. *Behavioral Neuroscience, 112,* 293–303.
Ragozzino, M. E., Wilcox, C., Raso, M., & Kesner, R. P. (1999a). Involvement of the medial prefrontal cortex subregions in strategy switching. *Behavioral Neuroscience, 113,* 32–41.
Ragozzino, M. E., Detrick, S., & Kesner, R. P. (1999b). Involvement of the prelimbic-infralimbic areas of the rodent prefrontal cortex in behavioral flexibility for place and response learning. *Journal of Neuroscience, 19,* 4585–4594.
Reep, R. L., Chandler, H. C., King, V., & Corwin, J. V. (1994). Rat posterior parietal cortex: Topography of cortico-cortical and thalamic connections. *Experimental Brain Research, 100,* 67–84.
Rogers, J. L., & Kesner, R. P. (2007). Hippocampal-parietal cortex interactions: Evidence from a disconnection study in the rat. *Behavioural Brain Research, 179,* 19–27.
Rolls, E. T. (1996). A theory of hippocampal function in memory. *Hippocampus, 6,* 601–620.
Rolls, E. T., & Kesner, R. P. (2006). A computational theory of hippocampal function, and empirical tests of the theory. *Progress in Neurobiology, 79,* 1–48.
Rolls, E. T., & Treves, A. (1998). *Neural networks and brain function.* Oxford: Oxford University Press.
Save, E., & Moghaddam, M. (1996). Effects of lesions of the associative parietal cortex on the acquisition and use of spatial memory in egocentric and allocentric navigation tasks in the rat. *Behavioral Neuroscience, 110,* 74–85.
Save, E., & Poucet, B. (2000). Involvement of the hippocampus and associative parietal cortex in the use of proximal and distal landmarks for navigation. *Behavioural Brain Research, 109,* 195–206.
Save, E., Buhot, M.-C., Foreman, N., & Thinus-Blanc, C. (1992). Exploratory activity and response to a spatial change in rats with hippocampal or posterior parietal cortical lesions. *Behavioural Brain Research, 47,* 113–127.
Schacter, D. L. (1987). Implicit memory: History and current status. *Journal of Experimental Psychology: Learning, Memory, and Cognition, 13,* 501–518.
Schacter, D. L., & Tulving, E. (Eds.). (1994). *Memory systems 1994.* Cambridge: MIT Press.
Schoenbaum, G., Chiba, A. A., & Gallagher, M. (1999). Neural encoding in orbitofrotal cortex and basolateral amygdala during olfactory discrimination learning. *Journal of Neuroscience, 19,* 1876–1884.
Seamans, J. K., Floresco, S. B., & Phillips, A. G. (1995). Functional differences between the prelimbic and anterior cingulate regions of the rat prefrontal cortex. *Behavioural Neuroscience, 109,* 1063–1073.
Shapiro, M. L., & Olton, D. S. (1994). Hippocampal function and interference. In D. L. Schacter & E. Tulving (Eds.), *Memory systems 1994* (pp. 141–146). Cambridge: MIT Press.
Shaw, C., & Aggleton, J. P. (1993). The effects of fornix and medial prefrontal lesions on delayed non-matching-to-sample by rats. *Behavioural Brain Research, 54,* 91–102.
Spear, N. F. (1976). Retrieval of memories: A psychobiological approach. In W. K. Estes (Ed.), *Handbook of learning and cognitive processes (Vol. 4). Attention and memory.* Hillsdale: Erlbaum.
Squire, L. R. (1994). Declarative and nondeclarative memory: Multiple brain systems supporting learning and memory. In D. L. Schacter & E. Tulving (Eds.), *Memory systems 1994* (pp. 203–231). Cambridge: MIT Press.

Squire, L. R., Stark, C. E., & Clark, R. E. (2004). The medial temporal lobe. *Annual Review Neuroscience, 27,* 279–306.

Tees, R. C. (1999). The effects of posterior parietal cortex and posterior temporal cortical lesions on multimodal spatial and nonspatial competencies in rats. *Behavioural Brain Research, 106,* 55–73.

Tulving, E. (1983). *Elements of episodic memory*. Oxford: Clarendon.

Underwood, B. J. (1969). Attributes of memory. *Psychology Review, 76,* 559–573.

Uylings, H. B. M., & van Eden, C. G. (1990). Qualitative and quantitative comparison of the prefrontal cortex in rat and in primates, including humans. *Progress in Brain Research, 85,* 31–61.

Van Groen, T., & Wyss, J. M. (1990). The connections of presubiculum and parasubiculum in the rat. *Brain Research, 518,* 227–243.

Vertes, R. P. (2006). Interactions among the medial prefrontal cortex, hippocampus and midline thalamus in emotional and cognitive processing in the rat. *Neuroscience, 142,* 1–20.

Weeden, C. S. S., Hu, N. J., Ho, L. U. N., & Kesner, R. P. (2014). The role of the ventral dentate gyrus in olfactory pattern separation. *Hippocampus, 24,* 553–559.

Whishaw, I. Q., Tomie, J., & Kolb, B. (1992). Ventrolateral prefrontal cortex lesions in rats impair the acquisition and retention of a tactile-olfactory configural task. *Behavioral Neuroscience, 106,* 597–603.

Winocur, G. (1991). Functional dissociation of the hippocampus and prefrontal cortex in learning and memory. *Psychobiology, 19,* 11–20.

Winocur, G., & Eskes, G. (1998). Prefrontal cortex and caudate nucleus in conditional associative learning: Dissociated effects of selective brain lesions in rats. *Behavioral Neuroscience, 112,* 89–101.

Wise, S. P., Murray, E. A., & Gerfen, C. R. (1996). The frontal cortex-basal ganglia system in primates. *Critical Reviews in Neurobiology, 10,* 317–356.

Witter, M. P., Groenewegen, H. J., Lopes da Silva, F. H., & Lohman, A. H. (1989). Functional organization of the extrinsic and intrinsic circuitry of the parahippocampal region. *Progress in Neurobiology, 33,* 161–253.

Part I
Hippocampal Processes

Chapter 2
How Does the Hippocampus Support the Spatial and Temporal Attributes of Memory?

Howard Eichenbaum, Robert Komorowski, Christopher J. MacDonald, Benjamin J. Kraus and Jonathan Robitsek

In 1987, Kesner and DiMattia proposed that progress toward our understanding of memory could be improved by fragmenting memory into attributes that characterize the structural organization of memory, including space, sensory-perception, time, response, and affect. They assigned to the hippocampus a key role in the organization of memories in both space and time, and later, Kesner (1990) proposed that "the interaction between spatial and temporal attributes can provide an external context for situations." In support of this proposal, Kesner cited existing models of the hippocampus as involved in a spatial mapping of contexts (O'Keefe and Nadel 1978) and as forming a representation of temporal context (Rawlins 1985; see also Olton 1986). At that time there was compelling evidence of hippocampal neuronal activity that signaled spatial representations—place cells—and many studies, including key experiments by Kesner and his colleagues, had demonstrated critical hippocampal involvement in spatial memory. Furthermore, Kesner argued that the hippocampus is essential in supporting the temporal attributes of memory, showing that hippocampal lesions impair memory for the order of arms visited in a radial arm maze (Kesner and Novak 1982).

H. Eichenbaum (✉) · B. J. Kraus
Center for Memory and Brain, Boston University, 2 Cummington Mall, Boston, MA 02215, USA
e-mail: hbe@bu.edu

R. Komorowski
Picower Institute for Learning and Memory, Massachusetts Institute of Technology, 77 Massachusetts Ave, Cambridge, MA 02139, USA

C. J. MacDonald
Picower Institute for Learning and Memory, RIKEN-MIT Center for Neural Circuit Genetics, Massachusetts Institute of Technology, 43 Vassar St., Cambridge, MA 02139, USA

B. J. Kraus
The MathWorks, Inc., 3 Apple Hill Road, Natick, MA, USA

J. Robitsek
Department of Surgery, Jamaica Hospital Medical Center, 8900 Van Wyck Expressway Jamaica, NY 11418, USA

P. A. Jackson et al. (eds.), *The Neurobiological Basis of Memory*,
DOI 10.1007/978-3-319-15759-7_2

One could argue that memory within the radial maze task has essential spatial as well as temporal attributes, thus confounding a demand for spatial memory with that of temporal organization. But, many additional experimental studies by Kesner and his colleagues have shown that the hippocampus is also required in a variety of tasks that contain a memory delay and in memory for the order of nonspatial stimuli (reviewed in Kesner and Hunsaker 2010). Perhaps most compelling were experiments that examined whether rats could remember unique sequences of odors, and compared their ability to remember temporal order with that for odor discrimination (Kesner et al. 2002) and for recognition of the odor stimuli that had appeared within the list (Fortin et al. 2002). In the tests of memory for order, rats initially were rewarded for sampling each of a list of five odors. A few minutes later, on the order test, they were presented two nonadjacent odors from the list and were required to choose the less recently experienced odor to obtain another reward. Rats performed well above chance on temporal order memory, and better when the lag between previously presented items was larger. Rats with selective hippocampal damage were impaired in memory for temporal order at all lags, and performance was above chance only for the largest lag. By contrast, on tests of odor discrimination and on the recognition tests, rats with hippocampal damage performed as well as normal rats; and the selective impairment in order memory compared to intact item memory was striking even when overall accuracy in normal animals was matched between tasks.

These findings indicate that the hippocampus is essential in processing the temporal organization per se, independent of the memories for the items themselves, which was intact following hippocampal damage. There is a large literature on the ability of rats to time intervals, some of which indicate a role for the hippocampus in the perception of time and memory for duration (e.g., Meck et al. 1984; Jacobs et al. 2013; reviewed in MacDonald 2014). In addition, several other brain areas have been implicated in the capacity to time intervals, so it is likely that the hippocampus utilizes temporal information from many sources in supporting its role in the temporal organization of memories (Mauk and Buonomano 2004; Buhusi and Meck 2005; Yin and Troger 2011; MacDonald 2013).

Here we consider why and how the hippocampus is involved in both the spatial and temporal attributes of memory organization. One possibility is that these attributes are supported separately by anatomically distinct subfields within the hippocampus. Some of Kesner's work supports this idea. For example, in one particularly important study, Kesner et al. (2005) tested rats with selective CA1, CA3, or control lesions on a task in which animals were taught associations between an object and an odor that were separated by a 10 s delay; they called this the object–trace–odor association task. The animals learned that if object A was presented before the delay, then a cup of sand would contain a food reward if it was scented with odor 1 (but not with odor 2). Conversely, if object B was presented first, then a cup of sand would contain a food reward if it was scented with odor 2 (but not odor 1). Memory was measured by a briefer latency to approach the scented cup on rewarded pairings (A-1 and B-2) than on non-rewarded pairings (A-2 and B-1). In control rats, the latency to approach rewarded cups gradually decreased over daily training sessions of 12 trials each. Rats with selective CA1 lesions showed no sign of acquiring the associations, even after extensive training, whereas rats with CA3 lesions acquired the task just as rapidly as normal control animals.

The results of this study were surprising not only because a difference between the lesion groups was observed but also because the difference was so stark. The CA1 group did not learn at all and the CA3 group performed entirely normally. These findings stand in striking contrast to the findings of another by Gilbert and Kesner (2003), where rats learned associations between a particular object or odor and their locations in specific places in an open field. Normal rats learned the object–place and odor–place problems at about the same rate as in the object–trace–odor association task. However, in contrast to those findings, selective lesions of CA3 impaired acquisition of object–place and odor–place associations, whereas CA1 lesions did not. Indeed, in the case of odor–place associations, CA3 lesioned animals showed no learning, whereas animals with CA1 lesions performed normally, a pattern of results opposite to the pattern found in the authors' more recent study. Thus, CA1 and CA3 each appeared to make unique contributions, respectively, to temporal and spatial attributes of memory. These findings are difficult to reconcile with the close serial anatomical connections between CA3 and CA1, but are consistent with other evidence of differential effects of selective lesions to these subfields (reviewed in Manns and Eichenbaum 2005). Yet, other studies have continued to provide compelling evidence that CA1 may play an especially important role when associations demand bridging a substantial temporal gap (Farovik et al. 2010).

On the other hand, in contrast to a clear separation of temporal from spatial coding within CA1, a major line of evidence suggesting that CA1 also processes spatial information is the prominent observation of spatial coding by place cells in area CA1. This prominent finding raises the question: Do hippocampal neurons also encode temporal attributes of memory? Temporal coding by CA1 neurons is much less studied than their role in spatial information processing, but recently, several experiments have reported temporal coding by neurons in area CA1. Here we present evidence that CA1 neurons encode both the spatial and temporal attributes of memories. Supporting Kesner's intuition that spatial and temporal attributes are organizing features of the context of memories, we will argue that spatial and temporal organization are prominent attributes of hippocampal neural networks that support memory.

How Memories are Represented in Space

Following on earlier studies of spatial and nonspatial firing properties of hippocampal neurons (e.g., Wood et al. 1999; reviewed in Eichenbaum et al. 1999; Eichenbaum 2004), in recent studies aimed at examining the mechanisms by which hippocampal networks represent memories in spatial contexts, we recorded the activity of CA1 principal neurons in rats performing a task that requires them to remember

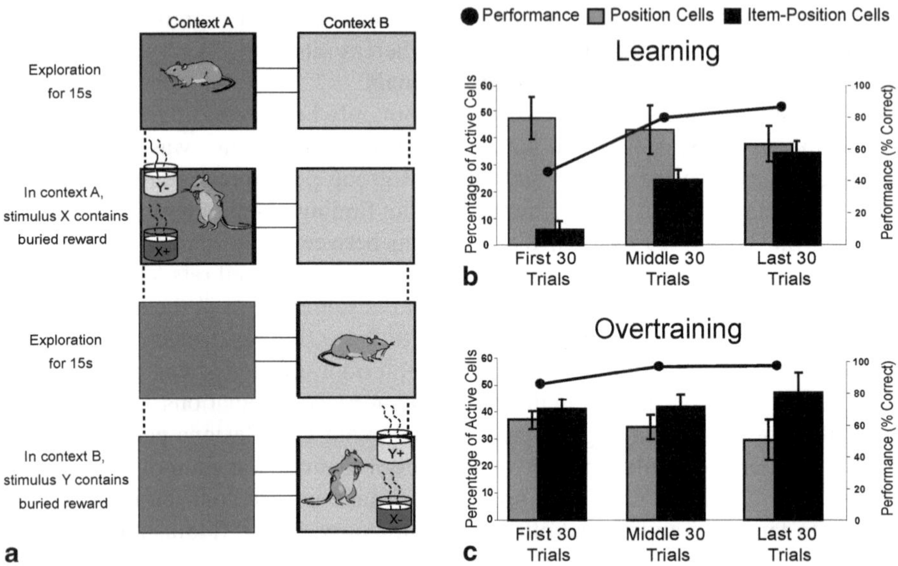

Fig. 2.1 Hippocampal neurons develop item–place representations in parallel with learning what happens where. **a** Object–context association task. The two contexts (represented by different shadings) differed in their flooring and wallpaper. The stimulus items (X or Y) differed in odor and in the medium that filled the pots. Items with a plus contained reward, whereas those with a minus did not, each depending upon the spatial context. **b** Changes in proportions of *Item-Position* and *Position cells* in learning vs. **c** overtraining sessions. (Data from Komorowski et al. 2009)

the differential reward associations of objects when they are presented in different places (Komorowski et al. 2009, 2013). In these experiments rats moved between environmental contexts that differed in visual, textural, and olfactory cues. On each trial, rats were initially allowed time to orient to the environment; then they were presented with two cups that were distinguished by both their odors and their digging media. In one environmental context (A), one of the stimuli (X) had a buried reward and the other stimulus (Y) did not, whereas in the other environmental context, the contingency was reversed (Y was baited and X was not; Fig. 2.1a). Therefore, the rat had to learn which of the two stimuli had been rewarded within each environment. We found that rats required several training sessions to acquire an initial problem of this type, but a subsequent second problem with new stimuli and new environmental contexts was typically acquired in the middle of a single 100-trial training session. This rapid learning allowed us to track the firing patterns of single neuron during the course of training on the second problem. We could therefore examine how neuronal firing patterns in the hippocampus might encode the relevant object–context associations.

We focused on the firing rates of hippocampal principal cells in areas CA1 and CA3 for a 1-s period surrounding when the rats sampled the stimuli during each trial. Earlier in training, we found that a large percentage of neurons fired when animals sampled either stimulus in a particular location in one of the two environments

(Fig. 2.1b; first 30 trials). These likely correspond to so-called place cells which fire when rats occupy a location in their environment. Some of these cells maintained the same place-specific firing patterns throughout training. At this stage, the firing patterns of virtually none of the cells distinguished the stimuli. However, as the animals acquired the context guided object association task, some neurons began to fire selectively during the sampling of one of the objects in one of the contexts and these cells continued to exhibit conjunctive object and place specificity after learning (Fig. 2.1b; middle 30 trials). The magnitude of item–context representation was robust in that, by the end of the training session, the proportion of hippocampal neurons that fired selectively during the sampling of one of the objects in a particular place or context equaled that of place cells (Fig. 2.1b; last 30 trials). This conjunctive object and place representation remained strong throughout recording sessions in which animals were highly overtrained on the task (Fig. 2.1c). Thus, a large percentage of hippocampal neurons developed representations of task-relevant object and place associations, and their evolution was closely correlated with learning those associations. Furthermore, subsequent analyses showed that the conjunctive representations developed from preexisting spatial representations into enhanced activations when particular objects were sampled in specific locations. Conversely, the representation of the objects alone was minimal throughout learning and the representation of places where any object was sampled, although strong, remained unchanged throughout training. These and other (Moita et al. 2003; Manns and Eichenbaum 2009) findings strongly suggest that the development of conjunctive object and location representations within the hippocampus underlies memories for items in the places where they occur.

Memories in Space and Time

Kesner and colleagues suggested that the entire hippocampus is engaged when a task demands both spatial and temporal attributes of memory (Hunsaker et al. 2006). In recent years, recordings of hippocampal neurons in animals performing tasks that require memory for spatial sequences have provided insights into how spatial and temporal attributes are integrated by hippocampal neuronal activity.

In addition to representation of elapsed time as a regularity of experiences, there is substantial evidence that hippocampal neuronal ensembles encode the order of events in sequence memories as revealed in studies showing that hippocampal neural ensembles "replay" sequences of place cell activations that occurred during previous experiences. The earliest studies on sequence replay by hippocampal neural ensembles focused on the tendency of place cells that fired in order during behavior to also fire in the same order when animals subsequently slept (Wilson and McNaughton 1994). Since then, numerous studies have reported forward and reverse replay of place cell sequences, both when animals are asleep and during periods of quiet wakefulness (see Karlsson and Frank 2009). Furthermore, when rats are engaged in vicarious trial and error of maze choices, hippocampal neurons replay firing sequences that reflect possible paths of response choices (Johnson and Redish 2007). And place cell sequences anticipate paths to be taken even in open fields (Pfeiffer and Foster 2013). Conversely, interfering with hippocampal replays retards learning of critical choices in spatial memories, but not the general skills of performance in the maze (Jadhav et al. 2012). In addition, hippocampal replays are synchronized with cortical replays, consistent with the view that sequence replays reflect a temporal organization involved in remembering and memory consolidation (Ji and Wilson 2007).

In a particularly striking recent study linking place cell replay with learning, Singer et al. (2013) recorded from CA1 and CA3 principal cells in rats performing a spatial alternation task in a "W" shaped maze. They examined neuronal activity during local field potential events known as sharp wave ripples (SWR), in which several earlier reports have shown a speeded "replay" of neuronal firing sequences that had occurred in earlier experiences. Specifically, their analyses focused on SWRs when the rat was relatively still while outbound on the center arm, heading toward the critical choice between the left or right arm as having the next reward. During these SWR events, they identified replays as co-activations of place cell activity that typically occurred during actual runs toward the left or right goals. They found that more replays occurred preceding subsequent correct choices than incorrect choices, and in the latter, the likelihood of replay was at chance level. In addition, there were usually multiple replays at these times, corresponding to both the correct and incorrect choice paths. Also, replays were common early in learning but no longer appeared when rats had mastered the task. Thus, associated with the course of learning, the hippocampus replays alternative paths just before a critical choice between those paths is made, and the occurrence of replay increases the accuracy of the subsequent choice.

The findings by Singer et al. (2013) showing that the hippocampus replays multiple alternative memories build on many earlier observations about hippocampal replay, including, in particular, that hippocampal neural ensembles replay both recent paths and paths not recently taken (Gupta et al. 2010). Also, the occurrence of replays is greater after novel experiences and correlates with memory performance (Dupret et al. 2010). And replays of alternative paths have also been observed when rats investigate possible choices during vicarious trial and error at a critical decision point (Johnson and Redish 2007). Here the trial-by-trial prediction of accuracy by the proportion of replays of alternative paths suggests that hippocampal replay reflects the retrieval of multiple relevant memories that can be evaluated to guide the correct subsequent choice, and this is of particular value early in learning.

The findings on hippocampal replay and its association with memory are paralleled by several observations on trajectory dependent activity of place cells (reviewed in Shapiro et al. 2006). In these studies, rats traverse overlapping routes through a maze and a typical observation is distinct place cell firing sequences for each route, including different firing patterns when the rat is traversing the overlapping part of different routes. In our first study of this phenomenon, rats were trained on the classic spatial T-maze alternation task in which successful performance depends on distinguishing left- and right-turn episodes to guide each subsequent choice (Wood et al. 2000). We reasoned that, if hippocampal neurons encode each sequential behavioral event within one type of episode, then neuronal activity at locations that overlap in left-to-right and right-to-left turn trials should vary according to the route currently under way. Indeed, virtually all cells that were active as the rat traversed these common locations were differentially active on left-to-right versus right-to-left trials. Although most cells exhibited similar quantitative differentiation of trial types, other cells fired exclusively on one type of trial. Similar results have subsequently been observed in several versions of this task (Bower et al. 2005; Ferbinteanu and Shapiro 2003; Frank et al. 2000; Griffin et al. 2007; Lee et al. 2006; Ainge et al. 2007; Pastalkova et al. 2008; for review, see Shapiro et al. 2006; but not all versions of the task Lenck-Santini et al. 2001; Bower et al. 2005). Furthermore, these observations are consistent with recent results in animals and humans showing that hippocampal neuronal activity captures sequential events that compose distinct memories (Ginther et al. 2011; Paz et al. 2010). These findings suggest a reconciliation of the current controversy about spatial navigation and episodic memory views of hippocampal function: Place cells represent the series of places where events occur in sequences that compose distinct memories.

Similar to the findings of Singer et al. (2013) on replays, trajectory-dependent activity of place cells is also strongly linked to memory performance, as its occurrence both prior to a memory delay and during memory retrieval predicts subsequent trial-by-trial memory accuracy (Robitsek et al. 2013). In that study, we first trained rats on the continuous spatial alternation task used in the Wood et al. (2000) study then, on subsequent recording sessions, recorded CA1 principal neurons as rats performed separate blocks of trials on the continuous alternation and on a delayed alternation version where they were constrained at the start of the common segment of the maze. Performance during delayed alternation was approximately 70 % correct, allowing a comparison of firing properties during accurate trials and errors when the animal ran on trajectories from left-to-left or right-to-right (Fig. 2.2). We found hippocampal place cells that fired when the rat traversed locations throughout the maze and their activity predicted accuracy of subsequent choices. In particular, we found that many place cells that fired at locations just before the delay were strongly activated in advance of subsequent correct choices, whereas the same cells fired much less or not at all in advance of errors. For example, the cell in Fig. 2.2a fires robustly as the rat approaches the end of the left return arm on correct but not error trials and the cells in Fig. 2.2b and c fire strongly as the rat is in the midst of the right return arm on correct trials, and much less on errors. Also, many of the cells that fired selectively associated with retrieval of left-to-right or right-to-left trials as the rat traversed the common segment of the maze also fired strongly in advance of correct choices but less so or not at all in advance of errors. For example, the cell in Fig. 2.2d fired robustly as the animal traverses the stem on correct left-to-right trials, much less so on right-to-left trials, and hardly fired on errors. Figures 2.2e and f show cells that fired at different locations on the common maze segment most strongly on correct left-to-right trials and slightly less on correct right-to-left trials, and did not fire on either type of error. The combined evidence on replay and trajectory-dependent firing strongly suggest that the activity of place cells in spatial

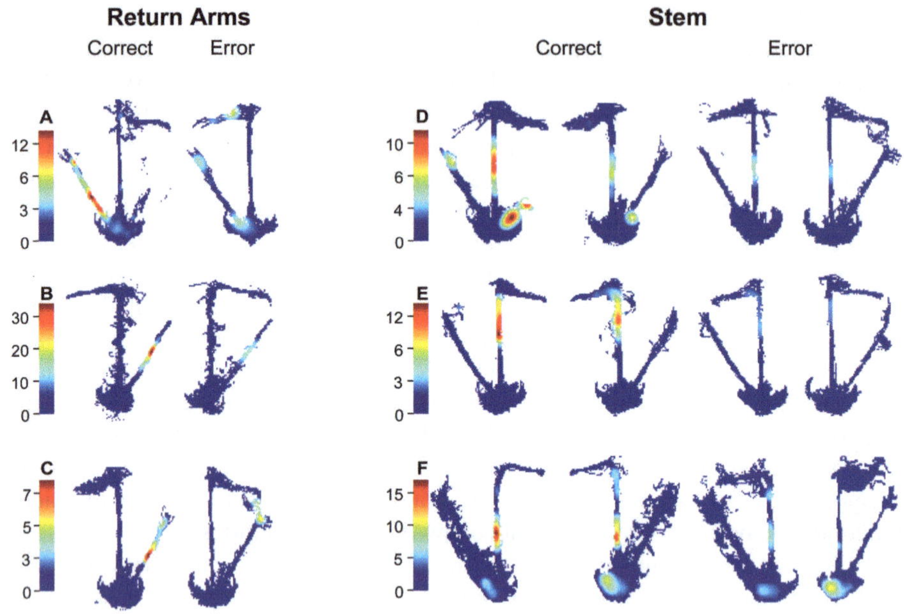

Fig. 2.2 CA1 neurons signal subsequent accurate memory on a spatial alternation task. **a–c** Cells that fired differentially as rats traversed different parts of the maze arm just prior to the memory delay. **d–f** Cells that fired differentially as rats traversed different parts of the maze common to both routes through the maze. See text for description. (Data from Robitsek et al. 2013)

memory tasks reflects the encoding and retrieval of sequences of places traversed that compose the memories of routes taken.

Do Hippocampal Neurons Represent the Temporal Attributes of Experience, Independent of Spatial Coding?

While there is an extensive literature on the spatial firing properties of hippocampal neurons, much less attention has been paid to how time itself is represented in the hippocampus, despite substantial evidence of hippocampal involvement in the temporal organization of memory (reviewed in Eichenbaum 2013). Recently, evidence has emerged showing that hippocampal neuronal networks compose a gradually changing representation of the flow of time, independent of explicitly identifiable locations or specific events that might directly drive sequential neural activations. Furthermore, the temporal signal has been dissociated from potential confounds of moving through space as well as self-generated movement cues (path integration) that could underlie an apparent temporal modulation of neural activity, as discussed in the interpretation of several experiments below.

The initial evidence of gradually changing temporal context representations in the hippocampus came in a study in which ensembles of CA1 neurons were recorded as rats performed the above-described task wherein rats encode and remember unique sequences of odors (Kesner et al. 2002; Fortin et al. 2002). The firing patterns of CA1 ensembles gradually evolved over entire recording sessions. Moreover, within those sessions, CA1 ensemble representations gradually changed even over a few minutes in which individual sequences were encoded, and the extent of ensemble change during the sequence of odor sampling events predicted subsequent success in remembering the order of odors experienced on each trial (Manns et al. 2007). Consistent with this observation, Naya and Suzuki (2011) observed that, when monkeys perform a task where they bridge a delay between two visual stimuli, hippocampal neural ensembles represent the evolving temporal context between the stimulus events.

As the Manns et al. (2007) task involved unique memories on each trial, it could not be determined whether distinct evolving temporal context representations are generated for specific memories. However, Pastalkova et al. (2008) recorded the activity of hippocampal (CA1) neurons as rats ran in a running wheel in between trials in a spatial alternation task and observed that different hippocampal ensemble sequences were associated with different subsequent memory choices and, when the animals made errors, these sequences were disrupted. Although Pastalkova et al. (2008) referred to these neurons as "episode cells," we prefer to call them "time cells" because, just as place cells encode locations in a specific space, time cells encode moments in a specific period of experience. The populations of time cells observed in Pastalkova's study likely reflect the repetition of ensemble firing patterns that gradually changed in the Manns et al. (2007) study.

The phenomenon of time cells was further examined using a nonspatial task developed by Kesner et al. (2005) that identified the hippocampal CA1 region as necessary for rats to learn distinct sequences in which an object and an odor were separated by a 10 s temporal gap (Fig. 2.3a). In this version of the task, rats moved through three sections of a linear maze, each of which composed a key phase in a sequence of events. Each trial began with the presentation of one of two objects that the rat investigated for a short period. Then the rat was confined in a small area for 10 s, after which it was presented with one of two odors mixed into common playground sand. Each odor was paired with one of the objects, such that if the odor followed the correctly paired object then the rat could dig in the sand for a buried reward. Conversely, the rat obtained no reward for digging when the odor followed the object with which it was not paired. Critically, the object–delay–odor sequences were presented repeatedly during each testing session, so the rats had to remember across the delay the object that had started the trial in order to respond appropriately to the odor at the end of the trial. As described above, rats with lesions of the CA1 region show no evidence of learning these object–odor sequences (Kesner et al. 2005). Conversely, rats with CA3 lesions learn the sequences with a time-course that is comparable to control rats. Taken together, these results are consistent with a selective role for the CA1 in representing a temporally extended sequence of events to compose a distinct experience.

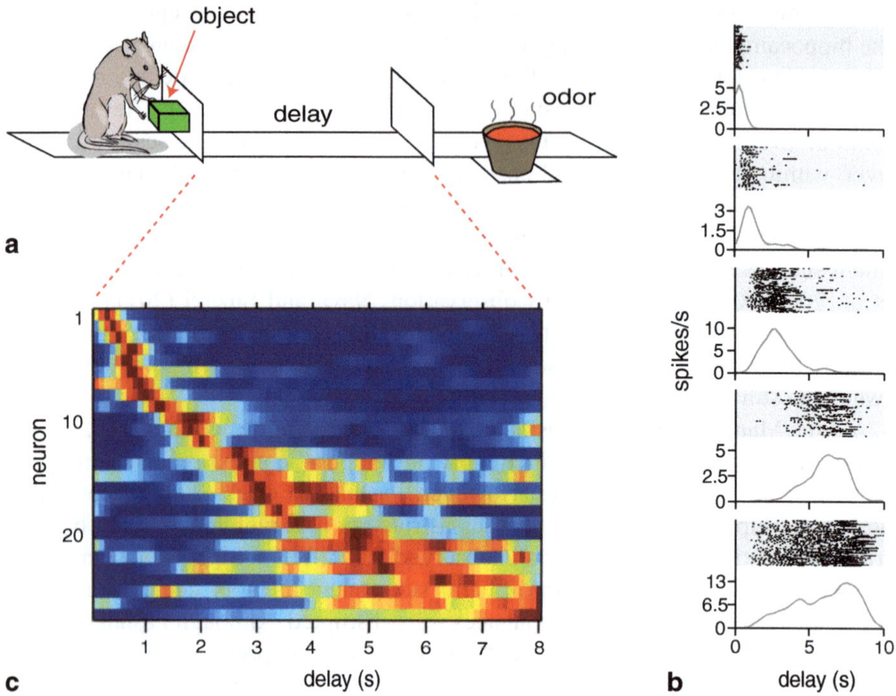

Fig. 2.3 **a** The trial structure for object–delay–odor sequences. **b** Each panel shows a raster plot and peri-event time histogram illustrating neural activity of a time cell during the delay period. **c** Normalized firing rates of 26 neurons recorded simultaneously during the delay period. Each row represents the activity pattern of a single neuron. (Data from MacDonald et al. 2011)

To explore the nature of the hippocampal representation supporting performance in this task, MacDonald et al. (2011) adapted the task and examined activity from large ensembles of hippocampal CA1 neurons monitored simultaneously. Many neurons activated during presentation of the object or odor and often fired differently depending on the object that started the trial, indicating that the hippocampus distinguished the key events composing each object–odor sequence. Most striking, nearly half of the cells that were recorded activated during the delay period, and the period of activity of each cell was typically selective for a specific moment (Fig. 2.3b). To better illustrate the temporal signature of these cells, Fig. 2.3c plots normalized firing patterns from an ensemble of cells recorded simultaneously during the delay. It is readily apparent that the cells activated in sequence, and the overlap among their firing fields bridged the delay. Importantly, time cells distinguished the object starting the trial, which is consistent with a function in integrating the object with its paired odor across the delay. These results confirmed a robust temporally organized representation for a sequence of events in the hippocampus, highlighted by cells that bridged the delay and composed the flow of time in a distinct memory.

Could temporal signals reflected in the activity of time cells be confounded with a reliable sequence of behaviors or a sequence of locations occupied during the delay? MacDonald et al. (2011) performed a detailed statistical analysis of the firing patterns of neurons and found that, while many of these cells also represented the spatial location and ongoing behavior during the delay, these factors did not account for the timing signal reflected in the activity of these cells. Thus, while many of these cells did incorporate information about spatial and behavioral events into the neural representation of the delay period, the temporal signal encoded by time cells was independent of the rat's location and movements.

Another alternative explanation of these findings is that hippocampal neurons integrated the path of movement animals took during the delay phase of the task (McNaughton et al. 1996). In the McDonald et al. (2011) and the Pastalkova et al. (2008) studies, as well as another study that observed time cells during the delay periods in a delayed spatial task (Gill et al. 2011), the rats were in motion over the entirety of the key delay periods. Therefore, the distance moved and time elapsed were entirely confounded during the periods when time cells were observed, and other studies have reported that hippocampal neurons can signal the accumulated linear distance that a rat has moved from a reference point (Gothard et al. 1996; Redish et al. 2000). Thus, it was unclear whether hippocampal neurons can signal the flow of time independent of self-generated cues that may support path integration (McNaughton et al. 2006). To address this issue, MacDonald et al. (2013) eliminated movement-related variables altogether by developing a head-fixed preparation for rats and recorded hippocampal CA1 activity while their memory was tested using an odor delayed matching to sample task. Each trial began with the presentation of a sample odor, followed by a fixed 2–5 s delay period, then presentation of a test odor. The restrained rats were rewarded with water for licking at a lick spout if the test odor matched the sample odor, but were not rewarded for licking when a nonmatching test odor was presented. This task was similar to the object–delay–odor sequence memory task in that there were a small number of highly repeated sequences that composed each combination of sample and test odors, and on each trial the rat had to remember the sample odor across the delay period to identify a target odor sequence.

Many hippocampal neurons activated at brief moments in sequence during the delay period. Therefore, even in head-fixed rats, hippocampal CA1 neurons segmented the delay period into discrete temporal units that reflected the flow of time within the trial. Moreover, many time cells were temporally modulated during the delay specifically following presentation of a particular odor that started the trial (Fig. 2.4a). Furthermore, most time cells contributed to a representation of only one odor memory while others contributed to more than one odor memory representation, though rarely to all four (Fig. 2.4a, b). In the latter case, some of these cells fired around the same time during delay following different odors, typically at different rates. Other cells had distinct temporal firing patterns after different sample odor presentations. Thus, each sample odor was represented during the delay by a largely distinct temporally organized ensemble of time cells. These data indicate that different neural ensembles activate in sequence over extended intervals to

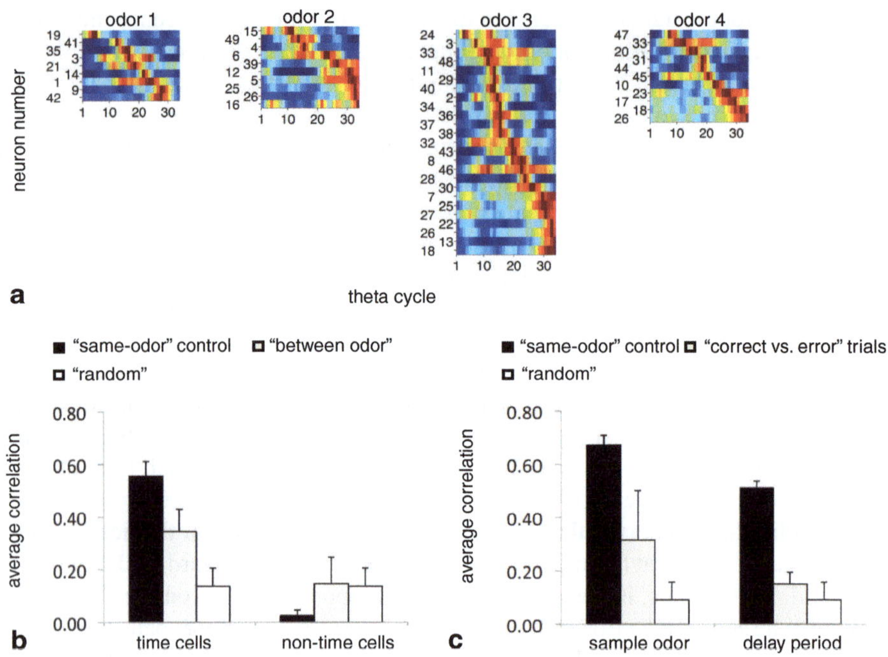

Fig. 2.4 Odor memory representations during the delay for each sample-odor defined trial type involved largely distinct, temporally organized neural ensemble activity. **a** Normalized firing rates over the delay for time cells (numerically labeled) for each of 4 sample-odor defined trial types in rat 5. **b** Average correlation coefficient between ensemble vectors for each trial type against the population vector for same set of neurons in all other trial types ("between-odor"). As one control, the average correlation coefficient between subsets of trials (even vs. odd) that began with the same odors is shown ("same-odor" control). As a second control, the average correlation coefficient between independent, randomly rearranged population vectors is shown ("random"). **c** For ensembles of cells that were temporally modulated in the sample odor or delay period, shown is the average correlation coefficient between populations vectors from correct trials that began with the same odor and error trials that began with the same odor ("correct vs. error" trials). The average correlation for the "same-odor" and "random" conditions are also shown. (From MacDonald et al. 2013)

compose the flow of time in specific odor memories. Moreover, the overlap among the different odor memories, embodied in cells that fire at the same or different rate at comparable moments during the delay, is consistent with the crucial role of the hippocampus in linking together different experiences (Eichenbaum et al. 1999; Eichenbaum 2004). Finally, these memory-specific, temporally organized representations predicted accurate memory performance, such that while ensemble representations were reliable during the sample and delay periods on successful trials, there was significantly less reliability during the sample phase and loss of the representation during the delay phase of error trials (Fig. 2.4c).

While the just described study revealed a temporal signal under conditions where head location was fixed and movement prevented, time cell firing patterns during movement could reflect path integration rather than elapsed time. To address this possibility, Kraus et al. (2013) recorded from multiple hippocampal neurons as rats

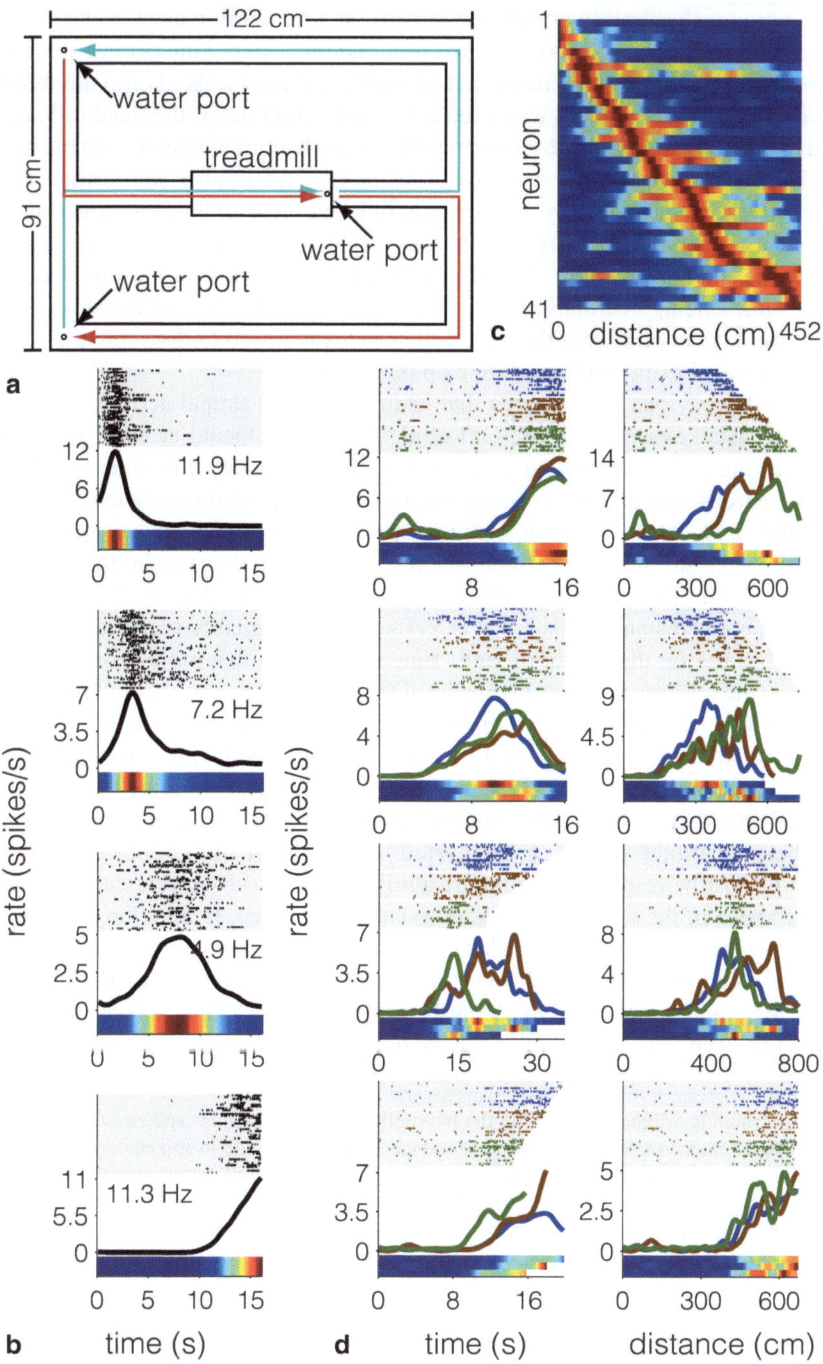

Fig. 2.5 Hippocampal activity during stationary treadmill running: temporal integration versus path integration. **a** Diagram of the figure-eight maze indicating the dimensions and location of the water ports and treadmill. *Cyan line* indicates *right-to-left* alternation; *red line* indicates *left-to-right* alternation. **b** Firing patterns of four different example neurons active during stationary

ran continuously in place at different speeds on a treadmill placed in the stem of a figure-eight maze (Fig. 2.5a). On each trial, the rats entered the central stem of the maze from one of two directions (left or right), and then walked onto the treadmill where they received a small water reward. After a short delay, the treadmill accelerated to a speed randomly chosen from within a predetermined range, and the rats ran in place until the treadmill stopped automatically and another small water reward was delivered. Subsequently, the animals finished the trial by turning in the direction opposite from their entry into the stem (spatial alternation) to arrive at a water port at the end of a goal arm. To distinguish behavior, location, time, and distance as factors influencing neuronal activity, behavior, and the location of the animal on the maze were "clamped," and the treadmill speed was varied to decouple the distance the rat traveled from its elapsed time on the treadmill.

As with previous experiments that examined hippocampal activity during task delays (Pastalkova et al. 2008; Gill et al. 2011; MacDonald et al. 2011), at each point during treadmill running a subset of hippocampal neurons fired, and the subset of neurons activated in a regular sequence that repeated during every treadmill run (Fig. 2.5b, c). In addition, running speed was systematically varied to allow post hoc analyses to separate the influences of time and distance on firing patterns, and to measure the extent to which each variable influenced firing. These analyses revealed both "distance cells," that is, cells that more reliably encoded the distance the rat has run on the treadmill, and "time cells," cells that more reliably encoded the time the rat has spent on the treadmill (Fig. 2.5d). The observation of "distance cells" in this task indicates that hippocampal neurons can integrate the length of a path even in the absence of visual flow usually associated with movement through space. Also, the presence of "distance cells" in this task indicates that these neurons are not driven entirely by network dynamics without the influence of either idiothetic or allothetic cues, as suggested by Pastalkova et al. (2008), because the neurons must be responding to the treadmill speed, or self-motion cues influenced by the speed of the treadmill, in order to encode distance. In addition, the observation of temporal modulation in addition to or without distance modulation indicates that these neurons are not exclusively driven by path integration but also by elapsed time (McNaughton et al. 1996, 2006; Etienne and Jeffery 2004). Thus, Kraus et al. (2013) showed that, when both of these dimensions are prominent, the hippocam-

---

treadmill running, aligned to the time the treadmill started. *Black lines* and *color bars* represent firing rate averaged over all runs. *Number* indicates peak firing rate in spikes per second (Hz). **c** Ensemble firing rate map showing all neurons active on the treadmill during a single session. Each row represents the normalized firing rate of one neuron, sorted by the peak firing time. In each row, *blue* represents no firing (zero spikes per second) and *red* represents peak firing for that particular neuron. **d** Examples shown in each row represent the activity from one neuron plotted both as a function of time since the treadmill started (*left column*) and distance traveled on the treadmill (*right column*). *Blue, brown, and green ticks* (*and tuning curves*) represent the slowest one third of runs, middle one third of runs, and fastest one third of runs, respectively. The rows in the raster plots in panels b and d are sorted with the slowest treadmill speed on top and fastest speed on the bottom. Note better alignment of the neural activity to time in the top two examples (time cells) and better alignment of neural activity to distance in the bottom two examples (distance cells). (Data from Kraus et al. 2013) pus represents both the distance traveled and time elapsed. Furthermore, a large fraction of hippocampal neurons combine information about these dimensions to varying extents, such that different neurons largely reflected distance or time and others equivalently reflected the combination of spatial and temporal dimensions, consistent with a unified representation of space and time attributes.

During treadmill running, when behavior and location were held relatively constant, time and distance predominated in their influence over the firing patterns of hippocampal neurons. However, other neurons, and many of the same neurons that were active on the treadmill, had place fields elsewhere on the maze, indicating that during other components of the task, where locations on the maze were important to task success, space was a strong influence over firing patterns of even the same neurons. These observations support the view that hippocampal neuronal activity reflects both the temporal and spatial regularities, along with other salient features of experience, consistent with a combined spatial-temporal organization of memories.

Conclusions

In 1987, Ray Kesner joined the then-prominent views of hippocampal function in spatial and temporal processing to propose that this brain area supported memory for the spatial–temporal context of memories. Many subsequent studies, including those of Kesner and his students, supported this idea, which we now recognize as a fundamental attribute of hippocampal dependent memory. Yet, most studies aimed to characterize the nature of information encoded by hippocampal neurons have focused solely on the spatial firing properties of hippocampal neurons and this has led to a separation between "navigation" (O'Keefe and Nadel 1978; Moser et al. 2008) and "memory" (Squire 2009) literatures on hippocampal function. However, the recent observations on temporal coding properties of hippocampal neurons, confirming Kesner's idea that the hippocampus also represents the temporal attributes of memories, offers a reconciliation of these views. The studies reviewed here show that the hippocampus is critical to memory for temporal organization independent of space, and the same neurons that are place cells when rats forage for food in open fields and traverse maze paths also fire sequentially when rats run in one location and when rats bridge gaps between remembered events independent of behavior and location. Furthermore, the hippocampus plays and replays sequences of place cell firings as a representation of spatial–temporal organization of memories. The combination of spatial and temporal organization can be considered fundamental to memory (Gallistel 1990).

These findings are examples of a growing set of studies that reveal a prominent role of the hippocampus in memory for temporal order in animals and humans, and provides a broad range of evidence for sequential activation of hippocampal neurons during memory retrieval of serial events in rats, monkeys, and humans. In particular, the existence of hippocampal "time cells" that encode moments in temporally extended memories, much as place cells encode locations in spatially extended environments, suggests that time, not place, is the fundamental dimension of hippocampal representation that is common to navigation and memory. Furthermore, recent evidence revealed temporal organization in hippocampal ensembles that exists prior to experiences, to which learning attaches specific memories (Dragoi and Tonegawa 2011). This observation of "preplay," which anticipates subsequent replay, suggests that temporal organization is primary, and may provide the scaffolding onto which spatial and nonspatial memories are hung. Combined with the other findings on time cells described above, these observations on temporal representation by hippocampal neurons offers considerable promise for a comprehensive understanding of the network mechanisms that underlie Kesner's prescient view on the spatial and temporal attributes of memory supported by the hippocampus.

References

Ainge, J. A., Tamosiunaite, M., Woergoetter, F., & Dudchenko, P. A. (2007). Hippocampal CA1 place cells encode intended destination on a maze with multiple choice points. *The Journal of Neuroscience, 27,* 9769–9779. doi:10.1523/JNEUROSCI.2011-07.2007.

Bower, M. R., Euston, D. R., & McNaughton, B. L. (2005). Sequential-context-dependent hippocampal activity is not necessary to learn sequences with repeated elements. *The Journal of Neuroscience, 25,* 1313–1323. doi:10.1523/JNEUROSCI.2901-04.2005.

Buhusi, C. V., & Meck, W. H. (2005). What makes us tick? Functional and neural mechanisms of interval timing. *Nature Reviews Neuroscience, 6,* 755–765. doi:10.1038/nrn1764.

Dragoi, G., & Tonegawa, S. (2011). Preplay of future place cell sequences by hippocampal cellular assemblies. *Nature, 469,* 397–401. doi:10.1038/nature09633.

Dupret, D., O'Neill, J., Pleydell-Bouverie, B., & Csicsvari, J. (2010). The reorganization and reactivation of hippocampal maps predict spatial memory performance. *Nature Neuroscience, 13,* 995–1002. doi:10.1038/nn.2599.

Eichenbaum, H. (2004). Hippocampus: Cognitive processes and neural representations that underlie declarative memory. *Neuron, 44,* 109–120. doi:10.1016/j.neuron.2004.08.028.

Eichenbaum, H. (2013). Memory on time. *Trends in Cognitive Sciences, 17,* 81–88. doi: 10.1016/j.tics.2012.12.007.

Eichenbaum, H., Dudchenko, P., Wood, E., Shapiro, M., & Tanila, H. (1999). The hippocampus, memory, and place cells: Is it spatial memory or a memory space? *Neuron, 23,* 209–226. doi:10.1016/j.neuron.2013.04.015.

Etienne, A. S., & Jeffery, K. J. (2004). Path integration in mammals. *Hippocampus, 14,* 180–192. doi:10.1002/hipo.10173.

Farovik, A., Dupont, L. M., & Eichenbaum, H. (2010). Distinct roles for dorsal CA3 and CA1 in memory for sequential nonspatial events. *Learning & Memory, 17,* 12–17. doi: 10.1101/lm.1616209.

Ferbinteanu, J., & Shapiro, M. L. (2003). Prospective and retrospective memory coding in the hippocampus. *Neuron, 40,* 1227–1239. doi:10.1016/S0896-6273(03)00752-9.

Fortin, N. J., Agster, K. L., & Eichenbaum, H. B. (2002). Critical role of the hippocampus in memory for sequences of events. *Nature Neuroscience, 5,* 458–462. doi:10.1038/nn834.

Frank, L. M., Brown, E. N., & Wilson, M. (2000). Trajectory encoding in the hippocampus and entorhinal cortex. *Neuron, 27,* 169–178. doi:10.1016/S0896-6273(00)00018-0.

Gallistel, C. R. (1990). *The organization of learning.* Cambridge: MIT Press.

Gibbon, J., Church, R. M., & Meck, W. H. (1984). Scalar timing in memory. *Annals of the New York Academy of Sciences, 423,* 52–77. doi:10.1111/j.1749-6632.1984.tb23417.x.

Gilbert, P. E., & Kesner, R. P. (2003). Localization of function within the dorsal hippocampus: the role of the CA3 subregion in paired-associate learning. *Behavioral Neuroscience, 117,* 1385–1394. doi:10.1037/0735-7044.117.6.1385.

Gill, P. R., Mizumori, S. J., & Smith, D. M. (2011). Hippocampal episode fields develop with learning. *Hippocampus, 21,* 1240–1249. doi:10.1002/hipo.20832.

Ginther, M. R., Walsh, D. F., & Ramus, S. J. (2011). Hippocampal neurons encode different episodes in an overlapping sequence of odors task. *The Journal of Neuroscience, 31,* 2706–2711. doi:10.1523/JNEUROSCI.3413-10.2011.

Gothard, K. M., Skaggs, W. E., McNaughton B. L. (1996) Dynamics of mismatch correction in the hippocampal ensemble code for space: interaction between path integration and environmental cues. *The Journal of Neuroscience, 16,* 8027–8040.

Griffin, A. L., Eichenbaum, H., & Hasselmo, M. E. (2007). Spatial representations of hippocampal CA1 neurons are modulated by behavioral context in a hippocampus-dependent memory task. *The Journal of Neuroscience, 27,* 2416–2423. doi: 10.1523/JNEUROSCI.4083-06.2007.

Gupta, A. S., van der Meer, M. A., Touretzky, D. S., & Redish, A. D. (2010). Hippocampal replay is not a simple function of experience. *Neuron, 65,* 695–705. doi:10.1016/j.neuron.2010.01.034.

Hunsaker, M. R., Thorup, J. A., Welch, T., & Kesner, R. P. (2006). The role of CA3 and CA1 in the acquisition of an object-trace-place paired associate task. *Behavioral Neuroscience, 120,* 1252–1256. doi:10.1037/0735-7044.120.6.1252.

Jacobs, N. S., Allen, T. A., Nguyen, N., & Fortin, N. J. (2013). Critical role of the hippocampus in memory for elapsed time. *The Journal of Neuroscience, 33,* 13888–13893. doi:10.1523/JNEUROSCI.1733-13.2013.

Jadhav, S. P., Kemere, C., German, P. W., & Frank, L. M. (2012). Awake hippocampal sharp-wave ripples support spatial memory. *Science, 336,* 1454–1458. doi:10.1126/science.1217230.

Ji, D., & Wilson, M. A. (2007). Coordinated memory replay in the visual cortex and hippocampus during sleep. *Nature Neuroscience, 10,* 100–107. doi:10.1038/nn1825.

Johnson, A., & Redish, A. D. (2007). Neural ensembles in CA3 transiently encode paths forward of the animal at a decision point. *The Journal of Neuroscience, 27,* 12176–12189. doi:10.1523/JNEUROSCI.3761-07.2007.

Karlsson, M. P., & Frank, L. M. (2009). Awake replay of remote experiences in the hippocampus. *Nature Neuroscience, 12,* 913–918. doi:10.1038/nn.2344.

Kesner, R. P. (1990). Learning and memory in rats with an emphasis on the role of the hippocampal formation. In: R. P. Kesner & D. S. Olton (Eds.), *Neurobiology of comparative cognition* (pp. 179–205). New Jersey: Erlbaum.

Kesner, R. P., & DiMattia, B. V. (1987). Neurobiology of an attribute model of memory. In: A. R. Morrison & A. N. Epstein (Eds.), *Progress in psychobiology and physiological psychology* (pp. 207–277). New York: Academic.

Kesner, R. P., & Hunsaker, M. R. (2010). The temporal attributes of episodic memory. *Behavioural Brain Research, 215,* 299–309. doi:10.1016/j.bbr.2009.12.029.

Kesner, R. P., & Novak, J. (1982). Serial position curve in rats: Role of the dorsal hippocampus. *Science, 218,* 173–175. doi:10.1126/science.7123228.

Kesner, R. P., Gilbert, P. E., & Barua, L. A. (2002). The role of the hippocampus in memory for the temporal order of a sequence of odors. *Behavioral Neuroscience, 116,* 286–290. doi:10.1037/0735-7044.116.2.286.

Kesner, R. P., Hunsaker M. R., & Gilbert, P.E. (2005). The role of CA1 in the acquisition of an object–trace–odor paired associate task. *Behavioral Neuroscience, 119,* 781–786. doi:10.1037/0735-7044.119.3.781

Komorowski, R. W., Manns, J. R., & Eichenbaum, H. (2009). Robust conjunctive item-place coding by hippocampal neurons parallels learning what happens where. *The Journal of Neuroscience, 29,* 9918–9929. doi:10.1523/JNEUROSCI.1378-09.2009.

Komorowski, R. W., Garcia, C. G., Wilson, A., Hattori, S., Howard, M. W., & Eichenbaum, H. (2013). Ventral hippocampal neurons are shaped by experience to represent behaviorally relevant contexts. *The Journal of Neuroscience, 33,* 8079–8087. doi:10.1523/JNEUROSCI.5458–12.2013.

Kraus, B. J., Robinson II, R. J., White, J. A., Eichenbaum, H., & Hasselmo, M. E. (2013). Hippocampal "time cells": Time versus path integration. *Neuron, 78,* 1090–1101. doi:10.1016/j.neuron.2013.04.015.

Lee, I., Griffin, A. L., Zilli, E. A., Eichenbaum, H., & Hasselmo, M. E. (2006). Gradual translocation of spatial correlates of neuronal firing in the hippocampus toward prospective locations. *Neuron, 51,* 639–650. doi:10.1016/j.neuron.2006.06.033.

Lenck-Santini, P. P., Save, E., & Poucet, B. (2001). Place-cell firing does not depend on the direction of turn in a Y-maze alternation task. *The European Journal of Neuroscience, 13,* 1055–1058. doi:10.1046/j.0953-816x.2001.01481.x.

MacDonald, C. J. (2014). Prospective and retrospective duration memory in the hippocampus: Is time in the foreground or background? *Proceedings of the Royal Society B: Biological Processes,* 2014 Jan 20; *369*(1637), 20120463.

MacDonald, C. J., Lepage, K. Q., Eden, U. T., & Eichenbaum, H. (2011). Hippocampal "time cells" bridge the gap in memory for discontiguous events. *Neuron, 71,* 737–749. doi:10.1016/j.neuron.2011.07.012.

MacDonald, C. J., Place, R., Carrow, S., & Eichenbaum, H. (2013). Distinct hippocampal time cell sequences represent odor memories in immobilized rats. *The Journal of Neuroscience, 33,* 14607–14616. doi:10.1523/JNEUROSCI.1537-13.2013.

Manns, J. R., & Eichenbaum, H. (2005). Time and treason to the trisynaptic teachings: theoretical comment on Kesner et Al. (2005). *Behavioral Neuroscience, 119,* 1140–1143. doi:10.1037/0735-7044.119.4.1140.

Manns, J. R., & Eichenbaum, H. (2009). A cognitive map for object memory in the hippocampus. *Learning and Memory, 16,* 616–624. doi:10.1101/lm.1484509.

Manns, J. R., Zilli, E. A., Ong, K. C., Hasselmo, M. E., & Eichenbaum, H. (2007) Hippocampal CA1 spiking during encoding and retrieval: Relation to theta phase. *Neurobiology of Learning and Memory, 87,* 9–20. doi:10.1016/j.nlm.2006.05.007.

Mauk, M. D., & Buonomano, D. V. (2004). The neural basis of temporal processing. *Annual Review of Neuroscience, 27,* 307–340. doi:10.1146/annurev.neuro.27.070203.144247.

McNaughton, B. L., Barnes, C. A., Gerrard, J. L., Gothard, K., Jung, M. W., Knierim, J. J., Kudrimoti, H., Qin, Y., Skaggs, W. E., Suster, M., & Weaver, K. L. (1996). Deciphering the hippocampal polyglot: The hippocampus as a path integration system. *The Journal of Experimental Biology, 199,* 173–185.

McNaughton, B. L., Battaglia, F. P., Jensen, O., Moser, E. I., & Moser, M. B. (2006). Path integration and the neural basis of the 'cognitive map'. *Nature Reviews Neuroscience, 7,* 663–678. doi:10.1038/nrn1932.

Moita, M. A. P., Rosis, S., Zhou, Y., LeDoux, J. E., & Blair, H. T. (2003). Hippocampal place cells acquire location-specific responses to the conditioned stimulus during auditory fear conditioning. *Neuron, 37,* 485–497. doi:10.1016/S0896-6273(03)00033-3.

Meck, W. H., Church, R. M., & Olton, D. S. (1984). Hippocampus, time, and memory. *Behavioral Neuroscience 98,* 3–22.

Moser, E. I., Kropff, E., & Moser, M. B. (2008). Place cells, grid cells, and the brain's spatial representation system. *Annual Review of Neuroscience, 31,* 69–89. doi:10.1146/annurev.neuro.31.061307.090723.

Naya, Y., & Suzuki, W. A. (2011). Integrating what and when across the primate medial temporal lobe. *Science, 333,* 773–776. doi:10.1126/science.1206773.

O'Keefe, J., & Nadel, L. (1978). *The Hippocampus as a cognitive map.* Oxford: Oxford University Press.

Olton, D. S. (1986). Hippocampal function and memory for temporal context. In: R. L. Isaacson & K. H. Pribram (Eds.), *The Hippocampus* (pp. 281–298). Philadelphia: Springer US.

Pastalkova, E., Itskov, V., Amarasingham, A., & Buzsáki, G. (2008). Internally generated cell assembly sequences in the rat hippocampus. *Science, 321,* 1322–1327. doi:10.1126/science.1159775.

Paz, R., Gelbard-Sagiv, H., Mukamel, R., Harel, M., Malach, R., & Fried, I. (2010). A neural substrate in the human hippocampus for linking successive events. *Proceedings of the Na-*

tional Academy of Sciences of the United States of America, 107, 6046–6051. doi:10.1073/pnas.0910834107.

Pfeiffer, B. E., & Foster, D. J. (2013). Hippocampal place-cell sequences depict future paths to remembered goals. *Nature, 497,* 74–79. doi:10.1038/nature12112.

Rawlins, J. N. P. (1985). Associations across time: The hippocampus as a temporary memory store. *Behavioral and Brain Sciences, 8,* 479–528. doi:10.1017/S0140525X00001291.

Redish, A. D., Rosenzweig, E. S., Bohanick, J. D., McNaughton, B. L., & Barnes, C. A. (2000). Dynamics of hippocampal ensemble activity realignment: time versus space. *The Journal of Neuroscience, 20,* 9298–9309.

Robitsek, J. R., White, J. A., & Eichenbaum, H. (2013). Place cell activation predicts subsequent memory. *Behavioural Brain Research, 254,* 65–72. doi:10.1016/j.bbr.2012.12.034.

Shapiro, M. L., Kennedy, P. J., & Ferbinteanu, J. (2006). Representing episodes in the mammalian brain. *Current Opinion in Neurobiology, 16,* 701–709. doi:10.1016/j.conb.2006.08.017.

Singer, A. C., Carr, M. F., Karlsson, M. P., & Frank, L. M. (2013). Hippocampal SWR activity predicts correct decisions during the initial learning of an alternation task. *Neuron, 77,* 1163–1173. doi:10.1016/j.neuron.2013.01.027.

Squire, L. R. (2009). Memory and brain systems: 1969–2009. *The Journal of Neuroscience, 29,* 12711–12716. doi:10.1523/JNEUROSCI.3575-09.2009.

Wilson, M. A., & McNaughton, B. L. (1994). Reactivation of hippocampal ensemble memories during sleep. *Science, 265,* 676–679. doi:10.1126/science.8036517.

Wood, E. R., Dudchenko, P. A., & Eichenbaum, H. (1999). The global record of memory in hippocampal neuronal activity. *Nature, 397,* 613–616. doi:10.1038/17605.

Wood, E. R., Dudchenko, P. A., Robitsek, R. J., & Eichenbaum, H. (2000). Hippocampal neurons encode information about different types of memory episodes occurring in the same location. *Neuron, 27,* 623–633. doi:10.1016/S0896-6273(00)00071-4

Yin, B., & Troger, A. B. (2011). Exploring the 4th dimension: Hippocampus, time, and memory revisited. *Frontiers in Integrative Neuroscience, 5,* Article 36. doi:10.3389/fnint.2011.00036.

Chapter 3
Space, Time, and the Hippocampus

Lara M. Rangel, Laleh K. Quinn and Andrea A. Chiba

The "Tapestry of Memory" (Fig. 3.1)

> What is the basic constituent of a memory? What is lost when we say that we have (alas) forgotten? (Underwood 1977)

Benton Underwood's (1969) notion of memory asserted that memory was composed of many attributes, or different types of information. Building upon this idea, Kesner first proposed (1980) that all memories are composed of a set of six salient features or attributes: space, time, affect, sensory perception, response, and language (in humans). Each experience would incorporate a specific and unique combination of attributes, and would be supported by neural processes. This attribute-based model of memory greatly advanced memory research in two important ways. First, the model defined memory as a distributed neural process. This definition asserted that memory, by necessity, could not be accomplished by a single brain region, but would instead require the integration of multiple memory systems thought, traditionally, to act independently. Investigations of this multidimensional model of memory from a neurobiological approach would, therefore, rely heavily on both the anatomy of individual brain regions for examining their individual contributions to memory, and upon the connections between regions for their cooperative function at a systems level. Second, in theory, the inclusion of multiple attributes

A. A. Chiba (✉)
Department of Cognitive Science and Program in Neuroscience,
Temporal Dynamics of Learning Center, University of California, San Diego,
9500 Gilman Drive, La Jolla, CA 92093-0515, USA
e-mail: achiba@ucsd.edu

L. M. Rangel
NSF Cognitive Rhythms Collaborative, College of Arts and Sciences Psychology,
Boston University, 2 Cummington St., Boston, MA 02135, USA

L. K. Quinn
Department of Cognitive Science, University of California, San Diego, 9500 Gilman Drive,
La Jolla, CA 92093-0515, USA

P. A. Jackson et al. (eds.), *The Neurobiological Basis of Memory*,
DOI 10.1007/978-3-319-15759-7_3

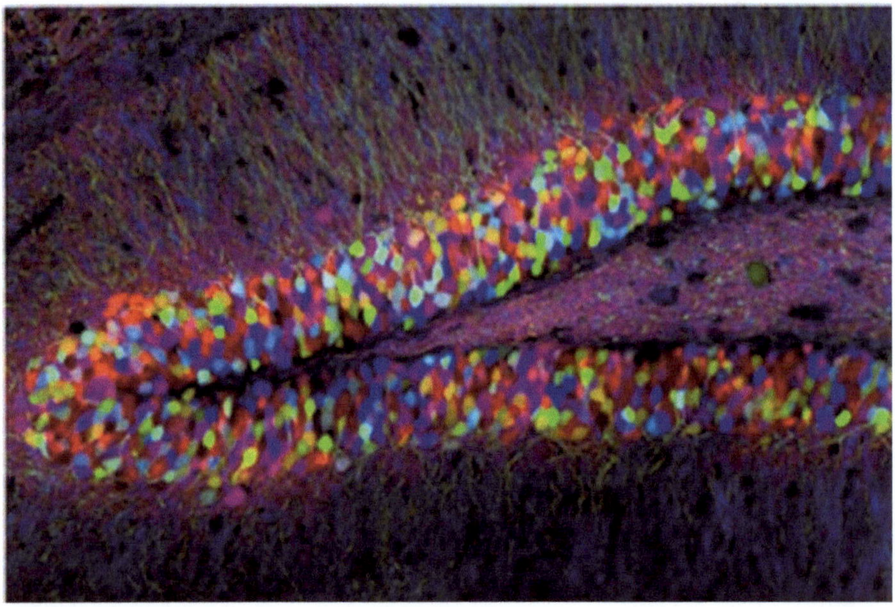

Fig. 3.1 This image shows the dentate gyrus of the hippocampus, a region of frequent discussion in this chapter. This photomicrograph was taken from a "Brainbow" transgenic mouse that allows distinction between neighboring neurons through color. (Photograph: J Livet (now Institut de la Vision, Paris), J W Lichtman, and J R Sanes (Harvard University))

would increase the dimensionality of a given experience, thus capturing the brain's computational ability to increase similarity or reduce interference between multiple experiences. The incorporation of multidimensional information processing into the function of a given brain region would dramatically affect computational models and experimental tests of memory processes, particularly in the hippocampus.

A primary assertion of the attribute-based model of memory is that functional circuits of the brain support attributes. In order for an organism to represent memory for an attribute, incoming sensory information must be encoded and temporarily stored within a neural system. In representing a spatial attribute, for example, the encoding and temporary storage of specific stimuli representing spatial locations, directions, and distances, which may or may not be independent of the subject's own body schema, must occur. A temporal attribute represents the occurrence of an episode in time, separating the episode from past or future episodes, as well as coding the duration of the episode.

The inclusion of both sensory perception and motoric (i.e., response) functions as essential to memory processes places this active account of memory within the realm of modern embodied cognition, requiring the intrinsic and positional state of an organism to be part of the initial processing of memory. Within this realm, the attribute of affect can be experiential or retrospective in that it involves the encoding and temporary storage of reinforcement contingencies that result in positive or negative emotional experiences in the visceral sense, which could subsequently be categorized. The interaction between memory for individual attributes, as a function of the activity of various neurobiological regions and their processes, combine to represent a unique memory.

Additionally, the attribute model accounts for differential processing of information by incorporating interconnected memory systems. For example, a data or an event-based memory system that emphasizes encoding of incoming information, combined with an expectancy or knowledge-based system that emphasizes top-down processing, allow for fluid use of previous knowledge in interpreting incoming information. Kesner emphasizes (Kesner 1990) that most situations require multiple such memory systems with disproportionate involvement of a system or two at any particular time. The theory is deeply rooted in the anatomy of the system, with an early understanding that the connectivity maps of the brain (an early embrace of the basis of modern "connectomics") are absolutely essential to the patterns of neural activity and the content of the ultimate recollection. Taking this perspective, memory is labile from the outset, and memories rely on the timing and availability of activation at the moment of recollection. Remember a time, for example, when you recalled an event (perhaps, a conversation with someone), without remembering when it occurred and you proceeded to reconstruct the context in order to remember the time of occurrence. This exemplifies the way in which the availability of a particular attribute can lead to an aggregation of the memory. From the perspective of memory processes, this can also be the point at which interference is reduced and the memory is effectively separated from other similar memories. Thus, memory is a multidimensional, distributed process.

The Functional Anatomy of Spatial and Temporal Memory Attributes

Whereas the theory concludes that behavioral or psychological processes are supported by brain function, the mapping of structure to function has taken an important turn towards a processing account of memory. Such an account acknowledges the important fact that the way in which behavioral or psychological functions are supported is reliant on a principled account of a brain that bears no obligation to function according to the psychological labels that are imposed on it. As such, a careful parsing of the *computational processes* subserved by the neural architecture is explored with respect to their ultimate role in mediating behavioral function (Kesner and Rolls 2001; Rolls and Kesner 2006).

Reducing Interference by Separating Attributes

The architecture of the hippocampus both constrains and allows for the separating or linking of specific types of information in the service of memory. For example, original computational models describing the hippocampal circuit endowed the dentate gyrus (DG) with the ability to pattern separate (McClelland et al. 1995;

O'Reilly and McClelland 1994; Rolls 2010; Treves and Rolls 1992). The idea of pattern separation addresses the requirement that there must be a mechanism to reduce interference of input patterns in order to form separate representations that will be transmitted to downstream targets. Pattern separation in such networks is based on the notion of orthogonalization. The DG in the rat has approximately 1 million neurons. It has more principal cells than the upstream entorhinal cortex (EC) and downstream CA3 combined (van Strien et al. 2009). A network with the anatomical properties contained in the EC–DG circuit is ideally situated to achieve highly disparate (non-overlapping) outputs, as the number of nodes in the DG network is orders of magnitude higher than the number of input nodes, thus allowing for a sparse and independent representation of overlapping inputs. If not true orthogonalization, the anatomy of the EC–DG circuit suggests that the dentate would at least act as a sparsifying network that encodes inputs in a non-distributed manner. The encoding of an experience containing multiple attributes would, thus, create a highly unique pattern of activity in DG, with each attribute acting as an additional means to separate or reduce interference of the memory from other experiences. In support of this hypothesis, experimental evidence has shown that DG neurons create more distinct representations of experiences at the single cell and population level than other subregions of the hippocampus (Deng et al. 2013; Leutgeb et al. 2007; Neunuebel and Knierim 2014; Rangel and Eichenbaum 2013).

Reinstating Memories from Linked Attributes

Early models also proposed that through Hebbian learning, or repeated experience, disparate attributes could be linked together in the CA3 region of the hippocampus such that incomplete features of a memory could reinstate the full original experience, a process called *pattern completion* (McClelland et al. 1995; O'Reilly and McClelland 1994). Specifically, repeated experience would strengthen the synaptic connections among activated neurons within CA3 through a long-term potentiation (LTP)-like mechanism, and partial or noisy activation would utilize CA3 recurrent collaterals to recruit linked neurons. The encoding of multiple attributes in this system therefore provides more avenues from which to reconstruct existing links or associations. Thus, attributes make memories more distinct in one hippocampal subregion, and more similar through acquired associations in another (Fig. 3.2).

Space and Time in Context

Space and Place Cells

The Kesner attribute model proposes that the rich architecture of the hippocampus supports the ongoing processing of space and time (Kesner et al. 1989). The contemporary accounts of the function of the hippocampus were entrenched in the

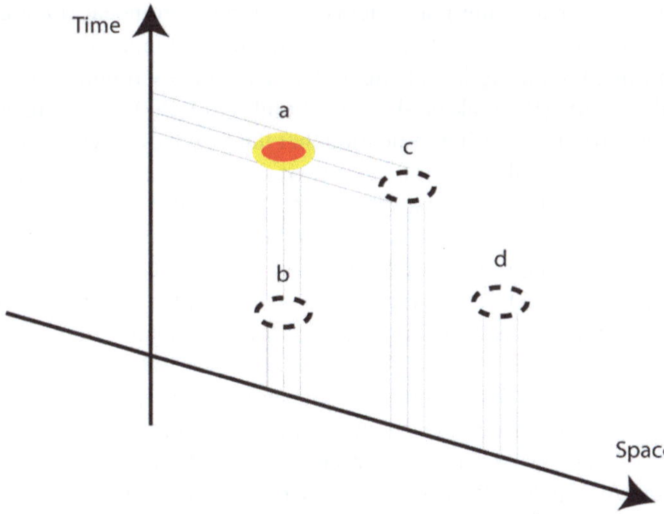

Fig. 3.2 Using both spatial and temporal attributes to link and separate new memories: event *b* occurs in the same place but at a different time than *a*, event *c* occurs in a different place but at nearly the same time as *a*, and event *d* occurs at a different time and place from *a*

powerful discovery of "place-cells" by O'Keefe and Dostrovsky (O'Keefe and Dostrovsky 1971) and the proposition that the hippocampus was the locus of the spatial "Cognitive Map," or our innate knowledge of space (O'Keefe and Nadel 1978). Spatial cognition in this case refers to a perception of the external world that is readily available and usable in every organism. The process of cognitive mapping, theoretically achieved by the hippocampus, can be described as:

> … a construct that encompasses those processes that enable people to acquire, code, store, recall, and manipulate information about the nature of their spatial environment. It refers to the attributes and relative locations of people and objects in the environment, and is an essential component in the adaptive process of spatial decision-making such as finding a safe and quick route to and from work, locating potential sites for a new house or business, and deciding where to travel on a vacation trip. Cognitive processes are not constant, but undergo change with age or development and use or learning. (O'Keefe and Nadel 1978)

If spatial cognition relies upon a cognitive map that is an innate ability in all organisms, then spatial information should be available in the hippocampus during novel exposures to spatial environments and prior to any learning. Indeed, single cells in the hippocampus demonstrate spatially specific activity in the form of *place fields* during even the first few minutes of novel exposures to an environment (Kentros et al. 1998). These cells additionally demonstrate large coverage of spatial environments at predictable spatial resolution along the septo-temporal axis (Kjelstrup et al. 2008). The hippocampus, thus, has the means to provide a spatial construct at the single cell level.

Place cell activity over the course of familiarity with a new environment is additionally reflective of increasing perception of space. A large body of experimental evidence suggests that stable place fields are highly contingent upon experience. Although these cells demonstrate place specific activity immediately, they remain unstable and flexible during initial encounters with an environment before demonstrating stable fields (Bostock et al. 1991; Frank et al. 2006; Kentros et al. 2004; Rowland et al. 2011). As further indication that the activity of these cells is linked to spatial perception, spatially specific activity of these cells is closely associated with animal movement and perceived location, rather than absolute allocentric location. Specifically, the firing rate of place cells as an animal travels through its place field can be heavily modulated by speed, direction, and trajectory (Frank et al. 2000; McNaughton et al. 1983). Moreover, rotations of spatial cues surrounding an environment cause predictable shifts in place field location relative to the degree of cue rotation (Lenck-Santini et al. 2005; Poucet et al. 2000). Taken together, these findings suggest that place cells not only provide an internal representation of a spatial environment, but also encode these features in a behaviorally meaningful manner.

If place cells enable an internal spatial representation of the world, then testing the extent to which spatial firing properties of place cells account for learned features of an environment and changing behavioral conditions can help determine their ultimate contribution to learning and behavior. Experimental evidence has demonstrated that even cells with stable place fields can demonstrate changes in firing rate or location when fields are in close proximity to changing components of an environment (Lenck-Santini et al. 2005; Rivard et al. 2004). Moreover, their long-term stability has been correlated with spatial learning performance, and their instability in aging is correlated with a decline in spatial learning ability (Kentros et al. 2004; Shen et al. 1997). Thus, in addition to providing a flexible spatial representation of an environment, in a way that is behaviorally meaningful to an organism, these place cells may be utilized and perhaps required for specific types of spatial learning.

Lesions of the hippocampus result in a long-term inability to encode episodic memories, or the conscious knowledge of specific personal experience (Milner et al. 1998; Rosenbaum et al. 2000, 2005; Tulving 2002). Rats and humans with hippocampal lesions or inactivations maintain an ability to navigate novel environments and perform spatial learning tasks over time. In rats, lesions of the hippocampus impair performance in spatial learning tasks (Morris et al. 1990; Olton and Papas 1979; Olton et al. 1978). Over a significantly longer period of time, however, successful performance in these tasks can be achieved, suggesting secondary mechanisms for forming spatial associations with additional knowledge (DiMattia and Kesner 1988). This is consistent with other studies demonstrating that the behavior of rats with hippocampal lesions reflects maintained perceptual learning of their spatial environment (Jackson-Smith et al. 1993). Moreover, rats with fimbria/fornix lesions that demonstrate spatial learning impairment, still exhibit the presence of place cell activity, suggesting that place cells alone are insufficient for spatial knowledge (Whishaw et al. 1995). In clinical research, humans with lesions of the hippocampus maintain sufficient spatial orientation to demonstrate an ability to navigate their current living environments (Milner et al. 1968). These results suggest that in the absence of a hippocampus, an internal spatial map is available for use, and according to the attribute model this is likely supported by the parietal cortex (Chiba et al. 2002). It is thus possible that the hippocampus is not necessary for all spatial processing per se, but rather the knowledge of the map's appropriate utility.

Yet, it is clear that humans and rodents without a hippocampus lack an explicit perception of changing environmental conditions. Whereas in the rat it is difficult to claim the presence of conscious spatial knowledge, previous research has used the ability to generate decisions (i.e., "declare") based on appropriate knowledge of spatial learning contingencies as evidence of declarative memory (DeCoteau and Kesner 2000). Even though rats with hippocampal lesions can perform spatial learning tasks over long periods of time, they demonstrate inflexibility in their ability to adapt to changing spatial conditions (Jacobson et al. 2011). This is in high contrast to control rats, which instead demonstrate faster learning and adaptation to changing environmental conditions with increased experience (Tse et al. 2007). Humans with hippocampal lesions demonstrate a similar inflexibility in being able to update their perception of changing environmental conditions. This is coupled by an inability to consciously recall the utility of their current spatial surroundings (Milner et al. 1968).

Thus, in both rats and humans, the ability to assess the appropriate utility of a map is impoverished without a hippocampus, despite the availability of spatial knowledge in other areas of the brain. Indeed, cells in other areas of the brain, such as parietal cortex, demonstrate place specific firing in a manner analogous to place cells in the hippocampus but with additional properties allowing knowledge of spatial routes and position (Nitz 2009), that may account for the maintenance of spatial perception following hippocampal lesions (Chiba et al. 2002). Here, both systems are privy to the spatial code of the EC (Leutgeb et al. 2005), providing a map of the local spatial topography of the environment.

It has been hypothesized that the function of the hippocampal spatial map is to serve as a lattice for memory, on which episodes can be superimposed (Burgess et al. 2002; de Pontes et al. 2005). Kesner's attribute model set forth the convergence of the spatiotemporal code as the defining feature of an episode. This view was influenced by Milner's (Milner and Penfield 1956) early work with HM and exemplified his foresight in developing a model that could account both for the human amnestic syndrome (the inability to code new memories in both space and time) that arises from hippocampal damage, and the obligatory spatial code of the hippocampal architecture. Other current theories of the hippocampus suggest that the hippocampus creates relationships between important features across experiences and is thus essential for both spatial and nonspatial, or relational memory (Cohen et al. 1997). Kesner's model specifically endows the CA3 "autoassociative network" with the capacity to form arbitrary or relational associations with the space (Kesner 2013). Both viewpoints assert that spatial encoding does not exist in isolation from the encoding of other attributes and that memories include the relationships between space and other dimensions that together compose rich contextual knowledge.

Time and Time Cells

The concept that time serves as an organizing principle for memory is age-old but not antiquated. Aristotle established, the "principle of contiguity" as one of his "Laws of Association" based on the general finding that recall of an item is facilitated by the presentation or recall of another item that occurred close in time to the target item (Aristotle and Barnes 1984). The role of temporal context in memory has since been extensively studied. After writing "Attributes of Memory," Underwood elaborated on the role of temporal context and order in his book, "Temporal Codes for Memory." There, he too emphasized that those items occurring in close temporal proximity were more likely to be conjoined whereas those occurring with greater temporal distance were more likely to be distinguishable. Shortly thereafter, Kesner designed a variety of tasks that paralleled contemporary human experiments, such as list-learning experiments, for use with rats. He demonstrated that serial position effects were constant across species and that both retroactive and proactive interference were present in rat models of list learning in which different places or maze arms represented the elements in the list (Kesner and Novak 1982; Kametani and Kesner 1989; Kesner et al. 1989). Since then, many studies have utilized the rat model organism to further investigate the neural substrates and underlying mechanisms of temporal order deficits and the encoding of temporal sequences (Allen et al. 2014; Howard et al. 2005; Howard and Kahana 2002). The creation of associations between temporally proximal events contributes greatly to episodic memory formation in humans. Specifically, when subjects bias their retrieval strategy to rely on temporal associations, they perform better on episodic recall tasks (Sederberg et al. 2010).

The role of the hippocampus in time and temporal order is likely to be fundamental to the role of the hippocampus in coding episodes that are essential to an individual's ongoing autobiography. The role of the hippocampus with respect to time is complex and occurs in a series of nested timescales from the short duration of spike timing within oscillatory neural circuits to the construal of time with respect to place and context. With respect to memory, time has been studied regarding the basic substrates of neuronal coding (firing rates, spike-time dependent plasticity, and rhythmic oscillatory activity), the sequential order of events (including succession, temporal order, and relative recency), and memory for the duration of events including memory for intervals or time periods between events (Jackson-Smith et al. 1993; Pastalkova et al. 2008). Just as spatial memories require organized associations between spatial features of an environment, the encoding of events in time also requires a temporal organization that can account for similar and distinct temporal features. Experimental evidence suggests that single cells in the hippocampus can demonstrate reliable, and temporally selective, sequential activity during a given length of time in a manner similar to place cell activity for distinct spatial locations over a given spatial environment (MacDonald et al. 2011; Munn and Bilkey 2011). These cells, labeled *time cells*, are some of the first evidence to suggest that the hippocampus may have a temporal organization for episodes that is very similar to its characterized mechanism for spatial organization of environments. More importantly, these findings are evidence for single cell encoding of a temporal dimension or attribute.

Conjoining Space and Time

To better understand why brain regions such as the hippocampus would need to encode temporal as well as spatial features into memory, it is important to know that space and time are linked and inseparable in nature. In fact, the physical dimensions of space and time are often considered together and referred to as *spacetime*, whereby they do not have separate existences. In extreme cases such as the realm of relativistic physics, this can mean that events occurring billions of years ago at the farthest reaches of the universe can occur simultaneously with your thoughts as you look up at the light from that event shining in the earth's sky. In our everyday lives, it can mean that the changes we observe in places we have known since childhood are measured by, and are indicators of the passage of time. In the latter case, which refers to our own egocentric view of the universe, we realize that memories of places ("where") from our childhood are dependent upon an inseparable temporal ("when") component. Since space and time are inseparable in nature, it should be tested whether the classical separation of space and time in the brain and in the encoding of memories is an artificial division. Although place and time cells provide a mechanism through which spatial and temporal associations can be made along two separate dimensions, it remains to be tested how associations are encoded across these dimensions. How are spatial features of memory linked to events in time? The unity of these dimensions could be accomplished by place cells dependent upon temporal features or time cells with spatial contingencies. To this end, it has been demonstrated that the temporal organization of place cell activity with respect to the phase of theta (4–12 Hz) frequency oscillations in the hippocampus is related to a rat's movement through space, a phenomenon that has been termed phase precession (Dragoi and Buzsáki 2006; Skaggs et al. 1996; Tsodyks et al. 1996). This phenomenon is observed in each of the subregions of the hippocampus. Additionally, sharp wave ripple events in CA3 and CA1 in the hippocampus elicit activity resembling the sequential firing of place cells during task behavior, providing evidence for the encoding of temporal order for spatial experiences by the network (Davidson et al. 2009; Gupta et al. 2010; Jadhav et al. 2012). Both phenomena are promising evidence that the hippocampus represents both spatial and temporal information together at short-time scales.

Conjoining Space and Time in the DG

A subregion within the hippocampal formation, the DG, has been hypothesized to combine both spatial and temporal dimensions at the single cell level (Aimone et al. 2009). The DG is the only subregion of the hippocampus that demonstrates neurogenesis, or the continuous birth of new neurons, throughout adulthood. Adult-born neurons are born in the subgranular zone of the DG granule cell layer and demonstrate a characteristic development before becoming mature functional granule cells. Importantly, immature adult-born neurons exhibit a transient period of both intrinsic and synaptic hyperexcitability that is due to low membrane capacitance and less synaptic inhibition, respectively (Esposito et al. 2005; Laplagne et al. 2006, 2007; Piatti et al. 2006). This transient physiological difference between mature and immature granule cells may yield a unique role for adult-born cells in temporal encoding. Computational models demonstrate that temporally proximal events occurring within the transient period of hyperexcitability for a set of adult-born neurons elicit activity from common immature cells in an otherwise sparse firing mature network (Aimone et al. 2006). There then exists a similarity in DG output for temporally proximal events that does not exist for events separated in time, a *temporal pattern integration*, that is the direct result of adult-born neuron physiology during development (Aimone et al. 2006, 2009). This temporal pattern integration can then ultimately link disparate features in the spatial dimension through close proximity along the temporal dimension.

The transient period of hyperexcitability in adult-born neuron development is also a critical period for regulation of their survival and activity. Although a majority of adult-born cells die before becoming mature granule cells, exposure to learning paradigms or enriching environments during this critical period can greatly enhance their survival and bias their activity toward input received during their development (Aimone et al. 2006; Tashiro et al. 2007). This has led to the prediction that surviving adult-born neurons provide dedicated and selective activity to temporally proximal events during their development and thus can create new outputs from the dentate that are temporally distinct. Adult neurogenesis in the DG can therefore provide a mechanism for an additional type of pattern separation, a *temporal pattern separation*, through the continuous contribution of new temporal dimensions to distinguish between similar events separated in time (Aimone et al. 2011). Recent behavioral studies have demonstrated that the ability to make accurate spatiotemporal order judgments relies on the integrity of the DG. In fact, after lesions of the DG or the selective elimination of postnatal neurogenesis, rats cannot disambiguate the order of presentation of two spatial locations that were visited contiguously (Kesner et al. 2014; Morris et al. 2013). At the single cell level, one would predict that place cells in DG separate similar or identical spatial locations that are far apart along the temporal dimension by exhibiting spatial activity that is dependent upon time. To this end, it was shown that single cells, and even place cells in the DG of the hippocampus, demonstrate activity that is temporally selective (Rangel et al. 2014). Specifically, temporal separation between different experiences created a more distinct population code for each experience than experiences with no temporal separation, and manipulations to reduce levels of adult neurogenesis increased the similarity of responses to the different experiences. Cells in the DG thus support the integration of both spatial and temporal information through activity that is selective to both space and time, revealing the relationship between these dimensions by encoding experiences as distinct events in *spacetime* (Fig. 3.3).

A Multidimensional Hippocampus

In addition to the DG, the integration of space and time has recently been shown to exist in the CA1 region of the hippocampus. Populations of CA1 place cells, but not CA3 place cells, demonstrate different patterns of activity across days with increasing temporal intervals between cell recordings (Mankin et al. 2012). This implies that subregions of the hippocampus may integrate time and space according to different timescales. It remains an open question, however, how the spatiotemporal coding observed in DG contributes to or interacts with the spatiotemporal coding observed in downstream CA1. The encoding of both spatial and temporal dimensions at the single cell level in the DG and downstream hippocampal subregions can provide a mechanism for a more complete theory of how the hippocampus accomplishes associations between complex features of memories. In relativistic physics, *spacetime* describes everything in the universe as events that occur in space

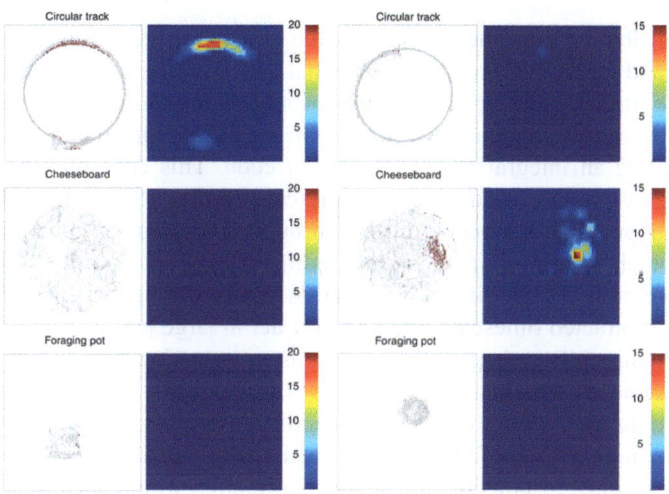

Fig. 3.3 Separating exposure to three different behavioral contexts during training (*a circular track, an open-field cheeseboard, and a square foraging pot*), resulted in place cells with activity selective for only one of three contexts during test exposures to all three contexts in the same day. This temporal selectivity was reduced in groups with shorter temporal separations between contexts and in groups with decreased levels of adult neurogenesis. (Taken from Rangel et al. 2014)

and time. The utility of combining these dimensions is that highly disparate locations in the universe can be linked in time, and highly overlapping locations can become more distinct in time. In other words, *spacetime* has the ability to reveal the relationship of events along both dimensions. In the DG, the combination of these two dimensions means that associations can be made between spatial and temporal features of events. Multidimensional activity in the hippocampus may thus be a mechanism through which the hippocampus accounts for relationships across spatial and nonspatial features of memories.

The hippocampus is thus more than a cognitive map, and more equivalent to a multidimensional terrain well suited for the demands of the attribute-based approach to memory. The large advantage of this multidimensional view is that it removes the hippocampus from the constraint of encoding complex features of memories along a single dimension. Instead, single cells are given the ability to reveal the relationships across spatial, temporal, and perhaps other dimensions. By acknowledging that these dimensions exist in the hippocampus, we can begin to examine the exact dynamics of the relationships between these dimensions and determine the rules, if any, regarding how these relationships manifest themselves in the activity of single cells. As the attribute-based model of memory suggests, these studies support the idea that space is so integrally linked with other dimensions that it would be difficult and potentially unmeaningful to examine it as encoded separately in the brain.

Conclusion

Kesner's attribute model theorizes that time and space are conjoined in the hippocampus and implicates the DG as essential to separating events that occur close in time. We further describe how the hippocampus, and more specifically the DG, may have the ability to create distinct spatial representations that also incorporate time, revealing an integrated *spatiotemporal* code. This code may be useful for segregating events that occur on long timescales. Here, spatiotemporal coding of contextual inputs may be accomplished through the continual generation of new neurons, which, due to their transient window of hyperexcitability and plasticity, allow for preferential encoding of information present during that temporal window. Thus, on a protracted timescale, the DG may act in large part as a sparsifying network, and temporally orthogonalize inputs, as computationally predicted (Aimone et al. 2009; Rangel et al. 2014).

By defining memory as composed of multiple complex features, the attribute model of memory first and foremost described memory as a systems level computation. The structures responsible for memory formation would need to have mechanisms for forming associations across features and for using different features to make memories distinct. This approach will continue to provide an inspirational framework for incorporating computational, behavioral, systems, cellular, and molecular level approaches towards investigating how rich contextual information aggregates to form our recollections.

Had there been an ageless observer at the sparkling moment of the creation of the egg—or of the hen—we would be no better off than we are today, for I am sure the observer would have soon forgotten which came first. Underwood 1977

Acknowledgments *Recollections of the Kesner Lab*

My (Chiba's) memories of Ray Kesner's lab in the context of graduate school surround the time of exciting theoretical advances, pushing the attribute model from a static to an active processing model. Daily candid exchanges were inspired by Ray's openness to creatively and rigorously testing, rather than simply supporting his theories. Ray's approach provided a platform for learning across several different labs working on similar questions. His genius for behavioral design and effervescence was contagious and as such all of us from that era inherited a portion of his passion and made his science part of our own. To our post-docs and students, there was nothing more inspiring than their first meal with Ray who is particularly facile at using restaurant condiments to represent all physical aspects of an experiment. The prize of the meal was the napkin covered with newly designed experiments to test the question of the evening. Each of us aspired to take at least some small aspect of Ray back to the lab. To Ray, we owe our intrinsic satisfaction from beautiful science; this is what makes a scientist for life and across many venues. What rich and perplexing lives he has given us. Thank you, Ray!

I (Chiba) also wish to acknowledge the late Dr. William H. Saufley II, a student of Underwood's, who instilled my early desire to pursue science and directed me towards Ray's Chapter in Learning and Memory: A Biological View (Eds. J L Martinez and R P Kesner 1986). This eye-catching book illuminated the path to Ray's lab.

References

Aimone, J. B., Wiles, J., & Gage, F. H. (2006). Potential role for adult neurogenesis in the encoding of time in new memories. *Nature Neuroscience, 9*(6), 723–727. doi:nn1707 [pii] 10.1038/nn1707.

Aimone, J. B., Wiles, J., & Gage, F. H. (2009). Computational influence of adult neurogenesis on memory encoding. *Neuron, 61*(2), 187–202. doi:S0896-6273(08)01019-2 [pii] 10.1016/j.neuron.2008.11.026.

Aimone, J. B., Deng, W., & Gage, F. H. (2011). Resolving new memories: a critical look at the dentate gyrus, adult neurogenesis, and pattern separation. *Neuron 70*(4), 589–596. doi: 10.1016/j.neuron.2011.05.010.

Allen, T. A., Morris, A. M., Mattfeld, A. T., Stark, C. E. L., & Fortin, N. J. (2014). A Sequence of events model of episodic memory shows parallels in rats and humans. *Hippocampus, 24*, 1178–1188 doi:10.1002/hipo.22301.

Aristotle, & Barnes, J. (1984) *The complete works of Aristotle: The revised Oxford translation (Bollingen Series LXXI-2)* (Vol. 2). New Jersey: Princeton University Press.

Bostock, E., Muller, R. U., & Kubie, J. L. (1991). Experience-dependent modifications of hippocampal place cell firing. *Hippocampus, 1*(2), 193–205. doi:10.1002/hipo.450010207.

Burgess, N., Maguire, E. A., & O'Keefe, J. (2002). The human hippocampus and spatial and episodic memory. *Neuron 35*(4), 625–641.

Chiba, A. A., Kesner, R. P., & Jackson, P. A. (2002). Two forms of spatial memory: A double dissociation between the parietal cortex and the hippocampus in the rat. *Behavioral Neuroscience, 116*(5), 874–883.

Cohen, N. J., Poldrack, R. A., & Eichenbaum, H. (1997). Memory for items and memory for relations in the procedural/declarative memory framework. *Memory (Hove, England), 5*(1–2), 131–178. doi:10.1080/741941149.

Davidson, T. J., Kloosterman, F., & Wilson, M. A. (2009). Hippocampal replay of extended experience. *Neuron, 63*(4), 497–507. doi:10.1016/j.neuron.2009.07.027.

DeCoteau, W. E., & Kesner, R. P. (2000). A double dissociation between the rat hippocampus and medial caudoputamen in processing two forms of knowledge. *Behavioral Neuroscience, 114*(6), 1096–1108.

Deng, W., Mayford, M., & Gage, F. H. (2013). Selection of distinct populations of dentate granule cells in response to inputs as a mechanism for pattern separation in mice. *eLife, 2*, e00312. doi:10.7554/eLife.00312

de Pontes, J. C. A., Batista, A. M., Viana, R. L., & Lopes, S. R. (2005). Short-term memories with a stochastic perturbation. *Chaos, Solitons & Fractals 23*(5), 1689–1694.

DiMattia, B. D., & Kesner, R. P. (1988). Spatial cognitive maps: Differential role of parietal cortex and hippocampal formation. *Behavioral Neuroscience, 102*(4), 471–480.

Dragoi, G., & Buzsáki, G. (2006). Temporal encoding of place sequences by hippocampal cell assemblies. *Neuron, 50*(1), 145–157. doi:10.1016/j.neuron.2006.02.023.

Esposito, M. S., Piatti, V. C., Laplagne, D. A., Morgenstern, N. A., Ferrari, C. C., Pitossi, F. J., & Schinder, A. F. (2005). Neuronal differentiation in the adult hippocampus recapitulates embryonic development. *The Journal of Neuroscience, 25*(44), 10074–10086. doi:25/44/10074 [pii] 10.1523/JNEUROSCI.3114-05.2005.

Frank, L. M., Brown, E. N., & Wilson, M. (2000). Trajectory encoding in the hippocampus and entorhinal cortex. *Neuron, 27*(1), 169–178.

Frank, L. M., Brown, E. N., & Stanley, G. B. (2006). Hippocampal and cortical place cell plasticity: Implications for episodic memory. *Hippocampus, 16*(9), 775–784. doi:10.1002/hipo.20200.

Gupta, A. S., van der Meer, M. A. A., Touretzky, D. S., & Redish, A. D. (2010). Hippocampal replay is not a simple function of experience. *Neuron, 65*(5), 695–705. doi:10.1016/j.neuron.2010.01.034.

Howard, M. W., & Kahana, M. J. (2002). A distributed representation of temporal context. *Journal of Mathematical Psychology, 46*(3), 269–299. doi:10.1006/jmps.2001.1388.

Howard, M. W., Fotedar, M. S., Datey, A. V, & Hasselmo, M. E. (2005). The temporal context model in spatial navigation and relational learning: Toward a common explanation of medial temporal lobe function across domains. *Psychological Review, 112*(1), 75–116. doi:10.1037/0033-295X.112.1.75.

Jackson-Smith, P., Kesner, R. P., & Chiba, A. A. (1993). Continuous recognition of spatial and nonspatial stimuli in hippocampal-lesioned rats. *Behavioral and Neural Biology, 59*(2), 107–119.

Jacobson, T. K., Gruenbaum, B. F., & Markus, E. J. (2011). Extensive training and hippocampus or striatum lesions: Effect on place and response strategies. *Physiology & Behavior, 105*(3), 645–652. doi:10.1016/j.physbeh.2011.09.027.

Jadhav, S. P., Kemere, C., German, P. W., & Frank, L. M. (2012). Awake hippocampal sharp-wave ripples support spatial memory. *Science (New York, N.Y.), 336*(6087), 1454–1458. doi:10.1126/science.1217230.

Kametani, H., & Kesner, R. P. (1989). Retrospective and prospective coding of information: Dissociation of parietal cortex and hippocampal formation. *Behavioral Neuroscience, 103*, 84–89.

Kentros, C., Hargreaves, E., Hawkins, R. D., Kandel, E. R., Shapiro, M., & Muller, R. V. (1998). Abolition of long-term stability of new hippocampal place cell maps by NMDA receptor blockade. *Science (New York, N.Y.), 280*(5372), 2121–2126.

Kentros, C. G., Agnihotri, N. T., Streater, S., Hawkins, R. D., & Kandel, E. R. (2004). Increased attention to spatial context increases both place field stability and spatial memory. *Neuron, 42*(2), 283–295.

Kesner, R. P. (1980). An attribute analysis of memory: The role of the hippocampus. *Physiology Psychology, 8*, 189–197.

Kesner, R. P. (1990). New approaches to the study of comparative cognition. *NIDA Research Monograph, 97*, 22–36 (http://www.ncbi.nlm.nih.gov/pubmed/2247138).

Kesner, R. P. (2013). A process analysis of the CA3 subregion of the hippocampus. *Frontiers in Cellular Neuroscience, 7*, 78. doi:10.3389/fncel.2013.00078.

Kesner, R. P., & Novak, J. (1982). Serial position curve in rats: Role of the dorsal hippocampus. *Science, 218*, 173–174.

Kesner, R. P., & Rolls, E. T. (2001). Role of long-term synaptic modification in short-term memory. *Hippocampus, 11*(3), 240–250. doi:10.1002/hipo.1040.

Kesner, R. P., Adelstein, T. B., & Crutcher, K. A. (1989). Equivalent spatial location memory deficits in rats with medial septum or hippocampal formation lesions and patients with dementia of the Alzheimer's type. *Brain and Cognition, 9*(2), 289–300.

Kesner, R. P., Hui, X., Sommer, T., Wright, C., Barrera, V. R., Fanselow, M. S. (2014). The role of postnatal neurogenesis in supporting remote memory and spatial metric processing. *Hippocampus, 24*, 1663–1671. doi:10.1002/hipo.22346.

Kjelstrup, K. B., Solstad, T., Brun, V. H., Hafting, T., Leutgeb, S., Witter, M. P., et al. (2008). Finite scale of spatial representation in the hippocampus. *Science, 321*(5885), 140–143. doi:321/5885/140 [pii] 10.1126/science.1157086.

Laplagne, D. A., Esposito, M. S., Piatti, V. C., Morgenstern, N. A., Zhao, C., van Praag, H., et al. (2006). Functional convergence of neurons generated in the developing and adult hippocampus. *Plos Biology, 4*(12), e409. doi:06-PLBI-RA-0577R3 [pii] 10.1371/journal.pbio.0040409.

Laplagne, D. A., Kamienkowski, J. E., Esposito, M. S., Piatti, V. C., Zhao, C., Gage, F. H., & Schinder, A. F. (2007). Similar GABAergic inputs in dentate granule cells born during embryonic and adult neurogenesis. *European Journal Neuroscience, 25*(10), 2973–2981. doi:EJN5549 [pii] 10.1111/j.1460-9568.2007.05549.x.

Lenck-Santini, P.-P., Rivard, B., Muller, R. U., & Poucet, B. (2005). Study of CA1 place cell activity and exploratory behavior following spatial and nonspatial changes in the environment. *Hippocampus, 15*(3), 356–69. doi:10.1002/hipo.20060.

Leutgeb, S., Leutgeb, J. K., Moser, M. B., & Moser, E. I. (2005). Place cells, spatial maps and the population code for memory. *Current Opinion Neurobiology, 15*(6), 738–746. doi:S0959-4388(05)00152-2 [pii] 10.1016/j.conb.2005.10.002.

Leutgeb, J. K., Leutgeb, S., Moser, M. B., & Moser, E. I. (2007). Pattern separation in the dentate gyrus and CA3 of the hippocampus. *Science, 315*(5814), 961–966. doi:315/5814/961 [pii] 10.1126/science.1135801.

MacDonald, C. J., Lepage, K. Q., Eden, U. T., & Eichenbaum, H. (2011). Hippocampal "time cells" bridge the gap in memory for discontiguous events. *Neuron, 71*(4), 737–749. doi:10.1016/j.neuron.2011.07.012.

Mankin, E. A., Sparks, F. T., Slayyeh, B., Sutherland, R. J., Leutgeb, S., & Leutgeb, J. K. (2012). Neuronal code for extended time in the hippocampus. *Proc Natl Acad Sci U S A. 109*(47), 19462–19467. doi: 10.1073/PNAS.1214107109. Epub 2012 Nov 6. Erratum in: Proc Natl Acad Sci U S A. 2015, 112(10), E1169.

McClelland, J. L., McNaughton, B. L., & O'Reilly, R. C. (1995). Why there are complementary learning systems in the hippocampus and neocortex: Insights from the successes and failures of connectionist models of learning and memory. *Psychological Review, 102*(3), 419–457.

McNaughton, B. L., Barnes, C. A., & O'Keefe, J. (1983). The contributions of position, direction, and velocity to single unit activity in the hippocampus of freely-moving rats. *Experimental Brain Research, 52*(1), 41–49.

Milner, B., & Penfield, W. (1956). *The effect of hippocampal lesions on recent memory*. Transactions of the American Neurological Association, 42-8(80th Meeting).

Milner, B., Corkin, S., & Teuber, H. L. (1968). Further analysis of the hippocampal amnesic syndrome: 14-year follow-up study of H.M. *Neuropsychologia, 6*(3), 215–234. doi:10.1016/0028-3932(68)90021-3.

Milner, B., Squire, L. R., & Kandel, E. R. (1998). Cognitive neuroscience and the study of memory. *Neuron, 20*(3), 445–468.

Morris, R. G. M., Schenk, F., Tweedie, F., & Jarrard, L. E. (1990). Ibotenate lesions of hippocampus and/or subiculum: Dissociating components of allocentric spatial learning. *The European Journal of Neuroscience, 2*(12), 1016–1028 (http://www.ncbi.nlm.nih.gov/pubmed/12106063).

Morris, A. M., Curtis, B. J., Churchwell, J. C., Maasberg, D. W., & Kesner, R. P. (2013). Temporal associations for spatial events: The role of the dentate gyrus. *Behavioural Brain Research, 256*, 250–6. doi:10.1016/j.bbr.2013.08.021.

Munn, R. G. K., & Bilkey, D. K. (2011). The firing rate of hippocampal CA1 place cells is modulated with a circadian period. *Hippocampus, 22*, 1325–1337. doi:10.1002/hipo.20969.

Neunuebel, J. P., & Knierim, J. J. (2014). CA3 retrieves coherent representations from degraded input: Direct evidence for CA3 pattern completion and dentate gyrus pattern separation. *Neuron, 81*(2), 416–427. doi:10.1016/j.neuron.2013.11.017.

Nitz, D. (2009). Parietal cortex, navigation, and the construction of arbitrary reference frames for spatial information. *Neurobiology of Learning and Memory, 91*(2), 179–185. doi:10.1016/j.nlm.2008.08.007.

O'Keefe, J., & Dostrovsky, J. (1971). The hippocampus as a spatial map. Preliminary evidence from unit activity in the freely-moving rat. *Brain Research, 34*(1), 171–175.

O'Keefe, J., & Nadel, L. (1978). *The hippocampus as a cognitive map. Hippocampus* (Vol. 3, p. 570). Oxford: Oxford University Press.

O'Reilly, R. C., & McClelland, J. L. (1994). Hippocampal conjunctive encoding, storage, and recall: Avoiding a trade-off. *Hippocampus, 4*(6), 661–682. doi:10.1002/hipo.450040605.

Olton, D. S., & Papas, B. C. (1979). Spatial memory and hippocampal function. *Neuropsychologia, 17*(6), 669–682.

Olton, D. S., Walker, J. A., & Gage, F. H. (1978). Hippocampal connections and spatial discrimination. *Brain Research, 139*(2), 295–308.

Pastalkova, E., Itskov, V., Amarasingham, A., & Buzsáki, G. (2008). Internally generated cell assembly sequences in the rat hippocampus. *Science (New York, N.Y.), 321*(5894), 1322–1327. doi:10.1126/science.1159775.

Piatti, V. C., Esposito, M. S., & Schinder, A. F. (2006). The timing of neuronal development in adult hippocampal neurogenesis. *The Neuroscientist, 12*(6), 463–468. doi:12/6/463 [pii] 10.1177/1073858406293538.

Poucet, B., Save, E., & Lenck-Santini, P. P. (2000). Sensory and memory properties of hippocampal place cells. *Reviews in the Neurosciences, 11*(2–3), 95–111.

Rangel, L. M., & Eichenbaum, H. (2013). What's new is older. *eLife, 2*, e00605. doi:10.7554/eLife.00605.

Rangel, L. M., Alexander, A. S., Aimone, J. B., Wiles, J., Gage, F. H., Chiba, A. A., & Quinn, L. K. (2014). Temporally selective contextual encoding in the dentate gyrus of the hippocampus. *Nature Communications, 5*, 3181. doi:10.1038/ncomms4181.

Rivard, B., Li, Y., Lenck-Santini, P.-P., Poucet, B., & Muller, R. U. (2004). Representation of objects in space by two classes of hippocampal pyramidal cells. *The Journal of General Physiology, 124*(1), 9–25. doi:10.1085/jgp.200409015.

Rolls, E. T. (2010). A computational theory of episodic memory formation in the hippocampus. *Behavioural Brain Research, 215*(2), 180–196. doi:10.1016/j.bbr.2010.03.027.

Rolls, E. T., & Kesner, R. P. (2006). A computational theory of hippocampal function, and empirical tests of the theory. *Progress in Neurobiology, 79*(1), 1–48. doi:S0301-0082(06)00036-0.

Rosenbaum, R. S., Priselac, S., Köhler, S., Black, S. E., Gao, F., Nadel, L., & Moscovitch, M. (2000). Remote spatial memory in an amnesic person with extensive bilateral hippocampal lesions. *Nature Neuroscience, 3*(10), 1044–1048. doi:10.1038/79867.

Rosenbaum, R. S., Köhler, S., Schacter, D. L., Moscovitch, M., Westmacott, R., Black, S. E., et al. (2005). The case of K.C.: Contributions of a memory-impaired person to memory theory. *Neuropsychologia, 43*(7), 989–1021. doi:10.1016/j.neuropsychologia.2004.10.007.

Rowland, D. C., Yanovich, Y., & Kentros, C. G. (2011). A stable hippocampal representation of a space requires its direct experience. *Proceedings of the National Academy of Sciences, 108*(35), 14654–14658. doi:10.1073/pnas.1105445108.

Sederberg, P. B., Miller, J. F., Howard, M. W., & Kahana, M. J. (2010). The temporal contiguity effect predicts episodic memory performance. *Memory & Cognition, 38*(6), 689–699. doi:10.3758/MC.38.6.689.

Shen, J., Barnes, C. A., McNaughton, B. L., Skaggs, W. E., & Weaver, K. L. (1997). The effect of aging on experience-dependent plasticity of hippocampal place cells. *The Journal of Neuroscience, 17*(17), 6769–6782.

Skaggs, W. E., McNaughton, B. L., Wilson, M. A., & Barnes, C. A. (1996). Theta phase precession in hippocampal neuronal populations and the compression of temporal sequences. *Hippocampus, 6*(2), 149–172. doi:10.1002/(SICI)1098-1063(1996)6:2<149::AID-HIPO6>3.0.CO;2-K.

Tashiro, A., Makino, H., & Gage, F. H. (2007). Experience-specific functional modification of the dentate gyrus through adult neurogenesis: A critical period during an immature stage. *The Journal of Neuroscience, 27*(12), 3252–3259. doi:27/12/3252 [pii] 10.1523/JNEUROSCI.4941-06.2007.

Treves, A., & Rolls, E. T. (1992). Computational constraints suggest the need for two distinct input systems to the hippocampal CA3 network. *Hippocampus, 2*(2), 189–199. doi:10.1002/hipo.450020209.

Tse, D., Langston, R. F., Kakeyama, M., Bethus, I., Spooner, P. A., Wood, E. R., et al. (2007). Schemas and memory consolidation. *Science (New York, N.Y.), 316*(5821), 76–82. doi:10.1126/science.1135935.

Tsodyks, M. V., Skaggs, W. E., Sejnowski, T. J., & McNaughton, B. L. (1996). Population dynamics and theta rhythm phase precession of hippocampal place cell firing: A spiking neuron model. *Hippocampus, 6*(3), 271–280. doi:10.1002/(SICI)1098-1063(1996)6:3<271::AID-HIPO5>3.0.CO;2-Q.

Tulving, E. (2002). Episodic memory: From mind to brain. *Annual Review of Psychology, 53*(1), 1–25. doi:10.1146/annurev.psych.53.100901.135114.

Underwood, B. J. (1969). Attributes of memory. *Psychological Review, 76*(6), 559–573.

Underwood, B. J. (1977). *Temporal codes for memories: Issues and problems*. Hillsdale: Erlbaum.

Van Strien, N. M., Cappaert, N. L. M., & Witter, M. P. (2009). The anatomy of memory: An interactive overview of the parahippocampal-hippocampal network. *Nature Reviews. Neuroscience, 10*(4), 272–282. doi:10.1038/nrn2614.

Whishaw, I. Q., Cassel, J. C., & Jarrad, L. E. (1995). Rats with fimbria-fornix lesions display a place response in a swimming pool: A dissociation between getting there and knowing where. *The Journal of Neuroscience, 15*(8), 5779–5788.

Chapter 4
Pattern Completion and Pattern Separation Mechanisms in the Hippocampus

Edmund T. Rolls

Introduction

There is great interest in how pattern separation and pattern completion in the hippocampus contribute to its functions in memory and spatial function (Giocomo et al. 2011; Jezek et al. 2011; Leutgeb et al. 2007; McHugh et al. 2007; Nakashiba et al. 2012; Nakazawa et al. 2002, 2003; Wills et al. 2005), and among those who have made many contributions in this area are Ray Kesner and his colleagues (Hunsaker and Kesner 2008, 2013; Kesner 2007, 2013; Kesner et al. 2012; Rolls and Kesner 2006).

This chapter describes some of the different types of pattern separation and pattern completion in the hippocampal system, and the mechanisms that implement them. More comprehensive descriptions of my theory of hippocampal function, and of differences between the primate and rodent hippocampal neuronal representations and the implications for understanding human memory are provided elsewhere (Rolls 2008, 2010b, 2013; Rolls and Kesner 2006; Rolls and Xiang 2006; Kesner and Rolls 2015). The theory has been developed through many stages (Rolls 1987, 1989a, b, c, 1990a, b, 1991, 1995, 1996b, 2008, 2010b; Rolls and Deco 2010; Rolls and Kesner 2006; Rolls and Treves 1998; Treves and Rolls 1991, 1992, 1994), has as a predecessor developments made by David Marr (Marr 1971) (though he never identified the CA3 system as an autoassociation network), and has benefitted greatly from collaborations with many whose names appear below in the citations, including Alessandro Treves and Simon Stringer. The operation of pattern association networks, autoassociation networks, and competitive networks has been described elsewhere (Hertz et al. 1991; Rolls 2008; Rolls and Treves 1998; Rolls 2016).

E. T. Rolls (✉)
Department of Computer Science and Oxford Centre for Computational Neuroscience, University of Warwick, Coventry CV4 7AL, UK
e-mail: Edmund.Rolls@oxcns.org

P. A. Jackson et al. (eds.), *The Neurobiological Basis of Memory*,
DOI 10.1007/978-3-319-15759-7_4

Background to the Approach to Hippocampal Function

Event and Episodic Memory

The focus is on a fundamental property of episodic memory, the ability to store and retrieve the memory of a particular single event involving an association between items such as the place and the object or reward seen at that place. Episodic memory, in the sense of a series of linked events, requires this type of event memory, and could be implemented by linking together a series of events.

An event consists of a set of items that occur together, such as seeing a particular object or person's face in a particular place. An everyday example might be remembering where one was for dinner, who was present, what was eaten, what was discussed, and the time at which it occurred. The spatial context is almost always an important part of an episodic memory (Dere et al. 2008), and it may be partly for this reason that episodic memory is linked to the functions of the hippocampal system which is involved in spatial processing and memory. The ability to recall a whole memory from a partial cue is an important property of episodic memory and is referred to as completion.

Systems-Level Functions and Connections of the Primate Hippocampus

Any theory of the hippocampus must state at the systems level what is computed by the hippocampus. Some of the relevant evidence about the functions of the hippocampus in memory comes from the effects of damage to the hippocampus, the responses of neurons in the hippocampus during behavior, and the systems-level connections of the hippocampus, as described in more detail elsewhere (Rolls 2008, 2010b; Kesner and Rolls 2015; Rolls and Xiang 2006). Many of the memory functions are important in event or episodic memory, in which the ability to remember what happened where on typically a single occasion (or trial in a learning experiment) is important. It is suggested that an autoassociation memory implemented by the CA3 neurons enables event or episodic memories to be formed by enabling associations to be formed between spatial and other including object or reward representations, and for completion to then occur in recall from any part. This is different from pattern association memory in which a visual stimulus might become associated with a taste by associative synaptic modification. Later presentation of the visual stimulus would retrieve the taste representation. However, presentation of the taste would not retrieve the visual representation, and this is an important and fundamental difference between autoassociation and pattern association, as described in detail elsewhere (Rolls 2008, 2014, 2016; Rolls and Treves 1998).

Information stored in the hippocampus will need to be retrieved and affect other parts of the brain in order to be used. The information about episodic events recalled from the hippocampus could be used to help form semantic memories (Rolls 1989b, c, 1990a; Treves and Rolls 1994). For example, remembering many particular journeys could help to build a geographic cognitive map in the neocortex. The hippocampus and neocortex would thus be complementary memory systems, with the hippocampus being used for rapid, "on the fly," unstructured storage of information involving activity potentially arriving from many areas of the neocortex; while the neocortex would gradually build and adjust on the basis of much accumulating information, often recalled from the hippocampal unstructured store, the semantic representation (McClelland et al. 1995; Moscovitch et al. 2005; Rolls 1989b; Treves and Rolls 1994). The theory shows how information could be retrieved within the hippocampus, and how this retrieved information could enable the activity in neocortical areas that was present during the original storage of the episodic event to be reinstated, thus implementing recall, by using hippocampo-neocortical backprojections is described elsewhere (Rolls 1995, 1996b, 2008, 2010b; Treves and Rolls 1994; see Fig. 4.1).

To understand the functions of the primate hippocampus in event or episodic memory, it is necessary to understand from which other parts of the brain it receives information. Does it, for example, receive object as well as spatial information in terms of its connectivity? The primate hippocampus receives inputs via the entorhinal cortex (area 28) and the highly developed parahippocampal gyrus (areas TF and TH) as well as the perirhinal cortex from the ends of many processing streams of the cerebral association cortex, including the visual and auditory temporal lobe association cortical areas, the prefrontal cortex, and the parietal cortex (Amaral 1987; Amaral et al. 1992; Lavenex et al. 2004; Rolls 2008; Rolls and Kesner 2006; Suzuki and Amaral 1994b; Van Hoesen 1982; Witter et al. 2000b; see Fig. 4.1). The hippocampus is thus by its connections potentially able to associate together object and spatial representations. In addition, the entorhinal cortex receives inputs from the amygdala and the orbitofrontal cortex, which could provide reward-related information to the hippocampus (Carmichael and Price 1995; Pitkanen et al. 2002; Stefanacci et al. 1996; Suzuki and Amaral 1994a).

The primary output from the hippocampus to neocortex originates in CA1 and projects to subiculum, entorhinal cortex, and parahippocampal structures (areas TF-TH) as well as prefrontal cortex (Delatour and Witter 2002; van Haeften et al 2003; Van Hoesen 1982; Witter 1993; see Fig. 4.1), though there are other outputs (Kesner and Rolls 2015). These are the pathways that are likely to be involved in the retrieval of information from the hippocampus back to the neocortex.

The theory is a quantitative theory and the numbers of synapses on the different types of neuron is an important feature of the circuitry emphasized next.

Hippocampal Circuitry

Hippocampal circuitry (Amaral 1993; Amaral and Witter 1989; Andersen et al. 2007; Kondo et al. 2009; Lavenex et al. 2004; Naber et al. 2001; Storm-Mathiesen et al. 1990; Witter 2007; Witter et al. 2000b) is illustrated in Fig. 4.1.

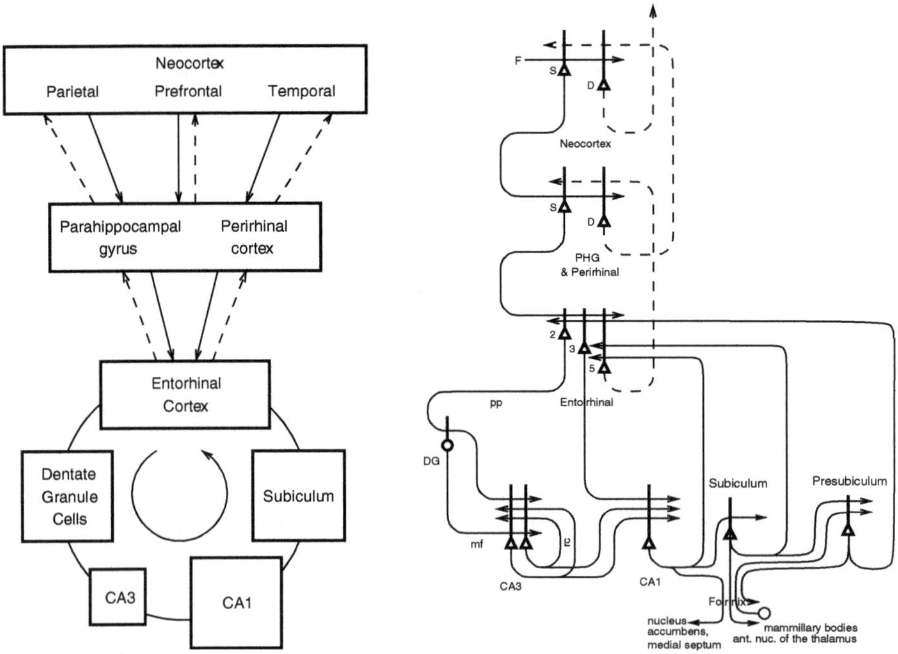

Fig. 4.1 Forward connections (*solid lines*) from areas of cerebral association neocortex via the parahippocampal gyrus and perirhinal cortex, and entorhinal cortex, to the hippocampus; and backprojections (*dashed lines*) via the hippocampal CA1 pyramidal cells, subiculum, and parahippocampal gyrus to the neocortex. There is great convergence in the forward connections down to the single network implemented in the CA3 pyramidal cells; and great divergence again in the backprojections. *Left*: block diagram. *Right*: more detailed representation of some of the principal excitatory neurons in the pathways. *D* Deep pyramidal cells, *DG* Dentate granule cells, *F* Forward inputs to areas of the association cortex from preceding cortical areas in the hierarchy, *mf* mossy fibers, *PHG* parahippocampal gyrus and perirhinal cortex, *pp* perforant path, *rc* recurrent collateral of the CA3 hippocampal pyramidal cells, *S* Superficial pyramidal cells, *2* pyramidal cells in layer 2 of the entorhinal cortex, *3* pyramidal cells in layer 3 of the entorhinal cortex. The *thick lines* above the cell bodies represent the dendrites

Projections from the entorhinal cortex layer 2 reach the granule cells (of which there are 10^6 in the rat) in the dentate gyrus (DG), via the perforant path (pp) (Witter 1993). The granule cells project to CA3 cells via the mossy fibers (mf), which provide a *sparse* but possibly powerful connection to the 3×10^5 CA3 pyramidal cells in the rat. Each CA3 cell receives approximately 46 mossy fiber inputs, so that the sparseness of this connectivity is thus 0.005 %. By contrast, there are many more—possibly weaker—direct perforant path inputs also from layer 2 of the entorhinal cortex onto each CA3 cell, in the rat of the order of 4×10^3. The largest number of synapses (about 1.2×10^4 in the rat) on the dendrites of CA3 pyramidal cells is, however, provided by the (recurrent) axon collaterals of CA3 cells themselves (rc) (see Fig. 4.2). It is remarkable that the recurrent collaterals are distributed to other CA3 cells largely throughout the hippocampus (Amaral et al. 1990; Amaral and Witter 1989, 1995; Ishizuka et al. 1990; Witter 2007), so that effectively the CA3 system

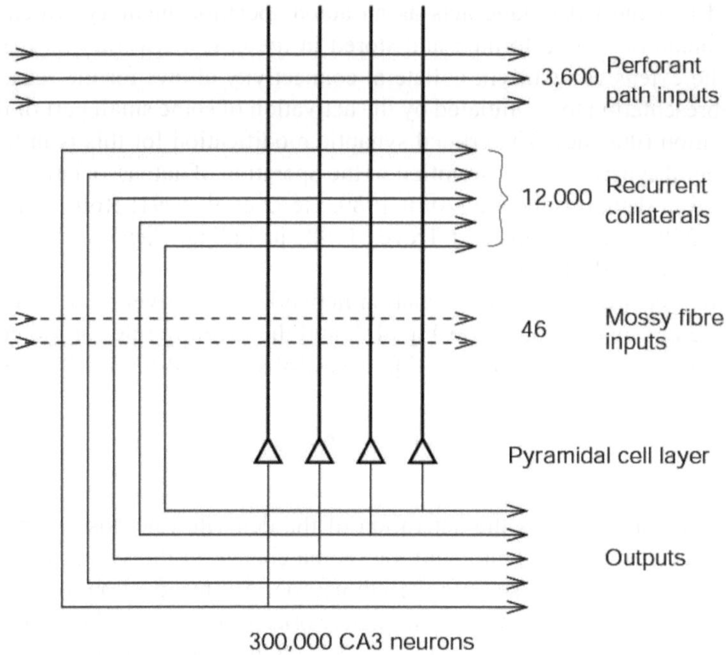

Fig. 4.2 The numbers of connections from three different sources onto each CA3 cell from three different sources in the rat. (After Rolls and Treves 1998; Treves and Rolls 1992)

provides a single network, with a connectivity of approximately 2% between the different CA3 neurons given that the connections are bilateral. The CA3–CA3 recurrent collateral system is even more extensive in macaques than in rats (Kondo et al. 2009). The neurons that comprise CA3, in turn, project to CA1 neurons via the Schaffer collaterals. In addition, projections that terminate in the CA1 region originate in layer 3 of the entorhinal cortex (see Fig. 4.1).

CA3 as an Autoassociation or Attractor Memory: Pattern Completion

Arbitrary Associations and Pattern Completion in Recall

Many of the synapses in the hippocampus show associative modification as shown by long-term potentiation, and this synaptic modification appears to be involved in learning (see Andersen et al. 2007; Lynch 2004; Morris 1989, 2003; Morris et al. 2003; Nakazawa et al. 2004; Nakazawa et al. 2003; Wang and Morris 2010). On the basis of the evidence summarized above, Rolls (1987, 1989a, b, c, 1990a, b, 1991) and others (Levy 1989; McNaughton 1991; McNaughton and Morris 1987) have suggested that the CA3 stage acts as an autoassociation memory which enables episodic memories to be formed and stored in the CA3 network, and that subsequently the extensive recurrent collateral connectivity allows for the retrieval of a whole representation to be initiated by the activation of some small part of the same representation (the cue). The crucial synaptic modification for this is in the recurrent collateral synapses. (A description of the operation of autoassociative networks is provided in detail elsewhere (Amit 1989; Hertz et al. 1991; Rolls 2010a; Rolls and Deco 2002, 2010; Rolls and Treves 1998) including *Memory, Attention, and Decision-Making* (Rolls 2008)).

The architecture of an autoassociation network is effectively that of the recurrent collateral synapses shown in Fig. 4.2, and the learning rule for the change in the synaptic weights is as shown in Eq. (4.1) (Rolls 2008; Rolls and Treves 1998).

$$\delta w_{ij} = k \cdot r_i \cdot r'_j \tag{4.1}$$

where k is a constant, r_i is the activation of the dendrite (the postsynaptic term), r'_j is the presynaptic firing rate, and δw_{ij} is the change in the synaptic weight w_{ij}. (w_{ij} refers to the j'th synapse onto the i'th neuron. An introduction to autoassociation, competitive, and pattern association networks is provided in the Appendices of *Memory, Attention and Decision-Making: A Unifying Computational Neuroscience Approach* (Rolls 2008).)

The hypothesis is that because the CA3 operates effectively as a single network, it can allow arbitrary associations between inputs originating from very different parts of the cerebral cortex to be formed. These might involve associations between information originating in the temporal visual cortex about the presence of an object, and information originating in the parietal cortex about where it is. I note that although there is some spatial gradient in the CA3 recurrent connections, so that the connectivity is not fully uniform (Ishizuka et al. 1990; Witter 2007), the network will still have the properties of a single interconnected autoassociation network allowing associations between arbitrary neurons to be formed, given the presence of many long-range connections which overlap from different CA3 cells, and the ability of attractor networks to operate with diluted connectivity shown in our computational studies prompted by this issue (Rolls 2012a; Rolls and Webb 2012; Treves 1990; Treves and Rolls 1991). It is very interesting indeed that in primates (macaques), the associational projections from CA3 to CA3 travel extensively along the longitudinal axis, and overall the radial, transverse, and longitudinal gradients of CA3 fiber distribution, clear in the rat, are much more subtle in the nonhuman primate brain (Kondo et al. 2009). The implication is that in primates, the CA3 network operates even more as a single network than in rodents.

Crucial issues include how many memories could be stored in this system (to determine whether the autoassociation hypothesis leads to a realistic estimate of the number of memories that the hippocampus could store); whether the whole of a memory could be completed from any part; whether the autoassociation memory can act as a short term memory, for which the architecture is inherently suited; and whether the system could operate with spatial representations, which are essentially continuous because of the continuous nature of space. These and related issues are considered in the remainder of "Storage Capacity" and in more detail elsewhere (Rolls 2008; Kesner and Rolls 2015).

Storage Capacity

We have performed quantitative analyses of the storage and retrieval processes in the CA3 network (Rolls 2012a; Rolls and Webb 2012; Treves and Rolls 1991, 1992; Webb et al. 2011). We have extended previous formal models of autoassociative memory (see Amit 1989) by analyzing a network with graded response units, so as to represent more realistically the continuously variable rates at which neurons fire, and with incomplete connectivity (Rolls et al. 1997b; Rolls and Webb 2012; Treves 1990; Treves and Rolls 1991; Webb et al. 2011). We have found that in general the maximum number p_{max} of firing patterns that can be (individually) retrieved is proportional to the number C^{RC} of (associatively) modifiable recurrent collateral (RC) synapses on to each neuron, by a factor that increases roughly with the inverse of the sparseness a of the neuronal representation. (Each memory is precisely defined in the theory: it is a set of firing rates of the population of neurons (which represent a memory) that can be stored and later retrieved, with retrieval being possible from a fraction of the originally stored set of neuronal firing rates.) The neuronal population sparseness a of the representation can be measured by extending the binary notion of the proportion of neurons that are firing to any one stimulus or event as

$$a = \left(\sum_{i=1,n} r_i / N\right)^2 / \sum_{i=1,n} \left(r_i^2 / N\right) \tag{4.2}$$

where r_i is the firing rate of the i'th neuron in the set of N neurons. The sparseness ranges from $1/N$, when only one of the neurons responds to a particular stimulus (a local or grandmother cell representation), to a value of 1.0, attained when all the neurons are responding to a given stimulus. Approximately,

$$p_{\max} \cong \frac{C^{RC}}{[a \ln(1/a)]} k \tag{4.3}$$

where k is a factor that depends weakly on the detailed structure of the rate distribution, on the connectivity pattern, etc., but is roughly in the order of 0.2–0.3 (Treves and Rolls 1991). For example, for $C^{RC}=12{,}000$ and $a=0.02$, p_{max} is calculated to be approximately 36,000. This analysis emphasizes the utility of having a sparse representation in the hippocampus, for this enables many different memories to be stored. (The sparseness a in this equation is strictly the population sparseness (Franco et al. 2007; Treves and Rolls 1991). The population sparseness a^p would be measured by measuring the distribution of firing rates of all neurons to a single stimulus at a single time. The single neuron sparseness or selectivity a^s would be measured by the distribution of firing rates to a set of stimuli, which would take a long time. The selectivity or sparseness a^s of a single neuron measured across a set of stimuli often takes a similar value to the population sparseness a^p in the brain, and does so if the tuning profiles of the neurons to the set of stimuli are uncorrelated (Franco et al. 2007). These concepts are elucidated by Franco, Rolls et al. (2007).) (I note that the sparseness estimates obtained by measuring early gene changes, which are effectively population sparsenesses, would be expected to depend greatly on the range of environments or stimuli in which these were measured. If the environment was restricted to one stimulus, this would reflect the population sparseness. If the environment was changing, the measure from early gene changes would be rather undefined, as all the populations of neurons activated in an undefined number of testing situations would be likely to be activated.)

In order for most associative networks to store information efficiently, heterosynaptic long-term depression (as well as LTP) is required (Fazeli and Collingridge 1996; Rolls 2008; Rolls and Deco 2002;Rolls and Treves 1990, 1998; Treves and Rolls 1991). Simulations that are fully consistent with the analytic theory are provided by Rolls (1995, 2012a), Simmen et al. (1996), and Rolls et al. (1997b).

A number of points that arise, including measurement of the total amount of information (in bits per synapse) that can be retrieved from the network, the computational definition of a memory, the computational sense in which CA3 is an attractor network, and the possible computational utility of memory reconsolidation, are treated elsewhere (Rolls 2008; Rolls and Kesner 2006). Here I note that given that the memory capacity of the hippocampal CA3 system is limited, it is necessary to have some form of forgetting in this store, or other mechanism to ensure that its capacity is not exceeded. (Exceeding the capacity can lead to a loss of much of the information retrievable from the network.) Heterosynaptic LTD could help this *forgetting*, by enabling new memories to overwrite old memories (Rolls 1996a, 2008). The limited capacity of the CA3 system does also provide one of the arguments that some transfer of information from the hippocampus to neocortical memory stores may be useful (see Treves and Rolls 1994). Given its limited capacity, the hippocampus might be a useful store for only a limited period, which might be in the order of days, weeks, or months. This period may well depend on the acquisition rate of new episodic memories. If the animal were in a constant and limited environment, then as new information is not being added to the hippocampus, the representations in the hippocampus would remain stable and persistent. These hypotheses have clear experimental implications, both for recordings from single neurons and for the gradient of retrograde amnesia, both of which might be expected to depend on whether the environment is stable or frequently changing. They show that the conditions under which a gradient of retrograde amnesia might be demonstrable would be when large numbers of new memories are being acquired, not when only a few memories (few in the case of the hippocampus being less than a few hundred) are being learned.

Recall and Completion

A fundamental property of the autoassociation model of the CA3 recurrent collateral network is that the recall can be symmetric, that is, the whole of the memory can be retrieved and completed from any part (Rolls 2008; Rolls and Kesner 2006; Rolls and Treves 1998). For example, in an object–place autoassociation memory, an object could be recalled from a place retrieval cue, and vice versa. Kesner et al. (2008) tested this using an object-cued spatial location recall task, and a spatial location-cued object recall task (developed from an episodic flavor–place paired-associate task (Day et al. 2003)). After rats were trained to a criterion of 80 % correct on 1 of the 2 tasks, they received either a dorsal CA3 lesion or a vehicle control lesion. Control animals continued performing well on both tasks. Rats with lesions to dorsal CA3 were impaired on both tasks and performed at chance but were able to perform a non-episodic version of the task as a control. These data provide evidence that CA3 mediates episodic learning of arbitrary associations as tested in the 1-trial object cue with spatial location recall task, and the spatial location cue with object recall task (Kesner et al. 2008).

In an object–place task, rats were trained in a study phase to learn in one trial an association between two flavors of food and two spatial locations (Day et al. 2003). During a recall test phase they were presented with a flavor which served as a cue for the selection of the correct location. They found that injections of an N-methyl-D-aspartate (NMDA) receptor blocker (AP5) or AMPA (α-amino-3-hydroxy-5-methyl-4-isoxazolepropionic acid)/kainate receptor blocker (CNQX) to the dorsal hippocampus prior to the study phase impaired encoding, but injections of AP5 prior to the test phase did not impair the place recall, whereas injections of CNQX did impair the place recall. The interpretation is that somewhere in the hippocampus NMDA receptors are necessary for learning one-trial odor–place associations, and that recall can be performed without further involvement of NMDA receptors.

Evidence that the CA3 system is not necessarily required during recall in a reference memory spatial task, such as the water maze spatial navigation for a single spatial location task, is that CA3 lesioned rats are not impaired during recall of a previously learned water maze task (Brun et al. 2002; Florian and Roullet 2004). However, if completion from an incomplete cue is needed, then CA3 NMDA receptors are necessary (presumably to ensure satisfactory CA3–CA3 learning) even in a reference memory task (Gold and Kesner 2005; Nakazawa et al. 2002). Thus, the CA3 system appears to be especially needed in rapid, one-trial object–place recall, and when completion from an incomplete cue is required (see further "Pattern Separation Performed By Dentate Granule Cells"). Especially important though in assessing the implications of all such tests is that the theory sets out how the system operates when large numbers of memories, in the order of thousands, are to be stored and retrieved, and this is difficult to test adequately in behavioral experiments. Effects found when the storage and retrieval of just a few memories are tested may not reflect well the operation of the system when it is heavily loaded, as it is expected to be when operating in the natural environment.

Evidence for pattern completion has been observed using imaging with voltage-sensitive dye in the CA3 region of a rat hippocampal slice. Following the induction of long-term potentiation from two stimulation sites activated simultaneously, stimulation at either of the two sites produced the whole pattern of activation that could be produced from both stimulation sites before LTP, thus demonstrating pattern completion in CA3 (Jackson 2013).

Continuous, Spatial, Patterns, and CA3 Representations

The fact that spatial patterns, which imply continuous representations of space, are represented in the hippocampus has led to the application of continuous attractor models to help understand hippocampal function. This has been necessary, because space is inherently continuous, because the firing of place and spatial view cells is approximately Gaussian as a function of the distance away from the preferred spatial location, because these cells have spatially overlapping fields, and because the theory is that these cells in CA3 are connected by Hebb-modifiable synapses. This specification would inherently lead the system to operate as a continuous attractor network. Continuous attractor network models have been studied by Amari (1977), Zhang (1996), Taylor (1999), Samsonovich and McNaughton (1997), Battaglia and Treves (1998), Stringer et al. (2002a, b, 2004), Stringer and Rolls (2002) and Rolls and Stringer (2005) (see Rolls 2008; Rolls and Deco 2002), and are described briefly next.

A "continuous attractor" neural network (CANN) can maintain the firing of its neurons to represent any location along a continuous physical dimension such as spatial view, spatial position, head direction, etc. It uses excitatory recurrent collateral connections between the neurons (as are present in CA3) to reflect the distance between the neurons in the state space of the animal (e.g., place or head direction). These networks can maintain the bubble or packet of neural activity constant for long periods wherever it is started to represent the current state (head direction, position, etc) of the animal, and are likely to be involved in many aspects of spatial processing and memory, including spatial vision. Global inhibition is used to keep the number of neurons in a bubble or packet of actively firing neurons relatively constant, and to help to ensure that there is only one activity packet.

Continuous attractor networks can be thought of as very similar to autoassociation or discrete attractor networks (Rolls 2008), and have the same architecture, as illustrated in Fig. 4.3. The main difference is that the patterns stored in a CANN are continuous patterns, with each neuron having broadly tuned firing which decreases with, for example, a Gaussian function as the distance from the optimal firing location of the cell is varied, and with different neurons having tuning that overlaps throughout the space. Such tuning is illustrated elsewhere (Rolls 2008; Rolls et al. 2002). For comparison, autoassociation networks normally have discrete (separate) patterns (each pattern implemented by the firing of a particular subset of the neurons), with no continuous distribution of the patterns throughout the space. A consequent difference is that the CANN can maintain its firing at any location in the

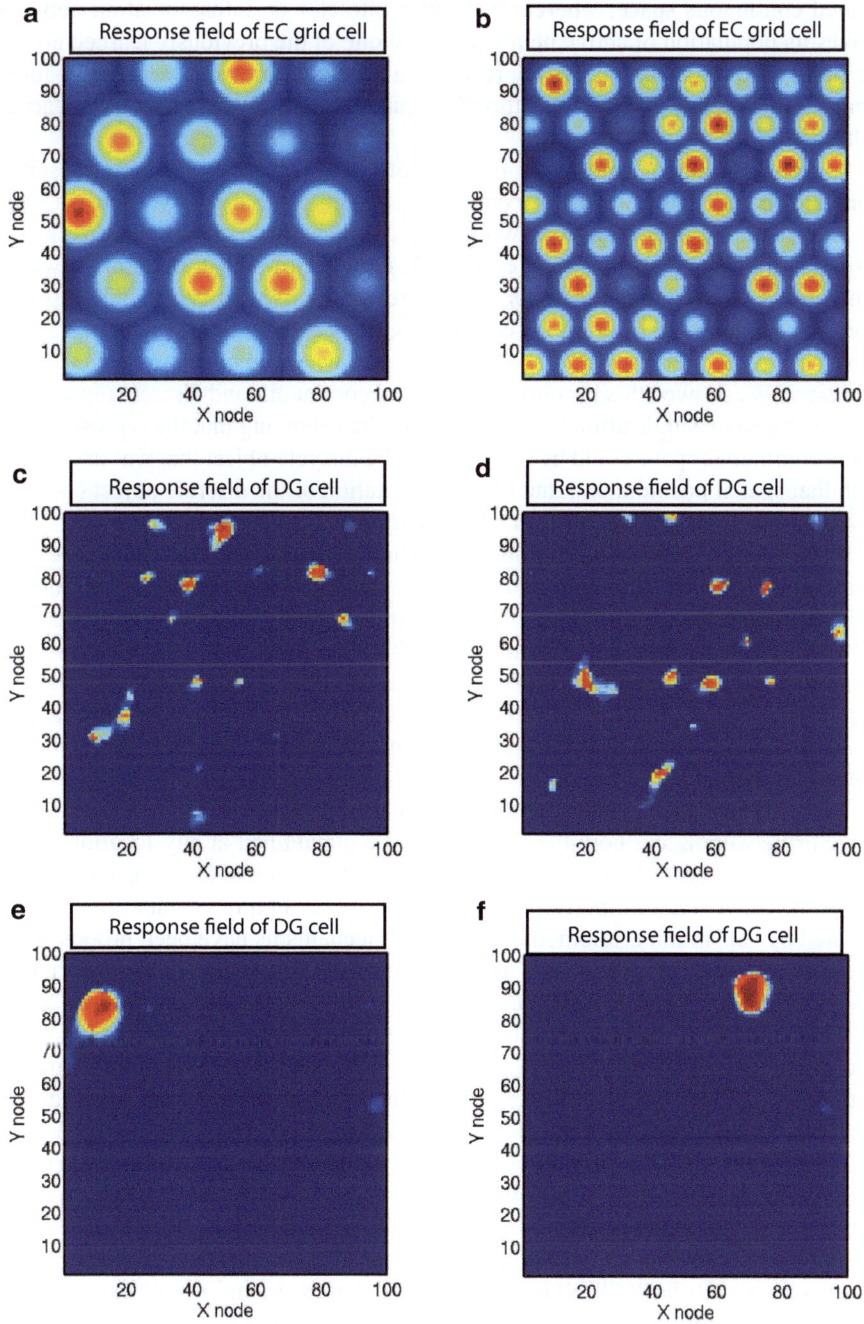

Fig. 4.3 Simulation of competitive learning in the dentate gyrus to produce place cells from the entorhinal cortex grid cell inputs. **a** and **b** Firing rate profiles of two entorhinal cortex (EC) grid cells with frequencies of 4 and 7 cycles. **c** and **d** Firing rate profiles of two dentate gyrus (DG) cells with no training using competitive learning. **e** and **f** Firing rate profiles of two dentate gyrus (DG) cells trained using competitive learning. (After Rolls et al. 2006.)

trained continuous space, whereas a discrete attractor or autoassociation network moves its population of active neurons toward one of the previously learned attractor states, and thus implements the recall of a particular previously learned pattern from an incomplete or noisy (distorted) version of one of the previously learned patterns.

Space is continuous, and object representations are discrete. If these representations are to be combined in for example an object–place memory, then we need to understand the operation of networks that combine these representations. Rolls, Stringer, and Trappenberg (Rolls et al. 2002) have shown that attractor networks can store both continuous patterns and discrete patterns, and can thus be used to store for example the location in (continuous, physical) space (e.g., the place "out there" in a room represented by spatial view cells) where an object (a discrete item) is present. We showed this by storing associated continuous and discrete representations in the same single attractor network, and then showing that the representation in the continuous space could be retrieved by the discrete object that was associated with that spatial position; and that the representation of the discrete object could be retrieved by providing the position in the continuous representation of space.

If spatial representations are stored in the hippocampus, the important issue arises in terms of understanding memories that include a spatial component or context of how many such spatial representations could be stored in a continuous attractor network. The very interesting result is that because there are in general low correlations between the representations of places in different maps or charts (where each map or chart might be of one room or locale), very many different maps or charts can be simultaneously stored in a continuous attractor network (Battaglia and Treves 1998).

We have considered how spatial representations could be stored in continuous attractor networks, and how the activity can be maintained at any location in the state space in a form of short-term memory when the external (e.g., visual) input is removed. However, a property of some spatial representations is that they can be updated by self-motion, idiothetic, input, and mechanisms have been proposed for how this could be achieved (Rolls and Stringer 2005; Samsonovich and McNaughton 1997; Stringer and Rolls 2006; Stringer et al. 2005, 2002a, b; Walters et al. 2013), including in the entorhinal cortex grid cell system (Giocomo et al. 2011; Kropff and Treves 2008; Zilli 2012). The ways in which path integration could be implemented in recurrent networks such as the CA3 system in the hippocampus or in related systems are described elsewhere (McNaughton et al. 2006; Samsonovich and McNaughton 1997; Stringer et al. 2002a, b), and have been applied to primate spatial view cells by Rolls and colleagues (Rolls and Stringer 2005; Stringer et al. 2004, 2005). Cognitive maps (O'Keefe and Nadel 1978) can be understood by the operations of these attractor networks, and how they are updated by learning and by self-motion (Rolls 2008). It has been argued that the bumpiness of the CA3 representation of space is more consistent with episodic memory storage, as argued in this chapter, than with spatial path integration using the CA3 system as a continuous attractor network implementing path integration (Cerasti and Treves 2013; Stella et al. 2013).

Perforant Path Inputs to CA3 Cells Perform Completion and Initiate Recall in CA3

By calculating the amount of information that would end up being carried by a CA3 firing pattern produced solely by the perforant path input and by the effect of the recurrent connections, we have been able to show (Treves and Rolls 1992) that an input of the perforant path type, alone, is unable to direct efficient information storage. Such an input is too weak, it turns out, to drive the firing of the cells, as the "dynamics" of the network is dominated by the randomizing effect of the recurrent collaterals. On the other hand, an autoassociative memory network needs afferent inputs to apply the retrieval cue to the network. We have shown (Treves and Rolls 1992) that the perforant path system is likely to be the one involved in relaying the cues that initiate retrieval in CA3. The concept is that to initiate retrieval, a numerically large input (the perforant path system, see Fig. 4.2) is useful so that even a partial cue is sufficient (see Eq. 17 of Treves and Rolls (1992)); and that the retrieval cue need not be very strong, as the recurrent collaterals (in CA3) then take over in the retrieval process to produce good recall (Rolls 2008; Treves and Rolls 1992). In this scenario, the perforant path to CA3 synapses operate as a pattern associator, the quantitative properties of which are described elsewhere (Rolls 2008, 2016; Rolls and Treves 1990, 1998). If an incomplete recall cue is provided to a pattern association network using distributed input representations, then most of the output pattern will be retrieved, and in this sense *pattern association networks do perform pattern generalization,* and this generalization performed at the perforant path synapses to CA3 cells helps in the completion produced by the recurrent collateral CA3–CA3 autoassociation process.

In contrast, during storage, strong signals, in the order of mV for each synaptic connection, are provided by the mossy fiber inputs to dominate the recurrent collateral activations, so that the new pattern of CA3 cell firing can be stored in the CA3 recurrent collateral connections (Rolls 2008; Treves and Rolls 1992).

The Dilution of the CA3 Recurrent Collateral Connectivity Enhances Memory Storage Capacity and Pattern Completion

Figure 4.2 shows that in the rat, there are approximately 300,000 CA3 neurons, but only 12,000 recurrent collateral synapses per neuron. The dilution of the connectivity is thus 12,000/300,000 = 0.04. The connectivity is thus not complete, and complete connectivity in an autoassociation network would make it simple, for the connectivity between the neurons would then be symmetric (i.e., the connection strength from any one neuron to another is matched by a connection of the same strength in the opposite direction), and this guarantees energy minima for the basins of attraction that will be stable, and a memory capacity than can be calculated (Hopfield 1982). We have shown how this attractor type of network can be extended to have similar properties with diluted connectivity, and also with sparse representations with graded firing rates (Rolls and Treves 1990; Treves 1990, 1991; Treves and Rolls 1991).

However, the question has recently been asked about whether there are any advantages to diluted autoassociation or attractor networks compared to fully connected attractor networks (Rolls 2012a). One biological property that may be a limiting factor is the number of synaptic connections per neuron, which is 12,000 in the CA3–CA3 network just for the recurrent collaterals (see Fig. 4.2). The number may be higher in humans, allowing more memories to be stored in the hippocampus than order 12,000. I note that the storage of large number of memories may be facilitated in humans because the left and right hippocampus appear to be much less connected between the two hemispheres than in the rat, which effectively has a single hippocampus (Rolls 2008). In humans, with effectively two separate CA3 networks, one on each side of the brain, the memory storage capacity may be doubled, as the capacity is set by the number of recurrent collaterals per neuron in each attractor network (Eq. 4.3). In humans, the right hippocampus may be devoted to episodic memories with spatial and visual components, whereas the left hippocampus may be devoted to memories with verbal/linguistic components, that is, in which words may be the part of the episode (e.g., who said what to whom and when) (Barkas et al. 2010; Bonelli et al. 2010; Sidhu et al. 2013).

The answer that has been suggested to why the connectivity of the CA3 autoassociation network is diluted (and why neocortical recurrent networks are also diluted), is that this may help to reduce the probability of having two or more synapses between any pair of randomly connected neurons within the network, which it has been shown greatly impairs the number of memories that can be stored in an attractor network, because of the distortion that this produces in the energy landscape (Rolls 2012a). In more detail, the hypothesis proposed is that the diluted connectivity allows biological processes that set up synaptic connections between neurons to arrange for there to be only very rarely more than one synaptic connection between any pair of neurons. If probabilistically there were more than one connection between any two neurons, it was shown by simulation of an autoassociation attractor network that such connections would dominate the attractor states into which the network could enter and be stable, thus strongly reducing the memory capacity of the network (the number of memories that can be stored and correctly retrieved), below the normal large capacity for diluted connectivity. Diluted connectivity between neurons in the cortex thus has an important role in allowing high capacity of memory networks in the cortex, and helping to ensure that the critical capacity is not reached at which overloading occurs leading to an impairment in the ability to retrieve any memories from the network (Rolls 2012a). The diluted connectivity is thus seen as an adaptation that simplifies the genetic specification of the wiring of the brain, by enabling just simple attributes of the connectivity to be specified (e.g., from a CA3 to another CA3 neuron chosen at random to specify the CA3 to CA3 recurrent collateral connectivity), rather than which particular neuron should connect to which other particular neuron (Rolls 2012a; Rolls and Stringer 2000). Consistent with this hypothesis, there are NMDA receptors with the genetic specification that they are NMDA receptors on neurons of a particular type, CA3 neurons (as shown by the evidence from CA3-specific vs. CA1-specific NMDA receptor knockouts) (Nakazawa et al. 2002, 2003, 2004; Rondi-Reig et al. 2001). A consequence is that the vector of output neuronal firing in the CA3 regions, that is, the number of CA3 neurons, is quite large (300,000 neurons in the rat). The large number of elements in this vector may have consequences for the noise in the system, as we will see below.

The dilution of the CA3–CA3 recurrent collateral connectivity at 0.04 may be greater dilution than that in a local neocortical area, which is in the order of 0.1 (Rolls 2008, 2012a). This is consistent with the hypothesis that the storage capacity of the CA3 system is at a premium, and so the dilution is kept to a low value (i.e., great dilution), as then there is lower distortion of the basins of attraction and hence the memory capacity is maximized (Rolls 2012a).

Pattern Separation of CA3 Cell Populations Encoding Different Memories

For the CA3 to operate with high capacity as an autoassociation or attractor memory, the sets of CA3 neurons that represent each event to be stored and later recalled need to be as uncorrelated from each other as possible. Correlations between patterns reduce the memory capacity of an autoassociation network (Kohonen 1977, 1984; Kohonen et al. 1981; Marr 1971), and because storage capacity is at a premium in an episodic memory system, there are several mechanisms that reduce the correlations between the firing of the population vectors of CA3 neuron firing each one of which represents a different event to be stored in memory. In the theoretical physics approach to the capacity of attractor networks, it is indeed assumed that the different vectors of firing rates to be stored are well separated from each other, by drawing each vector of firing at random, and by assuming very large (infinite) numbers of neurons in each pattern.

We have proposed that there are several mechanisms that help to achieve this pattern separation, namely the mossy fiber pattern separation effect produced by the small number of connections received by a CA3 neuron from mossy fibers which dominate the CA3 cell firing; the expansion recoding, and the sparse representation provided by the dentate granule cells that form the mossy fiber synapses; and the sparseness of the CA3 cell representation. Neurogenesis of dentate granule cells is a fifth potential contributor to achieving pattern separation of CA3 cell firing. The five factors are described next. Before this, it is remarked that some of this architecture may be special to the hippocampus, and not found in the neocortex, because of the importance of storing and retrieving large numbers of (episodic) memories in the hippocampus. The neocortex in contrast is more concerned with building new representations for which competitive learning is more important, and thus neocortical circuitry does not use a mossy fiber system to produce new random sets of neurons activated (Rolls 2008, 2016).

Pattern Separation and the Sparse Connectivity of the Mossy Fiber Inputs to CA3 Cells

We hypothesize that the mossy fiber inputs force efficient information storage by virtue of their strong and sparse influence on the CA3 cell firing rates (Rolls 1987, 1989b, c; Treves and Rolls 1992). (The strong effects likely to be mediated by the mossy fibers were also emphasized by McNaughton and Morris (1987) and McNaughton and Nadel (1990)). We (Rolls and Treves) (Rolls 1987, 1989b, 1989c, 1990b, 2008; Rolls and Treves 1998; Treves and Rolls 1992) hypothesize that the mossy fiber input appears to be particularly appropriate in several ways. First, the fact that mossy fiber synapses are large and located very close to the soma makes them relatively powerful in activating the postsynaptic cell. Second, the firing activity of dentate granule cells appears to be very sparse (Jung and McNaughton 1993; Leutgeb et al. 2007) and this, together with the small number of connections on each CA3 cell, produces a sparse signal, which can then be transformed into sparse firing activity in CA3 by a threshold effect. The hypothesis is that the mossy fiber sparse connectivity solution performs the appropriate function to enable learning to operate correctly in CA3 (Cerasti and Treves 2010; Treves and Rolls 1992). The perforant path input would, the quantitative analysis shows, not produce a pattern of firing in CA3 that contains sufficient information for learning (Treves and Rolls 1992) (see further Section "Perforant Path Inputs to CA3 Cells Perform Completion and Initiate Recall in CA3").

The particular property of the small number of mossy fiber connections onto a CA3 cell, approximately 46 (see Fig. 4.2), is that this has a *randomizing effect* on the representations set up in CA3, so that they are as different as possible from each other (Rolls 1989b, 1989c, 2008; Rolls and Kesner 2006; Rolls and Treves 1998; Treves and Rolls 1992). (This means, for example, that place cells in a given environment are well separated to cover the whole space.) The result is that any one event or episode will set up a representation that is very different from other events or episodes, because the set of CA3 neurons activated for each event is random. This is then the optimal situation for the CA3 recurrent collateral effect to operate, for it can then associate together the random set of neurons that are active for a particular event (e.g., an object in a particular place), and later recall the whole set from any part. It is because the representations in CA3 are unstructured, or random, in this way that large numbers of memories can be stored in the CA3 autoassociation system, and that interference between the different memories is kept as low as possible, in that they are maximally different from each other (Hopfield 1982; Rolls 2008; Rolls and Treves 1998; Treves and Rolls 1991).

The requirement for a small number of mossy fiber connections onto each CA3 neuron applies not only to discrete (Treves and Rolls 1992) but also to spatial representations, and some learning in these connections, whether associative or not, can help to select out the small number of mossy fibers that may be active at any one time to select a set of random neurons in the CA3 (Cerasti and Treves 2010). Any learning may help by reducing the accuracy required for a particular number of mossy fiber connections to be specified genetically onto each CA3 neuron. The optimal number of mossy fibers for the best information transfer from dentate granule cells to CA3 cells is in the order of 35–50 (Cerasti and Treves 2010; Treves and Rolls 1992). The mossy fibers also make connections useful for feedforward inhibition in CA3 (Acsady et al. 1998), which is likely to be useful to help in the sparse representations being formed in CA3.

On the basis of these and other points, we predicted that the mossy fibers may be necessary for new learning in the hippocampus, but may not be necessary for the recall of existing memories from the hippocampus (Rolls 2008; Rolls and Treves 1998; Treves and Rolls 1992). Experimental evidence consistent with this prediction about the role of the mossy fibers in learning has been found in rats with disruption of the dentate granule cells (Lassalle et al. 2000) (Pattern Separation Performed By Dentate Granule Cells).

We (Rolls and Kesner 2006) have hypothesized that nonassociative plasticity of mossy fibers (see Brown et al. 1989, 1990) might have a useful effect in enhancing the signal-to-noise ratio, in that a consistently firing mossy fiber would produce nonlinearly amplified currents in the postsynaptic cell, which would not happen with an occasionally firing fiber (Treves and Rolls 1992). This plasticity, and also learning in the dentate, would also have the effect that similar fragments of each episode (e.g., the same environmental location) recurring on subsequent occasions would be more likely to activate the same population of CA3 cells, which would have potential advantages in terms of economy of use of the CA3 cells in different memories, and in making some link between different episodic memories with a common feature, such as the same location in space. Consistent with this, dentate neurons that fire repeatedly are more effective in activating CA3 neurons (Henze et al. 2002).

As acetylcholine turns down the efficacy of the recurrent collateral synapses between CA3 neurons (Giocomo and Hasselmo 2007; Hasselmo et al. 1995), then cholinergic activation also might help to allow external inputs from the mossy fibers rather than the internal recurrent collateral inputs to dominate the firing of the CA3 neurons during learning, as the current theory proposes. If cholinergic activation at the same time facilitated LTP in the recurrent collaterals (as it appears to in the neocortex), then cholinergic activation could have a useful double role in facilitating new learning at times of behavioral activation (Giocomo and Hasselmo 2007; Hasselmo et al. 1995), when presumably it may be particularly relevant to allocate some of the limited memory capacity to new memories.

Pattern Separation and the Sparseness of the Firing of the Dentate Granule Cell Input Via the Mossy Fibers to CA3 Cells

The firing activity of dentate granule cells appears to be very sparse (Jung and McNaughton 1993; Leutgeb et al. 2007) and this, together with the small number of dentate mossy fiber connections on each CA3 cell, produces a sparse signal, which can then be transformed into sparse firing activity in CA3 by a threshold effect. The pattern separation mechanisms that enable the dentate to provide a sparse firing input to CA3 are described below.

Pattern Separation and the Large Number of Dentate Granule Cells Providing Inputs Via the Mossy Fibers to CA3 Cells

Expansion recoding can decorrelate input patterns, and this can be performed by a stage of competitive learning with a large number of neurons (Rolls 2008). A mechanism like this appears to be implemented by the dentate granule cells, which are numerous (1×10^6 in the rat, compared to 300,000 CA3 cells), have associatively modifiable synapses (required for a competitive network), and strong inhibition provided by the inhibitory interneurons. This may not represent expansion of numbers relative to the number of entorhinal cortex cells, but the principle of a large number of dentate granule cells, with competitive learning and strong inhibition through inhibitory interneurons, would produce a decorrelation of signals like that achieved by expansion recoding (Rolls 2008).

Sparseness of the CA3 Cell Representation and Pattern Separation

The firing of CA3 cells is relatively sparse, and this helps to decorrelate different population vectors of CA3 cell firing for different memories. (Sparse representations are more likely to be decorrelated with each other (Rolls 2008).) Evidence on the sparseness of the CA3 cell representation in rats includes evidence that CA3 cell ensembles may support the fast acquisition of detailed memories by providing a locally continuous, but globally orthogonal spatial representation, onto which new sensory inputs can rapidly be associated (Leutgeb and Leutgeb 2007). In the macaque hippocampus, in which spatial view cells are found (Georges-François et al. 1999; Robertson et al. 1998; Rolls et al. 1997a, 1998), for the representation of 64 locations around the walls of the room, the mean single cell sparseness a^s was 0.34, and the mean population sparseness $a^p$was 0.33 (Rolls 2008; Rolls and Treves 2011; Rolls et al. 1998). For comparison, the corresponding values for inferior temporal cortex neurons tuned to objects and faces were 0.77 (Franco et al. 2007; Rolls 2008; Rolls and Treves 2011); for taste and oral texture neurons in the insular cortex the population sparseness was 0.71; for taste and oral texture neurons in the orbitofrontal cortex was 0.61; and for taste and oral texture neurons in the orbitofrontal cortex was 0.81 (Rolls 2008; Rolls and Treves 2011). Thus, the evidence is that the hippocampal CA3/pyramidal cell representation is more sparse in macaques than in neocortical areas and the amygdala, and this is consistent with the importance in hippocampal CA3 of using a sparse representation to produce a large memory capacity.

Representations in the neocortex and in the hippocampus are often distributed with graded firing rates in the neuronal populations (Rolls and Treves 2011). The firing rate probability distribution of each neuron to a set of stimuli is often exponential or gamma (Rolls and Treves 2011). These graded firing rate distributed representations are present in the hippocampus, both for place cells in rodents and for spatial view cells in the primate (Georges-François et al. 1999; McNaughton et al. 1983; O' Keefe and Speakman 1987; O'Keefe 1979; Robertson et al. 1998; Rolls 2008; Rolls et al. 1997a, 1998; Rolls and Treves 2011). In processes in the brain such as memory recall in the hippocampus or decision-making in the cortex that are influenced by the noise produced by the close to random spike timings of each neuron for a given mean rate, the noise with this graded type of representation may be larger than with the binary firing rate distribution that is usually investigated (Rolls and Deco 2010). In integrate-and-fire simulations of an attractor decision-making network, we showed that the noise is indeed greater for a given sparseness of the representation for graded, exponential, than for binary firing rate distributions (Webb et al. 2011). The greater noise was measured by faster escaping times from the spontaneous firing rate state when the decision cues are applied, and this corresponds to faster decision or reaction times. The greater noise was also evident as less stability of the spontaneous firing state before the decision cues are applied. The implication is that spiking-related noise will continue to be a factor that influences processes such as decision-making, signal detection, short-term memory, and memory recall (including in the CA3 network) even with the quite large networks found in the cerebral cortex. In these networks there are several thousand recurrent collateral synapses onto each neuron. The greater noise with graded firing rate distributions has the advantage that it can increase the speed of operation of cortical circuitry (Webb et al. 2011). The graded firing rates also by operating in a nonlinear network effectively increase the sparseness of the representation, and this itself is a pattern separation effect (Webb et al. 2011).

Neurogenesis of Dentate Granule Cells to Provide New Representations in CA3 Uncorrelated with Previous CA3 Representations

If adult neurogenesis in the dentate gyrus does prove to be functionally relevant, its computational role could be to facilitate pattern separation for new patterns, by providing new dentate granule cells with new sets of random connections to CA3 neurons. Consistent with the dentate spatial pattern separation hypothesis (Rolls 1989b, c, 1996b; Treves and Rolls 1992, 1994), in mice with impaired dentate neurogenesis, spatial learning in a delayed non-matching-to-place task in the radial arm maze was impaired for arms that were presented with little separation, but no deficit was observed when the arms were presented farther apart (Clelland et al. 2009). Consistently, impaired neurogenesis in the dentate also produced a deficit for small spatial separations in an associative object-in-place task (Clelland et al. 2009).

The Direct Perforant Path to CA3 Cell Input: Poor at Pattern Separation and Forcing a New Memory Pattern into CA3 Cell Firing

It has been suggested that the feedforward connectivity from the entorhinal cortex via the perforant path to the CA3 neurons may act as a feedforward pattern association network that is more important than the CA3–CA3 recurrent collateral autoassociation system (Cheng 2013). The quantitative properties of pattern association networks are described elsewhere (Rolls 2008; Rolls and Treves 1990, 1998). If an incomplete recall cue is provided to a pattern association network using distributed input representations, then most of the output pattern will be retrieved, and in this sense *pattern association networks do generalize*. (As noted above, pattern association networks do not perform pattern completion, in that the unconditioned stimulus cannot recall the conditioned stimulus.) The analyses described in these sources shows that the capacity of pattern association networks (the maximum number of memories that can be stored and retrieved, here denoted by p_{max}) is approximately

$$p_{\max} \approx \frac{C^{\mathrm{PA}}}{[a_o \ln(1/a_o)]} \tag{4.4}$$

where C^{PA} is the number of feedforward associatively modifiable connections per neuron, and a_o is the sparseness of the representation in the output neurons of the pattern associator (Rolls 2008). Given that there are fewer feedforward (perforant path) synaptic connections onto CA3 neurons (3600) than recurrent synaptic connections between CA3 neurons (12,000 in the rat) (see Fig. 4.2), then the capacity of the feedforward system would be considerably smaller than that of the recurrent collateral CA3–CA3 system. (It is noted that the a_o of Eq. (4) would be the same number as the a of Eq. (3), as that is just the sparseness of the firing of the population of CA3 neurons. The number of perforant path synapses is sufficiently large that it can act as a retrieval cue for even an incomplete pattern so that the CA3–CA3 connections can then complete the retrieval, given that the recall signal for the perforant path pattern associator is proportional to the square root of the number of perforant path synapses, as shown by Eq. 17 of Treves and Rolls (1992).) The feedforward hypothesis (Cheng 2013) thus has a strong argument against it of storage capacity, which would be much less (approximately 3600/12,000) than that of the CA3–CA3 recurrent collateral system operating as an autoassociation memory. Another disadvantage of the feedforward hypothesis is that the attractor properties of the CA3–CA3 connections would be lost, and these potentially contribute to holding one or more items simultaneously active in short-term memory (Rolls 2008; Rolls et al. 2013), and providing a basis for temporal order memory as described in "Dilution of the CA3 Recurrent Collateral Connectivity Enhances Memory Storage Capacity and Pattern Completion." Another disadvantage is that we have been able to show (Treves and Rolls 1992) that an input of the perforant path type, alone, is unable to direct efficient information storage. Such an input is too weak, it turns out, to drive the firing of the cells, as the "dynamics" of the network is dominated by the randomizing effect of the recurrent collaterals. Another disadvantage of the feedforward hypothesis is that a pattern associator may not, with an incomplete cue, be able to recall the exact pattern that was stored, whereas an attractor network has the property that it can fall into an attractor basin that can reflect perfect retrieval of the memory (Rolls 2008; Rolls and Treves 1998).

Pattern Separation Performed by Dentate Granule Cells

The theory is that the dentate granule cell stage of hippocampal processing which precedes the CA3 stage acts as a competitive network in a number of ways to produce during learning the sparse yet efficient (i.e., nonredundant) representation in CA3 neurons that is required for the autoassociation implemented by CA3 to perform well (Rolls 1989b, c, 1990b; Kesner and Rolls 2015; Rolls et al. 2006; Treves and Rolls 1992). An important property for episodic memory is that the dentate by acting in this way would perform pattern separation (or orthogonalization) (Rolls 1989b; Kesner and Rolls 2015; Rolls et al. 2006; Treves and Rolls 1992), enabling the hippocampus to store different memories of even similar events, and this prediction has been confirmed (Gilbert et al. 2001; Goodrich-Hunsaker et al. 2008; Kesner et al. 2012; Leutgeb and Leutgeb 2007; McHugh et al. 2007; Rolls 2008; Rolls and Kesner 2006) ("Pattern Separation Performed By Dentate Granule Cells"). Consistently with this evidence for pattern separation by dentate granule cells, in rats small changes in the shape of the environment in which rats are exploring can substantially alter the activity patterns among place-modulated granule cells (Leutgeb et al. 2007).

As just described, the dentate granule cells could be important in helping to build and prepare spatial representations for the CA3 network. The actual representation of space in the primate hippocampus includes a representation of spatial view (Georges-François et al. 1999; Robertson et al. 1998; Rolls et al. 1997a, 1998; Rolls and Xiang 2006), whereas in the rat hippocampus it is of the place where the rat is. The representation in the rat may be related to the fact that with a much less developed visual system than the primate, the rat's representation of space may be defined more by the olfactory and tactile as well as distant visual cues present, and may thus tend to reflect the place where the rat is. However, the spatial representations in the rat and primate could arise from essentially the same computational process as follows (de Araujo et al. 2001; Rolls 1999). The starting assumption is that in both the rat and the primate, the dentate granule cells (and the CA3 and CA1 pyramidal cells) respond to combinations of the inputs received. In the case of the primate, a combination of visual features in the environment will, because of the fovea providing high spatial resolution over a typical viewing angle of perhaps 10–20 °, result in the formation of a spatial view cell, the effective trigger for which will thus be a combination of visual features within a relatively small part of space.

In contrast, in the rat, given the very extensive visual field subtended by the rodent retina, which may extend over 180–270 °, a combination of visual features formed over such a wide visual angle would effectively define a position in space that is a place (de Araujo et al. 2001).

The entorhinal cortex contains grid cells, which have high firing in the rat in a two-dimensional spatial grid as a rat traverses an environment, with larger grid spacings in the ventral entorhinal cortex (Fyhn et al. 2004; Hafting et al. 2005). This may be a system optimized for path integration (McNaughton et al. 2006) which may self-organize during locomotion with longer time constants producing more widely spaced grids in the ventral entorhinal cortex (Kropff and Treves 2008). How are the grid cell representations, which would not be suitable for association of an object or reward with a place to form an episodic memory, transformed into a place representation that would be appropriate for this type of episodic memory? I have proposed that this could be implemented by a competitive network (see Rolls 2008) in the dentate gyrus which operates to form place cells, implemented by each dentate granule cell learning to respond to particular combinations of entorhinal cortex cells firing, where each combination effectively specifies a place, and this has been shown to be feasible computationally (Rolls et al. 2006). The sparse representations in the dentate gyrus, implemented by the mutual inhibition through inhibitory interneurons and competitive learning, help to implement this "pattern separation" effect (Rolls 1989b, c, 2008; Rolls and Treves 1998). The investigations showed that learning in the perforant path to dentate granule cell representation, and the sparse representation in the dentate granule cells, are both important in the formation of place-like fields in dentate granule cells from the grid cells in the entorhinal cortex (Georges-François et al. 1999; Robertson et al. 1998; Rolls et al. 1997a; 1998). To illustrate this, Fig. 4.3 shows from these simulations the responses of the simulated grid cells (a, b), the dentate receptive fields formed by feedforward connections and a sparse representation in the dentate gyrus (c, d), and the dentate receptive fields formed when Hebbian synaptic modification and training is included in the feedforward connections to implement competitive learning (e, f). It is only with the full competitive learning that the dentate receptive fields self-organized to become small place-like receptive fields (Rolls et al. 2006) similar to those found in the rat dentate granule cells.

In primates, there is now evidence that there is a grid-cell like representation in the entorhinal cortex, with neurons having grid-like firing as the monkey moves the eyes across a spatial scene (Killian et al. 2012). Similar competitive learning processes may transform these entorhinal cortex "spatial view grid cells" into hippocampal spatial view cells, and may help with the idiothetic (produced in this case by movements of the eyes) update of spatial view cells (Robertson et al. 1998). The presence of spatial view grid cells in the entorhinal cortex of primates (Killian et al., 2012) is of course predicted from the presence of spatial view cells in the primate CA3 and CA1 regions (Georges-François et al. 1999; Robertson et al. 1998; Rolls 2008; Rolls et al. 1997a, 1998; Rolls and Xiang 2006). Further support of this type of representation of space being viewed "out there" rather than where one is located as for rat place cells is that cells in the human parahippocampal cortex with spatial view-like properties have now been described (Ekstrom et al. 2003).

CA1 Cells and Pattern Completion Prior to Hippocampo-Directed Recall to the Neocortex

The CA3 cells connect to the CA1 cells by the Schaeffer collateral synapses. The associative modifiability in this connection helps the full information present in CA3 to be retrieved in the CA1 neurons (Rolls 1995; Schultz and Rolls 1999; Treves 1995; Treves and Rolls 1994). Part of the hypothesis is that the separate subparts of an episodic memory, which must be represented separately in CA3 to allow for completion, can be combined together by competitive learning in CA1 to produce an efficient retrieval cue for the recall via the backprojection pathways to the neocortex of memories stored in the neocortex (Rolls 1989a, b, 1995, 1996b; Treves and Rolls 1994). Associative recall in the CA3 to CA1 feedforward connections is a prominent property which implements what amounts to pattern completion (Rolls 1995, 2008; Schultz et al. 2000), though for pattern associators this process is usually described as generalization (Rolls 2008).

Backprojections to the Neocortex, and Memory Retrieval from the Hippocampus Involving Pattern Completion

The need for information to be retrieved from the hippocampus to affect other brain areas was noted in the Introduction. The way in which this could be implemented via backprojections to the neocortex (Rolls 1995, 1996b, 2008; 2010b; Treves and Rolls 1994) is considered here in the context of recalling a complete memory representation in the complete set of cortical areas that provide inputs to the hippocampus (see Fig. 4.1).

It is suggested that the modifiable connections from the CA3 neurons to the CA1 neurons allow the whole episode in CA3 to be produced in CA1. The CA1 neurons would then activate, via their termination in the deep layers of the entorhinal cortex, at least the pyramidal cells in the deep layers of the entorhinal cortex (see Fig. 4.1). These entorhinal cortex layer 5 neurons would then, by virtue of their backprojections (Lavenex and Amaral 2000; Witter et al. 2000a) to the parts of cerebral cortex that originally provided the inputs to the hippocampus, terminate in the superficial layers (including layer 1) of those neocortical areas, where synapses would be made onto the distal parts of the dendrites of the (superficial and deep) cortical pyramidal cells (Rolls 1989a, b, c). The areas of cerebral neocortex in which this recall would be produced could include multimodal cortical areas (e.g., the cortex in the superior temporal sulcus which receives inputs from temporal, parietal, and occipital cortical areas, and from which it is thought that cortical areas such as 39 and 40 related to language developed), and also areas of unimodal association cortex (e.g., inferior temporal visual cortex). The backprojections, by recalling previous episodic events, could provide information useful to the neocortex in the building of new representations in the multimodal and unimodal association cortical areas, which by building new long-term and structured representations can be considered as a form of memory consolidation (Rolls 1989a, b, c; 1990a; b, 2008), or in organizing actions.

The hypothesis of the architecture with which this would be achieved is shown in Fig. 4.1. The feedforward connections from association areas of the cerebral neocortex (solid lines in Fig. 4.1), show major convergence as information is passed to CA3, with the CA3 autoassociation network having the smallest number of neurons at any stage of the processing. The backprojections allow for divergence back to neocortical areas. The way in which I suggest that the backprojection synapses are set up to have the appropriate strengths for recall is as follows (Rolls 1989a, b, c). During the setting up of a new episodic memory, there would be strong feedforward activity progressing toward the hippocampus. During the episode, the CA3 synapses would be modified, and via the CA1 neurons and the subiculum, a pattern of activity would be produced on the backprojecting synapses to the entorhinal cortex. Here, the backprojecting synapses from active backprojection axons onto pyramidal cells being activated by the forward inputs to entorhinal cortex would be associatively modified. A similar process would be implemented at preceding stages of neocortex, that is in the parahippocampal gyrus/perirhinal cortex stage, and in association cortical areas, as shown in Fig. 4.1.

The concept is that during the learning of an episodic memory, cortical pyramidal cells in at least one of the stages would be driven by forward inputs, but would simultaneously be receiving backprojected activity (indirectly) from the hippocampus which would by pattern association from the backprojecting synapses to the cortical pyramidal cells become associated with whichever cortical cells were being made to fire by the forward inputs. Then later on, during recall, a recall cue from perhaps another part of cortex might reach CA3, where the firing during the original episode would be completed. The resulting backprojecting activity would then, as a result of the pattern association learned previously, bring back the firing in any cortical area that was present during the original episode. Thus, retrieval involves reinstating the activity that was present in different cortical areas that was present during the learning of an episode. (The pattern association is also called heteroassociation, to contrast it with autoassociation. The pattern association operates at multiple stages in the backprojection pathway, as made evident in Fig. 4.1). If the recall cue was an object, this might result in recall of the neocortical firing that represented the place in which that object had been seen previously. As noted elsewhere in this chapter and by McClelland et al. (1995), that recall might be useful to the neocortex to help it build new semantic memories, which might inherently be a slow process and is not a part of the theory of recall.

A plausible requirement for a successful hippocampo-directed recall operation, is that the signal generated from the hippocampally retrieved pattern of activity, and carried backward toward neocortex, remain undegraded when compared to the noise due, at each stage, to the interference effects caused by the concurrent storage of other patterns of activity on the same backprojecting synaptic systems. That requirement is equivalent to that used in deriving the storage capacity of such a series of heteroassociative memories, and it was shown by Treves and Rolls (1991, 1994) that the maximum number of independently generated activity patterns that can be retrieved is given, essentially, by the same formula as (3) above where, however, a is now the sparseness of the representation at any given stage, and C is the average number of (back-)projections each cell of that stage receives from cells of the previous one. (k′ is a similar slowly varying factor to that introduced above.) If p is equal to the number of memories held in the hippocampal memory, it is limited by the retrieval capacity of the CA3 network, p_{max}. Putting together the formula for the latter with that shown here, one concludes that, roughly, the requirement implies that the number of afferents of (indirect) hippocampal origin to a given neocortical stage (C^{HBP}), must be $C^{HBP} = C^{RC} a_{nc} / a_{CA3}$, where C^{RC} is the number of recurrent collaterals to any given cell in CA3, the average sparseness of a representation is a_{nc}, and a_{CA3} is the sparseness of memory representations there in CA3.

The above requirement is very strong: even if representations were to remain as sparse as they are in CA3, which is unlikely, to avoid degrading the signal, C^{HBP} should be as large as C^{RC}, that is, 12,000 in the rat. If then C^{HBP} has to be of the same order as C^{RC}, one is led to a very definite conclusion: A mechanism of the type envisaged here could not possibly rely on a set of monosynaptic CA3-to-neocortex backprojections. This would imply that, to make a sufficient number of synapses on each of the vast number of neocortical cells, each cell in CA3 has to generate a disproportionate number of synapses (i.e., C^{HBP} times the ratio between the number of neocortical and that of CA3 cells). The required divergence can be kept within reasonable limits only by assuming that the backprojecting system is polysynaptic, provided that the number of cells involved grows gradually at each stage, from CA3 back to neocortical association areas (Treves and Rolls 1994) (cf. Fig. 4.1).

The theory of recall by the backprojections thus provides a quantitative account of why the cerebral cortex has as many backprojection as forward projection connections.

These concepts show how the backprojection system to neocortex can be conceptualized in terms of pattern completion, as follows. First, the information that is present when a memory is formed may be present in different areas of the cerebral cortex, for example of a face in a temporal cortex face area (Rolls 2012b), of a spatial location in a neocortical location area, and of a reward received in the orbitofrontal cortex (Rolls 2014). To achieve detailed retrieval of the memory, reinstatement of the activity during recall of the neuronal activity during the original memory formation may be needed. This is what the backprojection system described could achieve, and is a form of completion of the information that was represented in the different cortical areas when the memory was formed. Because such a wide set of different neocortical areas must be content addressed, a multistage feedback system is required, to keep the number of synapses per neuron in the backprojection pathways down to reasonable numbers. (Having CA1 directly address neocortical areas would require each CA1 neuron to have tens of millions of synapses with cortical neurons. That is the part of the computational problem solved by the multistage backprojection system shown in Fig. 4.1.) Second, the backprojection system with its series of pattern associators can each be thought of as performing a type of pattern completion.

Further aspects of the operation of the backprojecting systems are described elsewhere (Rolls 2008, 2016).

Tests of Pattern Separation and Pattern Completion

There is now a large literature on tests of pattern separation and pattern completion in the hippocampus (Giocomo et al. 2011; Hunsaker and Kesner 2008, 2013; Jezek et al. 2011; Kesner 2007, 2013; Kesner et al. 2012; Leutgeb et al. 2007; McHugh et al. 2007; Nakashiba et al. 2012; Nakazawa et al. 2002, 2003; Rolls and Kesner 2006; Wills et al. 2005), and a brief summary of some of the findings is provided next. An important point is that the theory (Rolls 1987, 1989a, b, c, 1990a, b, 1991, 1995, 1996b, 2008, 2010b, 2013; Rolls and Deco 2010; Kesner and Rolls 2015; Rolls and Treves 1998; Treves and Rolls 1991, 1992, 1994) is a quantitative theory of hippocampal function, and addresses how pattern separation and pattern completion are important in enabling the hippocampal system to operate up to capacity, which is in the order of tens of thousands of different memories. Some predictions from the theory may only hold when the system is well loaded, that is tested when the system is operating with thousands of memories, for then the pattern separation will be important. It is possible to test the predictions in simulations, where the system can be trained up to capacity (Rolls 1995, 2012a; Rolls et al. 1997b). In vivo, it may be useful to test the storage and recall of as many memories as possible, and in addition testing animals kept in environments where memories of the hippocampal type are needed may also help to test hypotheses in situations where the hippocampus has been at least moderately well loaded with many different memories.

Dentate Granule Cells

The theory predicts that pattern separation is performed by competitive learning by the dentate granule cells. Evidence consistent with this has been found neurophysiologically in the small sparsely encoded place fields of dentate neurons (Jung and McNaughton 1993; Leutgeb and Leutgeb 2007) and their reflection in CA3 neurons (Leutgeb and Leutgeb 2007). Further, and consistent with the theory, it has been shown that selective dentate lesions in rats (Gilbert and Kesner 2003; Gilbert et al. 2001; Goodrich-Hunsaker et al. 2008; Hunsaker and Kesner 2013; Kesner 2013; Rolls 2008; Kesner and Rolls 2015) or dentate NMDA receptor knockouts in mice (McHugh et al. 2007) impair spatial, object–place (or reward–place: Remembering where to find a reward) association tasks especially when the places are close together and require pattern separation before storage in CA3.

Mossy Fiber Inputs to CA3 and Learning

The theory predicts that the dentate granule cell mossy fiber system of inputs to the CA3 neurons is necessary to store spatial memories, but not to recall them. Lassalle et al. (2000) have obtained evidence consistent with this in rats with damage to the mossy fiber system (Lassalle et al. 2000), and there is further evidence consistent with this (Daumas et al. 2009; Lee and Kesner 2004; Rolls and Kesner 2006).

Perforant Path Inputs to CA3 and Recall

The theory predicts that the direct perforant path input from the entorhinal cortex to the CA3 cells (which bypasses the dentate granule cells) is involved in the recall of memory from the CA3 system, and Lee and Kesner (2004) have obtained evidence consistent with this in a Hebb–Williams maze recall task (Lee and Kesner 2004).

CA3 and Pattern Completion

The theory predicts that the CA3 system is especially important in object–place or reward–place tasks in which associations must be formed between any spatial location and any object (referred to as *arbitrary associations*). There is much evidence from subregion analyses involving disruption of CA3 that CA3 is necessary for arbitrary associations between places and objects or rewards (Gilbert and Kesner 2003; Hunsaker and Kesner 2013; Rolls and Kesner 2006). Similar impairments were obtained following deletion of CA3 NMDA receptors in mice in the acquisition of an odor–context paired associate learning task (Rajji et al. 2006). If place or time is not a component, associative tasks such as odor–object association are not impaired (Rolls and Kesner 2006), underlining the fact that the hippocampus is especially involved in episodic types of associative memory which typically involve place and/or time.

The theory predicts that the CA3 is especially important in object–place or reward–place *completion* tasks, in which associations must be completed from a part of the whole. It has been shown that if completion from an incomplete cue is needed, then CA3 NMDA receptors are necessary (presumably to ensure satisfactory CA3–CA3 learning) even in a reference memory task (Gold and Kesner 2005; Hunsaker and Kesner 2013; Nakazawa et al. 2002).

The theory predicts that the CA3 system is especially needed in *rapid, one-trial object–place, learning and recall.* It has been shown that hippocampal NMDA receptors (necessary for long-term potentiation to occur) are needed for one-trial flavor–place association learning, and that hippocampal AMPA/kainate receptors are sufficient for the recall, though the hippocampal subregion involved was not tested (Day et al. 2003). In subregion studies, Kesner and colleagues have shown that CA3 lesions produce chance performance on a one-trial object–place recall task (Kesner et al. 2008) and other object–spatial tasks (Kesner and Rolls 2001; Rolls and Kesner 2006). For example, CA3 lesions produced chance performance on both a one-trial object–place recall and place–object recall task (Kesner et al. 2008). This is evidence that CA3 supports arbitrary associations as well as episodic memory based on 1-trial learning. A control fixed visual conditional to place task with the same delay was not impaired, showing that it is recall after one-trial (or rapid, episodic) learning that is impaired (Kesner et al. 2008). CA3 NMDA receptors are as predicted by the theory necessary for rapid/one-trial spatial learning, as shown by a mouse knockout study by Nakazawa, Tonegawa and colleagues (Nakazawa et al. 2004, 2003; Tonegawa et al. 2003). We have shown that hippocampal CA3 neurons reflect the computational processes necessary for one-trial object–place event memory, used as a model for episodic memory (Rolls and Xiang 2006).

Another type of test of the autoassociation (or attractor) hypothesis for CA3 has been to train rats in different environments, for example, a square and a circular environment, and then test the prediction of the hypothesis that when presented with an environment ambiguous between these, hippocampal neurons will fall into an attractor state that represents one of the two previously learned environments, but not a mixture of the two environments. Evidence consistent with the hypothesis has been found (Wills et al. 2005). In a particularly dramatic example, it has been found that within each theta cycle, hippocampal pyramidal neurons may represent one or other of the learned environments (Jezek et al. 2011). This is an indication, predicted by Rolls and Treves (1998), that autoassociative memory recall can take place sufficiently rapidly to be complete within one theta cycle (120 ms), and that theta cycles could provide a mechanism for a fresh retrieval process to occur after a reset caused by the inhibitory part of each theta cycle, so that the memory can be updated rapidly to reflect a continuously changing environment, and not remain too long in an attractor state.

Evidence that the firing of hippocampal pyramidal cells in macaques is more sparse than in neocortical areas is described in “Sparseness of the CA3 Cell Representation and Pattern Separation.” This is consistent with the premium placed in the hippocampus for storing and retrieving large numbers of independent memories.

The theory predicts that if primates including humans can form an episodic memory in which objects or people are seen at particular locations even though the observer viewing the space has never been to those locations “out there” in space, there should be a neural system in CA3 that can support such associations between places “out there” in a scene and objects. Exactly this is provided by the spatial view neurons Rolls and colleagues have discovered that are present in primate CA3 (Georges-François et al. 1999; Robertson et al. 1998; Rolls et al. 1997a, 1998; Rolls and Xiang 2005, 2006; Rolls et al. 2005). Place cells will not do for this type of episodic memory (Rolls 2010b, 2013).

Recall Via CA1 to Neocortex: A Reverse Hierarchy of Pattern Associators Each Performing Pattern Completion

The theory shows quantitatively, analytically, how memories could be retrieved from the hippocampus to the neocortex (Treves and Rolls 1994), and this has been shown by simulation of the multistage hippocampal system including the entorhinal cortex, dentate, CA3, CA1, and return to the entorhinal cortex to recall the memory to be quantitatively realistic (Rolls 1995).

It has been shown that after learning in hippocampal-dependent tasks, neocortical representations may change (Schwindel and McNaughton 2011). Although this has been interpreted as the transfer of memories from the hippocampus to the neocortex (Schwindel and McNaughton 2011), it should be noted that if the hippocampal representation changes as a result of learning, then the altered representation in CA1 will, even with fixed synaptic connections back to neocortex, alter neocortical firing, with no learning or actual "transfer" involved. (This occurs whenever one vector of neuronal firing changes and influences another vector of neuronal firing through fixed connections.) It has also been suggested that the transfer of information from the hippocampus to the neocortex occurs especially during sleep (Marr 1971; Schwindel and McNaughton 2011). My own view is that during waking would be the best time to retrieve a memory from the hippocampus to the neocortex by using the hippocampus to retrieve the complete episodic memory from a fragment. The retrieval would reinstate the neocortical activity present when the event was originally learned. The retrieved information now present in the neocortex could then be used to build new semantic memories, for example, a narrative account of all the events that took place on one's fifth birthday party. During waking the building of semantic representations could be guided and organized by rational thought into useful semantic representations. To do this during sleep would run the risk of forming bizarre semantic representations of the type that we dream about during the unguided noise-driven stochastic firing during sleep (Rolls 2008; Rolls and Deco 2010). Further, the active recall during waking of memories from the hippocampus means that mainly relevant or useful memories would be retrieved from the hippocampus (not useless memories such as where one parked one's bicycle two weeks ago), and only these memories would tend to become incorporated into useful long-term semantic representations, allowing memories not retrieved from the hippocampus to be overwritten by new memories in the process of forgetting that involves using CA3 sets of neurons chosen at random for new episodic memories (Rolls 2008).

Many further tests of the theory are described elsewhere (Hunsaker and Kesner 2013; Kesner et al. 2012; Rolls 2008, 2010b; Rolls and Kesner 2006). The theory has recently been extended to temporal order memory and temporal pattern separation (Rolls 2010b, 2013), which are also related to hippocampal function (Hoge and Kesner 2007; Kesner et al. 2002; Kesner and Rolls 2015).

Acknowledgments Different parts of the research described here were supported by Program Grants from the Medical Research Council, by a Human Frontier Science Program Grant, by an EEC BRAIN grant, by the MRC Oxford Interdisciplinary Research Centre in Cognitive Neuroscience, by the Oxford McDonnell-Pew Centre in Cognitive Neuroscience, and by the Oxford Centre for Computational Neuroscience. The author has performed the experimental and theoretical work which is incorporated in some of the ideas presented here on the hippocampus with many colleagues, including Alessandro Treves, Simon Stringer, Ray Kesner, Robert Robertson, Pierre Georges-François, Shane O'Mara, and Alain Berthoz, and their contributions are sincerely acknowledged. Pdfs of some of the papers cited are available at www.oxcns.org.

References

Acsady, L., Kamondi, A., Sik, A., Freund, T., & Buzsaki, G. (1998). GABAergic cells are the major postsynaptic targets of mossy fibers in the rat hippocampus. *The Journal of Neuroscience, 18*(9), 3386–3403.

Amaral, D. G. (1987). Memory: Anatomical organization of candidate brain regions. In V. B. Mountcastle (Ed.), *Handbook of physiology. section 1, the nervous system* (Vol. V, higher functions of the brain, pp. 211–294). Washington DC: American Physiological Society.

Amaral, D. G. (1993). Emerging principles of intrinsic hippocampal organisation. *Current Opinion in Neurobiology, 3,* 225–229.

Amaral, D. G., & Witter, M. P. (1989). The three-dimensional organization of the hippocampal formation: A review of anatomical data. *Neuroscience, 31,* 571–591.

Amaral, D. G., & Witter, M. P. (1995). The hippocampal formation. In G. Paxinos (Ed.), *The rat nervous system* (pp. 443–493). San Diego: Academic.

Amaral, D. G., Ishizuka, N., & Claiborne, B. (1990). Neurons, numbers and the hippocampal network. *Progress in Brain Research, 83,* 1–11.

Amaral, D. G., Price, J. L., Pitkanen, A., & Carmichael, S. T. (1992). Anatomical organization of the primate amygdaloid complex. In J. P. Aggleton (Ed.), *The amygdala* (pp. 1–66). New York: Wiley-Liss.

Amari, S. (1977). Dynamics of pattern formation in lateral-inhibition type neural fields. *Biological Cybernetics, 27,* 77–87.

Amit, D. J. (1989). *Modeling brain function*. Cambridge: Cambridge University Press.

Andersen, P., Morris, R. G. M., Amaral, D. G., Bliss, T. V. P., & O'Keefe, J. (2007). *The hippocampus book.* London: Oxford University Press.

Barkas, L. J., Henderson, J. L., Hamilton, D. A., Redhead, E. S., & Gray, W. P. (2010). Selective temporal resections and spatial memory impairment: Cue dependent lateralization effects. *Behavioural Brain Research, 208*(2), 535–544. doi:10.1016/j.bbr.2009.12.035.

Battaglia, F. P., & Treves, A. (1998). Attractor neural networks storing multiple space representations: A model for hippocampal place fields. *Physical Review E, 58,* 7738–7753.

Bonelli, S. B., Powell, R. H., Yogarajah, M., Samson, R. S., Symms, M. R., Thompson, P. J., Duncan, J. S., et al. (2010). Imaging memory in temporal lobe epilepsy: Predicting the effects of temporal lobe resection. *Brain, 133*(Pt 4), 1186–1199. doi:10.1093/brain/awq006.

Brown, T. H., Ganong, A. H., Kairiss, E. W., Keenan, C. L., & Kelso, S. R. (Eds.). (1989). *Long-term potentiation in two synaptic systems of the hippocampal brain slice.* San Diego: Academic.

Brown, T. H., Kairiss, E. W., & Keenan, C. L. (1990). Hebbian synapses: Biophysical mechanisms and algorithms. *Annual Review of Neuroscience, 13,* 475–511.

Brun, V. H., Otnass, M. K., Molden, S., Steffenach, H. A., Witter, M. P., Moser, M. B., & Moser, E. I. (2002). Place cells and place recognition maintained by direct entorhinal-hippocampal circuitry. *Science, 296,* 2243–2246.

Carmichael, S. T., & Price, J. L. (1995). Limbic connections of the orbital and medial prefrontal cortex in macaque monkeys. *Journal of Comparative Neurology, 346,* 403–434.

Cerasti, E., & Treves, A. (2010). How informative are spatial CA3 representations established by the dentate gyrus? *PLoS Computational Biology, 6*(4), e1000759.

Cerasti, E., & Treves, A. (2013). The spatial representations acquired in CA3 by self-organizing recurrent connections. *Frontiers in Cellular Neuroscience, 7, 112.*

Cheng, S. (2013). The CRISP theory of hippocampal function in episodic memory. *Frontiers in Neural Circuits, 7,* 88.

Clelland, C. D., Choi, M., Romberg, C., Clemenson, G. D., Jr., Fragniere, A., Tyers, P., et al. (2009). A functional role for adult hippocampal neurogenesis in spatial pattern separation. *Science, 325*(5937), 210–213.

Daumas, S., Ceccom, J., Halley, H., Frances, B., & Lassalle, J. M. (2009). Activation of metabotropic glutamate receptor type 2/3 supports the involvement of the hippocampal mossy fiber pathway on contextual fear memory consolidation. *Learning and Memory, 16*(8), 504–507.

Day, M., Langston, R., & Morris, R. G. (2003). Glutamate-receptor-mediated encoding and retrieval of paired-associate learning. *Nature, 424,* 205–209.

de Araujo, I. E. T., Rolls, E. T., & Stringer, S. M. (2001). A view model which accounts for the spatial fields of hippocampal primate spatial view cells and rat place cells. *Hippocampus, 11,* 699–706.

Delatour, B., & Witter, M. P. (2002). Projections from the parahippocampal region to the prefrontal cortex in the rat: Evidence of multiple pathways. *The European Journal of Neuroscience, 15,* 1400–1407.

Dere, E., Easton, A., Nadel, L., & Huston, J. P. (Eds.). (2008). *Handbook of episodic memory.* Amsterdam: Elsevier.

Ekstrom, A. D., Kahana, M. J., Caplan, J. B., Fields, T. A., Isham, E. A., Newman, E. L., & Fried, I. (2003). Cellular networks underlying human spatial navigation. *Nature, 425,* 184–188.

Fazeli, M. S., & Collingridge, G. L. (Eds.). (1996). *Cortical plasticity: LTP and LTD.* Oxford: Bios.

Florian, C., & Roullet, P. (2004). Hippocampal CA3-region is crucial for acquisition and memory consolidation in Morris water maze task in mice. *Behavioural Brain Research, 154,* 365–374.

Franco, L., Rolls, E. T., Aggelopoulos, N. C., & Jerez, J. M. (2007). Neuronal selectivity, population sparseness, and ergodicity in the inferior temporal visual cortex. *Biological Cybernetics, 96,* 547–560.

Fyhn, M., Molden, S., Witter, M. P., Moser, E. I., & Moser, M. B. (2004). Spatial representation in the entorhinal cortex. *Science, 305,* 1258–1264.

Georges-François, P., Rolls, E. T., & Robertson, R. G. (1999). Spatial view cells in the primate hippocampus: Allocentric view not head direction or eye position or place. *Cerebral Cortex, 9,* 197–212.

Gilbert, P. E., & Kesner, R. P. (2003). Localization of function within the dorsal hippocampus: The role of the CA3 subregion in paired-associate learning. *Behavioral Neuroscience, 117,* 1385–1394.

Gilbert, P. E., Kesner, R. P., & Lee, I. (2001). Dissociating hippocampal subregions: Double dissociation between dentate gyrus and CA1. *Hippocampus, 11,* 626–636.

Giocomo, L. M., & Hasselmo, M. E. (2007). Neuromodulation by glutamate and acetylcholine can change circuit dynamics by regulating the relative influence of afferent input and excitatory feedback. *Molecular Neurobiology, 36*(2), 184–200.

Giocomo, L. M., Moser, M. B., & Moser, E. I. (2011). Computational models of grid cells. *Neuron, 71*(4), 589–603. doi:S0896-6273(11)00650-7[pii]10.1016/j.neuron.2011.07.023.

Gold, A. E., & Kesner, R. P. (2005). The role of the CA3 subregion of the dorsal hippocampus in spatial pattern completion in the rat. *Hippocampus, 15,* 808–814.

Goodrich-Hunsaker, N. J., Hunsaker, M. R., & Kesner, R. P. (2008). The interactions and dissociations of the dorsal hippocampus subregions: How the dentate gyrus, CA3, and CA1 process spatial information. *Behavioral Neuroscience, 122*(1), 16–26.

Hafting, T., Fyhn, M., Molden, S., Moser, M. B., & Moser, E. I. (2005). Microstructure of a spatial map in the entorhinal cortex. *Nature, 436,* 801–806.

Hasselmo, M. E., Schnell, E., & Barkai, E. (1995). Dynamics of learning and recall at excitatory recurrent synapses and cholinergic modulation in rat hippocampal region CA3. *The Journal of Neurosciences, 15,* 5249–5262.

Henze, D. A., Wittner, L., & Buzsaki, G. (2002). Single granule cells reliably discharge targets in the hippocampal CA3 network in vivo. *Nature Neuroscience, 5*(8), 790–795.

Hertz, J., Krogh, A., & Palmer, R. G. (1991). *An introduction to the theory of neural computation.* Wokingham: Addison-Wesley.

Hoge, J., & Kesner, R. P. (2007). Role of CA3 and CA1 subregions of the dorsal hippocampus on temporal processing of objects. *Neurobiology of Learning and Memory, 88*(2), 225–231.

Hopfield, J. J. (1982). Neural networks and physical systems with emergent collective computational abilities. *Proceedings of the National Academy of Science U S A, 79,* 2554–2558.

Hunsaker, M. R., & Kesner, R. P. (2008). Evaluating the differential roles of the dorsal dentate gyrus, dorsal CA3, and dorsal CA1 during a temporal ordering for spatial locations task. *Hippocampus, 18*(9), 955–964.

Hunsaker, M. R., & Kesner, R. P. (2013). The operation of pattern separation and pattern completion processes associated with different attributes or domains of memory. *Neuroscience and Biobehavioral Reviews, 37*(1), 36–58. doi:10.1016/j.neubiorev.2012.09.014.

Ishizuka, N., Weber, J., & Amaral, D. G. (1990). Organization of intrahippocampal projections originating from CA3 pyramidal cells in the rat. *Journal of Comparative Neurology, 295,* 580–623.

Jackson, M. B. (2013). Recall of Spatial Patterns Stored in a Hippocampal Slice By Long-Term Potentiation. *Journal of Neurophysiology*. doi:10.1152/jn.00533.2013.

Jezek, K., Henriksen, E. J., Treves, A., Moser, E. I., & Moser, M.-B. (2011). Theta-paced flickering between place-cell maps in the hippocampus. *Nature, 278,* 246–249.

Jung, M. W., & McNaughton, B. L. (1993). Spatial selectivity of unit activity in the hippocampal granular layer. *Hippocampus, 3,* 165–182.

Kesner, R. P. (2007). Behavioral functions of the CA3 subregion of the hippocampus. *Learning and Memory, 14*(11), 771–781. doi:10.1101/lm.688207.

Kesner, R. P. (2013). An analysis of the dentate gyrus function. *Behavioural Brain Research, 254,* 1–7. doi:10.1016/j.bbr.2013.01.012.

Kesner, R. P., & Rolls, E. T. (2001). Role of long term synaptic modification in short term memory. *Hippocampus, 11,* 240–250.

Kesner, R. P., & Rolls, E. T. (2015). A computational theory of hippocampal function, and tests of the theory: new developments. *Neuroscience and Biobehavioral Reviews 48,* 92–147.

Kesner, R. P., Gilbert, P. E., & Barua, L. A. (2002). The role of the hippocampus in memory for the temporal order of a sequence of odors. *Behavioral Neuroscience, 116,* 286–290.

Kesner, R. P., Hunsaker, M. R., & Warthen, M. W. (2008). The CA3 subregion of the hippocampus is critical for episodic memory processing by means of relational encoding in rats. *Behavioral Neuroscience, 122*(6), 1217–1225.

Kesner, R. P., Morris, A. M., & Weeden, C. S. S. (2012). Spatial, temporal, and associative behavioral functions associated with different subregions of the hippocampus. In T. R. Zentall & E. A. Wasserman (Eds.), *Oxford handbook of comparative cognition* (pp. 322–346). Oxford: Oxford University Press.

Killian, N. J., Jutras, M. J., & Buffalo, E. A. (2012). A map of visual space in the primate entorhinal cortex. *Nature, 491*(7426), 761–764. doi:10.1038/nature11587.

Kohonen, T. (1977). *Associative memory: A system theoretical approach*. New York: Springer.

Kohonen, T. (1984). *Self-organization and associative memory*. Berlin: Springer-Verlag.

Kohonen, T., Oja, E., & Lehtio, P. (1981). Storage and processing of information in distributed memory systems. In G. E. Hinton & J. A. Anderson (Eds.), *Parallel models of associative memory* (pp. 129–167). Hillsdale: Lawrence Erlbaum.

Kondo, H., Lavenex, P., & Amaral, D. G. (2009). Intrinsic connections of the macaque monkey hippocampal formation: II. CA3 connections. *Journal of Comparative Neurology, 515*(3), 349–377.

Kropff, E., & Treves, A. (2008). The emergence of grid cells: Intelligent design or just adaptation? *Hippocampus, 18*(12), 1256–1269.

Lassalle, J. M., Bataille, T., & Halley, H. (2000). Reversible inactivation of the hippocampal mossy fiber synapses in mice impairs spatial learning, but neither consolidation nor memory retrieval, in the Morris navigation task. *Neurobiology of Learning and Memory, 73,* 243–257.

Lavenex, P., & Amaral, D. G. (2000). Hippocampal-neocortical interaction: A hierarchy of associativity. *Hippocampus, 10,* 420–430.

Lavenex, P., Suzuki, W. A., & Amaral, D. G. (2004). Perirhinal and parahippocampal cortices of the macaque monkey: Intrinsic projections and interconnections. *Journal of Comparative Neurology, 472,* 371–394.

Lee, I., & Kesner, R. P. (2004). Encoding versus retrieval of spatial memory: Double dissociation between the dentate gyrus and the perforant path inputs into CA3 in the dorsal hippocampus. *Hippocampus, 14,* 66–76.

Leutgeb, S., & Leutgeb, J. K. (2007). Pattern separation, pattern completion, and new neuronal codes within a continuous CA3 map. *Learning and Memory, 14*(11), 745–757.

Leutgeb, J. K., Leutgeb, S., Moser, M. B., & Moser, E. I. (2007). Pattern separation in the dentate gyrus and CA3 of the hippocampus. *Science, 315*(5814), 961–966.

Levy, W. B. (1989). A computational approach to hippocampal function. In R. D. Hawkins & G. H. Bower (Eds.), *Computational models of learning in simple neural systems* (pp. 243–305). San Diego: Academic.

Lynch, M. A. (2004). Long-term potentiation and memory. *Psychological Review, 84,* 87–136.

Marr, D. (1971). Simple memory: A theory for archicortex. *Philosophical Transactions of the Royal Society of London. Series B, Biological Sciences, 262,* 23–81.

McClelland, J. L., McNaughton, B. L., & O'Reilly, R. C. (1995). Why there are complementary learning systems in the hippocampus and neocortex: Insights from the successes and failures of connectionist models of learning and memory. *Psychological Review, 102,* 419–457.

McHugh, T. J., Jones, M. W., Quinn, J. J., Balthasar, N., Coppari, R., Elmquist, J. K., et al. (2007). Dentate gyrus NMDA receptors mediate rapid pattern separation in the hippocampal network. *Science, 317*(5834), 94–99.

McNaughton, B. L. (1991). Associative pattern completion in hippocampal circuits: New evidence and new questions. *Brain Research Review, 16,* 193–220.

McNaughton, B. L., & Morris, R. G. M. (1987). Hippocampal synaptic enhancement and information storage within a distributed memory system. *Trends in Neurosciences, 10*(10), 408–415.

McNaughton, B. L., & Nadel, L. (1990). Hebb-Marr networks and the neurobiological representation of action in space. In M. A. Gluck & D. E. Rumelhart (Eds.), *Neuroscience and connectionist theory* (pp. 1–64). Hillsdale: Erlbaum.

McNaughton, B. L., Barnes, C. A., & O'Keefe, J. (1983). The contributions of position, direction, and velocity to single unit activity in the hippocampus of freely-moving rats. *Experimental Brain Research, 52,* 41–49.

McNaughton, B. L., Battaglia, F. P., Jensen, O., Moser, E. I., & Moser, M.-B. (2006). Path integration and the neural basis of the 'cognitive map.' *Nature Reviews Neuroscience, 7,* 663–678.

Morris, R. G. M. (1989). Does synaptic plasticity play a role in information storage in the vertebrate brain? In R. G. M. Morris (Ed.), *Parallel distributed processing: Implications for psychology and neurobiology* (pp. 248–285). Oxford: Oxford University Press.

Morris, R. G. (2003). Long-term potentiation and memory. *Philosophical Transactions of the Royal Society of London. Series B, Biological Sciences, 358,* 643–647.

Morris, R. G., Moser, E. I., Riedel, G., Martin, S. J., Sandin, J., Day, M., & O'Carroll, C. (2003). Elements of a neurobiological theory of the hippocampus: The role of activity-dependent synaptic plasticity in memory. *Philosophical Transactions of the Royal Society of London. Series B, Biological Sciences, 358,* 773–786.

Moscovitch, M., Rosenbaum, R. S., Gilboa, A., Addis, D. R., Westmacott, R., Grady, C., et al. (2005). Functional neuroanatomy of remote episodic, semantic and spatial memory: A unified account based on multiple trace theory. *Journal of Anatomy, 207,* 35–66.

Naber, P. A., Lopes da Silva, F. H., & Witter, M. P. (2001). Reciprocal connections between the entorhinal cortex and hippocampal fields CA1 and the subiculum are in register with the projections from CA1 to the subiculum. *Hippocampus, 11,* 99–104.

Nakashiba, T., Cushman, J. D., Pelkey, K. A., Renaudineau, S., Buhl, D. L., McHugh, T. J., et al. (2012). Young dentate granule cells mediate pattern separation, whereas old granule cells facilitate pattern completion. *Cell, 149*(1), 188–201. doi:S0092-8674(12)00157-2[pii]10.1016/j.cell.2012.01.046.

Nakazawa, K., Quirk, M. C., Chitwood, R. A., Watanabe, M., Yeckel, M. F., Sun, L. D., et al. (2002). Requirement for hippocampal CA3 NMDA receptors in associative memory recall. *Science, 297,* 211–218.

Nakazawa, K., Sun, L. D., Quirk, M. C., Rondi-Reig, L., Wilson, M. A., & Tonegawa, S. (2003). Hippocampal CA3 NMDA receptors are crucial for memory acquisition of one-time experience. *Neuron, 38,* 305–315.

Nakazawa, K., McHugh, T. J., Wilson, M. A., & Tonegawa, S. (2004). NMDA receptors, place cells and hippocampal spatial memory. *Nature Reviews Neuroscience, 5,* 361–372.

O'Keefe, J. (1979). A review of the hippocampal place cells. *Progress in Neurobiology, 13,* 419–439.

O'Keefe, J., & Nadel, L. (1978). *The hippocampus as a cognitive map.* Oxford: Clarendon Press.

O' Keefe, J., & Speakman, A. (1987). Single unit activity in the rat hippocampus during a spatial memory task. *Experimental Brain Research, 68,* 1–27.

Pitkanen, A., Kelly, J. L., & Amaral, D. G. (2002). Projections from the lateral, basal, and accessory basal nuclei of the amygdala to the entorhinal cortex in the macaque monkey. *Hippocampus, 12,* 186–205.

Rajji, T., Chapman, D., Eichenbaum, H., & Greene, R. (2006). The role of CA3 hippocampal NMDA receptors in paired associate learning. *The Journal of Neurosciences, 26*(3), 908–915.

Robertson, R. G., Rolls, E. T., & Georges-François, P. (1998). Spatial view cells in the primate hippocampus: Effects of removal of view details. *Journal of Neurophysiology, 79,* 1145–1156.

Rolls, E. T. (1987). Information representation, processing and storage in the brain: Analysis at the single neuron level. In J.-P. Changeux & M. Konishi (Eds.), *The neural and molecular bases of learning* (pp. 503–540). Chichester: Wiley.

Rolls, E. T. (1989a). Functions of neuronal networks in the hippocampus and cerebral cortex in memory. In R. M. J. Cotterill (Ed.), *Models of brain function* (pp. 15–33). Cambridge: Cambridge University Press.

Rolls, E. T. (1989b). Functions of neuronal networks in the hippocampus and neocortex in memory. In J. H. Byrne & W. O. Berry (Eds.), *Neural models of plasticity: Experimental and theoretical approaches* (pp. 240–265). San Diego: Academic.

Rolls, E. T. (1989c). The representation and storage of information in neuronal networks in the primate cerebral cortex and hippocampus. In R. Durbin, C. Miall, & G. Mitchison (Eds.), *The computing neuron* (pp. 125–159). Wokingham: Addison-Wesley.

Rolls, E. T. (1990a). Functions of the primate hippocampus in spatial processing and memory. In D. S. Olton & R. P. Kesner (Eds.), *Neurobiology of comparative cognition* (pp. 339–362). Hillsdale: L. Erlbaum.

Rolls, E. T. (1990b). Theoretical and neurophysiological analysis of the functions of the primate hippocampus in memory. *Cold Spring Harbor Symposia in Quantitative Biology, 55,* 995–1006.

Rolls, E. T. (1991). Functions of the primate hippocampus in spatial and non-spatial memory. *Hippocampus, 1,* 258–261.

Rolls, E. T. (1995). A model of the operation of the hippocampus and entorhinal cortex in memory. *International Journal of Neural Systems, 6,* 51–70.

Rolls, E. T. (1996a). Roles of long term potentiation and long term depression in neuronal network operations in the brain. In M. S. Fazeli & G. L. Collingridge (Eds.), *Cortical plasticity* (pp. 223–250). Oxford: Bios.

Rolls, E. T. (1996b). A theory of hippocampal function in memory. *Hippocampus, 6,* 601–620.

Rolls, E. T. (1999). Spatial view cells and the representation of place in the primate hippocampus. *Hippocampus, 9,* 467–480.

Rolls, E. T. (2008). *Memory, attention, and decision-making: A unifying computational neuroscience approach.* Oxford: Oxford University Press.

Rolls, E. T. (2010a). Attractor networks. *WIREs Cognitive Science, 1i,* 119–134.

Rolls, E. T. (2010b). A computational theory of episodic memory formation in the hippocampus. *Behavioural Brain Research, 205,* 180–196.

Rolls, E. T. (2012a). Advantages of dilution in the connectivity of attractor networks in the brain. *Biologically Inspired Cognitive Architectures, 1,* 44–54. doi: 10.1016/j.bica.2012.03.003.

Rolls, E. T. (2012b). Invariant visual object and face recognition: Neural and computational bases, and a model, VisNet. *Frontiers in Computational Neuroscience, 6,* 35, 1–70.

Rolls, E. T. (2013). A quantitative theory of the functions of the hippocampal CA3 network in memory. *Frontiers in Cellular Neuroscience, 7,* 98. doi:10.3389/fncel.2013.00098.

Rolls, E. T. (2014). *Emotion and decision-making explained.* Oxford: Oxford University Press.

Rolls, E. T. (2016). Cerebral cortex: Principles of operation. Oxford: Oxford University Press.

Rolls, E. T., & Deco, G. (2002). *Computational neuroscience of vision.* Oxford: Oxford University Press.

Rolls, E. T., & Deco, G. (2010). *The noisy brain: Stochastic dynamics as a principle of brain function.* Oxford: Oxford University Press.

Rolls, E. T., & Kesner, R. P. (2006). A computational theory of hippocampal function, and empirical tests of the theory. *Progress in Neurobiology, 79,* 1–48.

Rolls, E. T., & Stringer, S. M. (2000). On the design of neural networks in the brain by genetic evolution. *Progress in Neurobiology, 61,* 557–579.

Rolls, E. T., & Stringer, S. M. (2005). Spatial view cells in the hippocampus, and their idiothetic update based on place and head direction. *Neural Networks, 18,* 1229–1241.

Rolls, E. T., & Treves, A. (1990). The relative advantages of sparse versus distributed encoding for associative neuronal networks in the brain. *Network, 1,* 407–421.

Rolls, E. T., & Treves, A. (1998). *Neural networks and brain function.* Oxford: Oxford University Press.

Rolls, E. T., & Treves, A. (2011). The neuronal encoding of information in the brain. *Progress in Neurobiology, 95,* 448–490.

Rolls, E. T., & Webb, T. J. (2012). Cortical attractor network dynamics with diluted connectivity. *Brain Research, 1434,* 212–225.

Rolls, E. T., & Xiang, J.-Z. (2005). Reward-spatial view representations and learning in the hippocampus. *Journal of Neuroscience, 25,* 6167–6174.

Rolls, E. T., & Xiang, J.-Z. (2006). Spatial view cells in the primate hippocampus, and memory recall. *Reviews in the Neurosciences, 17,* 175–200.

Rolls, E. T., Robertson, R. G., & Georges-François, P. (1997a). Spatial view cells in the primate hippocampus. *European Journal of Neuroscience, 9,* 1789–1794.

Rolls, E. T., Treves, A., Foster, D., & Perez-Vicente, C. (1997b). Simulation studies of the CA3 hippocampal subfield modelled as an attractor neural network. *Neural Networks, 10,* 1559–1569.

Rolls, E. T., Treves, A., Robertson, R. G., Georges-François, P., & Panzeri, S. (1998). Information about spatial view in an ensemble of primate hippocampal cells. *Journal of Neurophysiology, 79,* 1797–1813.

Rolls, E. T., Stringer, S. M., & Trappenberg, T. P. (2002). A unified model of spatial and episodic memory. *Proceedings of the Royal Society of London B, 269,* 1087–1093.

Rolls, E. T., Xiang, J.-Z., & Franco, L. (2005). Object, space and object-space representations in the primate hippocampus. *Journal of Neurophysiology, 94,* 833–844.

Rolls, E. T., Stringer, S. M., & Elliot, T. (2006). Entorhinal cortex grid cells can map to hippocampal place cells by competitive learning. *Network: Computation in Neural Systems, 17,* 447–465.

Rolls, E. T., Dempere-Marco, L., & Deco, G. (2013). Holding multiple items in short term memory: A neural mechanism. *PLoS One, 8,* e61078.

Rondi-Reig, L., Libbey, M., Eichenbaum, H., & Tonegawa, S. (2001). CA1-specific N-methyl-D-aspartate receptor knockout mice are deficient in solving a nonspatial transverse patterning task. *Proceedings of the National Academy of Sciences of the United States of America, 98,* 3543–3548.

Samsonovich, A., & McNaughton, B. L. (1997). Path integration and cognitive mapping in a continuous attractor neural network model. *Journal of Neuroscience, 17,* 5900–5920.

Schultz, S., & Rolls, E. T. (1999). Analysis of information transmission in the Schaffer collaterals. *Hippocampus, 9,* 582–598.

Schultz, S., Panzeri, S., Rolls, E. T., & Treves, A. (2000). Quantitive model analysis of a Schaffer collateral model. In P. Hancock, P. Foldiak, & R. Baddeley (Eds.), *Information theory and the brain* (pp. 257–272,Chap. 14). Cambridge: Cambridge University Press.

Schwindel, C. D., & McNaughton, B. L. (2011). Hippocampal-cortical interactions and the dynamics of memory trace reactivation. *Progress in Brain Research, 193,* 163–177. doi:B978-0-444-53839-0.00011-9[pii]10.1016/B978-0-444-53839-0.00011-9.

Sidhu, M. K., Stretton, J., Winston, G. P., Bonelli, S., Centeno, M., Vollmar, C., et al. (2013). A functional magnetic resonance imaging study mapping the episodic memory encoding network in temporal lobe epilepsy. *Brain, 136,* 1868–1888. doi:10.1093/brain/awt099.

Simmen, M. W., Treves, A., & Rolls, E. T. (1996). Pattern retrieval in threshold-linear associative nets. *Network, 7,* 109–122.

Stefanacci, L., Suzuki, W. A., & Amaral, D. G. (1996). Organization of connections between the amygdaloid complex and the perirhinal and parahippocampal cortices in macaque monkeys. *Journal of Comparative Neurology, 375,* 552–582.

Stella, F., Cerasti, E., & Treves, A. (2013). Unveiling the metric structure of internal representations of space. *Frontiers in Neural Circuits, 7,* 81. doi:10.3389/fncir.2013.00081.

Storm-Mathiesen, J., Zimmer, J., & Ottersen, O. P. (Eds.). (1990). *Understanding the brain through the hippocampus* (Vol. 83). Oxford: Elsevier.

Stringer, S. M., & Rolls, E. T. (2002). Invariant object recognition in the visual system with novel views of 3D objects. *Neural Computation, 14,* 2585–2596.

Stringer, S. M., & Rolls, E. T. (2006). Self-organizing path integration using a linked continuous attractor and competitive network: Path integration of head direction. *Network: Computation in Neural Systems, 17,* 419–445.

Stringer, S. M., Rolls, E. T., Trappenberg, T. P., & Araujo, I. E. T. (2002a). Self-organizing continuous attractor networks and path integration. Two-dimensional models of place cells. *Network: Computation in Neural Systems, 13,* 429–446.

Stringer, S. M., Trappenberg, T. P., Rolls, E. T., & Araujo, I. E. T. (2002b). Self-organizing continuous attractor networks and path integration: One-dimensional models of head direction cells. *Network: Computation in Neural Systems, 13,* 217–242.

Stringer, S. M., Rolls, E. T., & Trappenberg, T. P. (2004). Self-organising continuous attractor networks with multiple activity packets, and the representation of space. *Neural Networks, 17,* 5–27.

Stringer, S. M., Rolls, E. T., & Trappenberg, T. P. (2005). Self-organizing continuous attractor network models of hippocampal spatial view cells. *Neurobiology of Learning and Memory, 83,* 79–92.

Suzuki, W. A., & Amaral, D. G. (1994a). Perirhinal and parahippocampal cortices of the macaque monkey—cortical afferents. *Journal of Comparative Neurology, 350*(4), 497–533.

Suzuki, W. A., & Amaral, D. G. (1994b). Topographic organization of the reciprocal connections between the monkey entorhinal cortex and the perirhinal and parahippocampal cortices. *Journal of Neuroscience, 14,* 1856–1877.

Taylor, J. G. (1999). Neural "bubble" dynamics in two dimensions: Foundations. *Biological Cybernetics, 80,* 393–409.

Tonegawa, S., Nakazawa, K., & Wilson, M. A. (2003). Genetic neuroscience of mammalian learning and memory. *Philosophical Transactions of the Royal Society of London. Series B, Biological Sciences, 358,* 787–795.

Treves, A. (1990). Graded-response neurons and information encodings in autoassociative memories. *Physical Review A, 42,* 2418–2430.

Treves, A. (1991). Dilution and sparse coding in threshold-linear nets. *Journal of Physics A, 24,* 327–335.

Treves, A. (1995). Quantitative estimate of the information relayed by Schaffer collaterals. *Journal of Computational Neuroscience, 2,* 259–272.

Treves, A., & Rolls, E. T. (1991). What determines the capacity of autoassociative memories in the brain? *Network, 2,* 371–397.

Treves, A., & Rolls, E. T. (1992). Computational constraints suggest the need for two distinct input systems to the hippocampal CA3 network. *Hippocampus, 2,* 189–199.

Treves, A., & Rolls, E. T. (1994). A computational analysis of the role of the hippocampus in memory. *Hippocampus, 4,* 374–391.

van Haeften, T., Baks-te-Bulte, L., Goede, P. H., Wouterlood, F. G., & Witter, M. P. (2003). Morphological and numerical analysis of synaptic interactions between neurons in deep and superficial layers of the entorhinal cortex of the rat. *Hippocampus, 13,* 943–952.

Van Hoesen, G. W. (1982). The parahippocampal gyrus. New observations regarding its cortical connections in the monkey. *Trends in Neuroscience, 5,* 345–350.

Walters, D. M., Stringer, S. M., & Rolls, E. T. (2013). Path integration of head direction: Updating a packet of neural activity at the correct speed using axonal conduction delays. *PLoS One, 8,* e58330.

Wang, S. H., & Morris, R. G. (2010). Hippocampal-neocortical interactions in memory formation, consolidation, and reconsolidation. *Annual Review of Psychology, 61,* 49–79, C41–44.

Webb, T., Rolls, E. T., Deco, G., & Feng, J. (2011). Noise in attractor networks in the brain produced by graded firing rate representations. *PLoS One, 6,* e23630.

Wills, T. J., Lever, C., Cacucci, F., Burgess, N., & O'Keefe, J. (2005). Attractor dynamics in the hippocampal representation of the local environment. *Science, 308,* 873–876.

Witter, M. P. (1993). Organization of the entorhinal-hippocampal system: A review of current anatomical data. *Hippocampus, 3,* 33–44.

Witter, M. P. (2007). Intrinsic and extrinsic wiring of CA3: Indications for connectional heterogeneity. *Learning and Memory, 14*(11), 705–713.

Witter, M. P., Naber, P. A., van Haeften, T., Machielsen, W. C., Rombouts, S. A., Barkhof, F., et al. (2000a). Cortico-hippocampal communication by way of parallel parahippocampal-subicular pathways. *Hippocampus, 10,* 398–410.

Witter, M. P., Wouterlood, F. G., Naber, P. A., & Van Haeften, T. (2000b). Anatomical organization of the parahippocampal-hippocampal network. *Annals of the New York Academy of Sciences, 911,* 1–24.

Zhang, K. (1996). Representation of spatial orientation by the intrinsic dynamics of the head-direction cell ensemble: A theory. *Journal of Neuroscience, 16,* 2112–2126.

Zilli, E. A. (2012). Models of grid cell spatial firing published 2005–2011. *Front Neural Circuits, 6,* 16. doi:10.3389/fncir.2012.00016.

Chapter 5
Pattern Separation: A Key Processing Deficit Associated with Aging?

Paul E. Gilbert, Heather M. Holden, David P. Sheppard and Andrea M. Morris

Pattern Separation and the Influence of Dr. Raymond Kesner

In a recent theoretical review entitled "A Tapestry of Memory," Dr. Raymond Kesner describes his Attribute Model of Memory as "a comprehensive view of memory organization based on multiple processes and multiple forms of memory representation and is based on the neurobiology of a multiple attribute, multiple process, tripartite system model of memory" (Kesner 2009, p. 3). Over the last 15 years of his career, Kesner focused on specific mnemonic processes associated with the event-based memory system with a particular emphasis on the hippocampus. In particular, he became interested in a process referred to as pattern separation. Pattern separation is hypothesized to serve as a mechanism for separating partially overlapping patterns of activation so that one pattern may be retrieved as separate from other similar patterns. A pattern separation mechanism may be critical for reducing potential interference among similar memory representations to enhance memory accuracy. A

P. E. Gilbert (✉) · D. P. Sheppard
Department of Psychology, San Diego State University, San Diego, CA, USA
e-mail: pgilbert@mail.sdsu.edu

P. E. Gilbert · H. M. Holden
Joint Doctoral Program in Clinical Psychology, San Diego State University—University of California San Diego, 6363 Alvarado Court, Suite 103, San Diego, CA 92120, USA

H. M. Holden
e-mail: hholden365@gmail.com

D. P. Sheppard
Department of Psychology, University of Houston, 126 Heyne Building, Suite 204, Houston, TX 77004-5022, USA
e-mail: dpsheppard@uh.edu

A. M. Morris
Department of Health Policy and Management, University of California, 650 Charles Young Dr. S. 31-269, Box 951772, Los Angeles, CA 90095, USA
e-mail: morrisandrea78@gmail.com

P. A. Jackson et al. (eds.), *The Neurobiological Basis of Memory*,
DOI 10.1007/978-3-319-15759-7_5

number of early theoretical and computational models suggested that the hippocampus supports pattern separation (Marr 1971; McNaughton and Nadel 1990; O'Reilly and McClelland 1994; Rolls 1996; Shapiro and Olton 1994; Treves and Rolls 1992). Kesner developed one of the first behavioral tasks used to demonstrate that lesions of the hippocampus impair spatial pattern separation (Gilbert et al. 1998). These theoretical and computational models also hypothesized that the dentate gyrus (DG) and CA3 subregions of the hippocampus may be particularly important for pattern separation (O'Reilly and McClelland 1994; Rolls 1996; Shapiro and Olton 1994). To test the predictions of these models, Kesner tested rats with neurotoxin-induced lesions of the DG or CA3 subregions on his spatial pattern separation task previously shown to be dependent on the hippocampus. The results provided support for the hypothesis that the DG (Gilbert et al. 2001) and CA3 (Gilbert and Kesner 2006) hippocampal subregions play a key role in spatial pattern separation. Over the last 15 years, Kesner and his colleagues have published numerous studies examining pattern separation for spatial information (Gilbert and Kesner 2006; Gilbert et al. 1998, 2001; Goodrich-Hunsaker et al. 2005; Hunsaker and Kesner 2008; Morris et al. 2012), temporal order of stimuli (Gilbert et al. 2001; Hunsaker et al. 2008; Kesner et al. 2002; Kesner and Hunsaker 2010; Kesner et al. 2010), olfactory stimuli (Kesner et al. 2011; Weeden et al. 2012), motor responses (Kesner and Gilbert 2006), scenes of visual objects (Gilbert and Kesner 2003), and reward magnitude (Gilbert and Kesner 2002). He has also published numerous theoretical models and review articles on pattern separation (Hunsaker and Kesner 2013; Kesner 2007, 2013a, b; Kesner et al. 2000; Kesner and Hopkins 2006; Rolls and Kesner 2006). The innovative behavioral studies conducted in the Kesner laboratory examining pattern separation have contributed greatly to our understanding of this process. In addition, his work has set the foundation for the recent behavioral investigations of age-related changes in pattern separation that will be reviewed in the present chapter.

In recent years, pattern separation has drawn considerable attention in the literature as an important mechanism for accurate memory formation and subsequent retrieval. Additional computational and theoretical models have been published detailing the role of the hippocampus in pattern separation (Kesner 2007; Myers and Scharfman 2009; Rolls 2010; Rolls and Kesner 2006). In addition, numerous researchers have shown that the DG and CA3 subregions of the hippocampus play a critical role in pattern separation in animal models using electrophysiological recordings (Leutgeb et al. 2007; McNaughton et al. 1989; Tanila 1999), neurotoxin-induced lesions and inactivations (Butterly et al. 2012; Gilbert and Kesner 2006; Gilbert et al. 2001; Goodrich-Hunsaker et al. 2008; Lee et al. 2005; McTighe et al. 2009; Morris et al. 2012), and genetic manipulations (Kubik et al. 2007; McHugh et al. 2007). Furthermore, studies using high-resolution functional magnetic resonance imaging (fMRI) have shown that the human hippocampus (Kirwan and Stark 2007; LaRocque et al. 2013; Motley and Kirwan 2012), and specifically the DG/CA3 subregions (Bakker et al. 2008; Lacy et al. 2011), are active during pattern separation tasks (see also reviews by Carr et al. 2010; Yassa and Stark 2011). Most recently, neuropsychological studies have shown that patients with hippocampal damage have deficits in pattern separation (Duff et al. 2012; Kirwan et al. 2012).

Age-related Changes in the Brain

Aging has been shown to result in both white matter and gray matter changes in various regions throughout the brain (Allen et al. 2005; Driscoll et al. 2009; Kennedy and Raz 2009; Ziegler et al. 2010); however, there has been particular focus in the literature on detrimental age-related changes in regions of the brain that support memory, including the hippocampus and surrounding medial temporal lobe structures (Allen et al. 2005; Driscoll and Sutherland 2005; Good et al. 2001; Raz et al. 2005; Walhovd et al. 2010). In aged rodents, a number of studies have reported preserved numbers of neurons in the hippocampus (Rapp and Gallagher 1996; Rapp et al. 1999; Rasmussen et al. 1996); however, others have reported decreased neuronal density (Driscoll et al. 2006). In addition, some studies have reported a lack of a relationship between hippocampal cell numbers and spatial learning deficits (Driscoll et al. 2006; Rapp and Gallagher 1996); however, hippocampal volume measured by MRI has been shown to correlate with water maze performance in aged rats (Driscoll et al. 2006). Since neuronal loss in the hippocampus alone is unlikely to account for the memory deficits observed in aged animals, it has been postulated that age-related memory decline may stem from functional changes in the hippocampus (Barnes 1994; Driscoll et al. 2006; Gallagher et al. 2010), localized synaptic loss (Wilson et al. 2006), and subregion-specific epigenetic and transcriptional changes in the hippocampus (Penner et al. 2011). In addition, age-related structural and functional changes have been reported in perforant path inputs to the DG from the entorhinal cortex (EC). The total number of contacts per neuron in the middle molecular layer of the DG (afferent EC fibers) was found to be significantly reduced in old rats (Geinisman et al. 1992; see also Smith et al. 2000). Perforant path connections to the DG in old rats were also found to be less excitable and required greater stimulation to achieve long-term potentiation compared to young rats (Burke and Barnes 2006).

Results of longitudinal studies in humans demonstrate that hippocampal and parahippocampal cortices exhibit decreased volumes as a function of increased age in non-demented older adults (Driscoll et al. 2009). The hippocampus has been reported to be particularly susceptible to age-related changes and this structure decreases in volume at a faster rate relative to other structures in the medial temporal lobe (Raz et al. 2004). In addition, the observed hippocampal volume loss has been reported to be a primary predictor of memory deficits in older adults (Kramer et al. 2007; Mungas et al. 2005). A recent longitudinal imaging study revealed that declines in episodic memory were associated with decreased hippocampal volume, as well as decreased activation in the left hippocampus, suggesting that structural and functional changes in the hippocampal formation are linked to memory decline (Persson et al. 2012). Small et al. (2002) reported that 60 % of an older adult sample had diminished MRI signal in at least one hippocampal subregion and this hippocampal dysfunction was associated with declines in memory ability. In addition, the authors demonstrated that DG dysfunction is associated with normal aging, whereas signal decline in the EC is indicative of a pathological process (see also Mueller et al. 2010). Although some studies have reported that the volume of the

EC is relatively resistant to aging (Mueller and Weiner 2009), other studies have reported that shrinkage of the EC is associated with poorer memory performance in older adults (Rodrigue and Raz 2004). Using ultrahigh-resolution microstructural diffusion tensor imaging, the perforant pathway has also been found to undergo significant structural changes with advanced age that related to memory function (Yassa et al. 2011b). As reviewed by Small et al. (2011), the DG has been reported to be particularly susceptible to age-related changes in both human (Small et al. 2002; Wu et al. 2008) and animal models (Patrylo and Williamson 2007; Small et al. 2004). In contrast, the pyramidal cells of the CA subregions are relatively less affected in aging (Small et al. 2004).

Pattern Separation and Aging

Wilson et al. (2006) proposed a model of neurocognitive aging, which suggests that age-related changes in the hippocampal processing circuit may account for some of the common episodic memory deficits experienced by many older adults. Based on a review of neurobiological and neurophysiological evidence, the authors suggest that subtle changes in each of the hippocampal subregions may lead to a functional reorganization of information processing in the aged hippocampus. Specifically, the DG receives less input and excitation from the EC via the perforant path, which may result in decreased pattern separation efficiency. The CA3 subregion also undergoes specific age-related changes, including decreased input from the EC and reduced ACh modulation. Reduced ACh input releases the CA3 auto-associative network from inhibition, causing this subregion to become entrenched in pattern completion—a mechanism that allows for completion of stored, familiar patterns given only partial cues (Kesner and Hopkins 2006). Collectively, the changes in the CA3 subregion may result in a strong bias toward retrieval of previously stored representations. The authors propose that the combination of a hypoactive DG and hyperactive CA3 in the aged hippocampus alters the balance of information processing, such that encoding of novel information (pattern separation) is attenuated due to interference from previously stored information (pattern completion). This functional reorganization may explain why older adults often have difficulty remembering new events whereas prior memories are relatively well preserved. In support of this model, Yassa et al. (2011) reported that age-related changes in perforant path integrity and changes in functional activity in the DG/CA3 network are associated with decreased pattern separation activity in older humans. These changes are suggested to increase reliance on retrieval of stored information at the expense of processing novel information (Yassa et al. 2011a).

Pattern Separation in Older Animals

Given the critical role of the DG subregion in supporting pattern separation and the susceptibility of this region to age-related neurobiological changes, recent studies have begun to examine a possible link between aging and efficiency of the pattern separation mechanism in rodents. A study published by Marrone et al. (2011) provided some of the first neurobiological insight into how age-related changes in the DG of rodents may affect pattern separation and spatial memory. The study used a marker of cellular activity (zif268/egr1) to examine granule cell activity in young and older animals during exploration of similar and dissimilar environments. The authors found that age-related changes in pattern separation correlated with a decreased ability of older animals to disambiguate similar contexts when performing a sequential spatial recognition task.

Another more recent study provides additional behavioral evidence that spatial pattern separation may be impaired in older rats (Gracian et al. 2013). Young and old rats were tested on a task developed by McDonald and White (1995) that was recently shown to be dependent on the DG hippocampal subregion (Morris et al. 2012). The rats were trained on a radial 8-arm maze to discriminate between a rewarded arm and a non-rewarded arm that were either adjacent to one another (high spatial interference) or separated by a distance of two arm positions (low spatial interference). The authors found that old rats committed significantly more errors compared to young rats on the adjacent condition. However, young and old rats committed similar numbers of errors in the separated condition. The authors concluded that decreased spatial pattern separation in old rats may impair performance in the adjacent condition, which involved greater spatial interference among distal cues. However, in the separated condition, when there was less overlap among distal cues and less need for pattern separation, performance improved in the older rats. Collectively, the aforementioned studies offer evidence that spatial pattern separation may become less efficient in rodents as a result of aging, presumably due to changes in the DG.

Studies have also provided some evidence that the reductions in neurogenesis observed in old animals (Kuhn et al. 1996) may be related to decreased hippocampal volume and impaired performance in hippocampal dependent tasks (Driscoll et al. 2006). Penner et al. (2011) suggest that age-related memory decline may stem from subregion-specific epigenetic and transcriptional changes in the hippocampus. Newborn neurons are reported to be involved in mnemonic processes such as pattern separation that are particularly dependent on the DG subregion (Aimone et al. 2010, 2011; Clelland et al. 2009; Creer et al. 2010; Deng et al. 2010; Luu et al. 2012; Sahay et al. 2011), whereas older DG cells may contribute to pattern completion (Nakashiba et al. 2012). Interventions that increase neurogenesis during adulthood may have clinical implications for reversing age-related impairments in pattern separation and associated DG dysfunction (Sahay et al. 2011). The development of such interventions may be particularly important given recent evidence in animals suggesting that pattern separation deficits may begin in middle age (Huxter et al. 2012). Creer et al. (2010) reported that voluntary running improved the ability of adult mice to discriminate between two spatially adjacent locations, suggesting an improvement in spatial pattern separation. In addition, this improvement was correlated with increased neurogenesis. Therefore, exercise may be a potential intervention to combat pattern separation deficits and decreased neurogenesis in adulthood. Unfortunately, voluntary running did not have similar effects on pattern separation or neurogenesis in very old mice (Creer et al. 2010). Given the aforementioned studies, the development of behavioral tasks sensitive to age-related changes in spatial pattern separation may have implications for future studies of neurogenesis in older animals.

Recent studies investigating age-related changes in visual object recognition have also provided evidence that pattern separation for visual object information may be impaired in aged rats (Burke et al. 2010, 2011) and monkeys (Burke et al. 2011). In a study by Burke et al. (2011), young and old rats were tested on a variant of the spontaneous object recognition task hypothesized to measure pattern separation. When the rats were tested on the task with objects that did not share any common features, both old and young rats showed an exploratory preference for the novel object. However, when the animals were tested using objects with overlapping features (presumably increasing the need for pattern separation); only young rats showed a preference for the novel object. In a second experiment, young and old monkeys were tested on an object discrimination task. When the objects were dissimilar, both young and old monkeys learned to choose the rewarded objects. However, when objects with overlapping features were used in the discriminations, old monkeys required more trials than young monkeys to learn the discriminations between the rewarded and non-rewarded objects. Given that the performance of the older animals was similar to that of animals with perirhinal cortex lesions (e.g. Bartko et al. 2007a; Bussey et al. 2003), the authors conclude that age-related changes in the perirhinal cortex may lessen the ability of aged animals to support visual object pattern separation (Burke et al. 2010, 2011, 2012). Continued efforts to investigate pattern separation in older animal models may provide a better understanding of the relationship between age-related changes in various brain regions and impaired pattern separation associated with aging.

Pattern Separation in Older Humans

Recent studies have also begun to examine the relationship between aging and decreased pattern separation efficiency in humans. Age-related changes in pattern separation ability have been demonstrated on tasks involving visual objects (Stark et al. 2013; Toner et al. 2009; Yassa et al. 2011), temporal order of items in a sequence (Tolentino et al. 2012), spatial locations (Holden et al. 2012; Stark et al. 2010), and perceptually related verbal stimuli (Ly et al. 2013). Toner et al. (2009) examined the performance of young and cognitively normal older adults on a continuous recognition paradigm developed by Kirwan and Stark (2007). Participants viewed pictures of everyday objects on a computer screen and were asked to make a judgment about whether or not they had seen each object previously in the task. Some of the objects were repeated across trials and some objects, referred to as lures, were similar but not identical to objects presented previously in the task. For each object, participants were asked to press a button to indicate whether the stimulus was: (1) new—the object had never been presented during the task, (2) old—the exact same object had been presented previously, or (3) similar—the object was similar, but not identical to one that had been presented previously during the task. This task was hypothesized to require pattern separation due to the highly overlapping object features of the lure items. Young adults significantly outperformed older adults in correct identification of lure items as similar, but there were no group differences in correct responses to new or repeated stimuli, suggesting that visual object pattern separation was less efficient in older adults (Toner et al. 2009).

In a more recent study, Yassa et al. (2011) used high-resolution fMRI to examine age-related neural changes in the human hippocampus whereas subjects performed the same task used by Toner et al. (2009). Behaviorally, the authors found a similar pattern of age-related impairment in the visual object pattern separation task. The study also included an additional experiment, which demonstrated that the behavioral pattern of activity maps onto the predictions of the model by Wilson et al. (2006). Specifically, older adults were found to require a larger degree of input dissimilarity before separation could occur. The results from the fMRI analyses revealed increased activity in the DG/CA3 subregions on trials that taxed pattern separation. On trials in which older adults were able to correctly identify lure stimuli as "similar," greater activation was observed in the DG/CA3 regions compared to when lure stimuli were incorrectly identified as "old." A subsequent study involving a similar incidental encoding behavioral task used high-resolution fMRI to reveal that representational rigidity (defined as the requirement for increased dissimilarity before stimuli can be orthogonalized) in the DG/CA3 regions of older adults was linked to deficits on the pattern separation task (Yassa et al. 2011). Using ultrahigh-resolution microstructural diffusion tensor imaging, the authors also found age-related changes in perforant path integrity that were inversely correlated with DG-CA3 representational rigidity in older adults. In addition, perforant path integrity was found to correlate with performance in the pattern separation task. The results provide further evidence for a reduction in pattern separation in DG/CA3 subregions of older adults. The findings reveal structural and functional deficits in the perforant path and the DG/CA3 subregions as potential contributors to pattern separation deficits associated with aging. The changes may result in a shift toward increased reliance on retrieval of stored information at the expense of processing novel information in older adults (Yassa et al. 2011).

In a recent study, Stark et al. (2013) used an incidental encoding version of the task described above to examine visual object pattern separation ability across lifespan. The study included cognitively normal adults divided into four age groups, ranging from 20 to 89 years of age. In the encoding phase of the task, participants were asked to make an indoor/outdoor judgment about pictures of everyday objects. In the subsequent recognition memory phase, participants were again presented with pictures of everyday objects and were asked to determine whether each object was new, old, or similar, using the same guidelines outlined for the continuous recognition task (Kirwan and Stark 2007). Recognition memory, measured by correct responses to repeated presentations of objects, did not differ across the four age groups. In contrast, as age increased, the ability to correctly identify lure objects as similar (pattern separation) declined in a linear fashion and leveled off around age 60. Performance was also examined as a function of the degree of mnemonic similarity among lure objects. The data revealed a systematic trend in which increased age was associated with a need for greater dissimilarity of lure objects to achieve accurate identification of the objects as similar. These results further support the hypothesis that visual object pattern separation efficiency declines with age.

Tolentino et al. (2012) examined the effects of temporal interference on sequence memory in young and nondemented older adults. Participants were presented with a sequence of eight circles at the end of each of the arms on a computerized version of a radial 8-arm maze. After the participant viewed the sequence, the radial 8-arm maze was presented with a circle at the end of two of the study phase arms. There were four possible temporal separations of 0, 2, 4, and 6 lags, which represented the number of circles in the original sequence that came between the two circles presented in the choice phase. The researchers hypothesized that circles closer together in the study phase sequence would result in increased interference and a greater need to temporally separate the items. This study involved two experiments, one with a new random sequence for each trial and one with a fixed sequence across trials. In the random sequence experiment, performance for both groups improved as the temporal lag increased and young adults outperformed older adults across all temporal lags. In the fixed sequence experiment, young adults performed significantly better than older adults on all temporal lags with the exception of the 6 lag, which involved the least amount of temporal interference. Both experiments demonstrated age-related deficits in temporal order memory as a function of increased interference. The authors postulated that temporal order memory is less efficient and more susceptible to interference in older adults, possibly due to impaired temporal pattern separation.

Age-related pattern separation deficits have also been demonstrated in memory for spatial location (Holden et al. 2012). Young adults and cognitively normal older adults performed a delayed match-to-sample task that involved manipulations of the degree of spatial interference. Participants were presented with a gray circle along a nonvisible horizontal line on a computer screen. After a short delay, two circles were presented simultaneously and the participant was asked to decide which circle was in the same location as the original gray circle. Distances of 0, 0.5, 1.0, and 1.5 cm separated the two choice circles. It was hypothesized that choice circles that were closer together would result in heightened interference and thus an increased need for pattern separation. Performance increased in both young and older adults as the distance between the two choice circles increased. However, young adults outperformed older adults, suggesting that spatial pattern separation was less efficient in aged individuals (see also Holden and Gilbert 2012).

In a recent study, Ly et al. (2013) sought to further elucidate the nature of age-related deficits in pattern separation by manipulating the type of interference. The authors were interested in understanding whether inefficient pattern separation in older adults is due to conceptual or perceptual interference and suggested that prior studies were unable to disentangle the two, due to the nature of the pictorial stimuli utilized. For this study, the researchers used verbal stimuli that were either phonologically similar (perceptual interference) or semantically similar (conceptual interference). The data revealed age-related deficits in pattern separation ability for perceptually related words, but no performance differences for conceptually related words. The authors proposed that perceptual recollection may be more sensitive to pattern separation deficits because it relies on item-specific information (e.g., item features and details), whereas conceptual recollection relies more on gist information. The results of this study suggest that not all types of memory are equally susceptible to interference and, more specifically, that age-related impairment in pattern separation may be specific to perceptual interference.

Variability in Pattern Separation Efficiency in Older Humans

Although the research reviewed thus far suggests that cognitive aging is associated with deficits in pattern separation, growing evidence also suggests that there may be individual differences among older adults in pattern separation efficiency. Stark et al. (2010) were the first to assess potential age-related variability in a task designed to measure spatial pattern separation. In this task, participants viewed pairs of pictures and were asked later to decide whether the pictures were in the same location or whether one of the pictures in the pair was in a different location. There were four possible conditions on the choice trial, one *same* condition (both pictures were in the same location) and three *different* conditions (one of the pictures in the pair had been moved). The *different* conditions were designated as *close, medium, and far*, representing the distance and angle from the original location. In the initial comparison of young and older adults, no group differences were found. However, when the older adult group was divided into an aged–impaired and aged–unimpaired group based on performance on a standardized auditory learning task, the young adults and aged–unimpaired groups performed significantly better than the aged–impaired group in the *different* trials that taxed spatial pattern separation. In an attempt to replicate these findings using a different paradigm to assess spatial pattern separation (described above), Holden et al. (2012) also divided older adults into impaired and unimpaired groups based on performance on standardized assessment of word learning. The pattern of deficits was remarkably similar to those of Stark et al. (2010). The group labeled older–impaired showed spatial pattern separation deficits relative to the young adults and older–unimpaired adults (Holden et al. 2012). The results of these two studies suggest that there may be individual differences in pattern separation deficits in the domain of spatial memory.

Evidence also suggests that there may be variability among older adults in visual object pattern separation. As discussed previously, Stark et al. (2013) utilized an incidental encoding task to examine pattern separation for visual object information. As part of this investigation, cognitively normal participants over 60 years of age were divided into aged–unimpaired and aged–impaired groups based on standardized list-learning task performance. These two groups of healthy older adults were compared to a group of individuals diagnosed with amnestic mild cognitive impairment (aMCI). The aged–unimpaired group outperformed both the aged–impaired group and aMCI group on trials that taxed visual object pattern separation, but there were no significant differences between the aged–impaired group and the aMCI group on these trials. In contrast, individuals with aMCI were impaired relative to both of the other groups on a measure of recognition memory, but there were no recognition memory differences between the aged–unimpaired and aged–impaired groups. In addition, when performance was examined as a function of the mnemonic similarity of lure objects, the correct identification of lures required greater object dissimilarity for aMCI individuals relative to the two older adult groups, as well as for the aged–impaired group relative to aged–unimpaired group. A previous study reported that when compared to cognitively normal older adults, individuals with aMCI were impaired in a continuous recognition task that taxed visual object pattern separation abilities and that the observed deficits were associated with structural and functional changes in the DG/CA3 region of the hippocampus (Yassa et al. 2010). The results of the recent study by Stark et al. (2013) suggest that it may be possible to further characterize impairment in mnemonic processes in older adults through specific patterns of impairment in individuals with aMCI (impaired recognition and pattern separation), cognitively normal individuals with subtle cognitive decline (intact recognition and impaired pattern separation), and those who are aging successfully (intact recognition and intact pattern separation).

Holden et al. (2013) also examined age-related variability in visual object pattern separation efficiency utilizing a task that involved intentional encoding (Toner et al. 2009; Yassa et al. 2011). Similar to previous studies that divided older adults into impaired and unimpaired groups (Holden et al. 2012; Stark et al. 2010, 2013), older adults were divided into two groups based on standardized verbal learning task performance. The data revealed that young adults and older–unimpaired adults outperformed older–impaired individuals when correctly identifying lure items as similar, suggesting that visual object pattern separation was less efficient only in this subset of older adults. All groups performed similarly in the correct identification of new and repeated stimuli, suggesting that the deficits were not due to general recognition memory impairment. The results of this study further support the idea that there may be individual variability in pattern separation ability among cognitively normal older adults and that this variability occurs across multiple domains, including memory for visual objects and spatial memory. In addition, the findings discussed above by Stark et al. (2013) and Yassa et al. (2010) provide evidence for a link between impaired pattern separation and a diagnosis of aMCI, which is a risk factor for the development of Alzheimer's disease (AD).

Is Memory Decline in Aging and Alzheimer's Disease Linked to Pattern Separation?

In the USA, AD is the most common cause of dementia in older adults and accounts for 60–80 % of dementia cases (Alzheimer's Association 2012). In the year 2012, an estimated 5.4 million Americans were diagnosed with AD; however, this number is projected to increase to 11–16 million by 2050 (Alzheimer's Association 2012). As a result of the aging "baby boom" generation and increasing longevity in the US population, the disease is a growing public health concern with costs estimated to reach $ 200 billion in 2012. Although a number of risk factors for AD have been discussed (e.g., diagnosis of mild cognitive impairment, family history of AD, genetics), one of the most well-documented risk factors for the disease is increasing age (Kamboh 2004). Therefore, a major aim of recent research has been to identify early indicators of cognitive dysfunction in older adults.

Age-related cognitive impairment has been documented in a variety of domains. However, one of the most commonly reported neurocognitive deficits associated with aging is memory decline. Although not all aspects of memory are equally affected by aging (e.g., source vs. item memory), some domains such as episodic memory appear to be particularly sensitive to age-related change. Episodic memory deficits have been well documented in older adults (Rand-Giovannetti et al. 2006) and are a prominent symptom of AD that may be detectable many years prior to disease onset (Bondi et al. 1999). Episodic memory impairment has also been documented in cognitively normal older adults who are at risk of AD by virtue of a diagnosis of mild cognitive impairment (Hodges et al. 2006) or genetic factors (Saunders et al. 1993). Episodic memory may rely on the functioning of the temporal and frontal lobes; however, the functional contributions of each cortical region can be dissociated (Kramer et al. 2005). The hippocampus may be important for memory accuracy, whereas the frontal lobes may be more important for decision-making and strategic aspects of episodic memory (Kramer et al. 2005). As discussed above, the hippocampus may support specific mnemonic processes, such as pattern separation, that may facilitate the encoding and subsequent retrieval of episodic memories to enhance memory accuracy. A key feature of episodic memory that differentiates it from other types of memory is that the elements of an episodic memory must be associated with a context to demarcate the episode in space and time. In addition, a pattern separation mechanism may be necessary to separate the elements of different episodic memories to avoid interference (Gilbert et al. 2001). The studies reviewed above provide evidence that less efficient pattern separation in older adults may contribute to age-related memory deficits, particularly in situations when interference is high. The identification of a key mnemonic processing deficit in pattern separation may result in behavioral interventions that structure daily living tasks to mitigate interference and potentially improve episodic memory in older adults.

Normal and pathological aging may have differential effects on subregions of the hippocampus. The DG subregion may be particularly susceptible to age-related changes in humans; however, there may be less impact on pyramidal cells in the CA subregions (Small et al. 2002). In contrast, the CA subregions may be more vulnerable to pathological changes associated with AD (Apostolova et al. 2010; Braak and Braak 1996; Price et al. 2001; Small et al. 2011; West et al. 2000). As mentioned previously, a primary goal in AD research is to identify risk factors and preclinical markers of the disease in older adults. Given the differential effects of normal aging and AD on the various subregions of the hippocampus, tasks that are sensitive to dysfunction in particular subregions, such as measures of pattern separation, may help to differentiate between cognitive impairment associated with normal aging and pathological changes associated with AD. In support of this idea, Stark et al. (2013) found unique patterns of performance in a visual object pattern separation task in individuals with aMCI, cognitively normal older individuals with subtle cognitive impairment, and cognitively normal older adults. In addition, another recent study utilized the continuous recognition memory task for visual objects (Kirwan and Stark 2007) used in previously mentioned aging studies (e.g. Toner et al. 2009; Yassa et al. 2011) to behaviorally examine pattern separation in individuals diagnosed with aMCI or mild AD (Ally et al. 2013). The authors also examined how performance changed as a function of the lag between the study and test objects. The data revealed that behavioral pattern separation rates decreased as a function of increasing lag between interfering objects in individuals diagnosed with aMCI. Performance of the aMCI group matched controls at the shortest lag of four interfering objects; however, the group performed comparably to the AD group at the largest lag of 40 interfering objects. The AD group was significantly impaired relative to controls across all lags. The data provide additional evidence for impaired visual object pattern separation associated with aMCI and offered some of the first behavioral evidence that pattern separation may be further impaired in those diagnosed with mild AD (Ally et al. 2013). Recent studies have begun to examine the relationship between standardized memory test performance and specific hippocampal subregion function (Brickman et al. 2011). Behavioral tasks that measure specific mnemonic processes, such as the previously reviewed pattern separation tasks, may be highly sensitive to subtle age-related changes. These tests may be used one day in conjunction with standardized neuropsychological measures to help differentiate normal aging and AD.

Pattern Separation Beyond the Hippocampus

Although most of the studies examining the neural substrates of pattern separation have focused on the DG/CA3 subregions, there is growing evidence that other regions of the brain may also support pattern separation (reviewed by Hunsaker and Kesner 2013; Yassa and Stark 2011). For example, researchers have reported that pattern separation may be facilitated by the CA1 hippocampal subregion for temporal order information (Gilbert et al. 2001; Hunsaker et al. 2008; Kesner and Hunsaker 2010; Kesner et al. 2010, 2011), the perirhinal cortex for visual object information (Barense et al. 2010; Bartko et al. 2007a, b; Burke et al. 2011; Gilbert and Kesner 2003), the piriform cortex for olfactory information (Barnes et al. 2008; Sahay et al. 2011; Wilson 2009; Wilson and Sullivan 2011), and the amygdala for reward value (Gilbert and Kesner 2002). Many of these regions of the brain undergo age-related change. For example, age-related functional changes have been observed in perirhinal cortex in rodents (Moyer and Brown 2006) and humans (Ryan et al. 2012). However, aging studies have reported that total neuron numbers in rodents (Rapp et al. 2002) and cortical volumes in humans (Insausti et al. 1998) are largely preserved in the perirhinal cortex. Although there is growing evidence to suggest that the human hippocampal subregions support pattern separation based on overlapping object features (Bakker et al. 2008; Kirwan and Stark 2007), there are data to suggest that perirhinal cortex may also play a role in pattern separation for visual objects. Rodent studies have shown that the perirhinal cortex may distinguish between visual objects with overlapping features to reduce feature ambiguity (Bartko et al. 2007a, b; Bussey et al. 2003, 2006; Gilbert and Kesner 2003; Norman and Eacott 2004). As discussed previously, data from the laboratory of Carol Barnes (Burke et al. 2010, 2011, 2012) provide evidence that age-related changes in the perirhinal cortex of rodents may impair pattern separation for visual objects. Therefore, functional changes in the perirhinal cortex of older animals and possibly humans may affect pattern separation for visual objects. As proposed by Burke et al. (2011), future studies should investigate whether the connections between the hippocampus and perirhinal cortex are necessary to support pattern separation. It is clear that additional research is needed to examine the relationship between age-related changes in brain regions outside of the hippocampus and pattern separation for various types of information. These studies are needed in animal models and also in humans using functional neuroimaging techniques. Although numerous computational and theoretical models have been published to describe potential neural mechanisms that may support pattern separation in the hippocampus, very little is known about pattern separation mechanisms in other brain regions. Therefore, future studies are needed to explore potential neural mechanisms for pattern separation beyond the hippocampus.

Conclusions

In conclusion, memory deficits have been well documented in older adults and may serve as an early indicator of MCI or AD in some individuals. Pattern separation may be a key mechanism for reducing interference among similar memory representations to enhance memory accuracy. Growing evidence suggests that brain regions critical to pattern separation, including the DG and CA3 hippocampal subregions and the perforant path input, may be particularly susceptible to adverse age-related changes. A growing literature indicates that pattern separation becomes less efficient as a result of normal aging in both humans and animal models. It is possible that this decreased pattern separation efficiency contributes to memory deficits, including episodic memory impairment, associated with aging. Given the evidence reviewed in the present chapter, it is clear that additional research is needed to examine the relationship between pattern separation and brain changes associated with aging and neurodegenerative disease. In addition, there is a need for additional research to examine this relationship in animal models. Through continued research we hope that new and innovative behavioral approaches and methodologies will be developed for future aging studies investigating: (1) episodic memory impairment, (2) hippocampal subregion specific epigenetic and transcriptional changes, (3) structural and functional changes in the hippocampus using neuroimaging techniques, and (4) the differentiation of preclinical markers of AD from those of normal aging. The findings may have important implications for studies in humans and translational studies in animal models to shed new light on processes that may contribute to hallmark age-related episodic memory deficits. Finally, we would like to acknowledge the work of Dr. Raymond Kesner and his significant contributions to our understanding of processes supported by the hippocampus such as pattern separation. The innovative behavioral tasks developed in his laboratory for use in rodents have set the foundation for many of the studies discussed in this review.

Acknowledgments A portion of the research reviewed in the chapter was supported by a National Institutes of Health Grant (#AG034202) from the National Institute on Aging awarded to Paul E. Gilbert. Paul Gilbert would like to thank and acknowledge Dr. Raymond Kesner for being an excellent mentor, collaborator, and friend. Andrea Morris also would like to thank Dr. Kesner for his mentorship, valuable scientific input, and advice. The authors thank Dr. Kesner for his contributions to the field and wish him the best in his retirement.

References

Aimone, J. B., Deng, W., & Gage, F. H. (2010). Adult neurogenesis: integrating theories and separating functions. *Trends in Cognitive Sciences, 14*(7), 325–337. doi:10.1016/j.tics.2010.04.003.

Aimone, J. B., Deng, W., & Gage, F. H. (2011). Resolving new memories: A critical look at the dentate gyrus, adult neurogenesis, and pattern separation. *Neuron, 70*(4), 589–596. doi:10.1016/j.neuron.2011.05.010.

Allen, J. S., Bruss, J., Brown, C. K., & Damasio, H. (2005). Normal neuroanatomical variation due to age: The major lobes and a parcellation of the temporal region. *Neurobiology of Aging, 26*(9), 1245–1260 (discussion 1279–1282). doi:10.1016/j.neurobiolaging.2005.05.023.

Ally, B. A., Hussey, E. P., Ko, P. C., & Molitor, R. J. (2013). Pattern separation and pattern completion in Alzheimer's disease: Evidence of rapid forgetting in amnestic mild cognitive impairment. *Hippocampus, 23*(12), 1246–1258. doi:10.1002/hipo.22162.

Alzheimer's Association. (2012). 2012 Alzheimer's disease facts and figures. *Alzheimer's & Dementia: The Journal of the Alzheimer's Association, 8*(2), 131–168. doi:10.1016/j.jalz.2012.02.001.

Apostolova, L. G., Hwang, K. S., Andrawis, J. P., Green, A. E., Babakchanian, S., Morra, J. H., Cummings, J. L., et al. (2010). 3D PIB and CSF biomarker associations with hippocampal atrophy in ADNI subjects. *Neurobiology of Aging, 31*(8), 1284–1303. doi:10.1016/j.neurobiolaging.2010.05.003.

Bakker, A., Kirwan, C. B., Miller, M., & Stark, C. E. L. (2008). Pattern separation in the human hippocampal CA3 and dentate gyrus. *Science (New York, N.Y.), 319*(5870), 1640–1642. doi:10.1126/science.1152882.

Barense, M. D., Henson, R. N. A., Lee, A. C. H., & Graham, K. S. (2010). Medial temporal lobe activity during complex discrimination of faces, objects, and scenes: Effects of viewpoint. *Hippocampus, 20*(3), 389–401. doi:10.1002/hipo.20641.

Barnes, C. A. (1994). Normal aging: Regionally specific changes in hippocampal synaptic transmission. *Trends in Neurosciences, 17*(1), 13–18.

Barnes, D. C., Hofacer, R. D., Zaman, A. R., Rennaker, R. L., & Wilson, D. A. (2008). Olfactory perceptual stability and discrimination. *Nature Neuroscience, 11*(12), 1378–1380. doi:10.1038/nn.2217.

Bartko, S. J., Winters, B. D., Cowell, R. A., Saksida, L. M., & Bussey, T. J. (2007a). Perceptual functions of perirhinal cortex in rats: zero-delay object recognition and simultaneous oddity discriminations. *The Journal of Neuroscience: The Official Journal of the Society for Neuroscience, 27*(10), 2548–2559. doi:10.1523/JNEUROSCI.5171-06.2007.

Bartko, S. J., Winters, B. D., Cowell, R. A., Saksida, L. M., & Bussey, T. J. (2007b). Perirhinal cortex resolves feature ambiguity in configural object recognition and perceptual oddity tasks. *Learning & Memory (Cold Spring Harbor, N.Y.), 14*(12), 821–832. doi:10.1101/lm.749207.

Bondi, M. W., Salmon, D. P., Galasko, D., Thomas, R. G., & Thal, L. J. (1999). Neuropsychological function and apolipoprotein E genotype in the preclinical detection of Alzheimer's disease. *Psychology and Aging, 14*(2), 295–303.

Braak, H., & Braak, E. (1996). Development of Alzheimer-related neurofibrillary changes in the neocortex inversely recapitulates cortical myelogenesis. *Acta Neuropathologica, 92*(2), 197–201.

Brickman, A. M., Stern, Y., & Small, S. A. (2011). Hippocampal subregions differentially associate with standardized memory tests. *Hippocampus, 21*(9), 923–928. doi:10.1002/hipo.20840.

Burke, S. N., & Barnes, C. A. (2006). Neural plasticity in the ageing brain. *Nature Reviews Neuroscience, 7*(1), 30–40. doi:10.1038/nrn1809.

Burke, S. N., Wallace, J. L., Nematollahi, S., Uprety, A. R., & Barnes, C. A. (2010). Pattern separation deficits may contribute to age-associated recognition impairments. *Behavioral Neuroscience, 124*(5), 559–573. doi:10.1037/a0020893.

Burke, S. N., Wallace, J. L., Hartzell, A. L., Nematollahi, S., Plange, K., & Barnes, C. A. (2011). Age-associated deficits in pattern separation functions of the perirhinal cortex: A cross-species consensus. *Behavioral Neuroscience, 125*(6), 836–847. doi:10.1037/a0026238.

Burke, S. N., Ryan, L., & Barnes, C. A. (2012). Characterizing cognitive aging of recognition memory and related processes in animal models and in humans. *Frontiers in Aging Neuroscience, 4,* 15. doi:10.3389/fnagi.2012.00015.

Bussey, T. J., Saksida, L. M., & Murray, E. A. (2003). Impairments in visual discrimination after perirhinal cortex lesions: Testing "declarative" vs. "perceptual-mnemonic" views of perirhinal cortex function. *The European Journal of Neuroscience, 17*(3), 649–660.

Bussey, T. J., Saksida, L. M., & Murray, E. A. (2006). Perirhinal cortex and feature-ambiguous discriminations. *Learning & Memory (Cold Spring Harbor, N.Y.), 13*(2), 103–105 (author reply 106–107). doi:10.1101/lm.163606.

Butterly, D. A., Petroccione, M. A., & Smith, D. M. (2012). Hippocampal context processing is critical for interference free recall of odor memories in rats. *Hippocampus, 22*(4), 906–913. doi:10.1002/hipo.20953.

Carr, V. A., Rissman, J., & Wagner, A. D. (2010). Imaging the human medial temporal lobe with high-resolution fMRI. *Neuron, 65*(3), 298–308. doi:10.1016/j.neuron.2009.12.022.

Clelland, C. D., Choi, M., Romberg, C., Clemenson, G. D., Jr, Fragniere, A., Tyers, P., Jessberger, S., et al. (2009). A functional role for adult hippocampal neurogenesis in spatial pattern separation. *Science (New York, N.Y.), 325*(5937), 210–213. doi:10.1126/science.1173215.

Creer, D. J., Romberg, C., Saksida, L. M., van Praag, H., & Bussey, T. J. (2010). Running enhances spatial pattern separation in mice. *Proceedings of the National Academy of Sciences of the United States of America, 107*(5), 2367–2372. doi:10.1073/pnas.0911725107.

Deng, W., Aimone, J. B., & Gage, F. H. (2010). New neurons and new memories: how does adult hippocampal neurogenesis affect learning and memory? *Nature Reviews Neuroscience, 11*(5), 339–350. doi:10.1038/nrn2822.

Driscoll, I., & Sutherland, R. J. (2005). The aging hippocampus: navigating between rat and human experiments. *Reviews in the Neurosciences, 16*(2), 87–121.

Driscoll, I., Howard, S. R., Stone, J. C., Monfils, M. H., Tomanek, B., Brooks, W. M., & Sutherland, R. J. (2006). The aging hippocampus: A multi-level analysis in the rat. *Neuroscience, 139*(4), 1173–1185. doi:10.1016/j.neuroscience.2006.01.040.

Driscoll, I., Davatzikos, C., An, Y., Wu, X., Shen, D., Kraut, M., & Resnick, S. M. (2009). Longitudinal pattern of regional brain volume change differentiates normal aging from MCI. *Neurology, 72*(22), 1906–1913. doi:10.1212/WNL.0b013e3181a82634.

Duff, M. C., Warren, D. E., Gupta, R., Vidal, J. P. B., Tranel, D., & Cohen, N. J. (2012). Teasing apart tangrams: Testing hippocampal pattern separation with a collaborative referencing paradigm. *Hippocampus, 22*(5), 1087–1091. doi:10.1002/hipo.20967.

Gallagher, M., Bakker, A., Yassa, M. A., & Stark, C. E. L. (2010). Bridging neurocognitive aging and disease modification: Targeting functional mechanisms of memory impairment. *Current Alzheimer Research, 7*(3), 197–199.

Geinisman, Y., deToledo-Morrell, L., Morrell, F., Persina, I. S., & Rossi, M. (1992). Age-related loss of axospinous synapses formed by two afferent systems in the rat dentate gyrus as revealed by the unbiased stereological dissector technique. *Hippocampus, 2*(4), 437–444. doi:10.1002/hipo.450020411.

Gilbert, P. E., & Kesner, R. P. (2002). The amygdala but not the hippocampus is involved in pattern separation based on reward value. *Neurobiology of Learning and Memory, 77*(3), 338–353. doi:10.1006/nlme.2001.4033.

Gilbert, P. E., & Kesner, R. P. (2003). Recognition memory for complex visual discriminations is influenced by stimulus interference in rodents with perirhinal cortex damage. *Learning and Memory (Cold Spring Harbor, N.Y.), 10*(6), 525–530. doi:10.1101/lm.64503.

Gilbert, P. E., & Kesner, R. P. (2006). The role of the dorsal CA3 hippocampal subregion in spatial working memory and pattern separation. *Behavioural Brain Research, 169*(1), 142–149. doi:10.1016/j.bbr.2006.01.002.

Gilbert, P. E., Kesner, R. P., & DeCoteau, W. E. (1998). Memory for spatial location: Role of the hippocampus in mediating spatial pattern separation. *The Journal of Neuroscience: The Official Journal of the Society for Neuroscience, 18*(2), 804–810.

Gilbert, P. E., Kesner, R. P., & Lee, I. (2001). Dissociating hippocampal subregions: Double dissociation between dentate gyrus and CA1. *Hippocampus, 11*(6), 626–636. doi:10.1002/hipo.1077.

Good, C. D., Johnsrude, I. S., Ashburner, J., Henson, R. N., Friston, K. J., & Frackowiak, R. S. (2001). A voxel-based morphometric study of ageing in 465 normal adult human brains. *NeuroImage, 14*(1 Pt 1), 21–36. doi:10.1006/nimg.2001.0786.

Goodrich-Hunsaker, N. J., Hunsaker, M. R., & Kesner, R. P. (2005). Dissociating the role of the parietal cortex and dorsal hippocampus for spatial information processing. *Behavioral Neuroscience, 119*(5), 1307–1315. doi:10.1037/0735-7044.119.5.1307.

Goodrich-Hunsaker, N. J., Hunsaker, M. R., & Kesner, R. P. (2008). The interactions and dissociations of the dorsal hippocampus subregions: How the dentate gyrus, CA3, and CA1 process spatial information. *Behavioral Neuroscience, 122*(1), 16–26. doi:10.1037/0735-7044.122.1.16.

Gracian, E. I., Shelley, L. E., Morris, A. M., & Gilbert, P. E. (2013). Age-related changes in place learning for adjacent and separate locations. *Neurobiology of Aging, 34*(10), 2304–2309. doi:10.1016/j.neurobiolaging.2013.03.033.

Hodges, J. R., Erzinçlioğlu, S., & Patterson, K. (2006). Evolution of cognitive deficits and conversion to dementia in patients with mild cognitive impairment: A very-long-term follow-up study. *Dementia and Geriatric Cognitive Disorders, 21*(5–6), 380–391. doi:10.1159/000092534.

Holden, H. M., & Gilbert, P. E. (2012). Less efficient pattern separation may contribute to age-related spatial memory deficits. *Frontiers in Aging Neuroscience, 4,* 9. doi: 10.3389/fnagi.2012.00009.

Holden, H. M., Hoebel, C., Loftis, K., & Gilbert, P. E. (2012). Spatial pattern separation in cognitively normal young and older adults. *Hippocampus, 22*(9), 1826–1832. doi:10.1002/hipo.22017.

Holden, H. M., Toner, C., Pirogovsky, E., Kirwan, C. B., & Gilbert, P. E. (2013). Visual object pattern separation varies in older adults. *Learning & Memory (Cold Spring Harbor, N.Y.), 20*(7), 358–362. doi:10.1101/lm.030171.112.

Hunsaker, M. R., & Kesner, R. P. (2008). Evaluating the differential roles of the dorsal dentate gyrus, dorsal CA3, and dorsal CA1 during a temporal ordering for spatial locations task. *Hippocampus, 18*(9), 955–964. doi:10.1002/hipo.20455.

Hunsaker, M. R., & Kesner, R. P. (2013). The operation of pattern separation and pattern completion processes associated with different attributes or domains of memory. *Neuroscience and Biobehavioral Reviews, 37*(1), 36–58. doi:10.1016/j.neubiorev.2012.09.014.

Hunsaker, M. R., Fieldsted, P. M., Rosenberg, J. S., & Kesner, R. P. (2008). Dissociating the roles of dorsal and ventral CA1 for the temporal processing of spatial locations, visual objects, and odors. *Behavioral Neuroscience, 122*(3), 643–650. doi:10.1037/0735-7044.122.3.643.

Huxter, J. R., Miranda, J. A., & Dias, R. (2012). The hippocampal physiology of approaching middle-age: Early indicators of change. *Hippocampus, 22*(9), 1923–1940. doi:10.1002/hipo.22027.

Insausti, R., Insausti, A. M., Sobreviela, M. T., Salinas, A., & Martínez-Peñuela, J. M. (1998). Human medial temporal lobe in aging: Anatomical basis of memory preservation. *Microscopy Research and Technique, 43*(1), 8–15. doi:10.1002/(SICI)1097-0029(19981001)43:13.0.CO;2-4.

Kamboh, M. I. (2004). Molecular genetics of late-onset Alzheimer's disease. *Annals of Human Genetics, 68*(Pt 4), 381–404. doi:10.1046/j.1529-8817.2004.00110.x.

Kennedy, K. M., & Raz, N. (2009). Aging white matter and cognition: Differential effects of regional variations in diffusion properties on memory, executive functions, and speed. *Neuropsychologia, 47*(3), 916–927. doi:10.1016/j.neuropsychologia.2009.01.001.

Kesner, R. P. (2007). A behavioral analysis of dentate gyrus function. *Progress in Brain Research, 163,* 567–576. doi:10.1016/S0079-6123(07)63030-1.

Kesner, R. P. (2009). Tapestry of memory. *Behavioral Neuroscience, 123*(1), 1–13. doi:10.1037/a0014004.

Kesner, R. P. (2013a). An analysis of the dentate gyrus function. *Behavioural Brain Research, 254,* 1–7. doi:10.1016/j.bbr.2013.01.012.

Kesner, R. P. (2013b). A process analysis of the CA3 subregion of the hippocampus. *Frontiers in Cellular Neuroscience, 7,* 78. doi:10.3389/fncel.2013.00078.

Kesner, R. P., & Gilbert, P. E. (2006). The role of the medial caudate nucleus, but not the hippocampus, in a matching-to sample task for a motor response. *The European Journal of Neuroscience, 23*(7), 1888–1894. doi:10.1111/j.1460-9568.2006.04709.x.

Kesner, R. P., & Hopkins, R. O. (2006). Mnemonic functions of the hippocampus: a comparison between animals and humans. *Biological Psychology, 73*(1), 3–18. doi:10.1016/j.biopsycho.2006.01.004.

Kesner, R. P., & Hunsaker, M. R. (2010). The temporal attributes of episodic memory. *Behavioural Brain Research, 215*(2), 299–309. doi:10.1016/j.bbr.2009.12.029.

Kesner, R. P., Gilbert, P. E., & Wallenstein, G. V. (2000). Testing neural network models of memory with behavioral experiments. *Current Opinion in Neurobiology, 10*(2), 260–265.

Kesner, R. P., Gilbert, P. E., & Barua, L. A. (2002). The role of the hippocampus in memory for the temporal order of a sequence of odors. *Behavioral Neuroscience, 116*(2), 286–290.

Kesner, R. P., Hunsaker, M. R., & Ziegler, W. (2010). The role of the dorsal CA1 and ventral CA1 in memory for the temporal order of a sequence of odors. *Neurobiology of Learning and Memory, 93*(1), 111–116. doi:10.1016/j.nlm.2009.08.010.

Kesner, R. P., Hunsaker, M. R., & Ziegler, W. (2011). The role of the dorsal and ventral hippocampus in olfactory working memory. *Neurobiology of Learning and Memory, 96*(2), 361–366. doi:10.1016/j.nlm.2011.06.011.

Kirwan, C. B., & Stark, C. E. L. (2007). Overcoming interference: an fMRI investigation of pattern separation in the medial temporal lobe. *Learning & Memory (Cold Spring Harbor, N.Y.), 14*(9), 625–633. doi:10.1101/lm.663507.

Kirwan, C. B., Hartshorn, A., Stark, S. M., Goodrich-Hunsaker, N. J., Hopkins, R. O., & Stark, C. E. L. (2012). Pattern separation deficits following damage to the hippocampus. *Neuropsychologia, 50*(10), 2408–2414. doi:10.1016/j.neuropsychologia.2012.06.011.

Kramer, J. H., Rosen, H. J., Du, A.-T., Schuff, N., Hollnagel, C., Weiner, M. W., Miller, B. L., et al. (2005). Dissociations in hippocampal and frontal contributions to episodic memory performance. *Neuropsychology, 19*(6), 799–805. doi:10.1037/0894-4105.19.6.7999.

Kramer, J. H., Mungas, D., Reed, B. R., Wetzel, M. E., Burnett, M. M., Miller, B. L., Weiner, M. W., et al. (2007). Longitudinal MRI and cognitive change in healthy elderly. *Neuropsychology, 21*(4), 412–418. doi:10.1037/0894-4105.21.4.412.

Kubik, S., Miyashita, T., & Guzowski, J. F. (2007). Using immediate-early genes to map hippocampal subregional functions. *Learning & Memory (Cold Spring Harbor, N.Y.), 14*(11), 758–770. doi:10.1101/lm.698107.

Kuhn, H. G., Dickinson-Anson, H., & Gage, F. H. (1996). Neurogenesis in the dentate gyrus of the adult rat: Age-related decrease of neuronal progenitor proliferation. *The Journal of Neuroscience: The Official Journal of the Society for Neuroscience, 16*(6), 2027–2033.

Lacy, J. W., Yassa, M. A., Stark, S. M., Muftuler, L. T., & Stark, C. E. L. (2011). Distinct pattern separation related transfer functions in human CA3/dentate and CA1 revealed using high-resolution fMRI and variable mnemonic similarity. *Learning & Memory (Cold Spring Harbor, N.Y.), 18*(1), 15–18. doi:10.1101/lm.1971111.

LaRocque, K. F., Smith, M. E., Carr, V. A., Witthoft, N., Grill-Spector, K., & Wagner, A. D. (2013). Global similarity and pattern separation in the human medial temporal lobe predict subsequent memory. *The Journal of Neuroscience: The Official Journal of the Society for Neuroscience, 33*(13), 5466–5474. doi:10.1523/JNEUROSCI.4293-12.2013.

Lee, I., Jerman, T. S., & Kesner, R. P. (2005). Disruption of delayed memory for a sequence of spatial locations following CA1- or CA3-lesions of the dorsal hippocampus. *Neurobiology of Learning and Memory, 84*(2), 138–147. doi:10.1016/j.nlm.2005.06.002.

Leutgeb, J. K., Leutgeb, S., Moser, M.-B., & Moser, E. I. (2007). Pattern separation in the dentate gyrus and CA3 of the hippocampus. *Science (New York, N.Y.), 315*(5814), 961–966. doi:10.1126/science.1135801.

Luu, P., Sill, O. C., Gao, L., Becker, S., Wojtowicz, J. M., & Smith, D. M. (2012). The role of adult hippocampal neurogenesis in reducing interference. *Behavioral Neuroscience, 126*(3), 381–391. doi:10.1037/a0028252.

Ly, M., Murray, E., & Yassa, M. A. (2013). Perceptual versus conceptual interference and pattern separation of verbal stimuli in young and older adults. *Hippocampus, 23*(6), 425–430. doi:10.1002/hipo.22110.

Marr, D. (1971). Simple memory: A theory for archicortex. *Philosophical Transactions of the Royal Society of London. Series B, Biological Sciences, 262*(841), 23–81.

Marrone, D. F., Adams, A. A., & Satvat, E. (2011). Increased pattern separation in the aged fascia dentata. *Neurobiology of Aging, 32*(12), 2317.e23–2317.e32. doi:10.1016/j.neurobiolaging.2010.03.021.

McDonald, R. J., & White, N. M. (1995). Hippocampal and nonhippocampal contributions to place learning in rats. *Behavioral Neuroscience, 109*(4), 579–593.

McHugh, T. J., Jones, M. W., Quinn, J. J., Balthasar, N., Coppari, R., Elmquist, J. K., Lowell, B. B., et al. (2007). Dentate gyrus NMDA receptors mediate rapid pattern separation in the hippocampal network. *Science (New York, N.Y.), 317*(5834), 94–99. doi:10.1126/science.1140263.

McNaughton, B. L., & Nadel, L. (1990). "Hebb-Marr networks and the neurobiological representation of action in space". In M. A. Gluck & D. E. Rumelhart (Eds.), *Neuroscience and connectionist theory* (pp. 1–63). Hillsdale: Erlbaum.

McNaughton, B. L., Barnes, C. A., Meltzer, J., & Sutherland, R. J. (1989). Hippocampal granule cells are necessary for normal spatial learning but not for spatially-selective pyramidal cell discharge. *Experimental Brain Research, 76*(3), 485–496.

McTighe, S. M., Mar, A. C., Romberg, C., Bussey, T. J., & Saksida, L. M. (2009). A new touchscreen test of pattern separation: effect of hippocampal lesions. *Neuroreport,* 20(9), 881–885. doi: 10.1097/WNR.0b013e32832c5eb2.

Morris, A. M., Churchwell, J. C., Kesner, R. P., & Gilbert, P. E. (2012). Selective lesions of the dentate gyrus produce disruptions in place learning for adjacent spatial locations. *Neurobiology of Learning and Memory, 97*(3), 326–331. doi:10.1016/j.nlm.2012.02.005.

Motley, S. E., & Kirwan, C. B. (2012). A parametric investigation of pattern separation processes in the medial temporal lobe. *The Journal of Neuroscience: The Official Journal of the Society for Neuroscience, 32*(38), 13076–13085. doi:10.1523/JNEUROSCI.5920-11.2012.

Moyer, J. R., Jr., & Brown, T. H. (2006). Impaired trace and contextual fear conditioning in aged rats. *Behavioral Neuroscience, 120*(3), 612–624. doi:10.1037/0735-7044.120.3.612.

Mueller, S. G., & Weiner, M. W. (2009). Selective effect of age, Apo e4, and Alzheimer's disease on hippocampal subfields. *Hippocampus, 19*(6), 558–564. doi:10.1002/hipo.20614.

Mueller, S. G., Schuff, N., Yaffe, K., Madison, C., Miller, B., & Weiner, M. W. (2010). Hippocampal atrophy patterns in mild cognitive impairment and Alzheimer's disease. *Human Brain Mapping, 31*(9), 1339–1347. doi:10.1002/hbm.20934.

Mungas, D., Harvey, D., Reed, B. R., Jagust, W. J., DeCarli, C., Beckett, L., Mack, W. J., et al. (2005). Longitudinal volumetric MRI change and rate of cognitive decline. *Neurology, 65*(4), 565–571. doi:10.1212/01.wnl.0000172913.88973.0d.

Myers, C. E., & Scharfman, H. E. (2009). A role for hilar cells in pattern separation in the dentate gyrus: a computational approach. *Hippocampus, 19*(4), 321–337. doi:10.1002/hipo.20516.

Nakashiba, T., Cushman, J. D., Pelkey, K. A., Renaudineau, S., Buhl, D. L., McHugh, T. J., Rodriguez Barrera, V., et al. (2012). Young dentate granule cells mediate pattern separation, whereas old granule cells facilitate pattern completion. *Cell, 149*(1), 188–201. doi:10.1016/j.cell.2012.01.046.

Norman, G., & Eacott, M. J. (2004). Impaired object recognition with increasing levels of feature ambiguity in rats with perirhinal cortex lesions. *Behavioural Brain Research, 148*(1–2), 79–91.

O'Reilly, R. C., & McClelland, J. L. (1994). Hippocampal conjunctive encoding, storage, and recall: Avoiding a trade-off. *Hippocampus, 4*(6), 661–682. doi:10.1002/hipo.450040605.

Patrylo, P. R., & Williamson, A. (2007). The effects of aging on dentate circuitry and function. *Progress in Brain Research, 163,* 679–696. doi:10.1016/S0079-6123(07)63037-4.

Penner, M. R., Roth, T. L., Chawla, M. K., Hoang, L. T., Roth, E. D., Lubin, F. D., Sweatt, J. D., et al. (2011). Age-related changes in Arc transcription and DNA methylation within the hippocampus. *Neurobiology of Aging, 32*(12), 2198–2210. doi:10.1016/j.neurobiolaging.2010.01.009.

Persson, J., Pudas, S., Lind, J., Kauppi, K., Nilsson, L.-G., & Nyberg, L. (2012). Longitudinal structure-function correlates in elderly reveal MTL dysfunction with cognitive decline. *Cerebral Cortex (New York, N.Y.: 1991), 22*(10), 2297–2304. doi:10.1093/cercor/bhr306.

Price, J. L., Ko, A. I., Wade, M. J., Tsou, S. K., McKeel, D. W., & Morris, J. C. (2001). Neuron number in the entorhinal cortex and CA1 in preclinical Alzheimer disease. *Archives of Neurology, 58*(9), 1395–1402.

Rand-Giovannetti, E., Chua, E. F., Driscoll, A. E., Schacter, D. L., Albert, M. S., & Sperling, R. A. (2006). Hippocampal and neocortical activation during repetitive encoding in older persons. *Neurobiology of Aging, 27*(1), 173–182. doi:10.1016/j.neurobiolaging.2004.12.013.

Rapp, P. R., & Gallagher, M. (1996). Preserved neuron number in the hippocampus of aged rats with spatial learning deficits. *Proceedings of the National Academy of Sciences of the United States of America, 93*(18), 9926–9930.

Rapp, P. R., Stack, E. C., & Gallagher, M. (1999). Morphometric studies of the aged hippocampus: I. Volumetric analysis in behaviorally characterized rats. *The Journal of Comparative Neurology, 403*(4), 459–470.

Rapp, P. R., Deroche, P. S., Mao, Y., & Burwell, R. D. (2002). Neuron number in the parahippocampal region is preserved in aged rats with spatial learning deficits. *Cerebral Cortex (New York, N.Y.: 1991), 12*(11), 1171–1179.

Rasmussen, T., Schliemann, T., Sørensen, J. C., Zimmer, J., & West, M. J. (1996). Memory impaired aged rats: No loss of principal hippocampal and subicular neurons. *Neurobiology of Aging, 17*(1), 143–147.

Raz, N., Gunning-Dixon, F., Head, D., Rodrigue, K. M., Williamson, A., & Acker, J. D. (2004). Aging, sexual dimorphism, and hemispheric asymmetry of the cerebral cortex: Replicability of regional differences in volume. *Neurobiology of Aging, 25*(3), 377–396. doi:10.1016/S0197-4580(03)00118-0.

Raz, N., Lindenberger, U., Rodrigue, K. M., Kennedy, K. M., Head, D., Williamson, A., Dahle, C., et al. (2005). Regional brain changes in aging healthy adults: General trends, individual differences and modifiers. *Cerebral Cortex (New York, N.Y.: 1991), 15*(11), 1676–1689. doi:10.1093/cercor/bhi044.

Rodrigue, K. M., & Raz, N. (2004). Shrinkage of the entorhinal cortex over five years predicts memory performance in healthy adults. *The Journal of Neuroscience: The Official Journal of the Society for Neuroscience, 24*(4), 956–963. doi:10.1523/JNEUROSCI.4166-03.2004.

Rolls, E T. (1996). A theory of hippocampal function in memory. *Hippocampus, 6*(6), 601–620. doi:10.1002/(SICI)1098-1063(1996)6:63.0.CO;2-J.

Rolls, E. T. (2010). A computational theory of episodic memory formation in the hippocampus. *Behavioural Brain Research, 215*(2), 180–196. doi:10.1016/j.bbr.2010.03.027.

Rolls, E. T., & Kesner, R. P. (2006). A computational theory of hippocampal function, and empirical tests of the theory. *Progress in Neurobiology, 79*(1), 1–48. doi:10.1016/j.pneurobio.2006.04.005.

Ryan, L., Cardoza, J. A., Barense, M. D., Kawa, K. H., Wallentin-Flores, J., Arnold, W. T., & Alexander, G. E. (2012). Age-related impairment in a complex object discrimination task that engages perirhinal cortex. *Hippocampus, 22*(10), 1978–1989. doi:10.1002/hipo.22069.

Sahay, A., Wilson, D. A., & Hen, R. (2011). Pattern separation: a common function for new neurons in hippocampus and olfactory bulb. *Neuron, 70*(4), 582–588. doi:10.1016/j.neuron.2011.05.012.

Saunders, A. M., Strittmatter, W. J., Schmechel, D., George-Hyslop, P. H., Pericak-Vance, M. A., Joo, S. H., Rosi, B. L., et al. (1993). Association of apolipoprotein E allele epsilon 4 with late-onset familial and sporadic Alzheimer's disease. *Neurology, 43*(8), 1467–1472.

Shapiro, M. L., and Olton, D. S. (1994). "Hippocampal function and interference". In D. L. Schacter & E. Tulving (Eds.), *Memory systems* (pp. 141–146). London: MIT.

Small, S. A., Tsai, W. Y., DeLaPaz, R., Mayeux, R., & Stern, Y. (2002). Imaging hippocampal function across the human life span: Is memory decline normal or not? *Annals of Neurology, 51*(3), 290–295.

Small, S. A., Chawla, M. K., Buonocore, M., Rapp, P. R., & Barnes, C. A. (2004). Imaging correlates of brain function in monkeys and rats isolates a hippocampal subregion differentially vulnerable to aging. *Proceedings of the National Academy of Sciences of the United States of America, 101*(18), 7181–7186. doi:10.1073/pnas.0400285101.

Small, S. A., Schobel, S. A., Buxton, R. B., Witter, M. P., & Barnes, C. A. (2011). A pathophysiological framework of hippocampal dysfunction in ageing and disease. *Nature Reviews Neuroscience, 12*(10), 585–601. doi:10.1038/nrn3085.

Smith, T. D., Adams, M. M., Gallagher, M., Morrison, J. H., & Rapp, P. R. (2000). Circuit-specific alterations in hippocampal synaptophysin immunoreactivity predict spatial learning impairment in aged rats. *The Journal of Neuroscience: The Official Journal of the Society for Neuroscience, 20*(17), 6587–6593.

Stark, S. M., Yassa, M. A., & Stark, C. E. L. (2010). Individual differences in spatial pattern separation performance associated with healthy aging in humans. *Learning & Memory (Cold Spring Harbor, N.Y.), 17*(6), 284–288. doi:10.1101/lm.1768110.

Stark, S. M., Yassa, M. A., Lacy, J. W., & Stark, C. E. L. (2013). A task to assess behavioral pattern separation (BPS) in humans: Data from healthy aging and mild cognitive impairment. *Neuropsychologia, 51,* 2442–2449. doi:10.1016/j.neuropsychologia.2012.12.014.

Tanila, H. (1999). Hippocampal place cells can develop distinct representations of two visually identical environments. *Hippocampus, 9*(3), 235–246. doi:10.1002/(SICI)1098-1063(1999)9:33.0.CO;2-3.

Tolentino, J. C., Pirogovsky, E., Luu, T., Toner, C. K., & Gilbert, P. E. (2012). The effect of interference on temporal order memory for random and fixed sequences in nondemented older adults. *Learning & Memory (Cold Spring Harbor, N.Y.), 19*(6), 251–255. doi:10.1101/lm.026062.112.

Toner, C. K., Pirogovsky, E., Kirwan, C. B., & Gilbert, P. E. (2009). Visual object pattern separation deficits in nondemented older adults. *Learning & Memory (Cold Spring Harbor, N.Y.), 16*(5), 338–342. doi:10.1101/lm.1315109.

Treves, A., & Rolls, E. T. (1992). Computational constraints suggest the need for two distinct input systems to the hippocampal CA3 network. *Hippocampus, 2*(2), 189–199. doi:10.1002/hipo.450020209.

Walhovd, K. B., Fjell, A. M., Dale, A. M., McEvoy, L. K., Brewer, J., Karow, D. S., Salmon, D. P., et al. (2010). Multi-modal imaging predicts memory performance in normal aging and cognitive decline. *Neurobiology of Aging, 31*(7), 1107–1121. doi:10.1016/j.neurobiolaging.2008.08.013.

Weeden, C.S.S., Hu, N.J., Ho, L.Y.N., & Kesner, R.P. (2012). The role of the ventral dentate gyrus in olfactory learning and memory. Neuroscience Meeting Planner, Program Number 397.17/EEE39.

West, M. J., Kawas, C. H., Martin, L. J., & Troncoso, J. C. (2000). The CA1 region of the human hippocampus is a hot spot in Alzheimer's disease. *Annals of the New York Academy of Sciences, 908,* 255–259.

Wilson, D. A. (2009). Pattern separation and completion in olfaction. *Annals of the New York Academy of Sciences, 1170,* 306–312. doi:10.1111/j.1749-6632.2009.04017.x.
Wilson, D. A., & Sullivan, R. M. (2011). Cortical processing of odor objects. *Neuron, 72*(4), 506–519. doi:10.1016/j.neuron.2011.10.027.
Wilson, I. A., Gallagher, M., Eichenbaum, H., & Tanila, H. (2006). Neurocognitive aging: Prior memories hinder new hippocampal encoding. *Trends in Neurosciences, 29*(12), 662–670. doi:10.1016/j.tins.2006.10.002.
Wu, W., Brickman, A. M., Luchsinger, J., Ferrazzano, P., Pichiule, P., Yoshita, M., Brown, T., et al. (2008). The brain in the age of old: the hippocampal formation is targeted differentially by diseases of late life. *Annals of Neurology, 64*(6), 698–706. doi:10.1002/ana.21557.
Yassa, M. A., & Stark, C. E. L. (2011). Pattern separation in the hippocampus. *Trends in Neurosciences, 34*(10), 515–525. doi:10.1016/j.tins.2011.06.006.
Yassa, M. A., Stark, S. M., Bakker, A., Albert, M. S., Gallagher, M., & Stark, C. E. L. (2010). High-resolution structural and functional MRI of hippocampal CA3 and dentate gyrus in patients with amnestic Mild Cognitive Impairment. *NeuroImage, 51*(3), 1242–1252. doi:10.1016/j.neuroimage.2010.03.040.
Yassa, M. A., Lacy, J. W., Stark, S. M., Albert, M. S., Gallagher, M., & Stark, C. E. L. (2011a). Pattern separation deficits associated with increased hippocampal CA3 and dentate gyrus activity in nondemented older adults. *Hippocampus, 21*(9), 968–979. doi:10.1002/hipo.20808.
Yassa, M. A., Mattfeld, A. T., Stark, S. M., & Stark, C. E. L. (2011b). Age-related memory deficits linked to circuit-specific disruptions in the hippocampus. *Proceedings of the National Academy of Sciences of the United States of America, 108*(21), 8873–8878. doi:10.1073/pnas.1101567108.
Ziegler, D. A., Piguet, O., Salat, D. H., Prince, K., Connally, E., & Corkin, S. (2010). Cognition in healthy aging is related to regional white matter integrity, but not cortical thickness. *Neurobiology of Aging, 31*(11), 1912–1926. doi:10.1016/j.neurobiolaging.2008.10.015.

Chapter 6
A Lifetime of Memories: Raymond Kesner's Contributions of the Attribute Model in Understanding Amnesia

Naomi J. Goodrich-Hunsaker and Ramona O. Hopkins

In the vanguard of learning and memory research, Raymond P. Kesner embraced translational research early in his career. He was not only interested in elucidating the neural mechanisms underlying memory in rats, but his research also ultimately strove to test whether parallel mnemonic processes operate in rodents and humans. Although it would be virtually impossible to list all the significant findings Kesner has made throughout his successful career, this chapter will focus specifically on those key contributions where the theories and ideas elucidated in his studies in animals have been directly useful in furthering our understanding of human amnesia. Besides providing scientific findings, Kesner has foremost been a remarkable mentor. Some of the new research conducted by individuals that he has mentored will also be discussed, as Kesner's work will continue to impact the field of learning and memory for years to come through the research of his protégés.

N. J. Goodrich-Hunsaker (✉)
Department of Psychiatry and Behavioral Sciences, MIND Institute, University of California, Davis Medical Center, Sacramento, CA, USA
e-mail: njhunsaker@icloud.com

Department of Psychology, Brigham Young University, Provo, UT, USA

R. O. Hopkins
Psychology Department and Neuroscience Center, Brigham Young University, 1022 SWKT, Provo, UT 84062, USA
e-mail: mona_hopkins@byu.edu

Department of Medicine, Pulmonary and Critical Care Division, Intermountain Medical Center, Murray, UT, USA

P. A. Jackson et al. (eds.), *The Neurobiological Basis of Memory*,
DOI 10.1007/978-3-319-15759-7_6

The Attribute Model

At the most basic level, memory involves the encoding of information for storage, retaining the encoded information, and later recalling the stored information. More realistically though, it is known that memory is extremely complex in terms of the kind of information that is represented in the brain, the associated memory processes, and the distribution of memory across a variety of neural systems. In an attempt to capture the complexity of memory, Kesner proposed a comprehensive neurobiologically based model of memory, which has fundamentally guided most of his research throughout his career. The attribute model consists of a set of three fundamental memory systems, the attributes or domains that represent specific types of memory representations (e.g., space, time, etc.), the associated neuroanatomical substrates, and a variety of mnemonic processes (i.e., pattern separation). Partially based on Atkinson–Shiffrin's multi-store model of memory, Kesner proposed three systems of memory: event-based memory, knowledge-based memory, and rule-based memory. First, event-based memory is defined as a limited and fleeting memory representation of the world based on incoming sensory information. From moment to moment, sensory information (e.g., sight, sound, smells, touch, taste, etc.) floods in through sensory receptors and is further processed by the nervous system. All of this sensory information has the potential to become part of a memory representation. The second system, knowledge-based memory, is a more lasting memory representation. At this stage, the memory representation is about more than just remembering; it is about the strategic control of encoding and rehearsal, which depends heavily upon attention mechanisms and cognitive control. Knowledge-based memory also involves transferring the memory representation to permanent "storage" (i.e., long-term memory). Third, rule-based memory assumes that a mental representation has been processed (event-based memory) and integrated as a permanent memory representation (knowledge-based memory). Rule-based memory uses processes, strategies, and rules to maintain and manipulate information during decision-making and action (e.g., working memory).

Kesner proposes that each of these memory systems consist of the same domains or attributes of memory. The most important attributes of memory are: space, time, response, sensory-perception, affect, and language (in humans). This chapter will focus primarily on the spatial attribute, and we will highlight several key studies that show parallel spatial processes in rodents and humans on similar tasks, namely the water maze (Morris et al. 1982) and the radial arm maze (Olton and Samuelson 1976). It was through Kesner's theory and ideas and mentorship that these studies were made possible.

Tolman (1948) first described the cognitive map theory, which proposes a spatial coordinate system that systematically represents both the physical environment and individual's location within the environment. Since the discovery of place cells in the hippocampus of rats (O'Keefe and Dostrovsky 1971), animal models have been extensively used to study the underlying processes associated with spatial memory. Subsequent lesion studies have compared navigation using a cognitive map with navigation using a combination of cues and responses. One task, the Morris water maze, has provided a definitive paradigm establishing that hippocampal lesions produce a deficit in place learning but not in cue-response learning in rodents (Morris et al. 1982).

The water maze consists of a large pool of opaque water with a platform on which rats can stand to escape from the water. Several versions of the water maze are used including the visible platform, submerged hidden platform, and landmark platform maze. In the visible platform water maze, the platform is above the water. In the hidden platform water maze, the platform is submerged just below the surface of the opaque water. The location of the hidden platform is static across trials. In the landmark platform water maze, the hidden platform location is marked by a cue such as a light bulb hanging directly over the platform (i.e., the location of the hidden platform). The landmark platform changes between trials, but an object or cue always demarcates the platform location. Lesion studies in rodents contrasted the use of a spatial map for navigation to a hidden platform with navigation based upon specific responses to a hidden platform marked by a single cue (i.e., landmark platform). Rodents with hippocampal damage have deficits in place learning of the hidden platform location, but are not impaired for stimulus-response learning of the location of the visible landmark platform.

The adaptation of animal tasks into analogous tasks to evaluate human mnemonic processes has been slow due to the difficulty identifying individuals who have damage or lesions restricted to specific brain structures such as the hippocampus and lack of a suitable water maze for humans. It was many decades after the Morris water maze was first used in rodents that a virtual version of the maze was designed and the task was administered to amnesic participants with focal, bilateral hippocampal damage (Goodrich-Hunsaker et al. 2010). The amnesic participants were impaired in navigating to the hidden platform location when there was no nearby landmark available. The amnesic participants spent more time searching for the platform and took longer paths to reach the platform than did the control participants. These amnesic participants with hippocampal damage showed the same pattern of deficits as hippocampal-damaged rodents showed on the visible and hidden platform trials of the Morris water maze.

In addition to the Morris water maze, the radial arm maze is a well-researched paradigm used to assess spatial memory in rodents (Olton and Samuelson 1976). The radial arm maze consists of a center platform with several identical evenly distributed arms radiating outward. Distal, salient cues are placed at the ends of each arm, which provide spatial context for the maze environment. At the end of each arm, there is a food well that can be baited with a reward (e.g., Froot Loop, sucrose water, chocolate milk, etc.). In the traditional radial arm maze with eight arms, four of the arms are randomly baited and four arms are non-baited. Through trial and error, rodents learn to retrieve each of the four rewards without entering the non-baited arms or reentering a previously rewarded arm. Errors are categorized as reference memory errors (e.g., entering the non-baited arms) or working memory errors (e.g., reentering rewarded arms). The spatial role of the hippocampus in the radial arm maze is well supported by studies that find that rats with hippocampal lesions are impaired on the maze making more reference memory errors relative to control animals (see Barnes 1988; Jarrard 1991 for review). However, several other studies find that rats with hippocampal lesions make both working and reference memory errors relative to control animals (Jarrard 1993; Lee and Kesner 2003).

Again, it was decades later that a virtual reality version of the radial arm maze was developed and administered to amnesic participants with hippocampal damage (same participants as those tested in the Morris water maze). Goodrich–Hunsaker and Hopkins (2010) found that amnesic subjects with focal, bilateral hippocampal damage had significant spatial memory deficits in a virtual radial arm maze. Specifically, amnesic subjects consistently spent more time searching for and took longer paths to reach the four rewarded arms than did the control subjects. Amnesic subjects also made significantly more reference memory errors by entering into non-rewarded arms and working memory errors by reentering into an arm where they had previously entered. As an editorial by Sutherland states, these findings show that the "… human hippocampus plays a very similar role in spatial process as the one that is well described in rodents. As such it strengthens the notion that a considerable proportion of the multilevel analysis of neurobiology of spatial memory in the rat will apply in a straightforward manner to humans (Sutherland 2010)."

The 1980s

Raymond Kesner was already exploring the similarities of spatial and temporal attributes of memory in animals and humans early in his career. Kesner et al. (1987, 1989) assessed memory for a list of items at varying spatial locations in rats with small or large lesions of the medial septum, dorsal hippocampus, or nucleus basalis and comparing those studies with results in patients with Alzheimer's disease. By administering analogous tasks to both rodents and humans, not only were Kesner and colleagues able to determine possible neuroanatomical systems associated with impaired memory at the different stages of dementia in Alzheimer's disease, but also they showed that animal models could successfully be used to better understand the neural underpinnings of memory deficits associated with Alzheimer's disease.

For the animal task, Kesner et al. (1987, 1989) used the standard 8-arm maze apparatus. During the study phase, animals were presented a sequence of five arms. The sequence of arms was presented one arm at a time. During the test phase, animals were given a choice between an arm that had been part of the sequence and an arm that was not included in the sequence. Animals received a reward when they selected the arm that was one of the five arms in the sequence. After animals had been trained on the task, they were given small or large lesions of medial septum, small or large lesions of the dorsal hippocampus, small or large nucleus basalis lesions, or a sham operation. After animals recovered from surgery, they were retested. Results are discussed below.

In this same set of studies a similar paradigm was also administered to college students, normal elderly individuals with no dementia, elderly individuals with early

Alzheimer's dementia, and elderly individuals with moderate Alzheimer's dementia (Kesner et al. 1987, 1989). Instead of an 8-arm maze, participants were shown a 4-by-4 grid on a piece of paper. During the study phase, an X appeared in one of the 16 squares. Participants were instructed to pay attention to the locations of the Xs. In the test phase participants were presented with two Xs (one location presented in the list and a novel location) and were asked to determine which location they saw in the study phase (item recognition memory).

Overall results from these two studies (Kesner et al. 1987, 1989) show that the animals with small medial septum or dorsal hippocampus lesions and patients with mild Alzheimer's dementia were able to remember the spatial locations presented later in the sequence (recency), but had impaired memory for locations that occurred early in the sequence. Further, animals with large medial septum or dorsal hippocampus lesions had memory impairments for all items of the sequence. Patients with moderate Alzheimer's dementia were also impaired for all items. The above data show that not only could rodents serve as a valid model for memory impairments in individuals who had Alzheimer's dementia, but also that deficits in the rodent model scaled with lesion size thus providing insight into the progressive neurodegenerative effects associated with Alzheimer's disease (increasing memory impairment with disease progression) (Kesner et al. 1987, 1989).

The 1990s

Up to this point, Kesner's human memory research primarily involved individuals with Alzheimer's dementia who not only had damage to the hippocampus but also had damage to other neural regions (e.g., medial temporal lobe). As such, a human model of memory in which damage was restricted to the hippocampus was needed. In the 1990s, Kesner's lab (Hopkins et al. 1995a, b) published several studies that assessed memory for temporal information including memory for temporal distances in amnesic participants with hypoxic brain injury based on studies in rodents (Jackson et al. 1998). Research suggests that memory for temporal information is better when more items (temporal distance) occur between any two to be remembered items, and memory declines the closer the items are to each other on a list. Using the identical task to that used in individuals with Alzheimer's disease, amnesic subjects with selective hippocampal damage were impaired for remembering spatial locations as well as the order in which the locations were presented, but had intact recency (able to remember the last items on the list) for spatial locations compared to healthy matched controls. Similar findings were found for words, pictures, and abstract designs (Hopkins et al. 1995b). Thus began a series of momentous studies, which were important because memory deficits were finally being assessed in amnesic individuals with focal bilateral hippocampal damage with no known cellular damage to the parahippocampal cortex or other temporal lobe regions.

In another study, amnesic participants and controls were presented with a list of eight spatial locations (Xs) on a grid of 16 possible locations, which assessed memory for the temporal order of a sequence of the spatial locations (Xs on the grid). This task is slightly different from the above sequence-learning task and is based on a similar paradigm administered to rats (Chiba et al. 1994). During the study phase a series of random (novel) sequences of eight Xs appeared on a computer screen for a period of 5 s each. Participants were instructed to pay attention to the locations of the Xs and to the order in which they occurred. In the test phase participants were presented with two Xs that were in sequence and they were asked to determine which X occurred earlier in the study phase. Temporal distances (the number of items in the study phase that occurred between the two test items) of 0, 2, 4, and 6 were assessed. Relative to controls, the amnesic participants were impaired for all temporal distances, with some improvement at the greatest distance (Hopkins et al. 1995b). These results were similar to animals with dorsal hippocampal lesions (Chiba et al. 1994) that found chance performance for all temporal distances with marginal improvements for the greatest temporal distance. The above data indicate that the hippocampus in both rats and humans is essential for memory for temporal information and uses similar processes in both species.

Another way to evaluate memory for temporal information is to assess memory for duration using a delayed conditional discrimination procedure (Jackson et al. 1998). Rats learned that a black rectangle stimulus visible for 2 s indicated a positive (go) reinforcement trial for one object (a ball), but no (no-go) reinforcement trial for a different object (bottle). Whereas a longer duration of presentation of the black rectangle (8 s), now indicated a "no-go" trial for the ball, but indicated a reinforced "go" trial for the bottle. After rats learned the discrimination, they received large (dorsal and ventral) lesions of the hippocampus or cortical control lesions. The rodents with hippocampal lesions were impaired on this task compared to rats with cortical control lesions (Jackson et al. 1998). Memory for duration was also assessed in amnesic participants with hippocampal damage and healthy control participants on a task that varied the duration of exposure of an object for a delay of 1, 4, 8, 12, or 16 s. Amnesic participant's memory for duration was impaired at all delays except 1 s compared to normal controls (Kesner and Hopkins 2001). The studies illustrated above show that both rodents and humans with hippocampal damage exhibit impairments for processing temporal information including impairments in temporal order and duration.

The 2000s

Kesner and colleagues (for review see, Kesner et al. 2004; Rolls and Kesner 2006) have demonstrated experimentally using behavioral tasks in rodents that the hippocampus facilitates the rapid association of information coming from multiple neocortical regions in forming new memories, storing memories independently from each other, subsequent retrieval of that memory from partial cues, and flexibly applying stored memories to novel situations through mnemonic processes known as *pattern separation* and *pattern completion*. Pattern separation separately encodes features within an event in order to maintain the overall composition and organizes events in terms of their occurrence across time so that spatial events are separated from each other due to attenuation of spatial interference. If pattern separation mediated by the hippocampus were not possible, then interference during encoding of new information would reduce the ability to discriminate between present and past experiences in memory. Historically, it has been shown that pattern separation processes can occur with any incoming sensory/perceptual information (Marr 1971; O'Reilly and Rudy 2000, 2001; O'Reilly et al. 2011). The study of pattern separation processes has been mostly limited to spatial domains in rats and mice. For example, Gilbert et al. (2001) tested rats using a paradigm that measured short-term memory for spatial locations. The task was designed to assess rats' ability to spatially discriminate two spatial locations at various levels of difficulty (i.e., interference due to small to large distance between objects). During the sample phase, an object was placed over a baited food well in a specific spatial location on the cheeseboard. Rats were trained to displace the object to receive a reward. For the test phase, the same object was again placed over the baited food well (correct choice), but now there was a second identical object placed over a different non-baited food well (incorrect foil object). This second object was between 15 and 105 cm away from the correct object. Rats with the hippocampus, and specifically dentate gyrus lesions, were impaired at discriminating the original spatial location from a foil location when the spatial distance between the two locations was less than 150 cm. Results show that when there is a greater degree of similarity or overlap between items to be remembered that needs to be overcome, the hippocampus achieves this via a pattern separation processes.

Kesner also assessed spatial pattern separation processes in humans (Kesner and Hopkins 2006). The above task in rats was adapted and administered on a computer to amnesic participants with focal, bilateral hippocampal damage due to hypoxia. For the task, participants were presented with a dot on a computer screen and asked to remember the location of the dot. The dot was then followed by a delay of 5, 10, 20, or 30 s. During the test phase, two identical dots appeared on the screen, one in the correct location and a foil in an incorrect location. Participants were instructed to choose the dot in the original, correct location. The distance between the original correct location and foil locations was 2.75, 1.75, or 0.75 cm. Like the rodents with dorsal hippocampus lesions and specifically dorsal dentate gyrus lesions, amnesic participants with focal, bilateral hippocampal damage showed the greatest impairment compared to controls when there was a lot of spatial interference to overcome because the correct spatial location and the foil location were very close together (Kesner and Hopkins 2006). As the spatial interference decreased making the task of determining the correct location over the foil location easier, the hypoxic participants were less impaired. Again, these results replicate the pattern separation findings in rodents with dorsal hippocampus and dorsal dentate gyrus lesions.

Kesner was also interested in understanding *how* spatial pattern separation processes contributed to the formation of spatial representations. In order to study two key spatial representation features, Goodrich–Hunsaker et al. (2005, 2008) designed two new tasks, the "metric" (also called "coordinate") task, and the "topological" (also called "categorical") task. The metric task determined rodents' ability to discriminate very small changes in the spatial relationships between stimuli (i.e., two identical objects were either moved closer or further apart) without regard to the identity of the objects. During the sample phase of the task, two objects were placed on a platform (i.e., the "cheeseboard apparatus"). Rodents were given time to explore these objects. During the test phase, these two objects were either moved closer or further apart. Animals were again given time to reexplore the objects. Increased reexploration of these objects during the test phase was used as an indicator that the animals detected the change in object distance. Whether using the term *coordinate* (Kosslyn 1987) or *metric* (Gallistel 1990; Goodrich–Hunsaker et al. 2005, 2008; Poucet 1993), the aforementioned spatial relationship refers to the precise spatial location of items in a context and can be expressed in terms of distances and angles between objects. A coordinate representation is useful for generating an accurate mental representation of one's environment. The topological task measured rodents' ability to process the overall configuration of the items in a context and their general relationships to each other. During the sample phase of the task, four objects were placed in a square configuration on a cheeseboard. Rodents were given time to explore these objects. During the test phase, two of the four objects were transposed. Animals were again given time to reexplore the objects. Increase reexploration of the two objects that were transposed was used as an indicator that the animals detected the change. This second spatial relationship is referred to as *categorical* information in the human literature (Kosslyn 1987) and is called *topological* information in the animal literature (Gallistel 1990; Goodrich-Hunsaker et al. 2005, 2008; Poucet 1993). In the Goodrich–Hunsaker et al. (2005) study described above, rats with lesions restricted to the dorsal hippocampus did not detect the change in distance between objects, but were capable of detecting a transposition of objects. On the other hand, rats with lesions restricted to the parietal cortex were unable to detect the transposition of objects, but showed no deficits in detecting when the objects were moved closer together. This double dissociation shows that in rats the hippocampus mediates fine spatial memory ("metric/coordinate memory"), whereas the parietal cortex underlies topological spatial information processing ("topological/categorical memory"). The results of this study were further expanded in subsequent lesion studies to implicate the dentate gyrus, and not CA3 or CA1, in high-resolution spatial information processing via pattern separation (Goodrich–Hunsaker et al. 2008).

A similar paradigm was carried out in amnesic participants with focal, bilateral hippocampal atrophy due to hypoxic brain injury and control participants (Goodrich-Hunsaker 2009). During the study phase, participants were shown naturalistic scenes with two foreground objects. The two foreground objects (see Fig. 6.1) were placed such that there were four possible distances between them (10-, 12-, 14-, and 16-cm). Participants were instructed to remember the precise location of objects in the picture. During the test phase, participants were presented with the same naturalistic scene and were instructed to place the two foreground objects in their original locations (i.e., participants had to reconstruct the scene). Topological/categorical errors were recorded if a participant transposed the two objects

Fig. 6.1 Scene duplication task. Displayed is one of the 24 possible scenes. All scenes were in full color. Participants saw this scene for 5 s. After a 30 s inter-stimulus interval, participants were presented with the background and given the space shuttle and astronaut to place in their correct locations

during the recreation of the scene. In order to assess participant's spatial resolution of metric/ coordinate spatial relationships, the distance between the two foreground objects were measured and compared against the actual distance of 10-, 12-, 14-, or 16-cm. The hypoxic participants were able to recreate the overall schematic of the scenes. They never transposed the two foreground objects. These data suggest that categorical/topological spatial information was intact in hypoxic participants. In terms of coordinate/metric spatial information processing, the hypoxic participants consistently placed the two foreground objects approximately 15-cm apart, even though the distance between the two foreground objects varied over trials from 10-, 12-, 14-, or 16-cm. On the other hand, normal control participants showed no impairments in processing the subtle differences in distances between the two objects. When the two objects were originally 10-, 12-, 14-, or 16-cm apart, normal control participants accurately placed the two objects 10-, 12-, 14-, and 16-cm apart, respectively. The amnesic participants placed the two objects in nearly the same spatial location across all trials and therefore displayed similar low-resolution spatial representation as hippocampal lesioned rats (Goodrich-Hunsaker et al. 2005). By comparing cross-species performance on analogous tasks, these data support the idea that the hippocampus may be critical for separating highly similar events or objects in both rodents and humans.

More recently, Baumann et al. (2012) administered a comparable virtual reality version of the task. Participants saw a central landmark and were asked to remember either (1) how faraway the target item was from the central landmark (coordinate/ metric) or (2) the quadrant the target item was located in reference to the central landmark (categorical/topological). Using functional magnetic resonance imaging (fMRI), Baumann et al. (2012) found that the parietal lobe, in general, responded more strongly when participants were encoding the quadrant location (categorical/topological) and the hippocampal formation responded more strongly when participants were encoding the precise distance between the two objects (coordinate/metric). These results support the previous animal (Goodrich-Hunsaker et al. 2005, 2008) and human findings (Goodrich-Hunsaker 2009).

Another aspect of studying the features of a mental representation includes the ability to associate which objects or parts are associated with locations. It is not enough to know precise locations or the general aspect of shape, but it is also necessary to be able to pair the object or item with each location. Previous research has shown that the hippocampus (Cornu Ammonis fields, dentate gyrus, and subiculum) supports mechanisms of associative learning and memory that bind features connected with an event into an integrated memory trace by linking neuronal activation from multiple sensory modalities: sight, sound, smell, etc. (Brown and Aggleton 2001; Davachi et al. 2003; Eichenbaum 2000; Kesner et al. 2000; O'Reilly and Rudy 2001). In rodents, the Kesner lab has shown that hippocampal damage impairs acquisition of object–place associations (Gilbert and Kesner 2004) and odor–place associations (Gilbert and Kesner 2002). Until the Goodrich–Hunsaker et al. (2009) study, information regarding the contribution of the human hippocampus in associative memory was incomplete, as studies had used item pairs that involved only pictorial, verbal, and/or spatial information. Based on the rodent task design from Kesner's lab (Gilbert and Kesner 2002), amnesic and healthy comparison participants were tested on an odor–place associative task, an odor recognition task, and a place recognition task (Goodrich-Hunsaker et al. 2009). The odor–place associative task required subjects to associate 6 odors with 6 spatial locations on a board. The recognition tasks required subjects to identify the 6 odors and the 6 locations that were presented during the associative task. Amnesic participants were impaired for odor–place memory and place recognition, but not odor recognition compared with non-amnesic comparison participants. These results suggest that the human hippocampus is necessary for odor–place associative memory and spatial recognition memory. These data provide support for the idea that odor–place associative memory is mediated by the hippocampus in both humans and rodents.

Into the Future

Kesner's Attribute Model of memory has widespread applicability in understanding the neurobiological processes underlying learning and memory. Kesner's theories will continue to influence this field for decades to come. Although, this review has focused predominately on the attribute of space, Kesner's Attribute Model of memory proposes several additional attributes: response, sensory-perception, affect, and language (in humans). In fact, tasks in humans have assessed item and order recognition memory for words, sentences, geographical locations, and abstract pictures (Hopkins et al. 1995b; Johnson and Kesner 1997; Madsen and Kesner 1995) not described here, as there are not analogous tasks in rodents.

While we have only discussed a small part of Raymond Kesner's contributions to memory, we have focused on some important comparative contributions of the role of the hippocampus in memory in both rodents and humans. This chapter highlights several landmark contributions of Raymond Kesner to the field of memory. (1) The Attribute Model of memory has guided his research and provided testable ideas for a plethora of innovative studies. (2) Kesner developed innovative tasks for a numbers of attributes and memory processes. (3) Kesner has developed analogous tasks for humans based on the tasks developed for animals and vice versa. (4) He focused on memory processes, such as pattern separation and pattern completion using a variety of tasks. And importantly (5) Raymond Kesner is an outstanding mentor who has left the field a legacy of scientists that continue to expand and elaborate on the research and ideas that he brought to life for them.

References

Barnes, C. A. (1988). Spatial learning and memory processes: The search for their neurobiological mechanisms in the rat. *Trends in Neurosciences, 11*(4), 163–169. doi:10.1016/0166-2236(88)90143-9.

Baumann, O., Chan, E., & Mattingley, J. B. (2012). Distinct neural networks underlie encoding of categorical versus coordinate spatial relations during active navigation. *NeuroImage, 60*(3), 1630–1637. doi:10.1016/j.neuroimage.2012.01.089.

Brown, M. W., & Aggleton, J. P. (2001). Recognition memory: What are the roles of the perirhinal cortex and hippocampus? *Nature reviews. Neuroscience, 2*(1), 51–61. doi:10.1038/35049064.

Chiba, A. A., Kesner, R. P., & Reynolds, A. M. (1994). Memory for spatial location as a function of temporal lag in rats: Role of hippocampus and medial prefrontal cortex. *Behav Neural Biol, 61*(2), 123–131. doi:10.1016/S0163-1047(05)80065-2.

Davachi, L., Mitchell, J. P., & Wagner, A. D. (2003). Multiple routes to memory: Distinct medial temporal lobe processes build item and source memories. *Proceedings of the National Academy of Sciences of the United States of America, 100*(4), 2157–2162. doi:10.1073/pnas.0337195100.

Eichenbaum, H. (2000). A cortical-hippocampal system for declarative memory. *Nature reviews. Neuroscience, 1*(1), 41–50. doi:10.1038/35036213.

Gallistel, C. R. (1990). Representations in animal cognition: An introduction. *Cognition, 37*(1–2), 1–22. doi:10.1016/0010-0277(90)90016-D.

Gilbert, P. E., & Kesner, R. P. (2002). Role of the rodent hippocampus in paired-associate learning involving associations between a stimulus and a spatial location. *Behavioral neuroscience, 116*(1), 63–71. doi:10.1037/0735-7044.116.1.63.

Gilbert, P. E., & Kesner, R. P. (2004). Memory for objects and their locations: The role of the hippocampus in retention of object-place associations. *Neurobiology of learning and memory, 81*(1), 39–45. doi:10.1016/S1074-7427(03)00069-8.

Gilbert, P. E., Kesner, R. P., & Lee, I. (2001). Dissociating hippocampal subregions: Double dissociation between dentate gyrus and CA1. *Hippocampus, 11*(6), 626–636. doi:10.1002/hipo.1077.

Goodrich-Hunsaker, N. J. (2009). Human spatial memory: Features and processes. (Doctoral Dissertation). Retrieved from ProQuest Dissertations and Theses. (3353728)

Goodrich-Hunsaker, N. J., & Hopkins, R. O. (2010). Spatial memory deficits in a virtual radial arm maze in amnesic participants with hippocampal damage. *Behavioral neuroscience, 124*(3), 405–413. doi:10.1037/a0019193.

Goodrich-Hunsaker, N. J., Hunsaker, M. R., & Kesner, R. P. (2005). Dissociating the role of the parietal cortex and dorsal hippocampus for spatial information processing. *Behavioral neuroscience, 119*(5), 1307–1315. doi:10.1037/0735-7044.119.5.1307.

Goodrich-Hunsaker, N. J., Hunsaker, M. R., & Kesner, R. P. (2008). The interactions and dissociations of the dorsal hippocampus subregions: How the dentate gyrus, CA3, and CA1 process spatial information. *Behavioral neuroscience, 122*(1), 16–26. doi:10.1037/0735-7044.122.1.16.

Goodrich-Hunsaker, N. J., Gilbert, P. E., & Hopkins, R. O. (2009). The role of the human hippocampus in odor-place associative memory. *Chemical senses, 34*(6), 513–521. doi:10.1093/chemse/bjp026.

Goodrich-Hunsaker, N. J., Livingstone, S. A., Skelton, R. W., & Hopkins, R. O. (2010). Spatial deficits in a virtual water maze in amnesic participants with hippocampal damage. *Hippocampus, 20*(4), 481–491. doi:10.1002/hipo.20651.

Hopkins, R. O., Kesner, R. P., & Goldstein, M. (1995b). Item and order recognition memory in subjects with hypoxic brain injury. *Brain and cognition, 27*(2), 180–201. doi:10.1006/brcg.1995.1016.

Hopkins, R. O., Kesner, R. P., & Goldstein, M. (1995b). Memory for novel and familiar spatial and linguistic temporal distance information in hypoxic subjects. *Journal of the International Neuropsychological Society: JINS, 1*(5), 454–468. doi:10.1017/S1355617700000552.

Jackson, P. A., Kesner, R. P., & Amann, K. (1998). Memory for duration: Role of hippocampus and medial prefrontal cortex. *Neurobiology of learning and memory, 70*(3), 328–348. doi:10.1006/nlme.1998.3859.

Jarrard, L. E. (1991). On the neural bases of the spatial mapping system: Hippocampus vs. hippocampal formation. *Hippocampus, 1*(3), 236–239. doi:10.1002/hipo.450010304.

Jarrard, L. E. (1993). On the role of the hippocampus in learning and memory in the rat. *Behav Neural Biol, 60*(1), 9–26. doi:10.1016/0163-1047(93)90664-4.

Johnson, D. L., & Kesner, R. P. (1997). Comparison of temporal order memory in early and middle stage Alzheimer's disease. *Journal of Clinical and Experimental Neuropsychology: Official Journal of The International Neuropsychological Society, 19*(1), 83–100. doi:10.1080/01688639708403839.

Kesner, R. P., & Hopkins, R. O. (2001). Short-term memory for duration and distance in humans: Role of the hippocampus. *Neuropsychology, 15*(1), 58–68. doi:10.1037/0894-4105.15.1.58.

Kesner, R. P., & Hopkins, R. O. (2006). Mnemonic functions of the hippocampus: A comparison between animals and humans. *Biological Psychology, 73*(1), 3–18. doi:10.1016/j.biopsycho.2006.01.004.

Kesner, R. P., Adelstein, T., & Crutcher, K. A. (1987). Rats with nucleus basalis magnocellularis lesions mimic mnemonic symptomatology observed in patients with dementia of the Alzheimer's type. *Behavioral neuroscience, 101*(4), 451–456. doi:10.1037/0735-7044.101.4.451.

Kesner, R. P., Adelstein, T. B., & Crutcher, K. A. (1989). Equivalent spatial location memory deficits in rats with medial septum or hippocampal formation lesions and patients with dementia of the Alzheimer's type. *Brain and cognition, 9*(2), 289–300. doi:10.1016/0278-2626(89)90038-9.

Kesner, R. P., Gilbert, P. E., & Wallenstein, G. V. (2000). Testing neural network models of memory with behavioral experiments. *Current opinion in neurobiology, 10*(2), 260–265. doi:10.1016/S0959-4388(00)00067-2.

Kesner, R. P., Lee, I., & Gilbert, P. (2004). A behavioral assessment of hippocampal function based on a subregional analysis. *Reviews in the neurosciences, 15*(5), 333–351.

Kosslyn, S. M. (1987). Seeing and imagining in the cerebral hemispheres: A computational approach. *Psychological review, 94*(2), 148–175. doi:10.1037/0033-295X.94.2.148.

Lee, I., & Kesner, R. P. (2003). Time-dependent relationship between the dorsal hippocampus and the prefrontal cortex in spatial memory. *The Journal of Neuroscience: The Official Journal of The Society for Neuroscience, 23*(4), 1517–1523..

Madsen, J., & Kesner, R. P. (1995). The temporal-distance effect in subjects with dementia of the Alzheimer type. *Alzheimer Disease and Associated Disorders, 9*(2), 94–100. doi:10.1097/00002093-199509020-00006.

Marr, D. (1971). Simple memory: A theory for archicortex. *Philosophical Transactions of the Royal Society of London. Series B, Biological Sciences, 262*(841), 23–81. doi:10.1098/rstb.1971.0078.

Morris, R. G., Garrud, P., Rawlins, J. N., & O'Keefe, J. (1982). Place navigation impaired in rats with hippocampal lesions. *Nature, 297*(5868), 681–683. doi:10.1038/297681a0.

O'Keefe, J., & Dostrovsky, J. (1971). The hippocampus as a spatial map. Preliminary evidence from unit activity in the freely-moving rat. *Brain Research, 34*(1), 171–175. doi:10.1016/0006-8993(71)90358-1.

O'Reilly, R. C., & Rudy, J. W. (2000). Computational principles of learning in the neocortex and hippocampus. *Hippocampus, 10*(4), 389–397. doi:10.1002/1098-1063(2000)10:4<389::AID-HIPO5>3.0.CO;2-P.

O'Reilly, R. C., & Rudy, J. W. (2001). Conjunctive representations in learning and memory: Principles of cortical and hippocampal function. *Psychological review, 108*(2), 311–345. doi:10.1037/0033-295X.108.2.311.

O'Reilly, R. C., Bhattacharyya, R., Howard, M. D., & Ketz, N. (2011). Complementary learning systems. *Cognitive Science*. doi:10.1111/j.1551-6709.2011.01214.x.

Olton, D. S., & Samuelson, R. J. (1976). Remembrance of places passed: Spatial memory in rats. *Journal of Experimental Psychology: Animal Behavior Processes, 2*(2), 97–116. doi:10.1037/0097-7403.2.2.97.

Poucet, B. (1993). Spatial cognitive maps in animals: New hypotheses on their structure and neural mechanisms. *Psychological review, 100*(2), 163–182. doi:10.1037/0033-295X.100.2.163.

Rolls, E. T., & Kesner, R. P. (2006). A computational theory of hippocampal function, and empirical tests of the theory. *Progress in neurobiology, 79*(1), 1–48. doi:10.1016/j.pneurobio.2006.04.005.

Sutherland, R. J. (2010). Spatial memory: A Rosetta stone for rat and human hippocampal discourse: Theoretical comment on Goodrich-Hunsaker and Hopkins (2010). *Behavioral neuroscience, 124*(3), 434–436. doi:10.1037/a0019614.

Tolman, E. C. (1948). Cognitive maps in rats and men. *Psychological review, 55*(4), 189–208. doi:10.1037/h0061626.

Chapter 7
Resolving Interference: The Role of the Human Hippocampus in Pattern Separation

C. Brock Kirwan and Michelle I. Nash

Interference in Memory

The medial temporal lobe (MTL) is critically involved in memory for facts and events (Kesner 2009; Squire et al. 2004). Damage to MTL structures, including the hippocampus and surrounding cortex (perirhinal, entorhinal, and parahippocampal), results in profound anterograde amnesia and temporally graded retrograde amnesia (Scoville and Milner 1957; Squire et al. 1989). Computational models of MTL function commonly posit that the MTL cortex establishes representations of statistical regularities in the environment through repeated exposures whereas the hippocampus is capable of establishing rapid, distinct, and nonoverlapping representations (O'Reilly and Rudy 2000, 2001). Several computational models of hippocampal function posit that sparse connections within the hippocampus allow for the establishment of distinct memory representations through a process known as pattern separation (McClelland et al. 1995; Norman and O'Reilly 2003; O'Reilly and Rudy 2001; Rolls and Treves 1998). This ability to establish nonoverlapping representations is essential for effective episodic memory (Tulving 2002) and allows the system to avoid "catastrophic interference" where retrieving one memory representation cues the retrieval of many unrelated memory representations (McClelland et al. 1995; McCloskey and Cohen 1989).

Pattern completion is the complementary computational process to pattern separation whereby previously stored representations are retrieved when given a noisy or degraded cue. Pattern separation and pattern completion are not mutually exclu-

The views expressed in this article are those of the authors and do not reflect the official policy or position of the United States Air Force, Department of Defense, or the U.S. Government.

C. B. Kirwan (✉) · M. I. Nash
Department of Psychology and Neuroscience Center, Brigham Young University, 1001 SWKT, Provo, UT 84602, USA
e-mail: kirwan@byu.edu

M. I. Nash
e-mail: michelle.nash13@gmail.com

P. A. Jackson et al. (eds.), *The Neurobiological Basis of Memory*,
DOI 10.1007/978-3-319-15759-7_7

sive processes that occur in distinct neuroanatomical locations. Indeed, computational principles predict that they necessarily must occur in the same brain region to be useful. However, different brain regions are differentially biased toward either pattern separation or pattern completion. The hippocampus, and the dentate gyrus (DG) in particular, is proposed to be biased toward pattern separation whereas later stages of the hippocampus, including the CA3 and CA1 in addition to the MTL cortex, are proposed to be biased toward pattern completion.

Effective pattern separation allows one to resolve interference between potentially overlapping memories. Day-to-day events have a great deal of overlapping information (such as location, people present, or time of day), but an effective episodic memory system must be able to resolve this interference in establishing unique representations for each new event that are later able to be retrieved individually. For example, the episode of eating breakfast on any particular morning is encoded separately from other similar episodes despite the potential overlap in location, time of day, and actors present. Interference between information acquired at different times is thought to be a major source of forgetting (Crowder 1976; Keppel 1984). Consequently, there is a long tradition of studying the effects of interference on memory. An early account of interference rested on the evidence for consolidation of memories over time, and has thus been termed consolidation theory. Müller and Pilzecker (1900) observed that a list of items was better remembered if learning was followed by a quiescent period before testing than if followed by another period of mental activity, such as learning another list. This phenomenon, when new information interferes with the retrieval of older information, is termed retroactive interference (RI). A classic source of evidence for consolidation theory comes from the studies by Jenkins and Dallenbach (1924), in which participants learned a list of words either early in the day or late in the evening. Testing occurred 8 h later, after either a full night's sleep or 8 h of normal daily activity. Retention was better for the list following sleep, indicating that the intervening activity during the day interfered with the material learned in the morning.

It seems intuitive that greater RI would be achieved when the intervening material is more similar to the original material. However, when the intervening material is exactly similar (i.e., identical) to the original material, there is no RI (performance improves instead). Thus, there should be a U-shaped pattern of performance, as performance is unaffected (or even benefited) by dissimilar or identical material, but is harmed by similar material. This theory, known as the Skaggs–Robinson law (Robinson, 1927; Skaggs 1925), lost appeal because of failure to demonstrate the full theoretical curve in a single experiment (Slamecka and Ceraso 1977).

Interference theory was heavily influenced by Underwood's (1957) review of proactive interference (PI). In the case of PI, previously learned material is detrimental to the memory performance on a subsequently learned list. Underwood demonstrated by a review of the literature that over a 24-h period, participants showed a 75 % reduction in memory for a verbal list, and that most of this reduction could be accounted for by previous, massed learning in the laboratory setting. He went so far as to observe that RI probably had little to do with this forgetting. Thus, it seemed plausible that forgetting would be explained in terms of PI. However, critical for the phenomenon of PI is the spontaneous recovery of the previously learned material, and PI offers no mechanism for this spontaneous recovery other than the simple passage of time. McGeoch's (1932) famous objection to decay theory, that time per se does not cause memories to fade any more than it causes rust to form, also applies to spontaneous recovery (Crowder 1976). Furthermore, PI could not account for the findings of Jenkins and Dallenbach (1924) and others (e.g., Ekstrand 1967) regarding the protective nature of sleep on memory. Proactive interference, and interference theory in general, has since fallen out of favor as a research topic, (for review, see Wixted 2004), but interference is nevertheless, still regarded as "a primary source of forgetting in explicit memory" (Lustig and Hasher 2001, p. 618).

Although interference theory today is not as heavily researched as it was for the first three quarters of the last century, it has not disappeared completely. Recent theoretical and empirical work (Blank 2005) has continued to test the predictions of interference theory. Of particular interest are neuropsychological and neuroimaging studies that bear on the question of how the brain deals with interference. Neuropsychological studies have indicated that damage to the frontal lobe increases the susceptibility to PI (e.g., Shimamura et al. 1995). Consistent with this, a number of neuroimaging studies that investigated the effects of interference in a number of different interference paradigms have reliably shown frontal activity during encoding (Henson et al. 2002) and retrieval (Badre and Wagner 2005; Henson et al. 2002; Herrmann et al. 2001; King et al. 2005; LePage et al. 2005) of high-interference materials. This is consistent with an interpretation that PI is caused by failures in source monitoring, a process known to depend on the frontal lobes (Johnson et al. 1993). Interestingly, many of these studies of PI fail to demonstrate a modulation of MTL activity as a function of interference (however, see LePage et al. 2005), although neuropsychological studies of patients with MTL damage suggest that amnesic patients are more susceptible to interference. Traditional tests of PI contrast two conditions that both require explicit or declarative memory. For example, in the AB–AD paradigm participants learn a list of paired-associates, the AB list, followed after a variable delay by learning a second list with the same stimulus terms paired with new response terms, the AD list. Recalling the AB list and recalling the AD list are both likely to engage the hippocampus strongly. The lack of neuroimaging evidence for activity modulation in the MTL as a function of interference, therefore, may be due in part to the fact that the critical comparison between two conditions is known to activate the MTL. In support of this view, Henson et al. (2002) show a main effect of retrieval in the MTL when collapsing across conditions of high and low interference as compared to a control condition.

Thus, there is a long tradition in psychology of studying the effects of interference on memory, although the failure of interference theory to tell a cohesive story about the characteristics and limitations of long-term declarative memory has caused it to fall out of favor in current theories of cognitive psychology and cognitive neuroscience. Recently, however, there has been a resurgence of interest in interference as a driving factor of hippocampal activity due in part to the suggestion that one of the main functions of the hippocampus is to perform pattern separation for similar or overlapping stimuli (Yassa and Stark 2011).

Testing Predictions of Computational Models

Various computational models propose that the hippocampus uses sparse representations to reduce representational overlap (Burgess and O'Keefe 1996; Hasselmo and Wyble 1997; McNaughton and Morris 1987; Norman ad O'Reilly 2003; Rolls 1989; Rolls and Treves 1998). Using a sparse representation rather than extracting statistical regularities from the environment allows the hippocampus to better represent overlapping stimuli without interference, compared with the cortex. Thus, damage limited to the hippocampus should result in an increased susceptibility to interference. Furthermore, due to the dual demands of encoding and retrieval, as stimulus similarity increases pattern separation processes will give way to pattern completion, and previously stored representations will be activated.

Computational models commonly posit that pattern separation is accomplished as the DG relays a sparse, orthogonalized representation to the CA3 via mossy fiber projections. This representation can be retrieved by reactivating a subset of the original pattern through pattern completion. The CA3 may therefore, show evidence of both pattern separation and pattern completion depending on the task demands and the active afferents to the area (Guzowski et al. 2004). Pattern separation would be evidenced by distinct representations and low representational overlap, while high representational overlap would be consistent with pattern completion.

A number of studies have tested the predictions of computational models using rodent models, particularly in the spatial and temporal domains. For example, Gilbert and colleagues (Gilbert et al. 1998) have demonstrated that lesions to the DG in rats disrupted discrimination of near spatial locations while leaving discrimination performance intact for distant locations. Leutgeb and colleagues (S. Leutgeb et al. 2004) also examined the changes in firing characteristics of CA3 and CA1 in response to changes in the environment and showed that place fields in CA1 showed a great deal of overlap between similar environments. When tested in the same environment in different rooms, the active set of neurons overlapped almost as much as on repeat tests in the same environment in the same room. Place cells in CA3, however, showed distinct firing patterns in different rooms, even in the similar environments, with the overlap between the two rooms being no more than would be expected for independent firing. In this case, the CA3 neurons show a large amount of pattern separation (Guzowski et al. 2004), while CA1 neurons show evidence of pattern completion.

There are also limited electrophysiological data from human patients supporting a role of the hippocampus in pattern separation. These data come from studies of patients with pharmacologically intractable epilepsy who have been implanted with depth electrodes in the MTL to localize the focus of seizure onset (Fried et al. 2002; Fried et al. 1997). Quiroga and colleagues (Quiroga et al. 2005) reported evidence for neurons in the MTL that showed view invariant responses to familiar stimuli. These cells demonstrate the sparse representation that is predicted by the computational models outlined above; however, neurons throughout the MTL demonstrated this sparse representation. Further, this study did not explicitly manipulate interstimulus similarity or interference, so it is difficult to draw conclusions regarding pattern separation processes.

To date, the majority of studies investigating pattern separation processes in the human hippocampus have used functional neuroimaging and behavioral tests of healthy young adults, healthy older adults, and patients with limited hippocampal damage. The results of these studies, reviewed below, are largely consistent with the predictions of computational models. Specifically, they show that the DG is involved in pattern separation in a number of modalities.

Functional MRI and Pattern Separation

A number of descriptive models of MTL function suggest a functional distinction between the hippocampus and the adjacent MTL cortical areas. According to the various descriptive models, the hippocampal region underlies conjunctive (O'Reilly and Rudy 2000; Sutherland and Rudy 1989), associative (Brown and Aggleton 2001), or recollective (Yonelinas 2002) processing, while the adjacent cortex supports memory for single items (Brown and Aggleton 2001) or familiarity processes (Yonelinas 2002). Within this class of models, the distinction between hippocampal and cortical processing is qualitative rather than quantitative, although the proposed processes often can be couched in terms of computational processes. For example, configural representations may be established in the hippocampus through setting up distinct, pattern-separated representations of external stimuli (O'Reilly and Rudy 2000, 2001). Pattern separation is also necessary for episodic and source memory (e.g., remembering where one parked one's car from day to day), while pattern completion is necessary for recollective processing. In spite of this apparent mapping, computational and descriptive models of MTL function make different predictions. For example, according to the descriptive models, lesions restricted to the hippocampus should disproportionately affect relational memory, while not adversely affecting memory for single items. However, computational models (e.g., Norman and O'Reilly 2003) predict that in some cases (e.g., when inter-stimulus similarity or pattern separation demands are increased) selective hippocampal lesions will also impair item memory (see Holdstock et al. 2002a). We suggest that the results of many experiments meant to test the predictions of descriptive models of MTL function (e.g., the distinction between hippocampus-dependent recollection and cortex-dependent familiarity) can be better understood in terms of the underlying computational principles rather than qualitative psychological phenomena.

For example, Kirwan and Stark (unpublished observations) used functional MRI (fMRI) to examine the processes of encoding and later recalling stimulus pairings within the MTL. Fifteen participants performed an encoding and then a cued-recall task while undergoing fMRI scanning. Participants were first familiarized to a set of card-like stimuli that varied on three dimensions: the shape of the marking on the card, the color of the marking, and the background pattern of the card (Fig. 7.1a). Following stimulus familiarization, participants performed a series of study/test blocks in which they were first shown a series of card pairs and instructed to memorize the pair for a later memory test. Each study phase consisted of three pairs presented twice in a random order. Following the study phase, participants performed a cued-recall task in which they were shown one card from a pair and asked to recall one dimension of the paired card (shape, color, or background). Each dimension of each pair was tested on different trials in the test phase. Pairs that were well learned were dropped from subsequent study/test blocks while pairs that were not learned were retained thus maintaining a constant level of performance across study/test blocks. For the fMRI analysis, study trials were sorted into four "memory strength" bins according to subsequent performance on the test block with the lowest memory strength reflecting near-chance performance and the highest memory strength reflecting perfect performance. During encoding, fMRI data analysis revealed activity changes associated with subsequent memory strength in bilateral hippocampus (Fig. 7.1b–c). Activity in this region was highest while studying pairs that were subsequently categorized as low memory strength. Activity decreased in a linear fashion as subsequent memory performance increased (Fig. 7.1d). FMRI activity during cued recall test trials on which participants were correct (hits) increased with increasing memory strength, not only in the hippocampus but also in

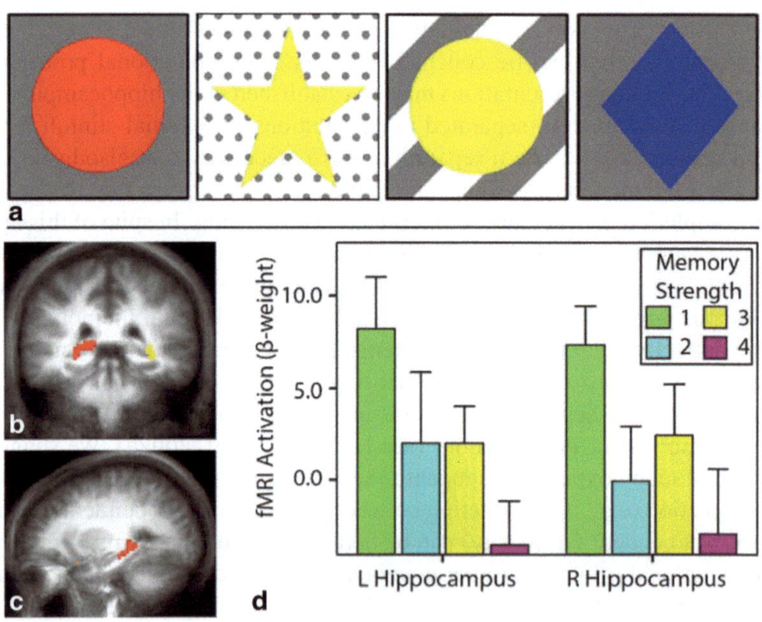

Fig. 7.1 Stimuli and fMRI results from a cued-recall paradigm. **a** Stimuli consisted of card-like stimuli that could vary in foreground shape, color, and background pattern. Subjects studied pairs of cards at encoding. At test, one card from the pair was presented and subjects were prompted to retrieve one aspect (shape, color, background) of the paired card. Pairs of stimuli were binned according to the number of features correctly retrieved into memory strength bins 1 (low) to 4 (high) (**b–c**), A contrast of encoding trials in memory strength bins 1 vs. 4 revealed activation in the left (red) and right (yellow) hippocampus where activity decreased in a linear fashion (**d**)

the adjacent cortical structures of the MTL. This pattern of results was also found in a distinct region of perirhinal cortex when just the miss trials were analyzed. While the pattern of results during the cued-recall trials is consistent with either a memory strength or recollection interpretation of MTL function (i.e., greater activity associated with the retrieval of more information or higher confidence responses), the pattern of results during encoding is more difficult to interpret. Subsequent analyses ruled out a novelty-detection response or a response as a function of the amount learned on each trial. One possible explanation is that pairs in the lowest memory strength bins were those with the highest pattern separation demands. Although the factorial combination of a limited number of stimulus features allowed us to test a relatively large stimulus set with limited behavioral responses, it also presented a unique challenge to participants. While overlap between pairs was intentionally limited, it could not be altogether eliminated. Interstimulus interference increased as more stimulus pairs with overlapping features were introduced into the experiment. Therefore, the pattern separation demands of encoding the stimulus pairs are potentially quite high. Here, we are operationally defining high pattern separation demands as any time the mnemonic demands of the task are high due to increased interstimulus overlap. This occurred in the current paradigm because the task demands required participants to attend to the three stimulus features (color, shape, background) and these features repeated in different combinations across stimulus pairs. This interpretation is supported by examination of a representative participant's behavioral performance (Fig. 7.2). Some pairs were learned quickly and the estimate for the probability for a subsequent correct response to that stimulus (our estimate of memory strength; see Law et al. 2005) increased rapidly. However, other stimulus pairs were not learned at all despite repeated testing. Further, these unlearned pairs seem to come later in the experiment as the potential for interference has built up with newer stimuli. Although the computational models of hippocampal function outlined above predict a high level of hippocampal activity due to pattern separation demands during the encoding phase of both preliminary experiments, the observed increases in hippocampal activity at time of encoding in this experiment cannot be attributed unequivocally to pattern separation mechanisms.

To explicitly test pattern separation processes in the MTL, Kirwan and Stark (2007) developed a continuous recognition paradigm that directly manipulated the similarity between stimuli in order to drive pattern separation demands. We hypothesized that a task that placed high demands on pattern separation processes would drive hippocampal activation. We further hypothesized that other MTL cortical regions would fail to show a distinction among stimulus types based on pattern separation demands. Rather than a standard study/test recognition memory paradigm, this study used a continuous recognition paradigm in which participants were required on each trial to encode stimuli to a sufficient threshold that they would be able to quickly and accurately recall the encoded information, compare it with current information, and determine whether or not they had previously encountered the information. Participants were shown either a series of objects (experiment 1) or a series of faces (experiment 2) while undergoing fMRI scanning. They were asked to determine whether each picture was new

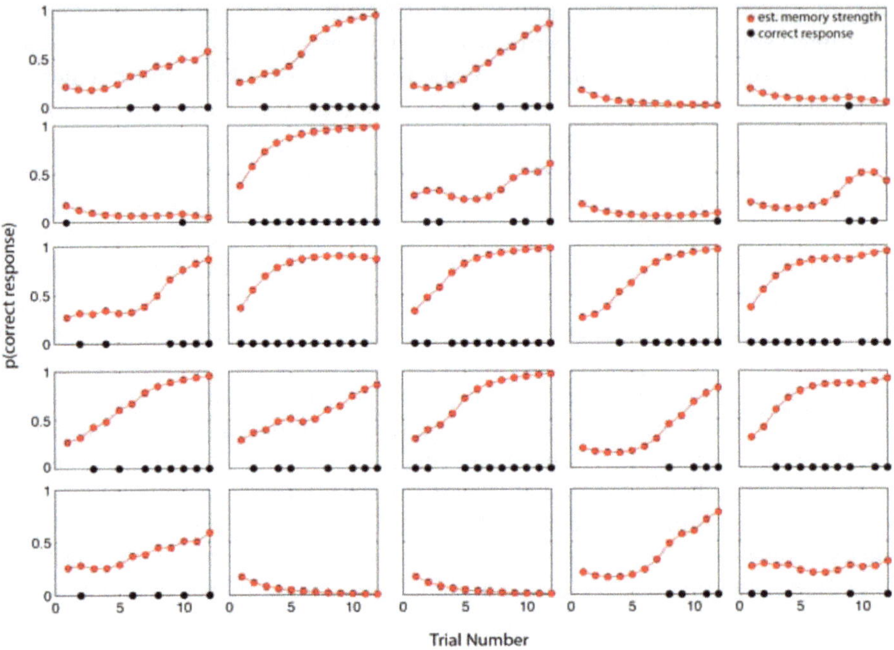

Fig. 7.2 Example performance on the paired-associates learning task. Subjects learned a series of pairs of cards, three pairs at a time. Each plot represents performance for one pair. Pairs were tested 12 times (four trials for each dimension of the test card). *Black dots* indicate correct trials (hits). *Red dots* indicate estimated memory strength for each pair. While some pairs were learned almost immediately, others were never learned

(never seen before), old (exact image as previously seen or "repeat"), or similar (similar to a previously seen image, but not exact or "lure"; Fig. 7.3). Repeated and lure items were separated from their first presentations by 10–40 items (or 25–100 s). This particular task places high demands on the participant and their ability to resolve memory interference and accurately perform pattern separation. The "recall to reject" strategy needed to successfully complete this task was supported by reaction time (RT) data; participants took longer to accurately identify a stimulus as "similar", presumably due to the process of recalling previously encoded information, comparing the current stimulus to the recalled stimuli, and then correctly identifying the stimulus as "similar". Furthermore, fMRI data from both experiments showed that the hippocampus was the only MTL structure to provide differentiated activation that distinguished the various trial types. Specifically, pattern of fMRI activity in the hippocampus was different for hits (correctly identifying a repeat stimulus as "old"), lure correct rejections (correctly identifying a similar stimulus as "similar"), and lure false alarms (incorrectly identifying a similar stimulus as "old"). This pattern of differentiation was not displayed in the parahippocampal gyrus.

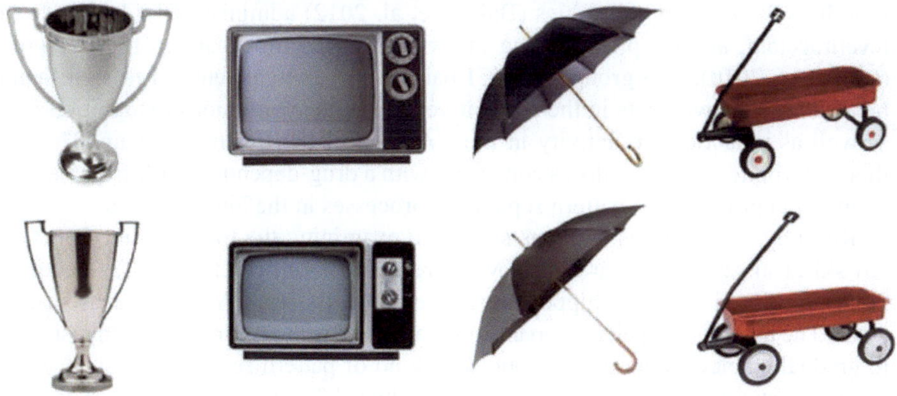

Fig. 7.3 Example target-lure stimulus pairs from the continuous recognition paradigm

Bakker and colleagues (Bakker et al. 2008) sought to differentially determine which subregions of the hippocampus are involved in pattern separation and which are involved in pattern completion. These authors again used a continuous recognition fMRI task consisting of new, repeated, and lure objects. However, instead of requiring participants to make an overt memory decision about each stimulus, they asked participants to identify whether an object was typically found indoors or outdoors. This incidental encoding task was used to remove any explicit memory confounds from the study and test the default hippocampal bias toward pattern separation or pattern completion. In addition, this methodology more closely resembled pattern separation and pattern completion tasks previously utilized in free exploration rodent studies. Using this indirect task, the authors hypothesized that lure stimuli would generate one of two types of activity within hippocampal subregions which could then be used to infer pattern separation or pattern completion processes. If lures generated a similar level of activity as new objects for a given subregion, it would be involved in pattern separation. In contrast, if lures generated a similar activity level as repeated objects a given subregion would be involved in pattern completion. Although high-resolution fMRI scans were performed in this study, resolution is still not sufficient enough to distinguish between CA3 and DG subregions of the hippocampus; therefore, these two regions were combined in the analyses. However, given that the DG projects predominantly to the CA3 region it is unproblematic to pair these regions together. Hippocampal activity inclined toward pattern separation in the CA3/DG subregions and activity favoring pattern completion was observed in the CA1 subregion as well as several other MTL regions. These results are in alignment with the prediction from computational models that the DG is primarily involved in pattern separation processes. DG activity has been observed to be elevated in amnestic mild cognitive impairment (aMCI), a putative precursor to Alzheimer's disease. This elevated activity may reflect either compensatory recruitment of additional neurocognitive resources or, alternately, aberrant activity that directly contributes to the behavioral memory impairment observed in aMCI. Bakker and colleagues (Bakker et al. 2012) administered a low dose of levetiracetam, an antiepileptic drug known to reduce hippocampal hyperactivity (Koh et al. 2010), to a group of aMCI patients. Drug treatment led to significant behavioral improvements in the explicit version of the continuous recognition task as well as reductions in activity in the CA3/DG of the treatment group. The authors interpreted these results as consistent with a drug-dependent shift from pattern completion processes to pattern separation processes in the hippocampus.

Recently, work in this area has shifted to examining the transfer function (Guzowski et al. 2004; Kumaran and Maguire 2009; Leutgeb et al. 2007; Leutgeb, 2008) in subregions of the hippocampus (i.e., CA3/DG vs. CA1) and other cortical areas. The hallmark of pattern separation is a step-like change in activity in response to gradual changes in stimulus input. This kind of pattern separation response has been demonstrated in place cell responses in the CA3 of rats in response to gradual changes across similar environments (Leutgeb et al. 2004). Pattern completion, on the other hand, should be evidenced by more gradual changes in output in response to gradual changes in input (Yassa and Stark 2011).

Lacy and colleagues (Lacy et al. 2011) used an incidental continuous recognition paradigm similar to Bakker et al. (2008). Based on normative similarity ratings, target-lure pairs were split into high- and low-similarity categories. Consistent with the findings of Bakker et al. (2008), activity in the CA3/DG was consistent with pattern separation processes whereas activity in the CA1 was more consistent with pattern completion processes. When examined separately, the difference between CA3/DG and CA1 was most pronounced for the high-similarity stimuli. Considering fMRI activation likely reflects input into an area (Logothetis et al. 2001), this finding is consistent with a differential transfer function between these regions where early stages of the hippocampus are more sensitive than later stages to small changes in input.

Rather than using a “mnemonic similarity” index, Motley and Kirwan (2012) explicitly manipulated target-lure similarity by rotating objects between study and test trials in a continuous recognition paradigm. These authors hypothesized that pattern-separation regions would have large differences in fMRI activity in response to small changes in stimulus similarity (i.e., small degree of rotation) and pattern-completion regions would have large differences only when stimulus similarity exceeded a threshold. They further hypothesized that intentional encoding would enhance the distinction between “old” and “similar” (i.e., rotated) stimuli and that top–down processing of task demands (dependent upon whether the task involved intentional encoding or incidental encoding) would enhance ventral stream inputs, both of which would aid hippocampal pattern separation. Individuals participated in either an intentional or incidental encoding pattern separation task. Individuals in the intentional paradigm were shown a series of objects and asked to identify them as either “new” (novel stimuli), “old” (previous seen stimuli), or “rotated” (previously seen stimuli from another angle). Independent behavioral testing indicated that rotations of 15°, 25°, 35° and 55° lead to a roughly linear change in behavioral performance from mostly false alarms (calling rotated stimuli “old”) to mostly correct rejections (calling rotated stimuli “rotated”). Participants in the incidental paradigm were shown the same stimuli as in the intentional paradigm, but were asked to identify objects as either "toy" or "not toy". The authors found that for the intentional encoding task, behavioral accuracy increased as the objects became more dissimilar in a linear fashion (i.e., as the rotation angle increased); however, activity in the hippocampus and posterior parahippocampal cortex was curvilinear, with a sharp change in fMRI activation from exact repeats to 15° rotations but a smaller change in activity between larger rotation differences. This pattern of large changes in fMRI activity in response to small changes in the input was taken as evidence of pattern separation processes. Task demands determined laterality in left and right medial temporal lobe. The authors found activity in the left hippocampus and posterior parahippocampal cortex consistent with pattern separation (i.e., large activity differences for small changes in the input) during the incidental encoding condition. In contrast, activity in the right hippocampus and parahippocampal cortex was consistent with pattern separation during the intentional encoding task. The finding of pattern-separation-like signals in the parahippocampal gyrus is consistent with rodent work that has shown a deficit in pattern separation for objects following perirhinal cortex damage (Kesner et al. 1993; 2001). The finding of lateral differences between the two tasks was interpreted in terms of semantic vs. spatial task demands in the incidental ("toy" or "not toy") and intentional ("old", "new", or "rotated") conditions. Furthermore, ventral stream activity depended upon the task encoding condition (i.e., whether the task involved intentional or incidental encoding). Specifically, there was decreased activation in the ventral stream during the incidental encoding task compared to intentional encoding, providing evidence for a top–down influence on hippocampal activity during the intentional condition. The authors speculated that the ventral stream inputs modulate information processing and are subsequently amplified by the hippocampus, which then conducts pattern separation processes.

Interference Following Hippocampal Damage

The computational models of hippocampal function outlined above predict a specific pattern of impairments when damage is limited to the hippocampus, namely a disproportionate impairment in pattern separation, behaviorally demonstrated as increased susceptibility to interstimulus interference. However, neuropsychological damage is rarely so selective as to affect only the hippocampus. Nevertheless, it is instructive to consider cases of amnesia caused by damage to structures including, or related to the hippocampus, as they demonstrate instances of hippocampal dysfunction. When considering cases of amnesia with a wide range of etiologies, it is apparent that amnesic patients are indeed more susceptible to interference than matched controls (for review, see Lustig and Hasher 2001).

The data supporting this claim come from a number of studies, each of which demonstrate significant PI (Kinsbourne and Winocur 1980; Mayes et al. 1987; Warrington and Weiskrantz 1974, 1976, 1978; Winocur and Moscovitch 1996) and RI (Winocur and Weiskrantz 1976) in amnesic patients. In each of these studies, patients and matched controls learned semantically related paired associates in the

AB–AD paradigm. Patient's performance, however, shows improvement relative to controls when the amount of interference is reduced, for example, when the AD list is not semantically related to the AB list (Winocur and Weiskrantz 1976), or the number of possible responses is limited (Kinsbourne and Winocur 1980; Warrington and Weiskrantz 1974, 1978). When the task has implicit memory instructions, i.e., when participants are given free-association instructions, controls' performance falls to the level of amnesics'. These data indicate that amnesic patients do not have the mechanisms available to resolve cases of high overlap (i.e., pattern separation) and must therefore rely on other memory mechanisms (i.e., cortical mechanisms) that are more prone to generalization and thus more susceptible to interference. This is not to say that amnesia per se is responsible for the increased susceptibility to interference (see Mayes and Downes 1997). Rather, the two are symptoms of damage to the MTL memory system.

Hippocampal amnesics are able to overcome this increased susceptibility to interference if their memory is probed in an appropriate way. Holdstock and colleagues (Holdstock et al. 2002b) describe the case of patient YR, who suffered selective adult-onset hippocampal damage. In tests of recognition memory, YR is unimpaired relative to matched controls for single items when tested in both forced choice and yes/no recognition formats (Holdstock et al. 2000; Mayes et al. 2002; Mayes et al. 2001). However, when target and lures were made more similar (i.e., interference was increased), YR's yes/no item recognition was impaired relative to controls (Holdstock et al. 2002b), indicating a pattern separation deficit (but see Bayley et al. 2008). Duff and associates (Duff et al. 2012) also demonstrated that when interstimulus similarity increased, patients with hippocampal damage were differentially impaired in a memory task relative to matched controls.

In a recent study, Kirwan and colleagues (Kirwan et al. 2012) hypothesized that memory-impaired patients with hippocampal damage would have pattern separation impairments which would present as an inability to correctly identify lure stimuli as similar compared to previously presented target stimuli; instead they speculated that these individuals would identify similar objects as "old". Three memory-impaired individuals with damage thought to be limited to the hippocampus and 11 controls performed baseline recognition memory tests for faces and objects. During the study phase, participants were shown a series of stimuli and asked to rate them as either "pleasant" or "unpleasant" and were told their memory of the stimuli would be tested later. After a brief delay, participants were shown a series of target (previously presented) and novel stimuli and were asked to identify which stimuli were "old" (i.e., targets) and which were "new". Following the baseline task, participants also completed a series of experimental tasks where interstimulus similarity was explicitly manipulated. The study phase of the experimental task also consisted of a series of stimuli (faces or objects), which the participants rated as either "pleasant" or "unpleasant". During the testing phase, participants were shown either repeat, similar, or novel stimuli and were asked to identify them as "old", "similar", or "new" respectively. In the face condition, similar stimuli consisted of the same individual as previously shown, but a different photograph of them in which some characteristic(s) differed (such as gaze direction, expression, hairstyle, clothes, etc.). In the object condition, similar stimuli consisted of the same type of object as previously shown, but a different specific example. Memory-impaired patients with hippocampal damage did not differ from controls in baseline memory recognition. However, patients were impaired compared to controls in the pattern separation conditions with significantly reduced "similar" responses to lure stimuli (corrected for overall "similar" response rates). Contrary to what was predicted, patients were not biased toward identifying similar objects as "old", rather they were more likely than controls to respond to similar objects as either "old" or "new" verses the correct response of "similar". The authors speculated that this response pattern indicates a pattern separation deficiency at the time of encoding for memory-impaired patients with hippocampal damage.

The preceding studies examined visual pattern separation abilities following damage to the hippocampus. According to Kesner's Attribute Model (Kesner 1991), the DG is involved in pattern separation of spatial locations. This conclusion is supported by a number of rodent lesion studies (e.g., Gilbert et al. 1998). Hopkins and Kesner (Hopkins and Kesner 1993; Kesner and Hopkins 2006) sought to extend these findings to spatial pattern separation in humans in a real world task. Patients with hippocampal atrophy due to hypoxia and matched controls performed a geographic distance task using cities on a map. In the study phase, participants were shown a series of cities on a map of New Brunswick and instructed to remember their locations. In the test phase, participants were given a pair of city names and asked which was further in a given direction (north, south, east, or west). The city pairs differed in the number of other cities in between, with separations of 0, 2, 4, and 6 intervening cities. Hypoxic patients were impaired relative to controls for all spatial distances. Taken together, these results indicate that damage to the hippocampus does result in pattern separation deficits for both objects and locations, as manifest by greater susceptibility to interstimulus interference.

Pattern Separation and the Aging Hippocampus

Memory impairments are one of the most common age-related cognitive complaints. One possible mechanism underlying age-related memory decline is a decrease in hippocampal integrity (Small et al. 2002). A number of studies have examined the effects of aging on pattern separation processes (see Holden and Gilbert 2012; Gilbert this volume). An emerging theme of these studies is that, similar to overall memory performance, there is a large degree of inter-subject variability in pattern separation performance among the aging population, even in the absence of any neurodegenerative disease.

Toner and colleagues (Toner et al. 2009) explored age-related changes in pattern separation utilizing a visual object continuous recognition task. They examined differences between young adults and nondemented older adults utilizing the previously discussed paradigm developed by Kirwan and Stark (2007). Since the lure items used are very similar but not identical to previously seen objects in this task, it was hypothesized that lures would produce increased interference and thus an increased demand on pattern separation. Older adults and younger adults performed similarly in their ability to distinguish new and old stimuli. However, younger adults significantly outperformed older adults in their ability to correctly identify lures as similar, with older adults more likely to identify lures as "old". The authors suggested that these results may come about due to age-related changes in hippocampal subregions resulting in less efficient pattern separation processes. Based on previous findings, they proposed the DG may be particularly vulnerable to age-related changes. The age-related differences in identifying lure objects may result from decreased integrity of the DG (Small et al. 2002), resulting in less DG dependent pattern separation. The authors speculate these age-related differences may result in a dominant pattern completion bias with lure stimuli incorrectly identified as old stimuli. In addition, they found the performance of older adults was significantly correlated with several neuropsychological tests involving the MTL. For example, number sequencing, a basic element involved in executive function tasks, was correlated with older adults correctly identifying lures as "similar". Thus pattern separation tasks which can discriminate performance in hippocampal subregions may be useful in identifying age-related changes in these areas. In a follow-up analysis, Holden and colleagues (Holden et al. 2013) split older adults into impaired and unimpaired groups based on their performance on a separate neuropsychological memory test (the Hopkins Verbal Learning Test-Revised, HVLT-R). Older adults in the unimpaired group performed as well as younger controls on the pattern separation task while those in the impaired group tended toward pattern completion (i.e., were more likely to call the lures "old" than "similar").

In another study, Holden and associates (Holden et al. 2012) used a delayed match-to-sample varying spatial location task to study pattern separation differences between young and older adults. During the first phase of this task a circle appeared in one of 18 possible locations on a screen. This was followed by a brief delay during which participants looked away from the screen and read a random letter sequence. Afterwards, two circles appeared on the screen and the participants had to identify which circle matched the location of the previously displayed circle. The authors found that young adults had significantly greater performance scores than older adults. Furthermore there was a significant linear effect of spatial separation; performance for both young and older adults increased as the spatial distance between the circles increased. To rule out the possibility of general memory deficits related to normal aging as the contributing factor to the performance differences, older adults were split into impaired and unimpaired groups based on their HVLT-R delayed recall scores. Results showed the original findings between young and older adults remained after accounting for verbal memory impairment. The authors concluded that although general memory decline cannot be ruled out, it is more likely that the differences between young and older adults is likely due to less efficient spatial pattern separation as a result of normal aging processes. They further elaborated that this inefficiency is likely due to age-related changes in the DG and CA3 subregions of the hippocampus.

Other studies have also explored spatial pattern separation abilities in older adults. For example, Stark and colleagues (Stark et al. 2010) modified a rodent spatial pattern separation task (Gilbert et al. 1998) for use within the human population. However, rather than only varying the spatial separation of stimuli on one dimension (e.g., distance) they varied it on two dimensions, changing both the distance and the angle of two stimuli. They hypothesized that the extent of movement between the stimuli would tax the spatial pattern separation demands in humans in a manner similar to that observed within rodent research. Furthermore, they sought to explore potential variability in task performance among healthy aged adults. They had young and older adults complete a series of standardized neuropsychological tests, which included the Rey Auditory-Verbal Learning Task (RAVLT), and participate a spatial pattern separation task. During the study phase, participants were briefly shown pairs of pictures and asked to remember the location of these stimuli. During the test phase, participants were asked to identify whether the pictures were in the same or a different location. For the different location trials, only one of the pictures from the pair changed locations and this change occurred in either a small (close), moderate (medium), or large (far) amount in both x- and y-coordinates. The results showed no overall differences between young and older adults with a positive linear trend across the conditions (i.e., same, close, medium, and far). The authors divided the older adult participants into impaired and unimpaired groups based on their RAVLT delayed word learning performance scores. While all of the older participants scored within the normal range for their age those that scored more than one standard deviation below the young adult norms were placed in the impaired group. Impaired participants performed worse on the different location trails compared to the unimpaired participants in the spatial pattern separation task. The authors found that as RAVLT delayed recall scores increased, participants' performance on the different location trails increased. The authors concluded that spatial pattern separation processing is diminished in mildly impaired older adults and that the spatial pattern separation task utilized in this study may be a sensitive marker of memory variability in aged individuals.

In a high-resolution fMRI paradigm, Yassa and associates (Yassa et al. 2011a) assessed changes in hippocampal subregion activity in older adults. They hypothesized that older adults would exhibit a bias toward pattern completion rather than pattern separation; that older adults would show increased activity in CA3 and/or DG subregions of the hippocampus; and that older adults would need greater dissimilarities between stimuli in order to correctly identify previously experienced stimuli from novel stimuli. Experiment 1 consisted of the continuous recognition task previously used by Kirwan and Stark (2007). The authors compared the contrast of lure trials correctly identified as "similar" against lure trials incorrectly identified as "old" to examine young adult and old adult MTL activity. Consistent with their hypothesis, the authors found that older adults were more likely to identify lure items as "old" (false alarm) rather than "similar" (correct rejection), indicating a bias toward pattern completion. Furthermore, older adults only successfully identified 33 % of the lure trials as "similar" compared with young adults correctly identifying 59 % of the lure trials. They also found a significant increase in signal (the difference between correct rejections and false alarms) in the right CA3/DG region of the hippocampus during the first and second presentations. Experiment 2 involved a mnemonic similarity task similar to that used in experiment 1; however, these stimuli had been previously normed to generate mnemonic similarity ratings for each pair of similar stimuli. The lure items were then sorted based on the degree of mnemonic similarity. They found that larger degrees of differences between similar stimuli were necessary for older adults to engage in successful pattern separation and correctly identify lure stimuli as "similar" rather than "old", confirming the last portion of their hypothesis.

In a separate study, to test whether CA3/DG pattern separation signals would diminish with age as stimuli increased in similarity, Yassa and colleagues (Yassa et al. 2011b) examined high-resolution fMRI hippocampal activity profiles in young and older adults during pattern separation tasks with varied stimuli similarity. The authors predicted that changes in the CA3/DG functional network would be correlated with structural indicators as measured by ultrahigh-resolution microstructural diffusion tensor imaging (msDTI). Furthermore, they hypothesized that degraded perforant pathway input to the DG and CA3 subregions would be linked to age-related pattern separation impairments. They had young and older adults participate in an explicit recognition task out of the scanner and in an implicit fMRI recognition task. In addition, ultrahigh-resolution msDTI scans were performed on each participant. Just like many of the other continuous recognition pattern separation tasks used in other studies, during the explicit recognition task, participants were shown a series of novel, repeat, and lure stimuli and asked to identify each as "new", "old", or "similar". During the implicit recognition task, participants were scanned while shown a different series of novel, repeat, and lure stimuli; however, in this task they were asked to identify whether each item was an "indoor" or "outdoor" object. Based on previous mnemonic similarity ratings (Yassa et al. 2011a), the lure stimuli were analyzed according to their degree of similarity. When lure stimuli were very different they found no differences in CA3/DG activity between young and older adults. Alternatively, as stimuli became more similar, CA3/DG responses diminished in older adults, but remained high in young adults. The authors report that this pattern indicates a weakened pattern separation response to lure stimuli in older adults' CA3/DG hippocampal region and refer to this change as "representational rigidity" or the "requirement for increased dissimilarity before stimuli can be orthogonalized" (Yassa et al. 2011b, p. 8873). The extent of the representational rigidity predicted behavioral deficits in the discrimination task. There was also a correlation between the left CA3/DG gray matter functional rigidity and fractional anisotropy in this same region. The authors suggest that these results indicate that structural dendritic changes in the CA3/DG region may influence the functional impairments observed in older adults. Finally, they found correlations between the perforant path integrity and the amount of left CA3/DG rigidity. In addition, perforant pathway integrity was predictive of older adult performance on the behavioral discrimination task. They further found that CA3 rigidity was correlated with the functional pairing of the CA3/DG region with the entorhinal cortex. The authors speculated that signal reduction between the entorhinal cortex and the hippocampus may be related to the level of CA3/DG rigidity resulting in greater resistance to change. The authors conclude that age-related degradation in the perforant path and CA3/DG network bias this system toward pattern completion and may impact mnemonic deficits often observed in older adults.

Most of the aging pattern separation studies have used pictures of objects in their paradigms, which are typically both perceptually and conceptually similar to each other. Due to the inability of pictorial pattern separation paradigms to differentiate between conceptual and perceptual interference, Ly et al. (2013) utilized a verbal stimuli paradigm to explore these differences in young and older adults. They expanded upon the Deese-Roediger and McDermott (DRM) paradigm (Roediger and McDermott 1995) and created two recognition conditions, one testing perceptual similarity and the other testing conceptual similarity. Two groups of young and older adults were randomized to either the perceptual or conceptual condition. Words used in the perceptual condition were phonologically similar and shared at least the first phoneme (e.g., cork and corn) while words in the conceptual condition were similar in semantic or categorical meaning (e.g., bell and whistle). During the initial encoding phase of the experiment, participants were shown a list of words and asked to identify whether each word represented an indoor or outdoor object. This phase was followed by the recognition phase (either perceptual or conceptual) during which participants were shown a series of novel, repeat, and lures) words and asked to identify if each was "old" or "new". The authors found no significant differences between young and older adults on the conceptual task; however, there were differences on the perceptual task with increased false alarms in older adults. To address whether perceptual impairments in older adults were influencing the results the authors conducted another experiment utilizing a match-to-sample paradigm on a separate group of young and older adults. In this paradigm, participants were shown the same stimuli used in the perceptual condition except that words were yoked to their similar lures and separated by a visual noise stimulus to remove any sensory trace. Participants were asked to identify whether the second word was the same or different from the first word in each pair. They found no differences between young and older adults and concluded that age-related deficits in perceptual working memory did not account for the previously found older adult impairment on the perceptual task. Instead they suggest that this impairment results from proactive interference and likely is the result of a gist or false familiarity processing bias. They further propose that while conceptually similar stimuli are likely immune from a pattern completion bias, perceptually similar stimuli are susceptible to pattern separation failure.

Improving Pattern Separation

There are many conditions that result in impairments to hippocampal dependent pattern separation, including depression (Déry et al. 2013), aMCI (Bakker et al. 2012), aging (Holden et al. 2013; Stark et al. 2010), and hippocampal damage (Kirwan et al. 2012). Is it possible to improve pattern separation? Rodent studies have demonstrated that increasing neurogenesis in the DG leads to improved performance in spatial pattern separation tasks (Sahay et al. 2011). Déry et al. (2013) administered an exercise intervention to a population of healthy young adults who were previously relatively sedentary. Since exercise has been shown to increase neurogenesis (van Praag et al. 1999), the authors reasoned that the exercise intervention would result in increased neurogenesis in the hippocampus and subsequent improvement in pattern separation performance. A modified version of the object pattern separation task (Kirwan and Stark 2007) was administered prior to and following a 6-week exercise intervention. Pattern separation performance in the exercise group improved as a function of the increase in physical fitness. The authors took this result as evidence that increased neurogenesis leads to improved pattern separation performance. These results are tempered, however, by the inclusion of a clear outlier in the group who responded to the exercise intervention (see Déry et al. 2013 their Fig. 7.2) and will require replication.

Another possible way to improve pattern separation processing in the hippocampus is to increase the amount of norepinephrine (NE) available in the DG. The DG has a high concentration of NE receptors and may be modulated by noradrenergic activity in the locus coeruleus and the basal lateral amygdala (McGaugh 2002; Young and Kuhar 1980). Segal and colleagues (Segal et al. 2012) manipulated NE levels by having participants view a series of high-emotional-valence stimuli prior to encoding a series of objects. Participants performed a recognition memory test with targets, novel foils, and similar lures following a 15-min delay. NE levels were assessed prior to the emotional arousal phase, prior to encoding, and at the beginning and end of the delay period via saliva sample. There was a positive relationship between the change in NE levels and performance on the pattern separation task, indicating that NE may have a direct effect on pattern separation processing in the DG.

Finally, it appears that ongoing behavioral state may also affect pattern separation performance. Duncan et al. (2012) demonstrated that the trial preceding a lure stimulus in the continuous recognition pattern separation task influence performance on the lure trial. Lures preceded by a (correctly identified) novel stimulus were more likely to be correctly identified as “similar” than those preceded by a repeated stimulus. The authors manipulated the inter-trial interval and found that the effect was most pronounced with short intervals (500 ms) and was absent at longer intervals (1500 and 2500 ms). The authors speculate that encoding a new stimulus disposes the hippocampus toward pattern separation processes while the retrieval of a previously stored representation may dispose the hippocampus toward pattern completion processes. Further work is needed to elucidate the mechanisms underlying this effect.

Conclusions

Since event-based memories have a high degree of overlap, the medial temporal lobe memory system must perform pattern separation in order to avoid catastrophic interference at the time of retrieval. Here, we have reviewed evidence from animal models that support the predictions of neuroanatomically-based computational models, which propose that the sparse connections in the DG are especially suited for performing pattern separation. There is also an increasing amount of supporting evidence from fMRI and behavioral studies with a variety of populations, including healthy young adults, memory-impaired patients with limited hippocampal damage, and healthy older adults. These and other studies offer promising avenues for both describing and improving pattern separation processes.

References

Badre, D., & Wagner, A. D. (2005). Frontal lobe mechanisms that resolve proactive interference. *Cerebral Cortex, 15,* 2003–2012.

Bakker, A., Kirwan, C. B., Miller, M., & Stark, C. E. (2008). Pattern separation in the human hippocampal CA3 and dentate gyrus. *Science, 319*(5870), 1640–1642.

Bakker, A., Krauss, G. L., Albert, M. S., Speck, C. L., Jones, L. R., Stark, C. E., Yassa, M. A., Bassett, S. S., Shelton, A. L., & Gallagher, M. (2012). Reduction of hippocampal hyperactivity improves cognition in amnestic mild cognitive impairment. *Neuron, 74*(3), 467–474.

Bayley, P. J., Wixted, J. T., Hopkins, R. O., & Squire, L. R. (2008). Yes/no recognition, forced-choice recognition, and the human hippocampus. *Journal of Cognitive Neuroscience, 20*(3), 505–512.

Blank, H. (2005). Another look at retroactive and proactive interference: A quantitative analysis of conversion processes. *Memory, 13*(2), 200–224.

Brown, M. W., & Aggleton, J. P. (2001). Recognition memory: What are the roles of the perirhinal cortex and hippocampus? *Nature Reviews Neuroscience, 2,* 51–61.

Burgess, N., & O'Keefe, J. (1996). Neuronal computations underlying the firing of place cells and their rolei n navigation. *Hippocampus, 6,* 749–762.

Crowder, R. G. (1976). *Principles of learning and memory.* New York: Lawrence Erlbaum Associates.

Déry, N., Pilgrim, M., Gibala, M., Gillen, J., Wojtowicz, J. M., Macqueen, G., & Becker, S. (2013). Adult hippocampal neurogenesis reduces memory interference in humans: Opposing effects of aerobic exercise and depression. *Frontiers in Neurosciences, 7,* 66.

Duff, M. C., Warren, D. E., Gupta, R., Vidal, J. P., Tranel, D., & Cohen, N. J. (2012). Teasing apart tangrams: Testing hippocampal pattern separation with a collaborative referencing paradigm. *Hippocampus, 22*(5), 1087–1091.

Duncan, K., Sadanand, A., & Davachi, L. (2012). Memory's penumbra: Episodic memory decisions induce lingering mnemonic biases. *Science, 337*(6093), 485–487.

Ekstrand, B. R. (1967). The effect of sleep on memory. *Journal of Experimental Psychology, Applied, 75,* 64–72.

Fried, I., MacDonald, K., & Wilson, C. (1997). Single neuron activity in human hippocampus and amygdala during recognition of faces and objects. *Neuron, 18,* 753–765.

Fried, I., Cameron, K., Yashar, S., Fong, R., & Morrow, J. (2002). Inhibitory and excitatory responses of single neurons in the human medial temporal lobe during recognition of faces and objects. *Cerebral Cortex, 12,* 575–584.

Gilbert, P. E., Kesner, R. P., & DeCoteau, W. E. (1998). The role of the hippocampus in mediating spatial pattern separation. *Journal of Neuroscience, 18,* 804–810.

Guzowski, J. F., Knierim, J. J., & Moser, E. I. (2004). Ensemble dynamics of hippocampal regions CA3 and CA1. *Neuron, 44,* 581–584.

Hasselmo, M. E., & Wyble, B. (1997). Free recall and recognition in a network model of the hippocampus: Simulating effects of scopolamine on human memory function. *Behavioral Brain Research, 89,* 1–34.

Henson, R. N. A., Shallice, T., Josephs, O., & Dolan, R. J. (2002). Functional magnetic resonance imaging of proactive interference during spoken cued recall. *NeuroImage, 17,* 543–558.

Herrmann, M., Rotte, M., Grubich, C., Ebert, A. D., Schiltz, K., Munte, T. F., & Heinze, H. J. (2001). Control of semantic interference in episodic memory retrieval is associated with an anterior cingulate-prefrontal activation pattern. *Human Brain Mapping, 13,* 94–103.

Holden, H. M., & Gilbert, P. E. (2012). Less efficient pattern separation may contribute to age-related spatial memory deficits. *Frontiers in Aging Neuroscience, 4,* 9.

Holden, H. M., Hoebel, C., Loftis, K., & Gilbert, P. E. (2012). Spatial pattern separation in cognitively normal young and older adults. *Hippocampus, 22*(9), 1826–1832.

Holden, H. M., Toner, C., Pirogovsky, E., Kirwan, C. B., & Gilbert, P. E. (2013). Visual object pattern separation varies in older adults. *Learning & memory, 20*(7), 358–362.

Holdstock, J. S., Gutnikov, S. A., Gaffan, D., & Mayes, A. R. (2000). Perceptual and mnemonic matching-to-sample in humans: Contributions of the hippocampus, perirhinal and other medial temporal lobe cortices. *Cortex, 36,* 301–322.

Holdstock, J. S., Mayes, A. R., Roberts, N., Cezayirli, E., Isaac, C., O'Reilly, R. C., & Norman, K. A. (2002a). Under what conditions is recognition spared relative to recall after selective hippocampal damage in humans? *Hippocampus, 12,* 341–351.

Holdstock, J. S., Mayes, A. R., Roberts, N., Cezayirli, E., Isaac, C. L., O'Reilly, R. C., & Norman, K. A. (2002b). Under what conditions is recognition spared relative to recall after selective hippocampal damage in humans? *Hippocampus, 12,* 341–351.

Hopkins, R. O., & Kesner, R. P. (1993). Memory for temporal and spatial distances for new and previously learned geographical information in hypoxic subjects. Paper presented at the Society for Neuroscience Abstracts.

Jenkins, J. B., & Dallenbach, K. M. (1924). Oblivescence during sleep and waking. *American Journal of Psychology, 35,* 605–612.

Johnson, M. K., Hashtroudi, S., & Lindsay, D. S. (1993). Source monitoring. *Psychological Bulletin, 114*(1), 3–28.

Keppel, G. (1984). Consolidation and forgetting theory. In H. Weingartner & E. S. Parker (Eds.), *Memory consolidation: Psychobiology of cognition* (pp. 149–161). Hillsdale: Erlbaum.

Kesner, R. P. (1991). The role of the hippocampus within an attribute model of memory. *Hippocampus, 1*(3), 279–282.

Kesner, R. P. (2009). Tapestry of memory. *Behavioral Neuroscience, 123*(1), 1–13.

Kesner, R. P., & Hopkins, R. O. (2006). Mnemonic functions of the hippocampus: A comparison between animals and humans. *Biological Psychology, 73*(1), 3–18.

Kesner, R. P., Ravindranathan, A., Jackson, P., Giles, R., & Chiba, A. A. (2001). A neural circuit analysis of visual recognition memory: Role of perirhinal, medial, and lateral entorhinal cortex. *Learning & Memory, 8*(2), 87–95.

King, J. A., Hartley, T., Spiers, H. J., Maguire, E. A., & Burgess, N. (2005). Anterior prefrontal involvement in episodic retrieval reflects contextual interference. *NeuroImage, 28*(1), 256–267.

Kinsbourne, M., & Winocur, G. (1980). Response competition and interference effects in paired-associate learning by Korsakoff amnesics. *Neuropsychologia, 18,* 541–548.

Kirwan, C. B., & Stark, C. E. L. (2007). Overcoming interference: An fMRI investigation of pattern separation in the medial temporal lobe. *Learning and Memory, 14*(9), 625–6333.

Kirwan, C. B., Hartshorn, J. A., Stark, S. M., Goodrich-Hunsaker, N. J., Hopkins, R. O., & Stark, C. E. L. (2012). Pattern separation deficits following damage to the hippocampus. *Neuropsychologia, 50,* 2408–2414.

Koh, M. T., Haberman, R. P., Foti, S., McCown, T. J., & Gallagher, M. (2010). Treatment strategies targeting excess hippocampal activity benefit aged rats with cognitive impairment. *Neuropsychopharmacology, 35*(4), 1016–1025.

Kumaran, D., & Maguire, E. A. (2009). Novelty signals: A window into hippocampal information processing. *Trends in Cognitive Sciences, 13*(2), 47–54.

Lacy, J. W., Yassa, M. A., Stark, S. M., Muftuler, L. T., & Stark, C. E. (2011). Distinct pattern separation related transfer functions in human CA3/dentate and CA1 revealed using high-resolution fMRI and variable mnemonic similarity. *Learning & Memory, 18*(1), 15–18.

Law, J. R., Flanery, M. A., Wirth, S., Yanike, M., Smith, A. C., Frank, L. M., Suzuki, W. A., Brown, E. N., & Stark, C. E. L. (2005). Functional magnetic resonance imaging activity during the gradual acquisition and expression of paired-associate memory. *Journal of Neuroscience, 25*(24), 5720–5729.

LePage, M., Blondin, F., Achim, A. M., Menear, M., & Brodeur, M. (2005). The interfering effect of related events on recognition memory discriminability: A functional magnetic resonance imaging study. *Cognitive Brain Research, 22,* 429–437.

Leutgeb, S. (2008). Neuroscience. Detailed differences. *Science, 319*(5870), 1623–1624.

Leutgeb, S., Leutgeb, J. K., Treves, A., Moser, M., & Moser, E. I. (2004). Distinct ensemble codes in hippocampal areas CA3 and CA1. *Science, 305,* 1295–1298.

Leutgeb, J. K., Leutgeb, S., Moser, M., & Moser, E. I. (2007). Pattern separation in dentate gyrus and CA3 of the hippocampus. *Science, 315*(5814), 961–966.

Logothetis, N. K., Pauls, J., Augath, M., Trinath, T., & Oeltermann, A. (2001). Neurophysiological investigation of the basis of the fMRI signal. *Nature, 412,* 150–157.

Lustig, C., & Hasher, L. (2001). Implicit memory is not immune to interference. *Psychological Bulletin, 127*(5), 618–628.

Ly, M., Murray, E., & Yassa, M. A. (2013). Perceptual versus conceptual interference and pattern separation of verbal stimuli in young and older adults. *Hippocampus, 23*(6), 425–430.

Mayes, A. R., & Downes, J. J. (1997). What do theories of the functional deficit(s) underlying amnesia have to explain? *Memory, 5*(1–2), 3–36.

Mayes, A. R., Pickering, A., & Fairbairn, A. (1987). Amnesic sensitivity to proactive interference: Its relationship to priming and the causes of amnesia. *Neuropsychologia, 25*(1B), 211–220.

Mayes, A. R., Isaac, C. L., Downes, J. J., Holdstock, J. S., Hunkin, N. M., Montaldi, D., MacDonald, C., Cezayirli, E., & Roberts, J. N. (2001). Memory for single items, word pairs, and temporal order in a patient with selective hippocampal lesions. *Cognitive Neuropsychology, 18,* 97–123.

Mayes, A. R., Holdstock, J. S., Isaac, C., Hunkin, N., & Roberts, N. (2002). Relative sparing of item recognition memory in a patient with adult-onset damage limited to the hippocampus. *Hippocampus, 12,* 325–340.

McClelland, J. L., McNaughton, B. L., & O'Reilly, R. C. (1995). Why there are complementary learning systems in the hippocampus and neocortex: Insights from the successes and failures of connectionist models of learning and memory. *Psychological Review, 102,* 419–457.

McCloskey, M., & Cohen, N. J. (1989). Catastrophic interference in connectionist networks: The sequential learning problem. In G. H. Bower (Ed.), *The psychology of learning and motivation* (Vol. 24, pp. 109–164). San Diego: Academic.

McGaugh, J. L. (2002). Memory consolidation and the amygdala: A systems perspective. *Trends in Neurosciences, 25*(9), 456.

McGeoch, J. A. (1932). Forgetting and the law of disuse. *Psychological Review, 39,* 352–370.

McNaughton, B. L., & Morris, R. G. M. (1987). Hippocampal synaptic enhancement and information storage within a distributed memory system. *Trends in Neurosciences, 10*(10), 408–415.

Motley, S. E., & Kirwan, C. B. (2012). A parametric investigation of pattern separation processes in the medial temporal lobe. *The Journal of Neuroscience, 32*(38), 13076–13085.

Müller, G. E., & Pilzicker, A. (1900). Experimentelle Beiträge zur Lehre vom Gedächtnis [Experimental contributions to the science of memory]. *Zeitschrift für Psychologie, 1,* 1–300.

Norman, K. A., & O'Reilly, R. C. (2003). Modeling hippocampal and neocortical contributions to recognition memory: A complementary learning systems approach. *Psychological Review, 110*(4), 611–646.

O'Reilly, R. C., & Rudy, J. W. (2000). Computational principles of learning in the neocortex and hippocampus. *Hippocampus, 10,* 389–397.

O'Reilly, R. C., & Rudy, J. W. (2001). Conjunctive representations in learning and memory: Principles of cortical and hippocampal function. *Psychological Review, 108,* 311–345.

Quiroga, R. Q., Reddy, L., Kreiman, G., Koch, C., & Fried, I. (2005). Invariant visual representation by single neurons in the human brain. *Nature, 435,* 1102–1107.

Robinson, E. S. (1927). The "similarity" factor in retroaction. *American Journal of Psychology, 39,* 297–312.

Roediger, H. L., & McDermott, K. B. (1995). Creating false memories: Remembering words not presented in lis. *Journal of Experimental Psychology. Learning, Memory, and Cognition, 12*(4), 803–814.

Rolls, E. T. (1989). Functions of neuronal networks in the hippocampus and neocortex in memory. In J. H. Byrne & W. O. Berry (Eds.), *Neural models of plasticity: Experimental and theoretical approaches*. San Diego: Academic.

Rolls, E. T., & Treves, A. (1998). *Neural networks and brain function*. Oxford: Oxford University Press.

Sahay, A., Scobie, K. N., Hill, A. S., O'Carroll, C. M., Kheirbek, M. A., Burghardt, N. S., Fenton, A. A., Dranovsky, A., & Hen, R. (2011). Increasing adult hippocampal neurogenesis is sufficient to improve pattern separation. *Nature, 472*(7344), 466–470.

Scoville, W. B., & Milner, B. (1957). Loss of recent memory after bilateral hippocampal lesions. *Journal of Neurology, Neurosurgery and Psychiatry, 20,* 11–21.

Segal, S. K., Stark, S. M., Kattan, D., Stark, C. E., & Yassa, M. A. (2012). Norepinephrine-mediated emotional arousal facilitates subsequent pattern separation. *Neurobiology of learning and memory, 97*(4), 465–469.

Shimamura, A. P., Jurica, P. J., Mangels, J. A., Gershberg, F. B., & Knight, R. T. (1995). Susceptibility to memory interference effects following frontal lobe damage: Findings from paired-associate learning. *Journal of Cognitive Neuroscience, 7,* 144–152.

Skaggs, E. B. (1925). Further studies in retroactive inhibition. Psychological Monographs *34*(8), 1–60.

Slamecka, N. J., & Ceraso, J. (1977). Retroactive and proactive inhibition of verbal learning. In W. L. Mikulas (Ed.), *Psychology of learning: Readings*. Chicago: Nelson-Hall.

Small, S. A., Tsai, W. Y., DeLaPaz, R., Mayeux, R., & Stern, Y. (2002). Imaging hippocampal function across the human life span: Is memory decline normal or not? *Annals of Neurology, 51*(3), 290–295.

Squire, L. R., Haist, F., & Shimamura, A. P. (1989). The neurology of memory: Quantitative assessment of retrograde amnesia in two groups of amnesic patients. *Journal of Neuroscience, 9*(3), 828–839.

Squire, L. R., Clark, R. E., & Bayley, P. J. (2004). Medial temporal lobe function and memory. In M. Gazzaniga (Ed.), *The cognitive neurosciences* (3rd ed.). Cambridge: MIT Press.

Stark, S. M., Yassa, M. A., & Stark, C. E. (2010). Individual differences in spatial pattern separation performance associated with healthy aging in humans. *Learning & Memory, 17*(6), 284–288.

Sutherland, R. W., & Rudy, J. W. (1989). Configural association theory: The role of the hippocampal formation in learning, memory and amnesia. *Psychobiology, 17,* 129–144.

Toner, C. K., Pirogovsky, E., Kirwan, C. B., & Gilbert, P. E. (2009). Visual object pattern separation deficits in nondemented older adults. *Learning & Memory, 16*(5), 338–342.

Tulving, E. (2002). Episodic memory: From mind to brain. *Annual Review of Psychology, 53*(1), 1–25.

Underwood, B. J. (1957). Interference and forgetting. *Psychological Review, 64*(1), 49–60.

van Praag, H., Kempermann, G., & Gage, F. H. (1999). Running increases cell proliferation and neurogenesis in the adult mouse dentate gyrus. *Nature Neuroscience, 2*(3), 266–270.

Warrington, E., & Weiskrantz, L. (1974). The effect of prior learning on subsequent retention in amensic patients. *Neuropsychologia, 12,* 419–428.

Warrington, E., & Weiskrantz, L. (1978). Further analysis of the prior learning effect in amnesic patients. *Neuropsychologia, 16,* 169–177.

Winocur, G., & Moscovitch, M. (1996). Heightened interference on implicit, but not explicit, tests of negative transfer: Evidence from patients with unilateral temporal lobe lesions and normal old people. *Brain and Cognition, 30,* 44–58.

Winocur, G., & Weiskrantz, L. (1976). An investigation of paired-associate learning in amnesic patients. *Neuropsychologia, 14,* 97–110.

Wixted, J. (2004). The psychology and neuroscience of forgetting. *Annual Review of Psychology, 55,* 235–269.

Yassa, M. A., & Stark, C. E. (2011). Pattern separation in the hippocampus. *Trends in Neurosciences, 34*(10), 515–525.

Yassa, M. A., Lacy, J. W., Stark, S. M., Albert, M. S., Gallagher, M., & Stark, C. E. (2011a). Pattern separation deficits associated with increased hippocampal CA3 and dentate gyrus activity in nondemented older adults. *Hippocampus, 21*(9), 968–979.

Yassa, M. A., Mattfeld, A. T., Stark, S. M., & Stark, C. E. (2011b). Age-related memory deficits linked to circuit-specific disruptions in the hippocampus. *Proceedings of the National Academy of Sciences of the United States of America, 108*(21), 8873–8878.

Yonelinas, A. P. (2002). The nature of recollection and familiarity: A review of 30 years of research. *Journal of Memory and Language, 46,* 441–517.

Young, W. S., 3rd, & Kuhar, M. J. (1980). Noradrenergic alpha 1 and alpha 2 receptors: Light microscopic autoradiographic localization. *Proceedings of the National Academy of Sciences of the United States of America, 77*(3), 1696–1700.

Chapter 8
Dorsoventral Hippocampus: Subregional Importance in Anxiety and Olfactory Learning and Memory

Christy S. S. Weeden

The hippocampus (HPP) has long been associated with learning and memory. Testing and observations of Henry Molaison, or case study H. M., and his unfortunate loss of declarative memory following the bilateral removal of portions of his temporal lobe influenced scientists to focus on memory investigation (Scoville and Milner 1957). Associations between the HPP and emotional behavior have a long history as well: James Papez initially linked the limbic system, named by Broca (2011), to emotional behavior. It has also been shown that the HPP is indeed, as Aransi (see Sano 1997) initially posited, involved in olfactory memory. Although discrepancy over the finest details remains, evidence indicates multiple roles for the HPP that involve learning and memory, olfaction, anxiety, and stress-related processing. The evidence further indicates that a pattern of function has emerged along the dorsoventral axis to suggest that the dorsal subregions are important for spatial processes and the ventral subregions are important in learning and memory processes for olfactory information.

Anatomy

The hippocampal trisynaptic circuit involves a series of projections leading from cortical structures to different subregions: dentate gyrus (DG), CA3, and CA1. Sensory information arrives at the hippocampal system through superficial layers of the entorhinal cortex with a majority of information arriving at DG from superficial layer II. Additional outputs from superficial layer II synapse onto CA3 and

C. S. S. Weeden (✉)
Section on Neuroplasticity, National Institute of Mental Health, National Institutes of Health, Bldg 35, Rm 3C-911, 35 Lincoln Drive, MSC 3718, Bethesda, MD 20892-3718, USA
e-mail: christy.weeden@nih.gov; christy.weeden@gmail.com

P. A. Jackson et al. (eds.), *The Neurobiological Basis of Memory*,
DOI 10.1007/978-3-319-15759-7_8

information from superficial layer III projects onto CA1 (Hafting et al. 2005; Witter 1993). Although the entorhinal cortex sends information directly to each subregion, the main source of input arrives at the DG. An excitatory cellular layer in DG, the hilus, sends information through an inhibitory layer of interneurons that provides the basis for a recurrent pathway within DG (Witter 1993). Sparse, mossy fibers project from the DG onto CA3 pyramidal cells. There is also a recurrent feedback system within CA3, as pyramidal cells interact via recurrent fibers (Amaral and Witter 1995). Information from CA3 projects onto CA1 via Schaffer collateral system, completing the feed forward of information from entorhinal cortex superficial layer II to DG, CA3, and CA1 subregions of the HPP (Witter 1993). Efferent projections from the HPP originate in CA1 and to a lesser extent, CA3. The majority of information is sent from CA1 to the subiculum, where it is projected to the entorhinal cortex, and from there it is forwarded to parahippocampal structures (Witter 1993). A small amount of information is thought to project from CA3 to the septum, which feeds-forward to the subiculum. Information is projected from the subiculum to entorhinal cortex to complete the processing route (Amaral and Witter 1995).

Although dorsal and ventral (posterior and anterior, respectively, in humans) HPP share common connectivity with some structures, differentiation has been observed. For example, place fields of the dorsal CA1 subregion are more dense, smaller, and fire more reliably in particular locations than do their ventral counterparts (Jung et al. 1994). The physical features of these cells indicate that while ventral CA1 may have some spatial capabilities, the dorsal portion is more suited to process detailed spatial information in the environment. Using anterograde and retrograde tracing techniques, Van Groen and Wyss (1990) conducted a projection study which provides detailed information about heterogeneous connectivity and further indicates common connection patterns along the dorsoventral axis of the HPP, indicating that both dorsal and ventral hippocampal streams project to the subiculum, parasubiculum, entorhinal cortex, and lateral septal nucleus. However, dorsal HPP selectively projects to the retrosplenial and perirhinal cortices. In rats, the dorsal CA1 sends additional information to the subiculum that is further projected to mammillary and anterior thalamic nuclei where it then returns to the dorsal HPP (Dong et al. 2009). The retrosplenial cortex is involved in learning and memory and is associated with anterograde amnesia (Bowers et al. 1988). Of note, mammillary and anterior thalamic nuclei are known to house navigational neurons, which further indicate spatial components (Taube 2007). Perirhinal cortex receives highly processed visual and spatial information from all sense modalities and plays a critical role in learning and memory in humans, nonhuman primates, and rodents (Squire and Zola-Morgan 1991; Witter et al. 1989). Connections between such structures that specialize in navigation and those structures that receive key visual information about the environment further underscore that the dorsal HPP is well-suited to conduct visuospatial processing for learning and memory.

Trace projection methods also illustrate strong ventral HPP-specific projections to structures involved in odor learning and memory as well as anxiety behaviors. Specifically, there are strong connections between ventral HPP and the nucleus accumbens, amygdala, hypothalamus, and olfactory bulb (Van Groen and Wyss 1990). Connections from the ventral HPP to the nucleus accumbens suggest a role in conditioning:

the nucleus accumbens is thought to be the neural substrate of drug addiction and to exert powerful conditioning effects through dopamine regulation (Everitt and Robbins 2005). Evidence suggests that the amygdala contributes to fear and the ventral HPP contributes to anxiety. While these are two distinct and separate functions, many unknown and dangerous situations result in alternation between fear and anxiety, and therefore fear and anxiety can be experienced simultaneously in such situations (Gray and McNaughton 2000). The amygdala and ventral HPP have direct communication, as would be expected between structures that produce behaviors that are frequently displayed together (Van Groen and Wyss 1990). Fear-based emotions, conditioning, and learning are important roles of the amygdala. Chronic stress, such as that experienced by people diagnosed with posttraumatic stress disorder, has been associated with amygdala enlargement and hippocampal atrophy (Hughes and Shin 2011).

Connections to the hypothalamus allow hippocampal communication for many bodily functions, such as hormone regulation, hunger, and feeding motivation, as well as stress and emotion states through hypothalamic involvement in the hypothalamic–pituitary–adrenal axis (HPA) (Dedovic et al. 2009). The ventral HPP also has an exclusive connection to the prefrontal cortex not shared with the dorsal HPP. This unique connection allows direct, rapid information sharing with the most evolved, powerful part of the brain (Bannerman et al. 2004). Both the dorsal and ventral HPP regions project to structures associated with learning and memory, but each region is suited to different processes. Specifically, dorsal HPP connections with spatial and navigational centers equip the region with the capability to navigate through space while the ventral HPP shares connections with structures associated with emotion and executive function.

The olfactory bulb is a key component to olfaction and hippocampal connections endorse associations between odors and memories (Eichenbaum et al. 1989). HPP connects to the olfactory bulb through entorhinal cortex (Vanderwolf 1992). There are additional direct connections between the ventral HPP and olfactory bulb (Gulyás et al. 1998). Gourevitch et al. (2010) provide evidence to suggest that the olfactory bulb drives hippocampal, mostly ventral, beta oscillations (15–35 Hz) at the time of odor information processing. Within the same study, there was no indication that theta oscillations (6–12 Hz) were induced by the olfactory bulb, but high theta activity was observed between dorsal and ventral HPP during the period between a signal that a new odor was available and actual sniffing of the odor, or stimulus expectation. A particularly strong connection exists between the olfactory bulb, through entorhinal cortex, to the DG subregion that is activated when rats experience predator scents (Heale and Vanderwolf 1999). Amygdala lesions fail to impact olfactory bulb-DG beta activity patterns, providing support to the concept that the amygdala and ventral HPP differentially contribute. Specifically, predator scent appears to activate anxiety states, as evidenced by precise patterned beta activity that is generated at the point of the olfactory bulb that projects to the DG of the HPP (Heale and Vanderwolf 1999). To further distinguish between fear and anxiety states, several studies have posited that the presence of predator scent indicates only that a predator has been in the local environment at some previous point and may no longer be present, and thus predator odors may not be tapping directly into fear but rather anxiety over the *possibility* of their presence. Dominant beta activity in the olfactory bulb, coupled with matched oscillations in the HPP and strong ventral HPP projections, lends substantial evidence to a role for the ventral HPP in olfactory-based learning and memory processes, while also indicating further specification for a role in anxiety, and not fear, based systems.

Anatomical evidence also indicates distinct receptor differentiation along the dorsoventral axis of the HPP. For example, a study conducted to investigate receptor characteristics in hippocampal rat brain slices suggests that there are different subtypes of the N-Methyl-D-aspartate (NMDA) receptor within the DG (Pandis et al. 2006). Research based on induction of stress and long term potentiation (LTP) function underscores the complexity of neural potentiation because different methodologies have been able to show increased, decreased, and unchanged LTP within the DG in response to stress (Bramham et al. 1998; Kavushansky et al. 2006; Vouimba et al. 2004). Due to complex relationships between environmental stressors and anatomical differences, it remains unclear whether differential cellular firing characteristics extend along the entire dorsoventral axis in a similar manner for each hippocampal subregion. Evidence from in vivo animal recordings suggests that the dorsal region of the HPP may contain more complex and reliable firing for place fields than ventral HPP (Royer et al. 2010). It is difficult to elicit a large population of recordings from ventral DG, and as such it is difficult to make direct comparisons. Despite methodological differences across studies, results from *individual* studies provide substantial support for clear differences in dorsal and ventral cellular firing characteristics. As the field of electrophysiological recording advances and new techniques arise to examine difficult-to-study regions such as ventral DG, it may be possible to conduct studies in which dorsal and ventral areas of DG, CA3, and CA1 all receive similar treatments to monitor possible differential output.

Theories About Dorsal and Ventral Hippocampal Functions

O'Keefe and Nadel (1978) suggest that the HPP is a cognitive mapping structure. They provide a two-part model in which the "place" system recognizes information related to familiar sensory information and in contrast the "misplace" system actively detects added or subtracted information. The nature of detecting what is different gives rise to exploratory behavior that is the result of a mathematical computation. The cognitive map theory places a high importance on spatial information, which has enticed many researchers to investigate the HPP from a spatial memory angle. The resultant research has led to decades of discoveries about the relationship between hippocampal function and spatial learning and memory processes.

Rolls and Kesner (2006) provide an updated computational theory that promotes subregional specificity based on anatomical characteristics as well as behavioral evidence. They contend that the purpose of the HPP and its subregions is to encode sensory and spatial information, with the ability to recall an entire episode based on a partial cue of that original episode. According to the cognitive map theory, information is transformed in order to facilitate "place" and "misplace" systems, however, Rolls and Kesner (2006) emphasize that the computational model receives and retrieves that information in its relatively similar, originial form.

The computational model proposed by Rolls and Kesner (2006) is backed by empirical behavioral evidence in rodents and nonhuman primates, and even human evidence is provided in some cases. Hippocampal subregions have been investigated and assigned particular roles within learning and memory processes. Briefly, the DG is necessary for pattern separation, through mossy fibers that output onto CA3 pyramidal cells. Physical characteristics and autoassociative inputs of CA3 neurons indicate a central role in rapid encoding and "arbitrary associations where space is a component." CA3 autoassociative network properties also indicate importance in pattern completion, in which it is possible to retrieve an entire memory from one cue. CA1's projections to the neocortex suggest a role in consolidation processes. Additionally, Rolls and Kesner (2006) point to evidence of the CA1 subregion's involvement in temporal processing of information across a series of time. This computational model focuses on dorsal hippocampal processing, as much of the behavioral evidence involves dorsal subregional lesions. However, this theory does not necessarily disagree with ventral-focused models of hippocampal subregion accounts of function, especially if dorsal functions are viewed as indicators of possible ventral functions for nonspatial sensory information processing.

Links between fear and/or anxiety-provoking tasks with rodents as well as strong hippocampal and amygdala connectivity led many researchers to attribute impairments in anxiety measures to amygdala manipulation of portions of the HPP. However, Gray and McNaughton (2000) make a convincing case to suggest that the amygdala contributes to fear, but the HPP contributes separately to anxiety. They posit that the HPP is involved when goals overlap, especially when at least one of the goals is associated with a known, possibly dangerous outcome. Gray and McNaughton (2000) suggest that the HPP contributes information about the previously experienced or known dangerous component and presents it in a way that increases the saliency of the possible dangerous component. In this way of increasing saliency of possible dangerous outcomes the HPP contributes to conflict resolution between competing goals.

The cost of magnification of possible dangerous outcomes, however, creates an "anxiety state" (Gray and McNaughton 2000). Specifically, underscoring dangerous possibilities creates a feeling of anxiety, and when that anxiety level exists simultaneously as the HPP is involved in processing possible outcomes, the information is stored for future goal conflict situations. Gray and McNaughton (2000) suggest that this is the key way in which the HPP possesses both cognitive and emotional components. They further indicate that the cognitive information related to anxiety is managed in the HPP and that may be independent of learning and memory. Gray and McNaughton (2000) offer a robust presentation of evidence, with the majority of support found in studies that involve antianxiety drugs, as such drugs tend to act on the HPP rather than amygdala.

Moser et al. (1993) provided initial, direct evidence of heterogeneity across the dorsoventral axis of the HPP. While rodents with dorsal HPP lesions displayed the expected spatial learning impairments, ventral HPP lesions revealed minimal spatial deficits. Further investigation in a water maze revealed that although ventral-lesioned animals were impaired on retrieval when lesions were made after training, they were not impaired when training occurred after lesions (Moser and Moser 1998a). Lesions of the whole HPP resulted in deficits for both pre- and post-training conditions and dorsal hippocampal lesions resulted in similar responses (Moser and Moser 1998a). These findings led the researchers to postulate that spatial information is processed in the dorsal region of the HPP and that while the ventral HPP can be involved in encoding, it is not necessary for encoding or retrieval of spatial information, further implicating alternative roles for the ventral HPP (Moser and Moser 1998b).

Bannerman et al. (1999) suggest that the ventral HPP is involved in anxiety and the dorsal HPP is specialized for retrieval processes. Further, this group proposed a solution to questions about what specifically defines dorsal and ventral portions of the HPP when they suggested that the dorsoventral axis should be divided equally; septal and temporal poles each contain 50 % of the total volume of the HPP. While this may seem an arbitrary division, this landmark system can be easily identified by researchers across the field—and can serve to clear up many discrepancies due to differential anatomical definitions. Bannerman et al. (1999) support the idea that the dorsal HPP is preferentially involved in spatial learning and memory processes and that the ventral HPP is preferentially involved in processing anxiety. They support Gray and McNaughton's (2000) concepts regarding hippocampal activity in cognition and emotion, but further stipulate that the HPP is differentially involved, in that the dorsal and ventral components of the HPP separately process information (Bannerman et al. 2004).

Fanselow and Dong (2010) relegate "cold" cognitive functions to the dorsal HPP and "hot" functions of stress, emotion, and affect to the ventral HPP. They suggest that the lack of a specific definition of the dorsal and ventral boundaries may be the source of some discrepancies in previous research and go beyond the suggestion from Bannerman et al. (2004) that dorsoventral distinctions can be drawn at equal halves. Instead, Fanselow and Dong (2010) posit that more precise borders can be identified with gene expression techniques and they provide behavioral and anatomical evidence to further bolster that claim. Individual expression patterns do not outline specific borders, but when several are co-examined their grouped expression patterns clearly differentiate along the dorsoventral axis and also by subregion. Molecular gene expression methods inform based on particular, individual expression for proteins that carry out small activities that eventually translate to cell-wide characteristics, further defining and influencing structural functionality.

Hippocampus

The ventral HPP plays a seminal role in both behavioral and hormonal expressions of anxiety. Gray and McNaughton (2000) have suggested that fear and anxiety are two separate emotional states that are differentially controlled by the amygdala and

HPP, respectively. The elevated plus maze is commonly used to measure anxiety in rodents and can serve here to highlight differences in fear and anxiety processes. It consists of two arms that have high walls, which greatly reduce the possibility of falling, and two open arms that do not protect the animal from potentially falling off. Anxiety arises due to the given nature of rats' exploration habits to explore all areas of a new environment in combination with the potential hazard of falling off one of the open arms. Rats with ventral or whole hippocampal lesions are impaired on the elevated plus maze, as evidenced by spending considerably more time in open arms than controls (Kjelstrup et al. 2002). Interestingly, rats administered anti-anxiety medications, such as midazolam, are similarly impaired on the elevated plus maze (Kjelstrup et al. 2002). According to Gray and McNaughton (2000), there may be a mechanism within the HPP that increases the saliency of the potential danger of falling in order to resolve the conflict. The fact that ventral HPP lesions mimic the behavioral effects of antianxiety medications is strong support that the ventral HPP mediates anxiety.

Several other methods of measuring anxiety in rodent models support an important role for the ventral HPP in anxiety. For example, the concept of the open field apparatus is that when rodents are exposed to bright, open spaces without the option to escape to a more secluded and covered environment, they have been shown to release more fecal matter than others not exposed to the open environment (Walsh and Cummins 1976). Rats with ventral or whole hippocampal lesions produce significantly less fecal matter than controls and dorsally lesioned subjects (Bannerman et al. 2003). Following exposure to foot shocks, measures of a stress hormone, corticosterone, were found in significantly lower levels in ventral and hippocampal-lesioned subjects compared to dorsal and control-lesioned subjects. However, in a working memory version of the Morris water maze, dorsal and complete hippocampal-lesioned subjects were impaired when compared to ventral HPP and control-lesioned subjects (Bannerman et al. 2003). There are clear distinctions between dorsal and ventral hippocampal lesions such that the dorsal HPP is critically involved in behaviors associated with spatial learning and memory, and the ventral HPP is critically involved in expression of anxiety behaviors.

Fear-based behavior can be associated with an almost reflexive-like behavior of moving away from a source of danger. But when removal from the situation is not possible, defensive behaviors, such as crouching or freezing, can be observed. A series of experiments exploring defensive behaviors indicate that the ventral HPP is involved in defensive behaviors. Specifically, rats with ventral, but not dorsal, hippocampal lesions failed to display defensive behaviors when exposed to cat odors, but did perform like controls when exposed to an actual cat (Pentkowski et al. 2006). The researchers suggest that the behaviors exhibited in the presence of predator odor and an actual predator presence arise from different sources. When cat odor was available, it may have provided the possibility of danger, which should have increased anxiety. However, the presence of an actual predator (cat) may have elicited a direct fear-based response that would be mediated by the amygdala. Interestingly, the authors of the defense behavior study note that in a subsequent re-exposure to the cat-odor context, rats with ventral lesions increased sniffing and exploration (Pentkowski et al. 2006). It is important to note that whole, but not dorsal, HPP lesions impair nonspatial odor representations, which indicates the ventral HPP is the necessary structure (Levita and Muzzio 2010). Therefore, it may be that ventral HPP-lesioned subjects were impaired in making associations between the context and predator odor. Further, it is possible that rodents with ventral hippocampal lesions may have impairments in forming the concept that a predator had recently been in the environment (as evidenced by the odor), as evidence suggests that the ventral CA1 subregion is necessary for temporal associations involving odors (Hunsaker et al. 2008). Although there was a foot-shock-context component to the behavioral battery, the task was not odor based. In order to determine if impairments were caused by inhibited anxiety or failed odor-context-threat associations, an alternative paring condition in which presentation of a novel odor and exposure to the actual predator occur simultaneously should be considered. The odor could then be presented in both similar and different contexts, as well as the home-cage, in order to further determine whether a true association had been formed. Additionally, one treatment that pairs a novel odor with foot-shock and another treatment that pairs only context with foot shock may shed more light on the context versus anxiety confound.

Recent investigations support important roles for the ventral HPP that have been previously attributed to amygdala control over the region. Rodents with ventral hippocampal lesions were faster than controls to enter new portions on a successive alleys test, in which a subject travels through sections of a runway that provide progressively less-protective features against falling and danger in regard to brightness, platform width, and wall height (Treit et al. 1993). Animals with amygdala lesions behaved similar to controls, but did explore the open arms after injection of an antianxiety drug, diazepam. The change to impaired ability to demonstrate anxiety behaviors only after administering antianxiety drugs indicates that the amygdala is not critical for expression of anxiety. However, the amygdala is necessary for fear expression and does not require the ventral HPP to play a critical role (Decker et al. 1995).

Rats are typically slow to begin eating novel food, especially in a novel environment, yet rats with ventral HPP lesions were significantly faster to begin consuming novel foods in familiar or novel environments, and rats with amygdala lesions were significantly slower to begin consumption (McHugh et al. 2004). In a social interaction test, rats with ventral hippocampal lesions spent significantly longer amounts of time investigating unfamiliar rats, but showed reduced aggressive behaviors (McHugh et al. 2004). In contrast, subjects with amygdala lesions showed a reduction in both social investigation and aggression. Disinterest in social exploration is similar to "social indifference" that is characterized in Klüver Bucy syndrome (Baron-Cohen et al. 2000). The evidence here clearly supports differential roles for the amygdala and ventral HPP in social exploratory behaviors such that the amygdala is important for interest in others, as well as dominance behaviors. However, the ventral HPP may serve to reduce anxiety associated with novel encounters and as such may increase exploration with a reduction in displays of aggression. It is important to again consider the possibility that impairment in odor memory may be an alternative explanation for heightened social exploration (especially sniffing) in ventral HPP-lesioned subjects. It may also play a role in readiness to consume novel foods. However, this is not likely, because ventral HPP lesions leave spatial learning and memory mainly intact. Because rats with ventral hippocampal lesions were also significantly faster to consume foods in novel environments, the evidence suggests that the overall lack of anxiety associated with novel foods (and social encounters) is not due to impairments in olfactory processes moderated by the ventral HPP.

Fear conditioning methodology relies on nonverbal behaviors, such as freezing, to indicate that an association between at least one conditioned stimulus and an unconditioned stimulus, such as a conditioned tone and an unconditioned foot shock (Fanselow 1984). Evidence suggests that the ventral, not dorsal, HPP is involved in fear conditioned freezing (Kjelstrup et al. 2002). Only ventral lesions of the HPP resulted in impaired freezing behavior similar to whole HPP lesions, while dorsal lesions did not impair freezing behavior. Yoon and Otto (2007) provide evidence to support the role of the ventral HPP in both acquisition and retrieval of auditory trace fear conditioning. Additionally, they found that the dorsal HPP-lesioned subjects were also impaired if the lesions occurred after training (Yoon and Otto 2007). One possible explanation is that the dorsal HPP may be involved in, but not critical for, auditory trace fear conditioning. Dorsal HPP lesions that occurred pre-training eliminated the optional use of dorsal HPP, and the ventral HPP may have supplemented encoding of training for auditory trace fear conditioning. But when training occurred before lesions were made, there may have been some dorsal processing that would later be lost when lesions were conducted, hence retrieval may be limited and may explain failure by post-trained ventral-lesioned subjects to navigate to the platform location. Given the well-established role of the amygdala in fear, one may question these observations of ventral HPP impairment in fear situations. One might assume, and many did, the ventral HPP is subservient to the amygdala. However, as several theorists have noted, conditioned freezing behavior may not be indicative of fear per se, but rather an induced anxiety state, which would be supported by the ventral HPP, rather than the fear-based network within the amygdala.

Electrophysiological recordings of activity within dorsal and ventral HPP provide evidence to suggest key differences in receptors and binding affinities of neurotransmitters. Specifically, Gage and Thompson (1980) found a higher binding affinity for serotonin in the dorsal HPP, but a higher density of serotonin receptors in the ventral region, as well as a higher density of norepinephrine receptors. Differences in posterior (dorsal) and anterior (ventral) hippocampal activity is also evident in primates. Electrophysiological recordings indicate that the posterior region of the HPP in monkeys is significantly more active during a delay in a spatial task (Colombo et al. 1998).

A well-known imaging study performed by Maguire et al. (2000) demonstrates the powerful effects that spatial navigation of an environment can have on the human HPP. Structural magnetic resonance imaging (MRI) was used to investigate the HPP of London taxi drivers, whose job is notoriously difficult as the streets of London are individually unique and complex: Essentially a cognitive spatial map. Imaging revealed that the posterior (dorsal) portion of the HPP had increased, and did so in relation to years of experience, and that the anterior (ventral) HPP in this group was smaller than controls (Maguire et al. 2000). This is strong evidence for the effects of spatial navigation expertise and how it affects the HPP. For obvious reasons, rodents are the main source of lesion data pertaining to the HPP. And there are always concerns of translation to human HPP functions, especially because a much larger percentage of the brain of rodents is dedicated to the HPP. This study suggests that the human HPP is also dedicated, to some degree, to spatial learning and memory processes and indicates that differentiation across the dorsoventral axis of the HPP is likely present in humans, at least for a dorsal HPP dominance in spatial learning and memory functions.

Dentate Gyrus

Behavioral studies are important to hippocampal research in general because many lesion techniques, both permanent and reversible, provide a way to silence particular subregions in order to determine their vital contributions to hippocampal processing. Investigations have led to identification of roles for DG in pattern separation, CA3 in rapid encoding, pattern completion, and arbitrary associations as well as a role for CA1 in temporal processes. A subregional analysis of evidence along the dorsoventral axis of the HPP has elucidated certain patterns and themes.

Behavioral research indicates that dorsal DG is important in spatial pattern separation. For example, Gilbert et al. (2001) demonstrated that the dorsal DG is critical for separating highly similar spatial information, or patterns of spatial location, in a delayed, matching-to-sample task. Rats given dorsal DG lesions after acquisition of a spatial pattern separation task were unable to perform the task correctly when choice objects were located close together, but performed similar to controls when the objects were placed further apart. A lack of impairment when objects were spaced further apart implies that the dorsal DG is not critical for well-distinguished-in-space objects, but the impairment that resulted when objects were close together suggests that the dorsal DG is involved in pattern separation for partially overlapping spatial representations (Gilbert et al. 2001).

In order to circumvent a potential confound of pretest training, Goodrich-Hunsaker et al. (2008) exposed subjects to exploratory tasks that did not require pre-training, as behavioral measurements are based on rodent tendency to investigate what is new. After an initial exploration, objects in the environment were moved closer together, and exploratory behavior during another session was observed. Though all lesioned subjects displayed a small degree of impairment, dorsal DG-lesioned subjects were severely impaired. As previously mentioned, the DG subregion of the HPP is particularly suited to perform pattern separation due to the sparse nature of the mossy fiber system. The ventral portion of the DG remained intact in the above studies and is therefore unlikely to be critically involved in pattern separation for spatial information (Goodrich-Hunsaker et al. 2008). From a parallel processing point of view, the well-documented role of the dorsal DG in pattern separation of spatial information suggests that the ventral DG may be important for pattern separation of olfactory information.

A nonspatial, odor-based test that was previously used by Kesner et al. (2011) to demonstrate that the ventral HPP but not dorsal HPP plays an important role in olfactory learning and memory for highly similar odor information was used to test the role of the ventral DG in odor processes. Rats were trained to dig in sand filled cups laced with odorant stimuli and then were given either control or ventral DG lesions. Odorants were a series of aliphatic acids identical to one another except for the number of methyl/carbon chains, which increased by one for each odorant in the series. For each test, rats were presented with two odor-laced choice cups. A correct choice was demonstrated by digging in a sand cup laced with the same odorant presented during the sample phase. Once criterion performance rates were met (see Fig. 8.1), rats were given either ventral DG lesions or sham surgery. Rats with ventral DG lesions performed relative to controls with a short delay between sample and test phases (see Fig. 8.2). When delays were extended to 60 s, rats with

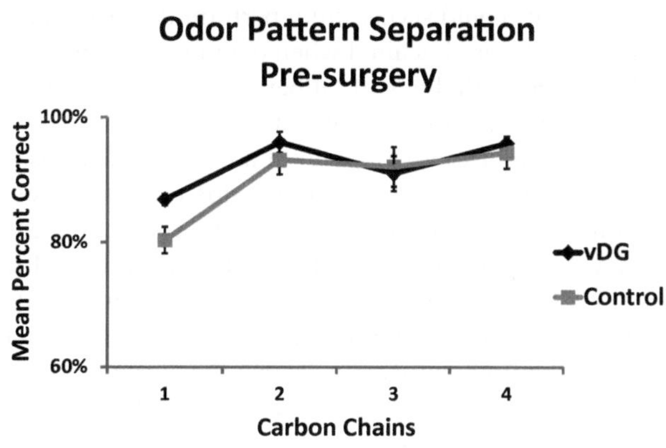

Fig. 8.1 Mean (±SE) performance of pre-lesion subjects for levels of odor separation. Data consist of performance levels at criteria, which was 80–90 % correct on the most recent block of 16 trials

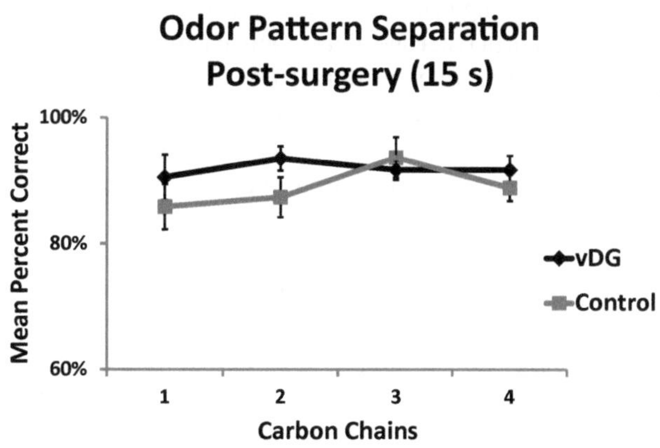

Fig. 8.2 Mean (±SE) performance of ventral DG and control rats for levels of odor separation with a delay of 15 s

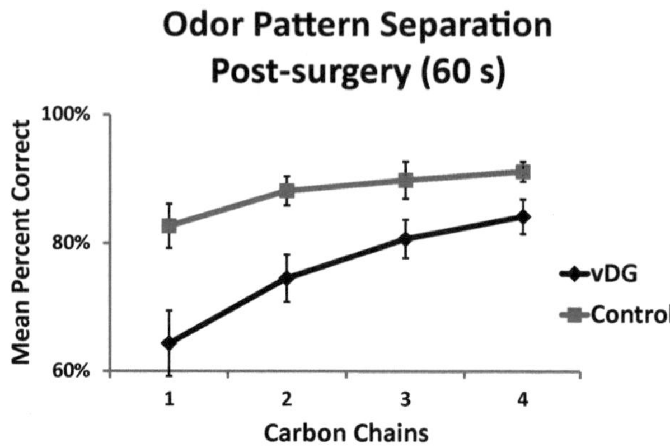

Fig. 8.3 Mean (±SE) performance of ventral DG and control rats for levels of odor separation with a delay of 60 s

ventral DG lesions were impaired relative to controls but only when choice odors were highly similar (one or two methyl group separations), however, they were not impaired when odorants were less similar (Weeden et al. 2012) (see Fig. 8.3). That lesioned rats could successfully perform the task when odor information was more distinct and were impaired when odorants were more similar compliments the dorsal DG spatial findings of Gilbert et al. (2001) and provides strength to a parallel processing role for the DG in pattern separation.

Investigations into the field of adult born neurons, or neurogenesis, have led to indirect comparisons that support a dorsoventral processing relationship for spatial information and anxiety. The process of development, organization and eventual survival of adult born neurons occurs in the DG of the HPP (Christie and Cameron 2006). The *Fmr1* knockout mouse, commonly employed as a mouse model of Fragile X Syndrome (FXS), has been associated with altered neurogenesis and has been shown to significantly reduce the survival rate of adult born neurons in the ventral, but not dorsal, DG (Eadie et al. 2009). *Fmr1* knockout mice exhibit less anxiety behaviors than controls on a series of anxiety tasks, such as the open field and elevated plus mazes. However, Eadie et al. (2009) did not observe impairments relative to controls for knockout mice in a spatial water maze task or subsequent reversal learning task. Subjects exposed to restraint stress were shown to have lower levels of corticosterone than controls in the same condition (Eadie et al. 2009). Higher survival rates of dorsal DG adult born neurons and successful completion of spatial tasks indicate minimal damage or dysfunction to the dorsal DG for *Fmr1* knockout mice. However, low neuron survival rates in the ventral DG, in combination with altered anxiety-based test performance and abnormally low corticosterone levels following stress suggest a significant level of impairment in the ventral DG.

Chronic stress has been associated with reduced hippocampal volume (Magariños et al. 1996). Chronic mild stress, over a period of weeks, has been shown to reduce ventral, but not dorsal, DG volume (Jayatissa et al. 2006). After initial chronic stress treatment, escitalopram, an antianxiety drug, was administered in one study. For those rodents indicating a positive response to the drug treatment, as measured by renewed consumption of sweetened water, reversal of ventral DG depletion was observed. Vulnerability of ventral DG cytogenesis to stressors is striking, and is further evidence of a role in anxiety and emotion, while the dorsal DG appears to have minimal interaction with the factors of chronic mild stress.

High resolution fMRI studies support pattern separation in the DG. Presentation of highly similar images resulted in brain activity patterns consistent with pattern separation and were found only in the DG/CA3 region of the medial temporal lobe in humans (Bakker et al. 2008). In a follow up study, images of objects were manipulated so that the level of similarity varied. Several regions indicated increased activation when larger changes were implemented, but only the DG/CA3 region provided a higher response when images were very similar, indicating detection of differences on a detailed scale, or pattern separation. Interestingly, older adults meeting the criteria of mild cognitive impairment perform poorly in pattern separation studies and show hyperactivity in the DG/CA3 region (Yassa et al. 2010). Participants displaying mild cognitive impairment were also shown to have significantly reduced hippocampal volume with the most severe deformities in the DG/CA3 region, which indicates the region is necessary for detection of differences in similar presentations, and thus pattern separation processes (Yassa et al. 2010).

CA3

It has been suggested that the anatomical features of the CA3 subregion of the HPP are suitable for rapid encoding, as the recurrent collaterals allow for new information to be processed (Hunsaker and Kesner 2012; Rolls et al. 1997). Lee and Kesner (2004) have shown that dorsal CA3-lesioned subjects are impaired during the onset of freezing behavior (but eventually did display freezing) during the intertrial interval following a foot-shock and tone acquisition session. The experiment authors interpret these observations as indications that the delay was the result of a lack of rapid encoding processing availability due to the CA3 lesion. Gold and Kesner (2005) have shown that the CA3 subregion is also important in pattern completion. After selecting one of five objects in a sample phase in a controlled environment where only four cues were present, rats were trained to select the same item on the test phase to retrieve a reward. Lesions were implemented post-training. CA3-lesioned rats were only impaired when a single cue was present, but performed similar to controls when more than one cue was available. Impairment when minimal cues are present may represent an inability to complete a pattern of spatial cues that are no longer present in order to locate the correct spatial location to make a selection (Gold and Kesner 2005).

Differences in receptors and receptor binding abilities indicate variations in synaptic communications and indeed, some characteristics of cellular firing patterns differentiate across the dorsoventral axis of the CA3. For example, place fields in dorsal CA3 have been shown to be more specific, with mostly a single place field association to one arm of a radial arm maze in a water-foraging task (Royer et al. 2010). In contrast, ventral CA3 place fields have been implicated as less consistent, often overlapping, in which one place field can be activated by multiple arms of a radial maze. In addition, reliability to consistently fire when in specific spatial locations was high in dorsal, but low in the ventral CA3 subregion (Royer et al. 2010). Overall, Royer et al. (2010) found fewer complex firings (consisting of multiple spike components) in ventral CA3 compared to dorsal CA3. The results are consistent with the view that dorsal HPP subregions are primarily involved in spatial functions.

Imaging evidence suggests the ventral CA3 plays an important role in olfactory memory functions. fMRI imaging coupled with non-echo planar imaging (EPI) revealed that neural activity shifted from the ventral toward the dorsal region of the HPP when mice had previous experience with an odor, such as with a foot shock (Kent et al. 2007). However, the experience-dependent effects were negated in mice lacking CA3 NMDA receptors. The evidence indicates support for a ventral CA3 role in forming associations between contexts and scents, which could be implicated as having parallel processes with the spatial associations of dorsal CA3.

CA1

Evidence suggests that the dorsal and ventral CA1 subregions of the HPP are differentially involved in temporal order processing for spatial and olfactory information (Kesner and Hunsaker 2010; Kesner et al. 2010). The dorsal CA1 subregion provides a critical role for learning and memory for sequences of spatial locations (Gilbert et al. 2001). It has been shown that dCA1 is involved, but not critically important in processing topological changes in objects (Goodrich-Hunsaker et al. 2008). Specifically, rats with CA1 lesions failed to demonstrate an increase in exploration for two of four objects that were switched from their original configuration. In the same report, Goodrich-Hunsaker et al. (2008) investigated object placement in which objects were moved to novel locations in a metric task. In this task, there was also a mild disruption of dorsal CA1-lesioned subject exploration compared to controls, though not as impaired as exploration in dDG-lesioned rats.

Hunsaker et al. (2008) conducted an investigation into the possibility of differential roles for the dorsal and ventral CA1 subregions within the scope of temporal processing. Lesions of the dorsal CA1 led to deficits in temporal processing for spatial locations in which objects in recent locations were explored more than objects in earlier locations within the sequence. Ventral CA1 lesions led to deficits in temporal processing of odors, in which rats displayed a preference to explore the most recent odorant. Additionally, dorsal CA1-lesioned rats demonstrated a preference for more recently displayed visual objects than those that were displayed earlier, while ventral CA1-lesioned subjects explored the more recent objects slightly more than they explored objects that were earlier in the sequence. Preference for novelty was confirmed in each of the three temporal tasks, which negates the question of preference for new versus old. The evidence provided by Hunsaker et al. (2008) contributes substantial support that parallel processes are at work across the dorsoventral axis of the CA1 subregion of the HPP. Specifically, CA1 is critically involved in temporal processing with parallel processes in the dorsal (spatial) and ventral (olfaction) poles.

Evidence suggests that dorsal and ventral CA1 subregions may also be differentially involved in trace fear conditioning. Rogers et al. (2006) investigated the dorsal and ventral CA1 in trace fear conditioning in which a 10 s trace was implemented between a tone and foot shock. After 24 h, subjects were tested for retrieval of contextual fear conditioning, and were tested again 48 h after acquisition for tone. Dorsal CA1-lesioned rats demonstrated acquisition of contextual fear conditioning but they showed diminished freezing behavior on a retention test for context 24 h later. Ventral CA1-lesioned rats also acquired fear conditioning but did not indicate retrieval for context on the follow up test, and were notably more impaired as evidenced by even less freezing behavior than rats with dorsal CA1 lesions. However, rats with ventral CA1 lesions showed impairment on trace interval and were somewhat impaired on tests of retention for tone at 48 h after acquisition while dorsal CA1-lesioned subjects did not display impairments on retrieval. There is a well-established literature on the importance of dorsal CA1 in temporal learning. It is possible that the CA1 subregion was not necessary to demonstrate acquisition of the tone-trace-shock pairings that occurred approximately 72 s apart. However, encoding of temporal information for a sequence of events over a longer period of time (such as consolidation) may, in fact, rely on the CA1 subregion of the HPP. This may explain why lesions to either dorsal or ventral CA1 would not result in immediate deficits in acquisition, but would imply impairment a day after initial exposure. The fact that both groups were not impaired on acquisition but were both impaired on context retrieval indicates that both dorsal and ventral CA1 are involved in temporal processing for sequential information. Additionally, ventral CA1 was impaired when presented with the tone 48 h after acquisition. It is also possible that there was an arbitrary association between foot shock and tone, without regard to temporal order, that would easily be evidenced by freezing. Controls and dorsal CA1 but not ventral CA1-lesioned rats displayed such freezing. Rather than being of a temporal nature, it is possible that lesioning of a portion of the ventral HPP reduced ability to express anxiety, which would be evidenced by freezing. This provides a possible explanation for the results, as interpreted along the lines of parallel processing, with anxiety behaviors being absent in ventral CA1-lesioned subjects, except during acquisition when many trials are presented, and dorsal CA1 showing impairment for temporal sequences of contextual, perhaps spatial, information, while not being impaired on tone trace retention because it is not of a spatial nature.

It was suggested previously that more complex patterns of electrophysiological activity in dorsal rather than ventral CA3 during exploration in live animals was support for a stronger, primary role for dorsal HPP in spatial encoding (Royer et al. 2010). One concern with this view is that perhaps anatomical differences in the cells make dorsal hippocampal cells better suited to produce complex spike patterns, regardless of the kind of information that is processed (spatial, temporal, olfactory, etc.). Gilbert et al. (1985) have observed an inverse relationship in firing patterns in which more complex firing patterns were observed in the ventral portion of CA1.

There are substantial differences between the studies that must be taken into account. Although more complex firing patterns were observed in dorsal CA3 cells during active exploration of an environment in vivo, the dominance of ventral CA1 firing patterns were solicited in vitro from prepared hippocampal slices in which the Schaffer collaterals were stimulated in dorsal and ventral regions (Gilbert et al. 1985; Royer et al. 2010). Another concern is that CA3 and CA1 cells have different assemblies and mechanisms. While a direct comparison cannot be made between CA3 and CA1 activity patterns, additional in vitro evidence demonstrates that it is possible to modify, and even invert the relationship of activity dominance. For example, ventral CA1 can produce synchronized activity at low frequencies that does not disrupt dorsal CA1 (Papatheodoropoulos and Kostopoulos 2000). And repeated ventral priming with a low frequency stimulus is able to increase the intensity of ventral activity to a higher degree than the dorsal portion (Papatheodoropoulos and Kostopoulos 2000). The ability to invert action potential dominance in vitro is a strong indicator that dorsal and ventral HPP take primary and secondary roles in different situations. A secondary role for dorsal CA1 would indicate that there may be at least one learning situation in which ventral CA1 is the primary or critical substrate.

Discussion

From speculations of a role in inhibition to a cognitive map, dorsal research has been the prominent hippocampal focus. Subregional analyses have, for the most part, clarified the roles of CA1, CA3, and DG within the HPP as they are related to spatial learning and memory processes. Ventral hippocampal roles have recently been investigated and are now beginning to be elucidated, resulting in less clarity and more questions about hippocampal functions. Behavioral studies provide substance for the claim that hippocampal subregions can be viewed as essentially separate structures along the dorsoventral axis (Fanselow and Dong 2010). The heterogeneity in dorsal and ventral HPP functions leaves the little-explored ventral HPP, at least from a subregional perspective, behind. Although it seems that ventral subregional investigations will also require decades of investigation to gain the level of knowledge that is known about dorsal subregions, one "shortcut" is to use the extant literature available about the dorsal HPP to guide further ventral research.

Over the last decade, Raymond Kesner has conducted a wealth of subregional behavioral investigations of the HPP for event-based memory processes (see Kesner 2009). Following his work, and that of others, new ventral studies that investigate roles in olfaction and anxiety have been conducted. For example, a behavioral lesion study was used to demonstrate the importance of the (dorsal) DG in spatial pattern separation inspired modification of an odor-based task to show that the ventral HPP, and later and more specifically, the ventral DG, is important for olfactory pattern separation (Cleland et al. 2002; Gilbert et al. 2001; Kesner et al. 2011; Weeden et al. 2012). Chiba et al. (1994) have further investigated temporal processing with respect to hippocampal function in a similar fashion. Specifically, they designed a radial maze paradigm and demonstrated that the HPP is important for temporal information processing for spatial information in rats (Chiba et al. 1994). In order to parse out the spatial and temporal components, Kesner et al. (2002) removed the spatial component by modifying the task to use odor-based cues. This task was then used to demonstrate that the CA1 subregion plays a critical role (Hunsaker et al. 2008). And finally, Kesner et al. (2010) used the odor-based paradigm to show that the ventral CA1, but not dorsal, subregion plays a critical role in temporal memory for odor information.

Such behavioral investigations, along with imaging, eletrophysiological, and anatomical evidence has provided a strong basis to claim that heterogeneity of hippocampal subregions follows a parallel processing construct. Together, the larger body of evidence reflects that the dorsal subregions of the HPP are important for spatial learning and memory and the ventral subregions of the HPP are important for olfactory learning and memory as well as for anxiety-based behaviors.

References

Amaral, D. G., & Witter, M. P. (1995). Hippocampal formation. In G. Paxinos (Ed.), *The rat nervous system* (pp. 443–493). San Diego: Academic.

Bakker, A., Kirwan, C. B., Miller, M., & Stark, C. E. L. (2008). Pattern separation in the human hippocampal CA3 and dentate gyrus. *Science (New York, N.Y.), 319*(5870), 1640–1642. doi:10.1126/science.1152882.

Bannerman, D. M., Yee, B. K., Good, M. A., Heupel, M. J., Iversen, S. D., & Rawlins, J. N. (1999). Double dissociation of function within the hippocampus: A comparison of dorsal, ventral, and complete hippocampal cytotoxic lesions. *Behavioral Neuroscience, 113*(6), 1170–1188.

Bannerman, D. M., Grubb, M., Deacon, R. M. J., Yee, B. K., Feldon, J., & Rawlins, J. N. P. (2003). ScienceDirect.com—behavioural brain research—ventral hippocampal lesions affect anxiety but not spatial learning. *Behavioural Brain Research, 139*(1–2), 197–213.

Bannerman, D. M., Rawlins, J. N. P., McHugh, S. B., Deacon, R. M. J., Yee, B. K., Bast, T., et al. (2004). Regional dissociations within the hippocampus—memory and anxiety. *Neuroscience & Biobehavioral Reviews, 28*(3), 273–283. doi:10.1016/j.neubiorev.2004.03.004.

Baron-Cohen, S., Ring, H. A., Bullmore, E. T., Wheelwright, S., Ashwin, C., & Williams, S. C. (2000). The amygdala theory of autism. *Neuroscience & Biobehavioral Reviews, 24*(3), 355–364.

Bowers, D., Verfaellie, M., Valenstein, E., & Heilman, K. M. (1988). Impaired acquisition of temporal information in retrosplenial amnesia. *Brain and Cognition, 8*(1), 47–66. doi:10.1016/0278-2626(88)90038-3.

Bramham, C. R., Southard, T., Ahlers, S. T., & Sarvey, J. M. (1998). Acute cold stress leading to elevated corticosterone neither enhances synaptic efficacy nor impairs LTP in the dentate gyrus of freely moving rats. *Brain Research, 789*(2), 245–255.

Chiba, A. A., Kesner, R. P., & Reynolds, A. M. (1994). Memory for spatial location as a function of temporal lag in rats: Role of hippocampus and medial prefrontal cortex. *Behavioral and Neural Biology, 61*(2), 123–131.

Christie, B. R., & Cameron, H. A. (2006). Neurogenesis in the adult hippocampus. *Hippocampus, 16*(3), 199–207. doi:10.1002/hipo.20151.

Cleland, T. A., Morse, A., Yue, E. L., & Linster, C. (2002). Behavioral models of odor similarity. *Behavioral Neuroscience, 116*(2), 222–231. doi:10.1037//0735-7044.116.2.222.

Colombo, M., Fernandez, T., Nakamura, K., & Gross, C. G. (1998). Functional differentiation along the anterior-posterior axis of the hippocampus in monkeys. *Journal of Neurophysiology, 80*(2), 1002–1005.

Decker, M. W., Curzon, P., & Brioni, J. D. (1995). Influence of separate and combined septal and amygdala lesions on memory, acoustic startle, anxiety, and locomotor activity in rats. *Neurobiology of Learning and Memory, 64*(2), 156–168. doi:10.1006/nlme.1995.1055.

Dedovic, K., Duchesne, A., Andrews, J., Engert, V., & Pruessner, J. C. (2009). The brain and the stress axis: The neural correlates of cortisol regulation in response to stress. *NeuroImage, 47*(3), 864–871. doi:10.1016/j.neuroimage.2009.05.074.

Dong, H.-W., Swanson, L. W., Chen, L., Fanselow, M. S., & Toga, A. W. (2009). Genomic-anatomic evidence for distinct functional domains in hippocampal field CA1. *Proceedings of the National Academy of Sciences of the United States of America, 106*(28), 11794–11799. doi:10.1073/pnas.0812608106.

Eadie, B. D., Zhang, W. N., Boehme, F., Gil-Mohapel, J., Kainer, L., Simpson, J. M., et al. (2009). Fmr1 knockout mice show reduced anxiety and alterations in neurogenesis that are specific to the ventral dentate gyrus. *Neurobiology of Disease, 36*(2), 361–373. doi:10.1016/j.nbd.2009.08.001.

Eichenbaum, H., Mathews, P., & Cohen, N. J. (1989). Further studies of hippocampal representation during odor discrimination learning. *Behavioral Neuroscience, 103*(6), 1207–1216.

Everitt, B. J., & Robbins, T. W. (2005). Neural systems of reinforcement for drug addiction: From actions to habits to compulsion. *Nature Neuroscience, 8*(11), 1481–1489. doi:10.1038/nn1579.

Fanselow, M. S. (1984). What is conditioned fear? *Trends in Neurosciences, 7*(12), 460–462.

Fanselow, M. S., & Dong, H.-W. (2010). Are the dorsal and ventral hippocampus functionally distinct structures? *Neuron, 65*(1), 7–19. doi:10.1016/j.neuron.2009.11.031.

Gage, F. H., & Thompson, R. G. (1980). Differential distribution of norepinephrine and serotonin along the dorsal-ventral axis of the hippocampal formation. *Brain Research Bulletin, 5*(6), 771–773. http://pdn.sciencedirect.com/science?_ob=MiamiImageURL&_cid=271056&_user=10843&_pii=0361923080902208&_check=y&_origin=article&_zone=toolbar&_coverDate=31-Dec-1980&view=c&originContentFamily=serial&wchp=dGLbVlk-zSkzk&md5=2857224f44d72adfdf9ea3facaa8bbee&pid=1-s2.0-0361923080902208-main.pdf&sqtrkid=0.21771737933158875. Accessed 12 June 2013.

Gilbert, M., Racine, R. J., & Smith, G. K. (1985). Epileptiform burst responses in ventral vs dorsal hippocampal slices. *Brain Research, 361*(1–2), 389–391.

Gilbert, P. E., Kesner, R. P., & Lee, I. (2001). Dissociating hippocampal subregions: A double dissociation between dentate gyrus and CA1. *Hippocampus, 11*(6), 626–636. doi:10.1002/hipo.1077.

Gold, A. E., & Kesner, R. P. (2005). The role of the CA3 subregion of the dorsal hippocampus in spatial pattern completion in the rat. *Hippocampus, 15*(6), 808–814. doi:10.1002/hipo.20103.

Goodrich-Hunsaker, N. J., Hunsaker, M. R., & Kesner, R. P. (2008). The interactions and dissociations of the dorsal hippocampus subregions: How the dentate gyrus, CA3, and CA1 process spatial information. *Behavioral Neuroscience, 122*(1), 16–26. doi:10.1037/0735-7044.122.1.16.

Gourevitch, B., Kay, L. M., & Martin, C. (2010). Directional coupling from the olfactory bulb to the hippocampus during a go/no-go odor discrimination task. *Journal of Neurophysiology, 103*(5), 2633–2641. doi:10.1152/jn.01075.2009.

Gray, J. A., & McNaughton, N. (2000). *The Neuropsychology of Anxiety* (2nd ed.). USA: Oxford University Press.

Gulyás, A. I., Tóth, K., McBain, C. J., & Freund, T. F. (1998). Stratum radiatum giant cells: A type of principal cell in the rat hippocampus. *The European journal of neuroscience, 10*(12), 3813–3822.

Hafting, T., Fyhn, M., Molden, S., Moser, M.-B., & Moser, E. I. (2005). Microstructure of a spatial map in the entorhinal cortex. *Nature, 436*(7052), 801–806. doi:10.1038/nature03721.

Heale, V. R., & Vanderwolf, C. H. (1999). Odor-induced fast waves in the dentate gyrus depend on a pathway through posterior cerebral cortex: Effects of limbic lesions and trimethyltin. *Brain Research Bulletin, 50*(4), 291–299.

Hughes, K. C., & Shin, L. M. (2011). Functional neuroimaging studies of post-traumatic stress disorder. *Expert Review of Neurotherapeutics, 11*(2), 275–285. doi:10.1586/ern.10.198.

Hunsaker, M. R., & Kesner, R. P. (2012). The operation of pattern separation and pattern completion processes associated with different attributes or domains of memory. *Neuroscience & Biobehavioral Reviews*. doi:10.1016/j.neubiorev.2012.09.014.

Hunsaker, M. R., Fieldsted, P. M., Rosenberg, J. S., & Kesner, R. P. (2008). Dissociating the roles of dorsal and ventral CA1 for the temporal processing of spatial locations, visual objects, and odors. *Behavioral Neuroscience, 122*(3), 643–650. doi:10.1037/0735-7044.122.3.643.

Jayatissa, M. N., Bisgaard, C., Tingström, A., Papp, M., & Wiborg, O. (2006). Hippocampal cytogenesis correlates to escitalopram-mediated recovery in a chronic mild stress rat model of depression. *Neuropsychopharmacology: Official Publication of the American College of Neuropsychopharmacology, 31*(11), 2395–2404. doi:10.1038/sj.npp.1301041.

Jung, M. W., Wiener, S. I., & McNaughton, B. L. (1994). Comparison of spatial firing characteristics of units in dorsal and ventral hippocampus of the rat. *The Journal of neuroscience: the official journal of the Society for Neuroscience, 14*(12), 7347–7356.

Kavushansky, A., Vouimba, R.-M., Cohen, H., & Richter-Levin, G. (2006). Activity and plasticity in the CA1, the dentate gyrus, and the amygdala following controllable vs. uncontrollable water stress. *Hippocampus, 16*(1), 35–42. doi:10.1002/hipo.20130.

Kent, K., Hess, K., Tonegawa, S., & Small, S. A. (2007). CA3 NMDA receptors are required for experience-dependent shifts in hippocampal activity. *Hippocampus, 17*(10), 1003–1011. doi:10.1002/hipo.20332.

Kesner, R. P. (2009). Tapestry of memory. *Behavioral Neuroscience, 123*(1), 1–13. doi:10.1037/a0014004.

Kesner, R. P., & Hunsaker, M. R. (2010). The temporal attributes of episodic memory. *Behavioural Brain Research, 215*(2), 299–309. doi:10.1016/j.bbr.2009.12.029.

Kesner, R. P., Gilbert, P. E., & Barua, L. A. (2002). The role of the hippocampus in memory for the temporal order of a sequence of odors. *Behavioral Neuroscience, 116*(2), 286–290.

Kesner, R. P., Hunsaker, M. R., & Ziegler, W. (2010). The role of the dorsal CA1 and ventral CA1 in memory for the temporal order of a sequence of odors. *Neurobiology of Learning and Memory, 93*(1), 111–116. doi:10.1016/j.nlm.2009.08.010.

Kesner, R. P., Hunsaker, M. R., & Ziegler, W. (2011). The role of the dorsal and ventral hippocampus in olfactory working memory. *Neurobiology of Learning and Memory, 96*(2), 361–366. doi:10.1016/j.nlm.2011.06.011.

Kjelstrup, K. G., Tuvnes, F. A., Steffenach, H.-A., Murison, R., Moser, E. I., & Moser, M.-B. (2002). Reduced fear expression after lesions of the ventral hippocampus. *Proceedings of the National Academy of Sciences of the United States of America, 99*(16), 10825–10830. doi:10.1073/pnas.152112399.

Lazar, R. M., & Mohr, J. P. (2011). Revisiting the contributions of Paul Broca to the study of aphasia. *Neuropsychology Review, 21*, 236–239. doi:10.1007/s11065-011-9176-8.

Lee, I., & Kesner, R. P. (2004). Differential contributions of dorsal hippocampal subregions to memory acquisition and retrieval in contextual fear-conditioning. *Hippocampus, 14*(3), 301–310. doi:10.1002/hipo.10177.

Levita, L., & Muzzio, I. A. (2010). Role of the hippocampus in goal-oriented tasks requiring retrieval of spatial versus non-spatial information. *Neurobiology of Learning and Memory, 93*(4), 581–588. doi:10.1016/j.nlm.2010.02.006.

Magariños, A. M., McEwen, B. S., Flügge, G., & Fuchs, E. (1996). Chronic psychosocial stress causes apical dendritic atrophy of hippocampal CA3 pyramidal neurons in subordinate tree shrews. *The Journal of Neuroscience: The Official Journal of the Society for Neuroscience, 16*(10), 3534–3540.

Maguire, E. A., Gadian, D. G., Johnsrude, I. S., Good, C. D., Ashburner, J., Frackowiak, R. S., et al. (2000). Navigation-related structural change in the hippocampi of taxi drivers. *Proceedings of the National Academy of Sciences of the United States of America, 97*(8), 4398–4403. doi:10.1073/pnas.070039597.

McHugh, S. B., Deacon, R. M. J., Rawlins, J. N. P., & Bannerman, D. M. (2004). Amygdala and ventral hippocampus contribute differentially to mechanisms of fear and anxiety. *Behavioral Neuroscience, 118*(1), 63–78. doi:10.1037/0735-7044.118.1.63.

Moser, M. B., & Moser, E. I. (1998a). Distributed encoding and retrieval of spatial memory in the hippocampus. *The Journal of Neuroscience: The Official Journal of the Society for Neuroscience, 18*(18), 7535–7542.

Moser, M. B., & Moser, E. I. (1998b). Functional differentiation in the hippocampus. *Hippocampus, 8*(6), 608–619. doi:10.1002/(SICI)1098-1063(1998)8:6<608::AID-HIPO3>3.0.CO;2-7.

Moser, E., Moser, M. B., & Andersen, P. (1993). Spatial learning impairment parallels the magnitude of dorsal hippocampal lesions, but is hardly present following ventral lesions. *The Journal of Neuroscience: The Official journal of the Society for Neuroscience, 13*(9), 3916–3925.

O'Keefe, J., & Nadel, L. (1978). *The hippocampus as a cognitive map.* Oxford: Clarendon Press.

Pandis, C., Sotiriou, E., Kouvaras, E., Asprodini, E., Papatheodoropoulos, C., & Angelatou, F. (2006). Differential expression of NMDA and AMPA receptor subunits in rat dorsal and ventral hippocampus. *Neuroscience, 140*(1), 163–175. doi:10.1016/j.neuroscience.2006.02.003.

Papatheodoropoulos, C., & Kostopoulos, G. (2000). Decreased ability of rat temporal hippocampal CA1 region to produce long-term potentiation. *Neuroscience Letters, 279*(3), 177–180.

Pentkowski, N. S., Blanchard, D. C., Lever, C., Litvin, Y., & Blanchard, R. J. (2006). Effects of lesions to the dorsal and ventral hippocampus on defensive behaviors in rats. *European Journal of Neuroscience, 23*(8), 2185–2196. doi:10.1111/j.1460-9568.2006.04754.x.

Rogers, J. L., Hunsaker, M. R., & Kesner, R. P. (2006). Effects of ventral and dorsal CA1 subregional lesions on trace fear conditioning. *Neurobiology of Learning and Memory, 86*(1), 72–81. doi:10.1016/j.nlm.2006.01.002.

Rolls, E. T., & Kesner, R. P. (2006). A computational theory of hippocampal function, and empirical tests of the theory. *Progress in Neurobiology, 79*(1), 1–48. doi:10.1016/j.pneurobio.2006.04.005.

Rolls, E. T., Treves, A., Foster, D., & Perez-Vicente, C. (1997). Simulation studies of the CA3 hippocampal subfield modelled as an attractor neural network. *Neural Networks, 10*(9), 1559–1569. doi:10.1016/S0893-6080(97)00092-0.

Royer, S., Sirota, A., Patel, J., & Buzsaki, G. (2010). Distinct representations and theta dynamics in dorsal and ventral hippocampus. *Journal of Neuroscience, 30*(5), 1777–1787. doi:10.1523/JNEUROSCI.4681-09.2010.

Sano, K. (1997). Hippocampus and epilepsy surgery. *Epilepsia, 38*(s6), 4–10. doi:10.1111/j.1528-1157.1997.tb00098.x

Scoville, W. B., & Milner, B. (1957). Loss of recent memory after bilateral hippocampal lesions. *Journal of Neurology, Neurosurgery, and Psychiatry, 20*(1), 11–21.

Squire, L. R., & Zola-Morgan, S. (1991). The medial temporal lobe memory system. *Science (New York, N.Y.), 253*(5026), 1380–1386.

Taube, J. S. (2007). The head direction signal: Origins and sensory-motor integration. *Annual Review of Neuroscience, 30,* 181–207. doi:10.1146/annurev.neuro.29.051605.112854.

Treit, D., Pesold, C., & Rotzinger, S. (1993). Noninteractive effects of diazepam and amygdaloid lesions in two animal models of anxiety. *Behavioral Neuroscience, 107*(6), 1099–1105.

Van Groen, T., & Wyss, J. M. (1990). Extrinsic projections from area CA1 of the rat hippocampus: Olfactory, cortical, subcortical, and bilateral hippocampal formation projections. *The Journal of Comparative Neurology, 302*(3), 515–528. doi:10.1002/cne.903020308.

Vanderwolf, C. H. (1992). Hippocampal activity, olfaction, and sniffing: An olfactory input to the dentate gyrus. *Brain Research, 593*(2), 197–208.

Vouimba, R.-M., Yaniv, D., Diamond, D., & Richter-Levin, G. (2004). Effects of inescapable stress on LTP in the amygdala versus the dentate gyrus of freely behaving rats. *The European Journal of Neuroscience, 19*(7), 1887–1894. doi:10.1111/j.1460-9568.2004.03294.x.

Walsh, R. N., & Cummins, R. A. (1976). The open-field test: A critical review. *Psychological Bulletin, 83*(3), 482–504.

Weeden, C. S. S., Hu, N. J., Ho, L. Y. N., & Kesner, R. P. (2012). The role of the ventral dentate gyrus in olfactory learning and memory. Presented at the Society for Neuroscience Abstracts, New Orleans.

Witter, M. P. (1993). Organization of the entorhinal-hippocampal system: A review of current anatomical data. *Hippocampus, 3* (Spec No), 33–44.

Witter, M. P., Van Hoesen, G. W., & Amaral, D. G. (1989). Topographical organization of the entorhinal projection to the dentate gyrus of the monkey. *The Journal of Neuroscience: The Official Journal of the Society for Neuroscience, 9*(1), 216–228..

Yassa, M. A., Stark, S. M., Bakker, A., Albert, M. S., Gallagher, M., & Stark, C. E. L. (2010). High-resolution structural and functional MRI of hippocampal CA3 and dentate gyrus in patients with amnestic mild cognitive impairment. *NeuroImage, 51*(3), 1242–1252. doi:10.1016/j.neuroimage.2010.03.040.

Yoon, T., & Otto, T. (2007). Differential contributions of dorsal vs. ventral hippocampus to auditory trace fear conditioning. *Neurobiology of Learning and Memory, 87*(4), 464–475. doi:10.1016/j.nlm.2006.12.006.

Part II
Memory System Interactions

Chapter 9
Self Regulation of Memory Processing Centers of the Brain

Sheri J. Y. Mizumori

Introduction

Our memories reflect the accumulation of our past experiences. They shape future decisions and determine what and how we learn over time. It should not be surprising, then, that many fundamental elements of memory (e.g., different associative algorithms, motivation, sensory, motor, attention, memory updating, and response selection) must work together to continuously guide experience-dependent and adaptive behaviors regardless of the nature of the type of currently active memory. Study of a variety of amnesic populations has illustrated that not only many regions of the brain play these important roles in memory, but also different brain areas do so for different reasons. Temporal lobe patients (the most famous of which is patient H. M.) show severe but select anterograde episodic memory impairment, while procedural memory remained intact (Bayley et al. 2005; Milner 2005). Patients suffering from basal ganglia dysfunction show selective impairment in habit learning and procedural memory (Knowlton et al. 1996; Yin and Knowlton 2006). Amygdala damage results in poor emotional regulation of memory (Adolphs et al. 2005; Paz and Pare 2013). Frontal patients suffer from inadequate working memory (Baddeley and Della Sala 1996; Goldman-Rakic 1996). These classic distinctions of the mnemonic consequences of damage to different brain areas in humans have been replicated in rodents by many, and the Kesner laboratory has been particularly successful at demonstrating not only double, but also often triple dissociations of functions of structures like the hippocampus, striatum, amygdala, and prefrontal cortex (e.g., Chiba et al. 2002; Gilbert and Kesner 2002; Kametani and Kesner 1989; Kesner et al. 1993; Kesner et al. 1989; Kesner and Williams 1995). Moreover, the often clever behavioral paradigms created in the Kesner laboratory over the years have been inspirational for generating more specific hypotheses about

S. J. Y. Mizumori (✉)
Department of Psychology, University of Washington, 3921 W. Stevens Wy NE, Seattle, WA 98195, USA
e-mail: mizumori@uw.edu

P. A. Jackson et al. (eds.), *The Neurobiological Basis of Memory*,
DOI 10.1007/978-3-319-15759-7_9

memory function that could be tested in human subjects (e.g., Hopkins et al. 1995, 2004; Kesner and Hopkins 2001, 2006).

Decades ago, Ray Kesner was perhaps the first to develop a broad theoretical model of memory that sought to explain many of its complexities. His attribute model of memory posits the existence of three basic types of memory (event, knowledge, and rule-based memories), each one of which incorporates similar and fundamental memory operations to establish and use their particular type of memory. Kesner has written many elegant reviews of his work (Kesner 1980, 1998, 2009; Kesner and DiMattia 1987; Kesner and Rogers 2004), and readers, if they have not done so already, should seek out those reviews to gain an appreciation for his most impressive programmatic, timely, and innovative research program. Kesner initially proposed his attribute model at a time when most studies on the neurobiological models of memory focused on simpler memory functions of a small number of brain regions. However, more recently, the development of new behavioral and neuroscience technologies has sparked the current, widespread, and strong interest in studying the multiple neural systems of the brain during complex memory function that involve learning and decision-making mechanisms. Thus, it is clear that the attribute model was and continues to be a visionary theoretical framework for studying brain mechanisms of memory, learning, and decision-making. Indeed, it is now generally accepted that as espoused by the attribute model, hierarchical sets of parallel and distributed neural networks mediate the complex and dynamic processes of simple and complex memories in the brain. Current challenges are to figure out how different networks interact, how behaviors come to guide memory operations, and how existing memories guide future learning and decisions.

Neurophysiological investigations of these memory-related brain regions both confirmed and challenged the multiple memory system view. Spatial and conjunctive context-dependent coding were identified in the hippocampus (O'Keefe and Dostrovsky 1971; O'Keefe and Nadel 1978) and this was consistent with the view that hippocampus mediates episodic memory (Tulving 2002). Response-related codes were found in the striatum (Eschenko and Mizumori 2007; Jog et al. 1999; Yeshenko et al. 2004), supporting the hypothesis that striatum mediate habit or response learning (Knowlton et al. 1996). Frontal cortical neurons remain active during delay periods (Goldman-Rakic 1995), a finding that one might expect from a brain region that is importantly involved in working memory (Fuster 2006, 2008, 2009). With time, however, additional studies began to show that these striking neural correlates of behavior were not so unique to the hippocampus, striatum, and frontal cortex. Egocentric movement-related firing by hippocampal interneurons was reported long ago (e.g., Vanderwolf 1969) but was largely unstudied until recently in favor of studying what was at the time the more intriguing place cells. Parietal cortical neurons also showed strong representations of behavioral responses (e.g., Fogassi et al. 2005; McNaughton et al. 1994). Delay cells were found in regions of the cortex other than the prefrontal cortex, for example, in somatosensory cortex (Meftah et al. 2009), parietal cortex (Snyder et al. 1997), frontal eye fields (Curtis et al. 2004), and less so in temporal cortex (Kurkin et al. 2011). The fact that the single unit evidence did not align directly with the lesion literature suggested that many regions of the brain use similar types of information during their mnemonic computations. However, since the single-unit data came from studies of rodents and primates that used different recording methods while subjects performed a diverse set of tasks, it became important to record from multiple memory-related brain structures as an animal performed a single task that required animals to switch between different memories to continue adaptive decision-making.

In the following section, we describe our efforts to address the issue of whether different brain regions mediate different memories because they represent different kinds of information. The last section of this chapter explores the hypothesis that the relative contribution of different brain areas to memory is driven by homeostatic neural mechanisms that insures proper self-regulation of a behavioral adaptation system that depends on the memory functions described in the attribute model.

Memory Specialization and the Brain

Given that different memory capacities exist across different brain structures, a major challenge has been to understand why those different brain areas make such specialized contributions to memory. The following describes investigations that tested the hypothesis that different brain areas represent different types of information and that this is responsible for their different memory capacities. Their specific focus here will be on a comparison of hippocampal and striatal neural representations as rats performed a hippocampal-dependent (spatial) task or a striatal-dependent (response) task. Importantly, these are tasks that show dissociable mnemonic involvement by the hippocampus and striatum in lesion and clinical investigations. Also, it is noted that the principle conclusions from these results should apply more broadly to an understanding of the relationship between other memory-related brain areas such as the amygdala and prefrontal cortex.

Are Memory Specializations Due to the Nature of the Information Represented by Neurons?

Dorsal striatal and dorsal hippocampal single-unit activity were recorded as rats performed a T-Maze task (Yeshenko et al. 2004; Eschenko and Mizumori 2007). One group of rats was trained to solve the first 10 trials of a recording session according to a spatial strategy and then the next 10 trials according to a response strategy. According to a spatial strategy, rats seek a location that had been previously associated with reward. A response strategy, on the other hand, requires rats to use the same egocentric response (i.e., right turn from the start location) to find food. Another group of rats ran 10 response trials followed by 10 spatial trials. Since these 20 trials were performed within one recording session, the same striatal and hippocampal neurons were recorded before, during, and after an experimenter-controlled switch in cognitive strategy (or memory). Also, since striatal neurons were often recorded simultaneously with hippocampal neurons, it was possible to compare directly activity in the two brain structures relative to the currently active memory, the accuracy of the choices made, and relative to the type of cognitive switch. Such switches included, not only changes from spatial to response strategy use (or vice versa) but also changes from one spatial memory (e.g., food is in the north location) to another spatial memory (e.g., food is in the south location), or from one response strategy (always turn right to find food) to another response strategy (always turn left to find food). Also, in some tests, the visual cues were altered to present another type of strategy (or memory) shift. Together, the inclusion of these different types of manipulations made it possible to see if neural responses in hippocampus or striatum were specific to a particular type of activated memory or cognitive change, or just cognitive change in general.

As had been reported in previous studies (e.g., McNaughton et al. 1983; O'Keefe and Dostrovsky 1971; Olton et al. 1978; Muller and Kubie 1987; Ranck 1973; Redish 1999; summarized in Mizumori 2008b; O'Keefe and Nadel 1978), hippocampal pyramidal cell discharge showed strong correlations with the location of an animal on the maze, while hippocampal (presumed) interneurons showed firing that was correlated with an animal's movement velocity (Eschenko and Mizumori 2007; Yeshenko et al. 2004). As expected from the results of striatal lesion studies, dorsal striatal neurons showed strong correlations with behavioral response parameters such as the rat's movement velocity and acceleration. Unexpectedly, location-selective neurons were also found in both medial and lateral sectors of dorsal striatum. (This pattern contrasts with an earlier report that a different type of neural representation of space, a rat's directional heading, is found in only the medial, not lateral, dorsal striatum, Ragozzino et al. 2001.) Most of the details of the properties of both the movement and location correlates of hippocampal and striatal neurons did not differ as a function of whether the rat was solving the task with a (hippocampal-dependent) spatial or (striatal-dependent) response strategy. This result indicates that both structures continuously and actively process similar types of information (although some of the details may vary) regardless of whether hippocampal-dependent or striatal-dependent memories are engaged.

Given that both spatial and response-related information are represented in hippocampus and striatum, it is possible that the distinct mnemonic contributions of these areas derive from differential sensitivities to changes in memory or contextual information. This hypothesis was also tested by Eschenko and Mizumori (2007) and Yeshenko et al. (2004) and the clear result was that both hippocampal and striatal spatial, and movement neural codes were sensitive to changes in the memory demands of the task regardless of how the change in memory was brought about (e.g., by changing choice strategies, the available cues, from spatial to response strategy, or vice versa). For example, velocity correlated firing by either striatal or hippocampal neurons showed significant changes in the magnitude of the velocity correlation after the cognitive demands of a task shifted. This was the case even though the actual behaviors and velocities exhibited by the rats were consistent in all phases of the test session. Interestingly, hippocampal movement codes responded most dramatically when rats shifted from one type of spatial strategy to another, whereas striatal movement codes responded similarly across all types of context shifts. Thus, details of a given context shift may differentially impact particular hippocampal neurons whereas the same could not be said for striatal networks. Perhaps striatum responds more generally than hippocampus to any type of context change, a conclusion that is consistent with the view that striatum is primarily responsive to changes in reinforcement conditions more generally. Striatal and hippocampal velocity codes per se then are not solely determined by the ongoing behavior of the animals, but rather it is determined, or gated, by memory.

Place-specific firing by striatum and hippocampus also showed sensitivity to changes in cognitive strategy (or activated memory) and this was evidenced by significant changes in place field location or in-field firing rates. In the future then it would be of interest to determine if the context-regulation of striatal neural codes derives from hippocampus by disabling hippocampus, while testing the striatal neural responses to a context change. Since the hippocampus is hypothesized to detect changes in context (e.g., Mizumori et al. 1999; Mizumori et al. 2007a; Mizumori and Jo 2013), one prediction is that striatal neurons will not respond to context manipulations without a proper functioning hippocampus. If, however, striatal context-sensitivity is not affected when hippocampal input goes off-line, then the context-dependent striatal codes may emerge from neocortical memory systems.

Do the Memory Specializations Within the Brain Reflect Compensation After Brain Damage?

Another consideration that can be used to explain the different memory consequences of hippocampal and striatal lesions is that their effects reflect the extent to which remaining brain areas can compensate for a particular type of lesion. If the intrinsic processing by the structure of interest is unique and essential for learning to take place, then no behavioral impairment should be observed if other neural circuits are compromised. Indeed, there is abundant evidence that under most conditions, stim ulus-response learning is not impaired following hippocampal lesions, presumably since striatal computations are sufficient to support such learning (e.g., Knowlton et al. 1996; McDonald and White 1993; Packard et al. 1989; Packard and McGaugh 1996; Yin and Knowlton 2006). This does not mean that the hippocampus does not normally play a role in stimulus-response performance; if hippocampus defines the context of the learning so that animals can quickly adapt when test conditions change, deficits in stimulus-response performance may be observed after hippocampal damage only if a context change is involved. In contrast, context-dependent learning is by definition complex and dynamic—a situation that the hippocampus seems uniquely qualified to handle. Thus, as shown in the literature, hippocampal, but not striatal, lesions result in selective context memory deficits.

Are Memory Specializations in the Brain Defined by the Coordination of Neural Networks Across Brain Structures?

It is possible that brain structures play different roles in memory because of task-dependent co-modulated neural activity across different brain structures at strategic times during task performance. Although much work remains to thoroughly test this hypothesis, there are initial indications that support this view. Ragozzino et al. (2001) recorded striatal head direction cells simultaneously with hippocampal place cells, and then compared their responses to different types of context shifts. It was found that when familiar cues were shifted, head directional preferences and place field locations shifted in a comparable fashion. In contrast, when rats were placed in new environments, the shift in head direction preferences and place fields appeared random relative to each other. This result suggests that memory (i.e., past experiences) can bias striatal and hippocampal neural responses in a coordinated fashion, and that without such memory guidance, the two structures function more independently.

Numerous laboratories report that specific neural population-based rhythms can be detected both within and between memory structures such as the hippocampus, striatum, or prefrontal cortex (DeCoteau et al. 2007a; Engel et al. 2001; Fell et al. 2001; Siapas et al 2005; Tabuchi et al. 2000; Varela et al. 2001; Womelsdorf et al. 2007). Hippocampal theta activity seems to coordinate the timing of spatial coding by hippocampal neurons (Gengler et al. 2005; O'Keefe and Recce 1993). Striatal theta oscillations have been shown to become entrained to the hippocampal theta rhythm (Berke et al. 2004; DeCoteau et al. 2007a) in a behaviorally-dependent fashion. Also, directly stimulating the striatum can induce hippocampal high frequency theta activity (Sabatino et al. 1985). When neural activity is disrupted in the striatum via D2 receptor antagonism, striatal modulation of high frequency hippocampal theta activity is reduced, motor and spatial/contextual information is not integrated, and task performance is impaired (Gengler et al. 2005). This is consistent with the idea that theta is important for sensory-motor integration (Hallworth and Bland 2004). It appears then that during goal directed navigation, hippocampal and striatal activity becomes increasingly coherent, and this pattern is dopamine dependent. Given its putative role in assessing the value of behavioral outcomes (e.g., Schultz and Dickinson 2000), dopamine may play an important role in biasing the relative strengths of hippocampal and striatal output signals according to the most effective mnemonic strategy (e.g., Mizumori et al. 2004).

Particularly intriguing is a finding common to both the hippocampus and striatum, and that is that synchronous neural activity occurs in specific task-relevant ways (e.g., Hyman et al. 2005; Jones and Wilson 2005), especially when rats engage in decision-making (e.g., Benchenane et al. 2010). For example, striatal theta is modified over the course of learning an egocentric T-maze task, increasing as the rat comes to regularly choose and initiate turn behavior (DeCoteau et al. 2007a, 2007b). Rats that learned the task developed an antiphase relationship between hippocampal and striatal theta oscillations, while rats that did not learn the task did not show this coherent theta relationship. This coherence has also been observed during striatal-dependent classical conditioning (Kropf and Kuschinsky 1993).

The possibility that dopamine contributes to the regulation of memory efficiency and strategies by coordinating ensemble neural activity in distant brain structures is intriguing given its role in decision-making: Coherent theta oscillations across distant brain structures can be enhanced with application of dopamine (Benchenane et al. 2010) and impaired by dopamine antagonism (Gengler et al. 2005). Functionally, this type of control by dopamine suggests that information about the saliency of reward may determine which brain systems become synchronized (and desynchronized), and this in turn informs the decision process about what information is used to update memories and which behaviors are selected.

The functional importance of neural oscillations in the gamma range (30–80 Hz) remains debated. However, since gamma oscillations tend to occur intermittently (i.e., "gamma bursts" of about 150–250 ms are followed by periods of desynchronous activity), information carried by the cells that participate in a gamma burst effectively become a clear and punctuate signal against a background of disorganized neural activity. For this reason, it has been suggested that gamma bursts represent a fundamental mechanism by which information becomes segmented and/or filtered within a structure, as well as a way to coordinate information across structures (Buzsaki 2006). Although theta and gamma frequencies are quite different (perhaps reflecting the type of information that each rhythm coordinates), there are many common physiological and behavioral relationships that suggest they are components of a coordinated and larger scale oscillatory network. For example, similar to theta rhythms, single unit responses that are recorded simultaneously with gamma oscillations have been found to have specific phase relationships to the gamma rhythm (e.g., Berke 2009; Kalenscher et al. 2010; van der Meer and Redish 2009). Also, it is hypothesized that gamma oscillations may effectively select salient information that can come to impact decisions, learning, and behavioral responses (e.g., Kalenscher et al. 2010; van der Meer and Redish 2009), since their appearance is often in relation to task-relevant events. Another similarity with the theta system is that the occurrence gamma oscillations appear to be at least in part regulated by the dopamine system (Berke 2009).

Are Memory Specializations in the Brain Defined by the Functional Significance of the Output Messages of Populations of Cells?

From the above discussion it is clear that there are widespread neural codes for spatial and response aspects of task performance across different brain areas, and that these are context- (or memory-) dependent. It is possible that neuromodulators such as dopamine bias the strengths of the output messages according to recent behavioral success. What then is the meaning of the efferent neural code at the population level? What neural mechanisms control this meaning, and how are the outputs of different memory processing areas of the brain coordinated to result in continuously adaptive behaviors? These are some of the big challenges that remain to be addressed before we can fully understand the neural systems basis of multiple memory function. What follows is a suggested approach to future investigations of these big challenges.

Predictive Memories and Adaptive Decisions

Known patterns of intrinsic neural connectivity indicate that each memory-related brain structure undoubtedly processes information in a somewhat unique way, and yet it is unclear if these differences are sufficiently unique to account for the documented specialized memory capacities of each brain area. There is, however, growing evidence that the output of different brain structures has a common goal for different kinds of information, and that is to relay the extent to which experience-based predictions relevant to optimal task performance are born out. In fact, an emerging view is that the brain evolved in large part to allow organisms to accurately predict the outcomes of events and behaviors (e.g., Buzsaki 2013; Buzsaki and Moser 2013; Llinas and Roy 2009; Mizumori and Jo 2013). In this way, organisms have been able to adapt to environments and societies of increasing complexity—a condition that required sophisticated mechanisms to make decisions and predictions in a dynamic and conditional environment. According to this view, the underlying neural mechanisms of predictions (and the assessment of their accuracy) are likely to be highly conserved across species (Adams et al. 2013; Watson and Platt 2008). This includes the ability to retain information over times of varying scales depending on the desired goal. Indeed, different brain areas are known to generate and retain sequences of information, an ability that can be accounted for by state-dependent changes in network dynamics (Mauk and Buonomano 2004), internally-generated oscillatory activity (Pastalkova et al. 2008), and/or dedicated "time cells" (Kraus et al. 2013).

Hippocampal Evaluation of the Accuracy of Predictions About Contextual Information

A *context discrimination hypothesis* (CDH) postulates that single hippocampal neurons provide multidimensional (context-defining) data to population-based network computations that ultimately determine whether expected contextual features of a situation have changed (e.g., Mizumori et al. 1999, 2000a, 2007a; Mizumori 2008a, b; Smith and Mizumori 2006a, b). Specifically, these hippocampal representations of spatial context information (O'Keefe and Nadel 138; Nadel and Payne 2002; Nadel and Wilner 1980) may contribute to a match-mismatch type of analysis that evaluates the present context according to how similar it is to the context that an animal expects to encounter based on past experiences (e.g., Anderson and Jeffery 2003; Gray 1982, 2000; Hasselmo 2005b; Hasselmo et al. 2002; Jeffery et al. 2004, Lisman and Otmakhova 2001; Manns et al. 2007a; Mizumori et al. 1999,

2000a; Smith and Mizumori 2006a, b; Nadel 2008; Vinogradova 1995). Detected mismatches can be used to identify novel situations, initiate learning-related neural plasticity mechanisms, and to distinguish different contexts in memory—functions that are necessary to define significant events or episodes. When a match is computed, the effect of hippocampal output could be to strengthen currently active memory networks located elsewhere in the brain (e.g., neocortex). Thus, hippocampus may play different mnemonic roles depending on whether or not contexts change.

In support of the CDH, disconnecting hippocampus by fornix lesions impairs context discrimination (Smith et al. 2004), and hippocampal lesions reduce animals' ability to respond to changes in a familiar environment (Good and Honey 1991; Save et al. 1992a, 1992b). Spatial novelty detection corresponds to selective elevation of the immediate early gene c-*fos* in hippocampus, and not in surrounding parahippocampal cortical regions (Jenkins et al. 2004). Also, as described above, hippocampal neurons show significantly altered firing patterns when rats experience spatial or nonspatial changes in a familiar environment (Eschenko and Mizumori 2007; Ferbinteanu and Shapiro 2003; Fyhn et al. 2002; Leutgeb et al. 2005a, 2005b; Moita et al. 2004; Muller and Kubie 1987; O'Keefe 1976; Puryear et al. 2006; Smith and Mizumori 2006b; Wood et al. 1999; Yeshenko et al. 2004). As an example, Smith and Mizumori (2006b) showed that hippocampal neurons develop context-specific responses, but only when rats were required to discriminate contexts. Discriminating neural responses were not observed when rats were allowed to randomly forage for the same amount of time. Further, Manns et al. (2007b) demonstrated that relative to match trials in an odor cue or object recognition task, CA1 neurons were preferentially discharged when animals experienced a nonmatch situation in these same tasks. Also consistent with the CDH, neuroimaging studies of human performance shows that hippocampus becomes differentially active during match and mismatch trials (Chen et al. 2011; Dickerson et al. 2011; Duncan et al. 2012a; Duncan et al. 2012b; Foerde and Shohamy 2011; Kuhl et al. 2010; Kumaran and Maguire 2007).

The detection of changes in context is fundamentally important for the continual selection of appropriate behaviors that optimize performance and learning in a variety of tasks (e.g., navigation-based learning, instrumental conditioning, or classical conditioning). Context discrimination engages and prepares cellular mechanisms for rapid and new learning at potentially important times (Paulsen and Moser 1998), as it is generally known that novelty detection increases attention and exploratory behaviors in a variety of tasks. Interestingly, hippocampal cell firing tends to occur during the "encoding phase" of the ongoing theta rhythm (Hasselmo 2005a), which is increased during exploratory and investigatory behaviors (Vanderwolf 1969). Thus, detection of a nonmatch situation can change the relationship between cell discharge and the local theta rhythm such that encoding functions are enhanced. Detection of matches, on the other hand, does not cause changes in the hippocampal neural activity profile, resulting in efferent messages that continue to retrieve/utilize the currently active memory network that drove the execution of recently successful responses. Context discrimination, then, can be viewed as being critical for the formation of new episodic memories because it leads to the separation in time and space, one meaningful event from the next. Such division of memories could facilitate long-term information storage according to memory schemas (Bethus et al. 2010; Tse et al. 2007).

Context discrimination, or the detection of a mismatch between expected and experienced context-specific information, is considered an example of an error in predicting the contextual details of the current situation, referred to as a *context prediction error*. Transmission of a context prediction error signal from hippocampus should inform distal brain areas that a change in the context has occurred. In this case, upon receipt of the context prediction error message, efferent midbrain structures may respond with changes in excitation or inhibition that are needed to evaluate the subjective value of the context prediction error signal (e.g., Humphries and Prescott 2010; Lisman and Grace 2005; Mizumori et al. 2004; Penner and Mizumori 2012a). On the other hand, a hippocampal signal indicating that there was no prediction error may enable plasticity mechanisms that facilitate the incorporation of new information into existing memory schemas (e.g., Bethus et al. 2010; Mizumori et al. 2007a, b; Tse et al. 2007). Thus, hippocampal context analysis become critical for the formation of new episodic memories not only because prediction signals provide a mechanism that separates in time and space one meaningful event from the next, but also because the outcome of the prediction error computation engages appropriate neuroplasticity mechanisms in efferent structures that promote subsequent adaptive decisions and memory.

Striatal Evaluation of the Accuracy of Predictions About Response Outcomes Analogous to hippocampus, the midbrain dopaminergic system also generates prediction error signals, but in this case the focus is on whether the outcome of goal-directed behaviors occur as predicted based on past experience (Bayer and Glimcher 2005; Hollerman et al. 1998; Hollerman and Schultz 1998; Mizumori et al. 2009; Stalnaker et al. 2012). In particular, it is thought that dopamine neurons transmit information about the subjective value of rewards in terms of reward prediction error signals (RPEs). RPEs are thought to initiate three distinct and parallel loops of information processing between striatum and neocortex as new associations become learned sufficiently to habitually drive behaviors (e.g., Alexander et al. 1986; Alexander and Crutcher 1990a, b; Haber 2003). Penner and Mizumori (2012b) recently summarized this vast literature: Information within the limbic loop flows between ventromedial prefrontal cortex with the ventral striatum (Alexander and Crutcher 1990a, b; Graybiel 2008; Graybiel et al. 1994; Pennartz et al. 2009; Voorn et al. 2004; Yin and Knowlton 2006) to mediate learning about the significance of previously neutral stimuli (i.e., as it occurs in Pavlovian learning). The associative loop involves the medial prefrontal cortex and the dorsomedial striatum to support action-outcome learning. The sensorimotor loop involves transmission between somatosensory and motor cortical areas with the dorsolateral striatum. The latter loop is suited for incremental sensory-motor learning that happens when new procedural memories are formed. It is hypothesized that the transformation of newly learned behaviors to habits occurs as a result of multiple iterations of information flow through these three information loops starting with the limbic loop, the associative loop, and then finally the sensorimotor loop. Importantly, information flow through these systems is thought to be continually informed about the expected values of goals via dopamine signaling from the ventral tegmental area (VTA) and/or the substantia nigra (SN; Horvitz 2002; Nicola et al. 2004; Schultz 2010). When performing well-learned habits, the striatum is particularly suitable to rapidly control behavior or to provide feedback about behaviors that led to prediction errors (Stalnaker et al. 2012) because of its rather unique pattern of reciprocal connections with sensory and motor cortical regions (Alexander and Crutcher 1990a; Groenewegen et al. 1999; Haber 2003), and because striatum can receive immediate feedback when goal outcomes are not what was expected. In this way, midbrain signals of errors in predicting rewards may initiate adjustments to future planned behaviors (Penner and Mizumori 2012b).

Sensory and Motor Predictions
In addition to hippocampus and striatum, various sensory-motor cortical and cerebellar areas have been reported to generate prediction errors when expected sensory or motor-related input does not match expectations (e.g., Scheidt et al. 2012; Tanaka et al. 2009). This sort of feedback permits temporally and spatially precise behavior adjustments based on past outcomes. Also, information about expected sensory and motor events can be used to plan future sensory expectations and specific anticipatory movements (e.g., Duhamel et al. 1992). Such prediction error mechanisms are thought to fine tune actions to optimize the chances of securing a desired goal.

Summary: Error Signaling in the Brain
The above description illustrates that the generation of neural responses that signal times when actual events or information do not match those expected based on past experiences (i.e., prediction error signals) is often observed across many brain areas. In fact, it has been suggested that the ability to predict behavioral outcomes has essentially driven the evolution of the entire brain (Llinas and Roy 2009). Such error signals allow organisms to appropriately refine movements and choices relative to their perceived utility or value, and thus ultimately determine future decisions and behavior (e.g., Doll et al. 2012; Schultz and Dickinson 2000; Walsh and Anderson 2012). At a cellular level, prediction error signals may elevate the level of excitability of efferent neurons such that they become more responsive to outcome signals. This greater neural responsiveness may enhance the temporal and spatial resolution of future neural responses, and this in turn may ultimately result in improved accuracy of future predictions. For example, if hippocampus detects a mismatch between expected and actual contextual features, it may generate an error signal that "alerts" striatal efferent structures so that they become more responsive to future rewards (Lisman and Grace 2005; Mizumori et al. 2000a, 2004, 2009; Penner and Mizumori 2012a, b; Schultz and Dickinson 2000). Midbrain-generated reward prediction error signals may destabilize cortical neural (memory) networks so that they become more readily updated with new information (Mizumori 2008a; Penner and Mizumori 2012b). The updated memory information can then be passed on to hippocampus in the form of the most up-to-date context expectations. This view of how error signals can inform future processing in other prediction regions

of the brain is consistent with the view that there is a high degree of interdependence across mnemonic structures regardless of the task (Mizumori et al. 2004; Yeshenko et al. 2004).

Homeostatic Regulation of Predictive Memory Systems

An interesting and often described feature of memory functions is the rapid and seemingly automatic nature of its processes, or changes in processing, when a significant feature of a memory task changes. A challenge for future research then is to understand the neural mechanisms of the apparent automaticity and high accuracy of prediction analyses. An intriguing possibility is that the seemingly automatic nature derives from self-regulatory, intrinsic synaptic mechanisms rather than (only) responses to external information. Such mechanisms may align with the principles of homeostasis similar to those described for self-regulation at synaptic and neural circuit levels (e.g., Marder and Goaillard 2006; Marder and Prinz 2003; Mizumori and Jo 2013; Turrigiano 1999, 2008, 2011; Turrigiano and Nelson 2004). That is, *homeostatic regulation could drive the automatic and continuous maintenance of the balance between stable and flexible processing that neural networks need to retain learned (stable or expectancy) information that can to be (flexibly and adaptively) updated as needed.*

Marder and Goaillard (2006) suggested that homeostatic neural plasticity may be nested: Calcium sensors may monitor neural firing rates, then up or down regulate the availability of glutamate receptors to ramp up or down firing rates toward an optimal firing rate set point. Groups of neurons or neural networks may sense changes in firing collectively to regulate experience-dependent population activity levels and patterns of activation. In this way homeostatic synaptic plasticity enables groups of neural circuits to find a balance between flexible and stable processing as needed to learn from experiences, and to be responsive to future changed inputs. While details of how networks of cells or their connections engage in homeostatic regulation remain to be discovered it is worth noting that homeostatic regulation at the neural systems level indeed occurs, as is evident from studies of brain development, as well as from studies of reactive or compensatory neuroplasticity mechanisms that occur in response to experience (e.g., sensorimotor learning; Froemke et al. 2007) or brain injury (e.g., brain trauma or addiction; Nudo 2011; Robinson and Kolb 2004). Although homeostatic neural plasticity mechanisms have not been used to account for complex learning, current theories of reinforcement- and context-based learning and memory commonly rely on the autoregulation of feedback loops between systems that assess the outcomes of choices and existing (episodic) memory systems.

If homeostatic regulation pertains to neural networks that underlie adaptive memory, it should be possible to identify key components including *variables* that are being monitored by *sensors* and then regulated by *controllers*. For complex memory systems, such a model likely contains multiple hierarchically organized loops of control similar to what was described by Buzsaki (2013). For example, iterative updating via interactions between hippocampus, the dopamine striatal system, and memory networks would allow context prediction errors to guide future adaptive behaviors. This is somewhat similar to the micro- macro-agent distinction recently described by Kurth-Nelson and Redish (2009). This process may be enabled by sensors which monitor cell excitability within each structure. In this case, changes in calcium flux appear to be an important part of the sensing system that determines the current level of firing. A change in firing rates (either higher or lower than the optimal level) is taken as an indicator of a mismatch between optimal and actual rates, and a *controller* mechanism becomes engaged to bring the firing rates back to the optimal levels. Such a natural restorative mechanism that responds to cellular detection of errors in prediction (as reflected in firing rate deviations) seems essential since it would be unhealthy for neurons to exist in a chronically excited or inhibited state.

It has been suggested that the prefrontal cortex serves in a controller role that maintains the optimal excitatory level in prediction regions of the brain (more details below). In particular prefrontal cortex is strategically situated to enable mechanisms that restore afferent firing rates to a predetermined 'set point' via its detailed excitatory and inhibitory extrinsic connections (as reviewed in Arnsten et al. 2012). In this way, prefrontal cortex may orchestrate and coordinate the level of neural excitability in different prediction error brain areas (e.g., hippocampus and striatum) according to homeostatic principles. Thus, prefrontal cortex biases the nature of the outputs of connected brain areas according to experience and recent outcomes of decisions (Fig. 9.1).

It should be noted that although it is reasonable to assume that the prefrontal cortex is a major controller of the impact of prediction error signaling in the brain, other sources of control of cell excitability may arise via direct interconnections between the multiple prediction detection areas of the brain. For example, a prediction error from the hippocampus could be transmitted to midbrain–striatal neurons along pathways that do not necessarily include the prefrontal cortex. Indirect support for this idea comes from observations that conditions that produce error messages in the hippocampus change reward responses of dopamine neurons (Jo et al. 2013; Puryear et al. 2010), phasic theta comodulation is observed between hippocampus and striatum (DeCoteau et al. 2007a) during decision tasks, and comodulation of neural activity has been reported between prefrontal cortex and parietal cortex (Diwadkar et al. 2000). Perhaps neuromodulators such as dopamine strengthen or weaken associations that led to the last correct or incorrect, respectively, decision or behavior.

In sum, homeostatic regulatory processes may contribute to the automatic and continuous self-regulatory nature of prediction error analysis, decision-making, and learning. Such a naturally adaptive mechanism may optimize the relative contribution of different types of prediction error signals to future decisions and actions according to the pattern of recent successes and failures in prediction.

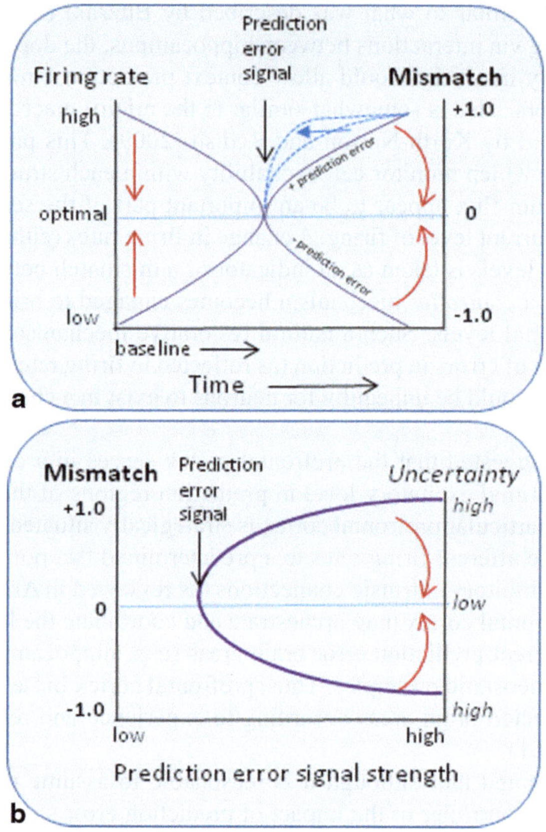

Fig. 9.1 **a** A homeostatic model of memory processing suggests that the primary goal of prefrontal cortex interactions with prediction centers of the brain (e.g., hippocampus and the midbrain-striatal area) is to maintain the baseline (tonic) firing rate of neurons within these centers at a set point level that is optimal for detecting future prediction errors. Modulating factors such as one's motivation or emotional state can elevate or reduce the baseline firing rates. It is postulated that the prefrontal cortex continually receives information from the prediction areas regarding the current population firing rates. If the baseline rates become elevated (e.g., due to stress) the prefrontal cortex is equipped to anatomically and physiologically restore firing rates (*red straight arrows*) to their optimal (baseline) levels. If the rates become too low (e.g., in depression), the prefrontal cortex should engage mechanisms to elevate firing to optimal levels over time. When a prediction error signal arrives in, or is generated by, a given prediction structure, firing rates can increase (in cases when prediction errors are positive) or decrease (when prediction errors are negative). The degree of rate increases or decreases scales to the degree of mismatch that is detected, and the slope of the increase or decrease in firing may vary between individuals and/or as a function of experience (*blue arrow*). **b** The prefrontal cortex is responsible for restoring the firing rates back to optimal levels and this reduces the uncertainty that was generated by the prediction error signals

Setting the Baseline from Which Prediction Error Signals Emerge: A Role for Hypothalamus, Amygdala, and the Prefrontal Cortex

Individual neurons face a continual barrage of excitatory inputs across tens of thousands of synaptic connections. Yet, neurons cannot maintain high levels of excitability and remain viable in the long term. Fortunately, individual neurons appear to be able to naturally and automatically engage mechanisms that control their level of excitability. This may occur by sensing and controlling the flow of various ions across cell membranes (e.g., Burrone et al. 2002; Turrigiano 2008; Turrigiano et al. 1998, see more detailed description below). Optimal levels of neuronal activity can be maintained also by achieving a relatively constant balance of excitatory and inhibitory synaptic input (e.g., Burrone et al. 2002). Together these factors define the baseline level of tonic activity from which phasic error signals are generated. Interestingly, the tonic level of cell excitability can be set according to the motivational state of an animal (Cagniard et al. 2006; Pecina et al. 2003; Puryear et al. 2010) suggesting that ones motivational state may play a significant role in determining the threshold for phasic neural and behavioral responding to unexpected events.

Motivational state information (e.g., signals of hunger or thirst) may arrive in prediction error structures (e.g. the hippocampus or striatum) and/or their controller (prefrontal cortex) via hypothalamic afferent systems. For example, lateral hypothalamus signals of hunger that reach brain areas that evaluate predictions may increase subsequent reward-responsiveness of efferent target neurons. Elevated responses to reward may reflect higher subjective values of the reward, an interpretation that is consistent with the biological needs of an animal. The amygdala, on the other hand, is thought to mediate a different motivational variable, and that is the emotional state of animals (Johansen et al. 2011). A message reflecting the current emotional state may emerge from the amygdala's role in associating cues with their aversive consequences (e.g., Chau and Galvez 2012; Paz and Pare 2013). Amygdala likely alters its neural activity in response to fear (Ciocchi et al. 2010; Haubensak et al. 2010; Li et al. 2013). Since the amygdala has direct excitatory effects on SN or VTA neurons (Lee et al. 2005; Zahm et al. 2011), fear-induced amygdala activation may increase the likelihood that dopamine neurons transition to a more excitable "up-state" (Wilson 1993; Wilson and Kawaguchi 1996) when hippocampal messages arrive. In this way, in urgent situations, animals can more readily assess the value of a changed context since transitioning to an "up-state" could make the dopamine cells respond more quickly to an input. This could be adaptive since responses can be implemented more quickly.

In addition to generally biasing the levels of neural excitability (which may translate to biasing the threshold for prediction error signaling), the amygdala may modulate prediction error-based learning efficiency on a trial by trial basis. For example, it is known that there is increased attention to cues or rewards that are unexpected or surprising based on past experiences (Pearce and Hall 1980; Rescorla and Wagner 1972). The dopamine system clearly plays a role in surprise-induced enhancement of learning (e.g., Schultz et al. 1997; Schultz and Dickinson 2000), and this may relate to transient influences of the amygdala on dopamine neurons since this prediction error-based learning effect is abolished in rats with amygdala disruption (Holland and Gallagher 2006) with no effects on the subsequent expression of surprise-induced enhanced learning (Lee et al. 2008). The amygdala and hypothalamus, then, may orchestrate information processing circuits/systems by ultimately setting the threshold for future error detection via direct connections to prediction error structures such as the hippocampus, striatum, sensory and motor cortex, and the cerebellum.

The prefrontal cortex can also be thought of as regulating the excitability state of neurons in predictive centers of the brain but for different reasons than both amygdala and hypothalamus. The prefrontal cortex is commonly thought to be important for holding information on-line in a working memory buffer (e.g., Arnsten et al. 2012; Fuster 2008). This function is considered essential to be able to select appropriate responses and/or for switching behavioral strategies (Ragozzino et al. 1999a; Ragozzino et al. 1999b; Young and Shapiro 2009), and this interpretation is consistent with findings that transient functional connections exist between the prefrontal cortex and the hippocampus or striatum especially when working memory is helpful for optimal behaviors. For example, hippocampal and prefrontal theta become co-modulated at times when animals make choices (e.g., Hyman et al. 2005; Shirvalkar et al. 2010), but not at other times during task performance. Co-activation of striatal and prefrontal activity has also been observed when working memory is required for accurate response selection (Levy et al. 1997; Scimeca and Badre 2012). Thus, the functional connections between striatum and prefrontal cortex, or between hippocampus and prefrontal cortex, can vary in strength and impact depending on the current task demands. Presumably this variation reflects the phasic task-dependent coordination of patterns of excitation and inhibition between prefrontal cortex and its efferent targets. Since the prefrontal cortex is thought to play a role in prediction analysis (e.g., Holyroyd et al. 2002), we suggest the possibility that its major contribution is to regulate efferent cell excitability according to recent behavioral outcomes. Indeed, Karlsson et al. (2012) recently showed that prefrontal cortical representations switch states of stability when conditions of greater uncertainty arise, that is, when response outcomes do not occur as predicted. Also, Merchant et al. (2011) suggest that prefrontal cortex exerts "top-down" control over parietal cortical responses in a match-to-sample task.

Prefrontal modulation is made possible due to the rather complex pattern of inhibitory and excitatory control over multiple types of efferent neurons (i.e., both interneurons and projection neurons) in efferent prediction brain areas (as reviewed in Arnsten 2011, Arnsten et al. 2012; Khan and Muly 2011), neurons that then could return information back to prefrontal cortex about their current activity state. Neocortex has indeed been shown to regulate the excitability states of subcortical neurons (e.g., Calhoon and O'Donnell 2013; Plenz and Arnsten 1996; Plenz and Kitai 1998). During baseline conditions, prefrontal cortex in particular may continually receive information about the current level of neural activity in target regions, and then use this afferent data to determine the extent and type of excitatory and inhibitory control needed to achieve optimal tonic activity within each of the multiple efferent prediction error systems. If the tonic activity becomes too low, for example, at times when there are no prediction error signals, prefrontal cortex may elevate the state of neural excitability so that the prediction cells are more responsive to future error signals, a feature that should increase the speed and accuracy of the error signaling. If, on the other hand, the baseline activity of a target region is higher than the optimal for the detection of prediction errors, further increasing the excitability of the cells may be detrimental for the cell's health and ability to produce clear error signals. In this case, it would be most adaptive if the prefrontal cortex lowered the level of excitability of its target cells so that optimal responsivity can be restored in the target region.

Recurrent neurocircuitry within the prefrontal cortex is thought to contribute to its working memory capacity (e.g., Arnsten et al. 2012), and as such this circuit is a clear candidate system to not only integrate error signals arriving from the different prediction error brain regions, but to also bias the thresholds and strengths of subsequent error-related signals from the brain regions that originally produced the error signal. The particular constellation of excitatory and inhibitory biases presumably will result in the most desired behavioral outcome.

In summary, at specific times when working memory is needed, the intrinsic recurrent neural circuits of the prefrontal cortex (Arnsten et al. 2012) may selectively and strategically exploit (differentially or in concert) its rich array of excitatory and inhibitory efferent connections to regulate the probability of neural firing in different prediction areas of the brain such that the relative responsiveness of different prediction brain regions changes in task-dependent ways. When prediction errors are detected and firing rates change, the prefrontal cortex may not only integrate the signal within its recurrent intrinsic circuitry, but it may have a key restorative function in efferent structures such that the firing rates return to a baseline tonic level that optimizes subsequent responsiveness to input. Thus, the prefrontal cortex may bias efferent neurons' ability to engage in, or efficiently use, prediction error analysis and hence their ability to adaptively guide future behaviors. This process may be a key factor that determines which prediction error-generating (i.e., memory) system ultimately controls the selection of future responses.

Final Comments

Memory research continues to evolve in complexity as new technologies become available. Ray Kesner's conceptualization of the cognitive and neurobiological underpinnings of memory was visionary in that it laid out a multidimensional neural systems view of memory that has provided guidance for a number of decades. Current challenges are to understand why different brain structures have select roles in memory, the nature of the interaction between brain structures, the control mechanisms for the interactions between brain areas, and the mechanisms by which memory functions, in all of its complexities, appear to be self-regulated and automatic. We offer a new hypothesis that the different memory regions of the brain make special contributions to memory because they assess the validity of different types of predictions that are needed for one to continually make adaptive choices and engage in adaptive and goal-directed behaviors. A homeostatic model of memory regulation is described to at least in part account for the seemingly automatic nature of memory function. Thank you Ray, for guiding the field to this very important and crucial time in memory research, one that promises exponential growth in the near future.

Acknowledgements This data described in this chapter were generated with grant support over the past 20 years (most recently NIH grant MH58755). The current conceptual frameworked evolved over that time, and they were based on discussions with a number of graduate students and postdoctoral students. Most recently, these included Wambura Fobbs, Katy Gill, Yong Sang Jo, Adria Martig, Sujean Oh, Corey Puryear, David Smith, Valerie Tryon, and Oxana Yeshenko.

References

Adams, R. A., Shipp, S., & Friston, K. J. (2013). Predictions not commands: Active interference in the motor system. *Brain Structure and Function, 218,* 611–643.

Adolphs, R., Tranel, D., & Buchanan T. W. (2005). Amygdala damage impairs emotional memory for gist but not details of complex stimuli. *Nature Neuroscience, 8,* 512–518.

Alexander, G. E., & Crutcher, M. D. (1990a). Functional architecture of basal ganglia circuits: Neural substrates of parallel processing. *Trends in Neuroscience, 13,* 266–271.

Alexander, G. E., & Crutcher, M. D. (1990b). Preparation for movement: Neural representations of intended direction in three motor areas of the monkey. *Journal of Neurophysiology, 64,* 133–150.

Alexander, G. E., DeLong, M. R., & Strick, P. L. (1986). Parallel organization of functionally segregated circuits linking basal ganglia and cortex. *Annual Reviews in Neuroscience,* 9, 357–381.

Anderson, M. I., & Jeffery, K. J. (2003). Heterogeneous modulation of place cell firing by changes in context. *Journal of Neuroscience, 23,* 8827–8835.

Arnsten, A. F. T. (2011). Prefrontal cortical network connections: Key site of vulnerability in stress and schizophrenia. *International Journal of Developmental Neuroscience, 29,* 21–223.

Arnsten, A. F. T., Wang, M. J., & Paspalas, C. D. (2012). Neuromodulation of thought: Flexibilities and vulnerabilities in prefrontal cortical network synapses. *Neuron, 76,* 223–239.

Baddeley, A., & Della Sala, S. (1996). Working memory and executive control. *Royal Society of London B Biological Sciences, 351,* 1397–1403.

Bayer, H. M., & Glimcher, P. W. (2005). Midbrain dopamine neurons encode a quantitative reward prediction error signal. *Neuron, 47,* 129–141.

Bayley, P.J., Frascino, J.C., & Squire, L.R. (2005). Robust habit learning in the absence of awareness and independent of the medial temporal lobe. *Nature, 436,* 550–553.

Benchenane, K., Peyrache, A., Khamassi, M., Tierney, P. L., Gioanni, Y., Battaglia, F. P., & Wiener, S. I. (2010). Coherent theta oscillations and reorganization of spike timing in the hippocampal- prefrontal network upon learning. *Neuron, 66,* 921–936.

Berke, J. D. (2009). Fast oscillations in cortical-striatal networks switch frequency following rewarding events and stimulant drugs. *European Journal of Neuroscience, 30,* 848–859.

Berke, J. D., Okatan, M., Skurski, J., & Eichenbaum, H. B. (2004). Oscillatory entrainment of striatal neurons in freely moving rats. *Neuron, 43,* 883–896.

Bethus, I., Tse, D., & Morris, R. G. (2010). Dopamine and memory: modulation of the persistence of memory for novel hippocampal NMDA receptor-dependent paired associates. *Journal of Neuroscience, 30,* 1610–1618.

Burrone, J., O'Bryne, M., & Murthy V. N. 2002. Multiple forms of synaptic plasticity triggered by selective suppression of activity in individual neurons. *Nature, 420,* 414–419.

Buzsaki, G. (2006). *Rhythms of the brain.* New York: Oxford Press.

Buzsaki, G. (2013). Time, space and memory. *Nature, 497,* 568–569.

Buzsaki, G., & Moser, E. I. (2013). Memory, navigation and theta rhythm in the hippocampal-entorhinal system. *Nature Neuroscience, 16,* 130–138.

Cagniard, B., Beeler, J. A., Britt, J. P., McGhee, D. S., Marinelli, M., & Zhuang, X. (2006). Dopamine scales performance in the absence of new learning. *Neuron, 51,* 541–547.

Calhoon, G. G., & O'Donnell, P. (2013). Closing the gate in the limbic striatum: Prefrontal suppression of hippocampal and thalamic inputs. *Neuron, 78,* 181–190.

Chau, L. S., & Galvez, R. (2012). Amygdala's involvement in facilitating associative learning-induced plasticity: A promiscuous role for hte amygdala in memory acquisition. *Frontiers in Integrative Neuroscience, 6,* 92.

Chen, J., Olsen, R. K., Preston, A. R., Glover, G. H., & Wagner A. D. (2011). Associative retrieval processes in the human medial temporal lobe: Hippocampal retrieval success and CA1 mismatch detection. *Learning and Memory, 18,* 523–528.

Chiba, A. A., Kesner, R. P., & Jackson, P. (2002). Two forms of spatial memory: A double dissociation between the parietal cortex and the hippocampus in the rat. *Behavioral Neuroscience, 116,* 874–883.

Ciocchi, S., Herry, C., Grenier, F., Wolff, S. B., Letzkus, J. J., Vlachos, I., Ehrlich, I., Sprengel, R., Deisseroth, K. I., Stadler, M. B., Muller, C., & Luthi, A. (2010). Encoding of conditioned fear in central amygdala inhibitory circuits. *Nature, 468,* 277–282.

Curtis, C. E., Rao, V. Y., & E'Esposito, M. (2004). Maintenance of spatial and motor codes during oculomotor delayed response tasks. *Journal of Neuroscience, 24,* 3944–3952.

DeCoteau, W. E., Thorn, C., Gibson, D. J., Courtemanche, R., Mitra, P., Kubota, Y., & Graybiel, A. M. (2007a). Learning-related coordination of striatal and hippocampal theta rhythms during acquisition of a procedural maze task. *Proceedings of the National Academy of Science U S A, 104,* 5644–5649.

DeCoteau, W. E., Thorn, C., Gibson, D. J., Courtemanche, R., Mitra, P., Kubota, Y., & Graybiel, A. M. (2007b). Oscillations of local field potentials in the rat dorsal striatum during spontaneous and instructed behaviors. *Journal of Neurophysiology, 97,* 3800–3805.

Dickerson, K. C., Li, J., & Delgado, M. R. (2011). Parallel contributions of distinct human memory systems during probabilistic learning. *Neuroimage, 55,* 266–276.

Diwadkar, V. A., Carpenter, P. A., & Just, M. A. (2000). Collaborative activity between parietal and dorso-lateral prefrontal cortex in dynamic spatial working memory revealed by fMRI. *Neuroimage, 12,* 85–99.

Doll, B. B., Simon, D. A., & Daw, N. D. (2012). The ubiquity of model-based reinforcement learning. *Current Opinion in Neurobiology, 22,* 1075–1081.

Duhamel, J. R., Colby, C. L., & Goldberg, M. E. (1992). The updating of the representation of visual space in parietal cortex by intended eye movements. *Science, 255,* 90–92.

Duncan, K., Ketz, N., Inati, S. J., & Davachi, L. (2012a). Evidence for area CA1 as a match/mismatch detector: A high-resolution fMRI study of the human hippocampus. *Hippocampus,* 22, 389–398.

Duncan, K., Sadanand, A., & Davachi, L. (2012b). Memory's penumbra: Episodic memory decisions induce lingering mnemonic biases. *Science, 337,* 485–487.

Engel, A. K., Fries, P., & Singer, W. (2001). Dynamic predictions: Oscillations and synchrony in top-down processing. *Nature Reviews in Neuroscience, 2,* 704–716.

Eschenko, O., & Mizumori, S. J.Y. (2007). Memory influences on hippocampal and striatal neural codes: Effects of a shift between task rules. *Neurobiology of Learning and Memory, 87,* 495–509.

Fell, J., Klaver, P., Lehnertz, K., Grunwald, T., Schaller, C., Elger, C. E., & Fernandez, G. (2001). Human memory formation is accompanied by rhinal-hippocampal coupling and decoupling. *Nature Neuroscience, 4,* 1259–1264.

Ferbinteanu, J., & Shapiro, M. L. (2003). Prospective and retrospective memory coding in the hippocampus. *Neuron, 40,* 1227–1239.

Foerde, K., & Shohamy, D. (2011). Feedback timing modulates brain systems for learning in humans. *Journal of Neuroscience, 31,* 13157–13167.

Fogassi, L., Ferrari, P. F., Gesierich, B., Rozzi, S., Chersi, F., & Rizzolatti, G. (2005). Parietal lobe: From action organization to intention understanding. *Science, 308,* 662–667.

Froemke, R. C., Merzenich, M. M., & Schreiner, C. E. (2007). A synaptic memory trace for cortical receptive field plasticity. *Nature, 450,* 425–429.

Fuster, J. M. (2006). The cognit: A network model of cortical representation. *International Journal of Psychophysiology, 60,* 125–132.

Fuster, J. M. (2008). *The prefrontal cortex* (4th ed.). London: Academic Press.

Fuster, J. M. (2009). Cortex and memory: Emergence of a new paradigm. *Journal of Cognitive Neuroscience, 21,* 2047–2072.

Fyhn, M., Molden, S., Hollup, S., Moser, M. B., & Moser E. (2002). Hippocampal neurons responding to first-time dislocation of a target object. *Neuron, 35,* 555–566.

Gengler, S., Mallot, H. A., & Holscher C. (2005). Inactivation of the rat dorsal striatum impairs performance in spatial tasks and alters hippocampal theta in the freely moving rat. *Behavioral Brain Research, 164,* 73–82.

Gilbert, P. E., & Kesner, R. P. (2002). The amygdala but not the hippocampus is involved in pattern separation based on reward value. *Neurobiology of Learning and Memory, 77,* 338–353.

Goldman-Rakic, P. S. (1995). Cellular basis of working memory. *Neuron, 14,* 477–485.

Goldman-Rakic, P. S. (1996). The prefrontal landscape: Implications of functional architecture for understanding human mentation and the central executive. *Philosophical Translation of the Royal Society of London B Biological Sciences, 351,* 1445–1453.

Good, M., & Honey, R. C. (1991). Conditioning and contextual retrieval in hippocampal rats. *Behavioral Neuroscience, 105,* 499–509.

Gray, J. A. (1982). *The neuropsychology of anxiety: An enquiry into the functions of the septo-hippocampal system*. Oxford: Oxford University Press.

Gray, J. A. (2000). *The neuropsychology of anxiety: An enquiry into the functions of the septo-hippocampal system* (2nd ed.). Oxford: Oxford University Press.

Graybiel, A. M. (2008) Habits, rituals, and the evaluative brain. *Annual Review in Neuroscience, 31,* 359–387.

Graybiel, A. M., Aosaki, T., Flaherty, A. W., & Kimura, M. (1994). The basal ganglia and adaptive motor control. *Science, 265,* 1826–1831.

Groenewegen, H. J., Galis-de Graaf, Y., & Smeets, W. J. (1999). Integration and segregation of limbic cortico-striatal loops at the thalamic level: An experimental tracing study in rats. *Journal of Chemical Neuroanatomy, 16,* 167–185.

Haber, S. N. (2003). The primate basal ganglia: Parallel and integrative networks. *Journal of Chemical Neuroanatomy, 26,* 317–330.

Hallworth, N. E., & Bland, B. H. (2004). Basal ganglia—hippocampal interactions support the role of the hippocampal formation in sensorimotor integration. *Experimental Neurology, 188,* 430–443.

Hasselmo, M. E. (2005a). What is the function of hippocampal theta rhythm?–Linking behavioral data to phasic properties of field potential and unit recording data. *Hippocampus, 15,* 936–949.

Hasselmo, M. E. (2005b). The role of hippocampal regions CA3 and CA1 in matching entorhinal input with retrieval of associations between objects and context: Theoretical comment on Lee et al. (2005). *Behavioral Neuroscience, 119,* 342–345.

Hasselmo, M. E., Hay, J., Ilyn, M., & Gorchetchnikov, A. (2002). Neuromodulation, theta rhythm and rat spatial navigation. *Neural Networks, 15,* 689–707.

Haubensak, W., Kunwar, P. S., Cai, H., Ciocchi, S., Wall, N. R., Ponnusamy, R., Biag, J., Dong, H. W., Deisseroth, K., Callaway, E. M., Fanselow, M. S., Lüthi, A., & Anderson, D. J. (2010). Genetic dissection of an amygdala microcircuit that gates conditioned fear. *Nature, 468,* 270–276

Holland, P. C., & Gallagher, M. (2006). Different roles for amygdala central nucleus and substantia innominate in the surprise-induced enhancement of learning. *Journal of Neuroscience, 26,* 3791–3797.

Hollerman, J. R., & Schultz, W. (1998). Dopamine neurons report an error in the temporal prediction of reward during learning. *Nature Neuroscience, 1,* 304–309.

Hollerman, J. R., Tremblay, L., & Schultz, W. (1998) Influence of reward expectation on behavior-related neuronal activity in primate striatum. *Journal of Neurophysiology, 80,* 947–963.

Holroyd, C. B., Coles, M. G., & Nieuwenhuis, S. (2002). Medial prefrontal cortex and error potentials. *Science, 29*(6), 1610–1611.

Hopkins, R. O., Kesner, R. P., & Goldstein, M. (1995). Memory for novel and familiar temporal spatial and linguistic temporal distance information in hypoxic subjects. *International Journal of Neuropsychology, 1,* 454–468.

Hopkins, R. O., Waldram, K., & Kesner, R. P. (2004). Sequences assessed by declarative and procedural tests of memory in amnesic patients with hippocampal damage. *Neuropsychologia, 42,* 1877–1886.

Horvitz, J. C. (2002). Dopamine gating of glutamatergic sensorimotor and incentive motivational input signals to the striatum. *Behavioral Brain Research, 137,* 65–74.

Humphries, M. D., & Prescott, T. J. (2010). The ventral basal ganglia, a selection mechanism at the crossroads of space, strategy, and reward. *Progress in Neurobiology, 90,* 385–417.

Hyman, J. M., Zilli, E. A., Paley, A. M., & Hasselmo, M. E. (2005). Medial prefrontal cortex cells show dynamic modulation with the hippocampal theta rhythm dependent on behavior. *Hippocampus, 15,* 739–749.

Jeffery, K. J., Anderson, M. I., Hayman, R., & Chakraborty, S. (2004). A proposed architecture for the neural representation of spatial context. *Neuroscience and Biobehavioral Review, 28,* 201–218.

Jenkins, T. A., Amin, E., Pearce, J. M., Brown, M. W., & Aggleton. J. P. (2004). Novel spatial arrangements of familiar visual stimuli promote activity in the rat hippocampal formation but not the parahippocampal cortices: A c-fos expression study. *Neuroscience, 124,* 43–52.

Jo, Y. S., Lee, J., & Mizumori, S. J. Y. (2013). Effects of prefrontal cortical inactivation on neural activity in the ventral tegmental area. *Journal of Neuroscience, 33,* 8159–8171.

Jog, M. S., Kubota, Y., Connolly, C. I., Hillegaart, V., & Graybiel, A. M. (1999). Building neural representations of habits. *Science, 286,* 1745–1749.

Johansen, J. P., Cain, C. K., Ostroff, L. E., & LeDoux, J. E. (2011). Molecular mechanisms of fear learning and memory. *Cell, 147,* 509–524.

Jones, M. W., & Wilson, M. A. (2005). Theta rhythms coordinate hippocampal-prefrontal interactions in a spatial memory task. *PLoS Biology,* 3, e402.

Khan, Z. U., & Muly, E. C. (2011). Molecular mechanisms of working memory. *Behavioral and Brain Research, 219,* 329–341.

Kalenscher, T., Lansink, C. S., Lankelma, J. V., & Pennartz, C. M. (2010). Reward-associated gamma oscillations in ventral striatum are regionally differentiated and modulate local firing activity. *Journal of Neurophysiology, 103,* 1658–1672.

Kametani, H., & Kesner, R. P. (1989). Retrospective and prospective coding of information: Dissociation of parietal cortex and hippocampal formation. *Behavioral Neuroscience, 103,* 84–89.

Karlsson, M. P., Tervo, D. G. R., & Karpova, A. Y. (2012). Network resets in medial prefrontal cortex mark the onset of behavioral uncertainty. *Science, 338,* 135–139.

Kesner, R. P. (1980). An attribute analysis of memory: The role of the hippocampus. *Physiology Psychology, 8,* 189–197.

Kesner, R. P. (1989). Retrospective and prospective coding of information: Role of the medial prefrontal cortex. *Journal of Experimental Brain Research, 74,* 163–167.

Kesner, R. P. (1998). Neurobiological views of memory. In J. L. Martinez & R. P. Kesner (Eds.), *The neurobiology of learning and memory* (pp. 361–416). San Diego: Academic.

Kesner, R.P. (2009). Tapestry of memory. *Behavioral Neuroscience, 123,* 1–13.

Kesner, R.P., & DiMattia, B.V. (1987). Neurobiology of an attribute model of memory. In A. R. Morrison & A. N. Epstein (Eds.), *Progress in psychobiology and physiological psychology* (pp. 207–277). New York: Academic.

Kesner, R. P., & Hopkins, R. O. (2001). Short-term memory for duration and distance in humans: Role of the hippocampus. *Neuropsychology, 15,* 58–68.

Kesner, R. P., & Hopkins, R. O. (2006). Mnemonic functions of the hippocampus: A comparison between animals and humans. *Biological Psychology, 73,* 3–18.

Kesner, R.P., & Rogers, L. (2004). An analysis of independence and interactions of brain substrates that subserve multiple attributes, memory systems, and underlying processes. *Neurobiology of Learning and Memory, 82,* 199–215.

Kesner, R. P., & Williams, J. M. (1995). Memory for magnitude of reinforcement: Dissociation between the amygdala and hippocampus. *Neurobiology of Learning and Memory, 64,* 237–244

Kesner, R. P., Farnsworth, G., & DiMattia, B. V. (1989). Double dissociation of egocentric and allocentric space following medial prefrontal and parietal cortex lesions in the rat. *Behavioral Neuroscience, 103,* 956–961.

Kesner, R. P., Bolland, B., & Dakis, M. (1993). Memory for spatial locations, motor responses, and objects: Triple dissociations among the hippocampus, caudate nucleus and extrastriate visual cortex. *Experimental Brain Research, 93,* 462–470.

Knowlton, B. J., Mangels, J. A., & Squire, L.R. (1996). A neostriatal habit learning system in humans. *Science, 273,* 1399–1402.

Kraus, B. J., Robinson II, R. J., White, J. A., Eichenbaum, H., & Hasselmo, M. E. (2013). Hippocampal "time cells": Time versus path integration. *Neuron, 78,* 1–12.

Kropf, W., & Kuschinsky, K. (1993). Conditioned effects of apomorphine are manifest in regional EEG of rats both in hippocampus and in striatum. *Naunyn-Schmiedeberg's Archives of Pharmacology, 347,* 487–493.

Kuhl, B. A., Shah, A. T., DuBrow, S., & Wagner AD. (2010). Resistance to forgetting associated with hippocampus-mediated reactivation during new learning. *Nature Neuroscience, 13,* 501–506.

Kumaran, D., & Maguire, E. A. (2007). Match-mismatch processes underlie human hippocampal responses to associative novelty. *Journal of Neuroscience, 27,* 8517–8524.

Kurkin, S., Akao, T., Shichinohe, N., Fukushima, J., & Fukushima, K. (2011). Neuronal activity in medial superior temporal area (MST) during memory-based smooth pursuit eye movements in monkeys. *Experimental Brain Research, 214,* 293–301.

Kurth-Nelson, Z., & Redish A. D. (2009). Temporal-difference reinforcement learning with distributed representations. *PLoS One, 4,* 37362.

Lee, H. J., Groshek, F., Petrovich, G. D., Cantalini, J. P., Gallagher, M., & Holland, P. C. (2005). Role of amygdalo-nigral circuitry in conditioning of a visual stimulus paired with food. *Journal of Neuroscience, 25,* 3881–3888.

Lee, H. J., Youn, J. M., Gallagher, M., & Holland, P. C. (2008). Temporally limited role of substantia nigra-central amygdala connections in surprise-induced enhancement of learning. *European Journal of Neuroscience, 27,* 3043–3049.

Leutgeb, J. K., Leutgeb, S., Treves, A., Meyer, R., Barnes, C. A., McNaughton, B. L., Moser, M. B., & Moser, E. I. (2005a). Progressive transformation of hippocampal neuronal representations in "morphed" environments. *Neuron, 48,* 345–358.

Leutgeb, S., Leutgeb, J. K., Barnes, C. A., Moser, E. I., McNaughton, B. L., & Moser, M. B. (2005b). Independent codes for spatial and episodic memory in hippocampal neuronal ensembles. *Science, 309,* 619–623.

Levy, R., Friedman, H. R., Davachi, L., & Goldman-Rakic, P. S. (1997). Differential activation of the caudate nucleus in primates performing spatial and nonspatial working memory tasks. *Journal of Neuroscience, 17,* 3870–3882.

Li, H., Penzo, M. A., Taniguchi, H., Kopec, C. D., Huang, Z. J., & Li, B. (2013). Experience-dependent modification of a central amygdala fear circuit. *Nature Neuroscience, 16,* 332–329.

Lisman, J. E., & Grace, A. A. (2005). The hippocampal-VTA loop: Controlling the entry of information into long-term memory. *Neuron, 46,* 703–713.

Lisman, J. E., & Otmakhova, N. A. (2001). Storage, recall, and novelty detection of sequences by the hippocampus: Elaborating on the SOCRATIC model to account for normal and aberrant effects of dopamine. *Hippocampus, 11,* 551–568.

Llinas, R. R., & Roy, S. (2009). The 'predictive imperative' as the basis for self-awareness. *Philosophical Transactions of the Royal Society B Biological Sciences, 364,* 1301–1307.

Manns, J. R., Howard, M. W., & Eichenbaum, H. (2007a). Gradual changes inhippocampal activity support remembering the order of events. *Neuron, 56,* 530–540.

Manns, J. R., Zilli, E. A., Ong, K. C., Hasselmo, M. E., & Eichenbaum, H. (2007b). Hippocampal CA1 spiking during encoding and retrieval: Relation to theta phase. *Neurobiology Learning Memory, 87,* 9–20.

Marder, E., & Goaillard, J. M. (2006). Variability, compensation and homeostasis in neuron and network function. *Nature Reviews Neuroscience, 7,* 563–574.

Marder, E., & Prinz, A. A. (2003). Current compensation in neuronal homeostasis. *Neuron, 37,* 2–4.

Mauk, M. D., & Buonomano, D. V. (2004). The neural basis of temporal processing. *Annual Reviews in Neuroscience, 27,* 307–340.

McDonald, R. J., & White, N. M. (1993). A triple dissociation of memory systems: Hippocampus, amygdala and dorsal striatum. *Behavioral Neuroscience, 107,* 371–388.

McNaughton, B. L., Barnes, C. A., & O'Keefe, J. (1983). The contributions of position, direction, and velocity to single unit activity in the hippocampus of freely-moving rats. *Experimental Brain Research, 52,* 41–49.

McNaughton, B. L., Mizumori, S. J., Barnes, C. A., Leonard, B. J., Marquis, M., & Green, E. J. (1994). Cortical representation of motion during unrestrained spatial navigation in the rat. *Cerebral Cortex, 4,* 26–39.

Meftah, E.-M, Bourgeon, S., & Chapman, C. E. (2009). Instructed delay discharge in primary and secondary somatosensory cortex within the context of a selective attention task. *Journal of Neurophysiology, 101,* 2649–2667.

Merchant H., Crowe, D. A., Robertson, M. S., Fortes, A. F., & Georgopoulos, A. P. (2011). Top-down spatial categorization signal from prefrontal to posterior parietal cortex in the primate. *Frontiers in Systems Neuroscience, 5,* 1–9.

Milner, B. (2005). The medial temporal-lobe amnesic syndrome. *Psychiatry Clinical North America, 28,* 599–611.

Mizumori, S. J. Y. (2008a). A context for hippocampal place cells during learning. In S. J. Y. Mizumori (Ed.), *Hippocampal place fields: Relevance to learning and memory* (pp. 16–43). New York: Oxford University Press.

Mizumori, S. J. Y. (2008b). *Hippocampal place fields: Relevance to learning and memory*. New York: Oxford University Press.

Mizumori, S. J. Y., & Jo, Y. S. (2013). Homeostatic regulation of memory systems and adaptive decisions. *Hippocampus, 23*(11), 1103–1124.

Mizumori, S. J. Y., Ragozzino, K. E., Cooper, B. G., & Leutgeb, S. (1999). Hippocampal representational organization and spatial context. *Hippocampus, 9,* 444–451.

Mizumori, S. J. Y., Cooper, B. G., Leutgeb, S., & Pratt, W. E. (2000a). A neural systems analysis of adaptive navigation. *Molecular Neurobiology, 21,* 57–82.

Mizumori, S. J. Y., Ragozzino, K. E., & Cooper, B. G. (2000b). Location and head direction representation in the dorsal striatum of rats. *Psychobiology, 28,* 441–462.

Mizumori, S. J. Y., Yeshenko, O., Gill, K., & Davis, D. (2004). Parallel processing across neural systems: Implications for a multiple memory systems hypothesis. *Neurobiology of Learning and Memory, 82,* 278–298.

Mizumori, S. J. Y., Smith, D.M., & Puryear, C.B. (2007a). Hippocampal and neocortical interactions during context discrimination: electrophysiological evidence from the rat. *Hippocampus, 17,* 851–862.

Mizumori, S. J. Y., Smith, D. M., & Puryear, C. B. (2007b). A role for place representation in episodic memory. In J. L. Martinez & R. P. Kesner (Eds.), *Neurobiology of learning and memory* (pp. 155–189). Second Edition Burlington Massachusetts: A: Academic.

Mizumori S. J. Y., Puryear, C. B., & Martig, A. K. (2009). Basal ganglia contributions to adaptive navigation. *Behavioural Brain Research, 199,* 32–42.

Moita, M. A., Rosis, S., Zhou, Y., LeDoux, J. E., & Blair, H. T. (2004). Putting fear in its place: Remapping of hippocampal place cells during fear conditioning. *Journal of Neuroscience, 24,* 7015–7023.

Muller, R. U., & Kubie, J. L. (1987). The effects of changes in the environment on the spatial firing of hippocampal complex-spike cells. *Journal of Neuroscience, 7,* 1951–1968.

Nadel, L. (2008). The hippocampus and context revisited. In S. J. Y. Mizumori (Ed.), *Hippocampal place fields: Relevance to learning and memory* (pp. 3–15). New York: Oxford University Press.

Nadel, L., & Payne, J. D. (2002). The hippocampus, wayfinding and episodic memory. In P. E. Sharp (Ed.), *The neural basis of navigation: Evidence from single cell recording* (pp. 235–248). Boston: Kluwer.

Nadel, L., & Wilner, J. (1980). Context and conditioning: a place for space. *Physiological Psychology, 8,* 218–228.

Nicola, S. M., Yun, I. A., Wakabayashi, K. T., & Fields, H. L. (2004). Firing of nucleus accumbens neurons during the consummatory phase of a discriminative stimulus task depends on previous reward predictive cues. *Journal of Neurophysiology, 91,* 1866–1882.

Nudo, R. J. (2011). Neural bases of recovery after brain injury. *Journal of Communications Disorder, 44,* 515–520.

O'Keefe, J. (1976). Place units in the hippocampus of the freely moving rat. *Experimental Neurology, 51,* 78–109.

O'Keefe, J., & Dostrovsky, J. (1971). The hippocampus as a spatial map. Preliminary evidence from unit activity in the freely-moving rat. *Brain Research, 34,* 171–175.

O'Keefe, J. & Nadel, L. (1978). *The hippocampus as a cognitive map.* Oxford University Press: Oxford.

O'Keefe, J., & Recce, M. L. (1993). Phase relationship between hippocampal place units and the EEG theta rhythm. *Hippocampus, 3,* 317–330.

Olton, D. S., Branch, M., & Best, P. J. (1978). Spatial correlates of hippocampal unit activity. *Experimental Neurology, 58,* 387–409.

Packard, M. G., & McGaugh, J. L. (1996). Inactivation of hippocampus or caudate nucleus with lidocaine differentially affects expression of place and response learning. *Neurobiology of Learning and Memory, 65,* 65–72.

Packard, M. G., Hirsh, R., & White, N. M. (1989). Differential effects of fornix and caudate nucleus lesions on two radial maze tasks: Evidence for multiple memory systems. *Journal of Neuroscience, 9,* 1465–1472.

Pastalkova, E., Itskov, V., Amarasingham, A., & Buzsaki, G. (2008). Internally generated cell assembly sequences in the rat hippocampus. *Science, 321,* 1322–1327.

Paulsen, O., & Moser, E. I. (1998). A model of hippocampal memory encoding and retrieval: GABAergic control of synaptic plasticity. *Trends in Neuroscience, 21,* 273–278.

Paz, R., & Pare, D. (2013). Physiological basis for emotional modulation of memory circuits by the amygdala. *Current Opinion in Neurobiology, 23,* 1–6.

Pearce, J. M., & Hall. G. (1980). A model for Pavlovian learning: Variations in the effectiveness of conditioned but not of unconditioned stimuli. *Psychological Review, 87,* 532–552.

Pecina, S., Cagniard, B., Berridge, K. C., Aldridge J. W., & Zhuang, X. (2003). Hyperdopaminergic mutant mice have higher 'wanting' but not 'liking' for sweet rewards. *Journal of Neuroscience, 23,* 9395–9402.

Pennartz, C. M., Berke, J. D., Graybiel, A. M., Ito, R., Lansink, C. S., van der Meer, M., Redish, A. D., Smith, K. S., & Voorn, P. (2009). Corticostriatal interactions during learning, memory processing, and decision making. *Journal of Neuroscience, 29,* 12831–12838.

Penner, M. R., & Mizumori, S. J. Y. (2012a). Age-associated changes in the hippocampal-ventral striatum-ventral tegmental loop that impact learning, prediction and context discrimination. *Frontiers in Aging Neuroscience, 4,* 1–12.

Penner, M. R., & Mizumori, S. J. Y. (2012b). Neural systems analysis of decision making during goal-directed navigation. *Progress in Neurobiology, 96,* 96–135.

Plenz, D., & Aertsen, A. (1996). Neural dynamics in cortex–striatum cocultures. I. Anatomy and electrophysiology of neuronal cell types. *Neuroscience, 70,* 861–891.

Plenz, D., & Kitai, S. T. (1998). Up and down states in striatal medium spiny neurons simultaneously recorded with spontaneous activity in fast-spiking interneurons studied in cortex-striatum-substantia nigra organotypic cultures. *Journal of Neuroscience, 18,* 266–283.

Puryear, C. B., King, M., & Mizumori, S. J. Y. (2006). Specific changes in hippocampal spatial codes predict spatial working memory performance. *Behavioral Brain Research, 169,* 168–175.

Puryear, C. B., Kim, M. J., & Mizumori, S. J. Y. (2010). Conjunctive encoding of reward and movement by ventral tegmental area neurons: Contextual control during adaptive spatial navigation. *Behavioral Neuroscience, 124,* 234–247. (Erratum in: Behav Neurosci. 2010 Jun;124(3):336).

Ragozzino, M. E., Wilcox, C., Raso, M., & Kesner, R. P. (1999a). Involvement of rodent prefrontal cortex subregions in strategy switching. *Behavioral Neuroscience, 113,* 32–41.

Ragozzino, M. E., Detrick, S., & Kesner, R. P. (1999b). Involvement of the prelimbic infralimbic areas of the rodent prefrontal cortex in behavioral flexibility for place and response learning. *Journal of Neuroscience, 19,* 4585–4594.

Ragozzino, K. E., Leutgeb, S., & Mizumori, S. J. Y. (2001). Conditional coupling of dorsal striatal head direction and hippocampal place representations during spatial navigation. *Experimental Brain Research, 139,* 372–376.

Ragozzino, M. E., Ragozzino, K. E., Mizumori, S. J. Y., & Kesner, R. P. (2002). The role of the dorsomedial striatum in behavioral flexibility for response and visual cue discrimination learning. *Behavioral Neuroscience, 116,* 105–115.

Ranck, J. B., Jr. (1973). Studies on single neurons in dorsal hippocampal formation and septum in unrestrained rats. I. Behavioral correlates and firing repertoires. *Experimental Neurology, 41,* 461–531.

Redish, A. D. (1999). *Beyond the cognitive map: From place cells to episodic memory.* Cambridge: MIT Press

Rescorla, R. A., & Wagner, A. R. (1972). A theory of Pavlovian conditioning: Variations in the effectiveness of reinforcement and nonreinforcement. In A. H. Black & W. F. Prokasy (Eds). *Classical conditioning II: Current research and theory* (pp. 64–99). New York: Appleton Century Crofts.

Robinson, T. E., & Kolb, B. (2004). Structural plasticity associated with exposure to drugs of abuse. *Neuropharmacology, 47*(Suppl 1), 33–46.

Sabatino, M., Ferraro, G., Liberti, G., Vella, N., & La Grutta, V. (1985). Striatal and septal influence on hippocampal theta and spikes in the cat. *Neuroscience Letters, 61,* 55–59.

Save, E., Buhot, M. C., Foreman, N., & Thinus-Blanc, C. (1992a). Exploratory activity and response to a spatial change in rats with hippocampal or posterior parietal cortical lesions. *Behavioral Brain Research, 47,* 113–127

Save, E., Poucet, B., Foreman, N., & Buhot, M. C. (1992b). Object exploration and reactions to spatial and nonspatial changes in hooded rats following damage to parietal cortex or hippocampal formation. *Behavioral Neuroscience, 106,* 447–456.

Scheidt, R. A., Zimbelman, J. L., Salowitz, N. M. G., Suminski, A. J., Mosier, K. M., Houk, J., & Simo, L. (2012). Remembering forward: Neural correlates of memory and prediction in human motor adaptation. *Neuroimage, 59,* 582–600

Schultz, W. (2010). Dopamine signals for reward value and risk: basic and recent data. *Behavior and Brain Function, 6,* 24.

Schultz, W., & Dickinson, A. (2000). Neuronal coding of prediction errors. *Annual Reviews in Neuroscience, 23,* 473–500.

Schultz, W., Dayan, P., & Montague, P. R. (1997). A neural substrate of prediction and reward. *Science, 275,* 1593–1599.

Scimeca, J. M., & Badre, D. (2012). Striatal contributions to declarative memory retrieval. *Neuron, 75,* 380–392.

Shirvalkar, P. R., Rapp, P. R., & Shapiro, M. L. (2010). Bidirectional changes to hippocampal theta-gamma comodulation predict memory for recent spatial episodes. *Proceedings of the National Academy of Sciences U S A, 107,* 7054–7059.

Siapas, A. G., Lubenov, E. V., & Wilson, M. A. (2005). Prefrontal phase locking to hippocampal theta oscillations. *Neuron, 46,* 141–151.

Smith, D. M., & Mizumori, S. J. Y. (2006a). Hippocampal place cells, context and episodic memory. *Hippocampus, 16,* 716–729.

Smith, D. M., & Mizumori, S. J. Y. (2006b). Learning-related development of context-specific neuronal responses to places and events: The hippocampal role in context processing. *Journal of Neuroscience, 26,* 3154–3163.

Smith, D. M., Wakeman, D., Patel, J., & Gabriel, M. (2004). Fornix lesions impair context-related cingulothalamic neuronal patterns and concurrent discrimination learning. *Behavioral Neuroscience, 118,* 1225–1239.

Snyder, L. H., Batista, A. P., & Andersen, R. A. (1997). Coding of intention in the posterior parietal cortex. *Nature, 386,* 167–170.

Stalnaker, T. A., Calhoon, G. G., Ogawa, M., Roesch, M. T., & Schoenbaum, G. (2012). Reward prediction error signaling in posterior dorsomedial striatum is action specific. *Journal of Neuroscience, 32,* 10296–10305.

Tabuchi, E. T., Mulder, A. B., & Wiener, S. I. (2000). Position and behavioral modulation of synchronization of hippocampal and accumbens neuronal discharges in freely moving rats. *Hippocampus, 10,* 717–728.

Tanaka, H., Sejnowski, T. J., & Krakauer, J. W. (2009). Adaptation to visuomotor rotation through interaction between posterior parietal and motor cortical areas. *Journal of Neurophysiology, 102,* 2921–2932.

Tse, D., Langston, R. F., Kakeyama, M., Bethus, I., Spooner, P. A., Wood, E. R., Witter, M. P., & Morris, R. G. (2007). Schemas and memory consolidation. *Science, 316,* 76–82.

Tulving, E. (2002). Episodic memory: From mind to brain. *Annual Reviews in Psychology, 53,* 1–25.

Turrigiano, G. G. (1999). Homeostatic plasticity in neuronal networks: The more things change, the more they stay the same. *Trends in Neuroscience, 22,* 221–227.

Turrigiano, G. G. (2008). The self-tuning neuron: Synaptic scaling of excitatory synapses. *Cell, 135,* 422–435.

Turrigiano, G. G. (2011). Homeostatic synaptic plasticity: Local and global mechanisms for stabilizing neuronal function. *Cold Spring Harbor Perspectives in Biology, 4a,* 005736.

Turrigiano, G. G., & Nelson, S. B. (2004). Homeostatic plasticity in the developing nervous system. *Nature Reviews in Neuroscience, 5,* 97–107.

Turrigiano, G. G., Leslie, K. R., Desai, N. S., Rutherford, L. C., & Nelson, S. B. (1998). Activity-dependent scaling of quantal amplitude in neocortical neurons. *Nature, 391,* 892–896.

van der Meer, M. A., & Redish, A. D. (2009). Low and high gamma oscillations in rat ventral striatum have distinct relationships to behavior, reward, and spiking activity on a learned spatial decision task. *Frontiers in Integrative Neuroscience, 3,* 9.

Vanderwolf, C. H. (1969). Hippocampal electrical activity and voluntary movement in the rat. *Electroencephalography Clinical Neurophysiology, 26,* 407–418.

Varela, F., Lachaux, J. P., Rodriguez, E., & Martinerie, J. (2001). The brainweb: Phase synchronization and large-scale integration. *Nature Reviews in Neuroscience, 2,* 229–239.

Vinogradova, O. S. (1995). Expression, control, and probably functional significance of the neuronal theta-rhythm. *Progress in Neurobiology, 45,* 523–583.

Voorn, P., Vanderschuren, L. J., Groenewegen, H. J., Robbins, T. W., & Pennartz, C. M. (2004) Putting a spin on the dorsal-ventral divide of the striatum. *Trends in Neuroscience, 27,* 468–474.

Walsh, M. M., & Anderson, J. R. (2012). Learning from experience: event-related potential correlates of reward processing, neural adaptation, and behavioral choice. *Neuroscience and Biobehaioral Reviews, 36,* 1870–1884.

Watson, K. K., & Platt, M. L. (2008). Neuroethology of reward and decision making. *Philosophical Transactions of the Royal Society of London B, 363,* 3825–3835.

Wilson, C. J. (1993). The generation of natural firing patterns in neostriatal neurons. *Progress in Brain Research, 99,* 277–298.

Wilson, C. J., & Kawaguchi, Y. (1996). The origins of two-state spontaneous membrane potential fluctuations of neostriatal spiny neurons. *Journal of Neuroscience, 16,* 2397–2410.

Womelsdorf, T., Schoffelen, J. M., Oostenveld, R., Singer, W., Desimone, R., Engel, A. K., & Fries, P. (2007). Modulation of neuronal interactions through neuronal synchronization. *Science, 316,* 1609–1612.

Wood, E. R., Dudchenko, P. A., & Eichenbaum H. (1999). The global record of memory in hippocampal neural activity. *Nature, 397,* 613–616.

Yeshenko, O., Guazzelli, A., & Mizumori, S. J. (2004). Context-dependent reorganization of spatial and movement representations by simultaneously recorded hippocampal and striatal neurons during performance of allocentric and egocentric tasks. *Behavioral Neuroscience, 118,* 751–769.

Yin, H. H., & Knowlton, B. J. (2006). The role of the basal ganglia in habit formation. *Nature Reviews in Neuroscience, 7,* 464–476.

Young, J. J., & Shapiro, M. L. (2009). Double dissociation and hierarchical organization of strategy switches and reversals in the rat PFC. *Behavioral Neuroscience, 123,* 1028–1035.

Zahm, D. S., Cheng, A. Y., Lee, T. J., Ghobadi, C. W., Schwartz, Z. M., Geisler, S., Parsely, K. P., Gruber, C., & Veh, R. W. (2011). Inputs to the midbrain dopaminergic complex in the rat, with emphasis on extended amygdala-recipient sectors. *Journal of Comparative Neurology, 19,* 3159–3188.

Chapter 10
Attribute Memory Model and Behavioral Neurophysiology of Memory

Inah Lee and Choong-Hee Lee

Kesner's Attribute Model

The Attribute Model for Memory

The attribute memory model was originally proposed by Kesner (Kesner 1980; Kesner and Hardy 1983) in an effort to develop a comprehensive memory theory that could cover both cognitive and neurobiological aspects. An event occurs, typically composed of different attributes according to the attribute model. The attributes in the model include sensory-perceptual, place, time, affect, and response. One may group these attributes into three categories: perceptual, cognitive, and response attributes. The attribute model proposes that different brain regions are involved as an attribute or a set of attributes is processed while an animal experiences an event. The model goes further and suggests that different brain regions form a distinct memory system (i.e., event-based, knowledge-based, and rule-based memory systems) depending on whether memory content is episodic or semantic in nature (Squire and Zola-Morgan 1991; Tulving 1972). For example, if a person parks his/her car at a certain location in a parking lot for a temporary visit and returns to find the car in a few hours, the parking lot memory being formed and retrieved in this case should be processed by the event-based memory system, according to the Kesner model. After parking at temporary locations in the lot daily for several days, if the person happens to find that he now has a permanent job there, which would let him park the car in the same location all the time, the memory representations originally processed by the event-based memory system should be processed by the knowledge-based

I. Lee (✉) · C.-H. Lee
Department of Brain and Cognitive Sciences, Laboratory for Behavioral Neurophysiology, Seoul National University, Rm 103A, Building 203 (SPC), Shilim-dong, Gwanak-gu, 151-747 Seoul, Republic of Korea
e-mail: inahlee@snu.ac.kr

C.-H. Lee
e-mail: choonghee.bcs@gmail.com

P. A. Jackson et al. (eds.), *The Neurobiological Basis of Memory*,
DOI 10.1007/978-3-319-15759-7_10

memory system. The knowledge-based memory system assumes not much change in the memory content and relatively constant ways in which memory is retrieved and used. The attribute model later added the third system, the rule-based memory system, but it is unclear how distinctions can be made at a conceptual level between knowledge-based and rule-based memory systems. That is, a rule is formed cognitively as various events are experienced, some key commonalities (as similar to the union of Venn diagrams) are extracted during those experiences, and become a common rule (i.e., knowledge of the way things should be dealt with).

The organizational scheme of the attribute memory model can be illustrated as shown in Fig. 10.1. As shown in the figure, the model requires both the attribute component and the memory system component as its major dimensions. The six attributes are associated with particular brain regions (sometimes with multiple ones) and those brain regions in turn belong to a particular memory system (event-based, knowledge-based, and rule-based memory systems). It is notable that the hippocampus and the prefrontal cortex (PFC) are the only brain regions that process multiple attributes (i.e., place, time, and language for the hippocampus; sensory-perceptual,

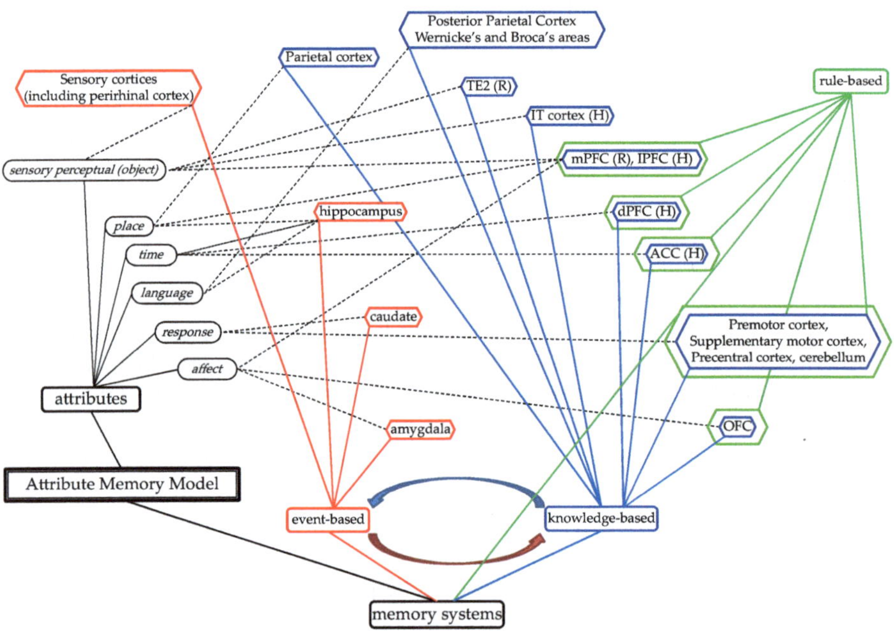

Fig. 10.1 The organizational scheme of the attribute memory model, showing brain regions and corresponding memory systems and attributes. Individual attributes (sensory-perceptual, place, time, language, response, and affect) and their associated brain regions are connected in dotted lines. The three memory systems (event-based, knowledge-based, and rule-based memory systems) and their associated brain areas are shown in different colors. *mPFC* medial prefrontal cortex, *lPFC* lateral prefrontal cortex, *dPFC* dorsal prefrontal cortex, *ACC* anterior cingulate cortex, *OFC* orbitofrontal cortex, *IT cortex* inferior temporal cortex. *H* human, *R* rodent. The *arrows* between the event-based and knowledge-based memory systems denote interactions between the two systems (*lower arrow*—updating the knowledge-based system by the event-based system; *upper arrow*—top-down influence of the knowledge-based system on the event-based memory system)

place, and affect for the medial PFC, or mPFC, in rodents (R) and lateral PFC, or lPFC, in primates and humans (H)) in the model as all other brain areas maintain one-to-one mapping with single attributes.

Attributes

The most critical dimension of Kesner's memory theory that makes the model unique is perhaps the *attribute*. Kesner categorizes the tasks the brain performs, according to which *attributes* (or type of information) play central roles in a certain task. For example, a task may require an animal to use sensory-perceptual and place information predominantly (e.g., Morris water maze) and the neural systems for processing matching attributes should be engaged in that case. There are six attributes that play major roles in Kesner's attribute theory (Fig. 10.1) and those are place, time, sensory-perception, language, response, and affect. The theory assumes interactions among these attributes depending on the nature of the task (Kesner 2009; Kesner and Rogers 2004), which should be the case because the sensory-perceptual attribute seems to be the first stage of information processing in the brain that may precede all the other attributes. Also, one can easily imagine the interaction between place and time attributes as the person in the parking-lot example above would experience as he/she navigates across space and time to find the car. Affect (e.g., valence, emotion, etc.) should also be interacting with other attributes. This naturally means that the brain areas associated with these attributes should interact with each other as an event memory is processed.

Memory Systems

The second dimension of Kesner's memory theory is the *memory system*. It is now widely accepted that there are multiple memory systems in the brain, and different theories have been put forth accordingly (Eichenbaum and Cohen 2001; McDonald and White 1993; Squire 2004; Squire and Zola-Morgan 1991; Tulving 1985). The attribute memory model categorizes the brain's memory systems into three according to the natural course of the development of mnemonic representations or to the order of engagement of different systems over time and experience. A task-based explanation may also be useful for understanding this aspect of the model. For example, when an animal first encounters a spatial memory task in an 8-arm maze, the novel experience of trying to find its way in the maze can be described as an "event." The brain can quickly form new memory representations that may last for several seconds to minutes for navigating across different arms of the maze, some of which may or may not last after the task is completed. The *event-based memory system* thus is needed when novel information associated with an event should be represented temporarily. One may imagine the same rat experiencing the task for multiple days. In that case, once novel and episodic information now becomes knowledge, and the *knowledge-based memory system* is in charge of making the rat get around in the maze. If the rat performs similar 8-arm maze tasks across different rooms and different mazes, then it may develop a general rule (or knowledge) that applies for most 8-arm maze situations by extracting and storing common principles. The *rule-based memory system,* according to Kesner's theory, integrates the information received from the event-based and knowledge-based memory systems by applying rules and strategies, but we would argue that it would be simpler to think of the rule-based memory system as a subcategory of the knowledge-based memory system.

Mnemonic Processes

The final dimension of Kesner's memory theory is composed of several mnemonic "processes" that work within and across the memory systems. The mnemonic processes included in his theory are mostly for explaining effective information processing such as for reducing interference for achieving specificity (e.g., pattern separation, information filtering, information selection, attention, etc.), maintaining information for different lengths of duration (e.g., short-term, intermediate-term, and long-term storage), and flexible generalization for information processing and action (e.g., pattern completion). The Kesner group, in particular, made significant contributions to experimentally testing computational models associating cognitive functions with specific subfields in the hippocampus. Specifically, computational modelers have proposed specific computations that might be performed by individual subfields of the hippocampus (Hasselmo and Wyble 1997; Marr 1971; Oreilly and Mcclelland 1994; Rolls 1996; Treves and Rolls 1994; Wallenstein et al. 1998). There are excellent reviews on this topic (Gluck et al. 2003; Kesner et al. 2000; Rolls and Kesner 2006; Yassa and Stark 2011) and it is beyond the scope of this chapter to review the hippocampal computational models in detail here. Briefly, on the basis of computational models, there had been theoretical predictions that the dentate gyrus (DG) was important for amplifying small differences in an environment so that similar contextual stimuli could be separately represented in the hippocampus, the function often called pattern separation in the literature (Marr 1971; Oreilly and Mcclelland 1994; Rolls 1996). The CA3 subfield had been suggested as a region in the hippocampus where autoassociative memory representations were formed and retrieved, even when partial or noisy cues were presented, the function often called as pattern completion (Marr 1971; Oreilly and Mcclelland 1994; Rolls 1996). Despite the existence of these models, the Kesner group found very little, if any, experimental evidence that could support those theoretical predictions in the mid 1990s. By developing and successfully implementing various lesion techniques for the individual subfields in the hippocampus, the Kesner group indeed produced some significant experimental evidence that could support computational models for hippocampal subfields. For example, rats with dorsal hippocampal lesions or DG specific lesions were impaired in spatial pattern separation under conditions of spatial similarity (Gilbert and Kesner 2006; Gilbert et al. 1998), rats with NMDA receptors blocked in CA3 were impaired in acquiring, but not retrieving, a spatial working memory task (Lee and Kesner 2002), the perforant path and mossy fibers seemed to perform different functions with respect to retrieval *versus* acquisition of spatial memory (Lee and Kesner 2004), to name a few. Electrophysiological studies confirmed some of these findings later (Lee et al. 2004; Leutgeb et al. 2007; Leutgeb et al. 2004), but it is worth mentioning that the Kesner group pioneered the experimental paradigms for testing computational models for hippocampal subfields by developing several key behavioral tasks and subfield-lesion methods.

It is also noteworthy that the computational processes examined in the hippocampus may not be unique to the hippocampus per se, but may universally be found in other brain regions including sensory and motor cortices. For example, computational processes such as pattern separation and completion have also been shown in the olfactory bulb and the piriform cortex in rodents and zebrafish (Barnes et al. 2008; Niessing and Friedrich 2010). The paper published in 2008 studied single-unit activities and their ensemble patterns responding to similar odor variations. By replacing or excluding one or two components out of ten chemical components that made up previously experienced odors, they found that neural ensembles in olfactory bulb showed decreased correlation (pattern separation) between similar but different odors, whereas ensemble activity in the anterior piriform cortex evoked by the odor with a component removed showed similar patterns to that of original odors (pattern completion). The Niessing and Friedrich study (2010) used two-photon calcium imaging of zebrafish olfactory bulb and showed that the small changes in odor composition evoked sudden changes in ensemble activity (pattern separation). These studies show that pattern separation and completion occur not exclusively in the hippocampus in the brain.

Attribute Memory Model and Memory System Interactions

Each memory system (i.e., event-based, knowledge-based, and rule-based memory systems) may work on tasks associated with different attributes according to the attribute theory. For example, the 8-arm maze task example above may require the place attribute to be processed by the event-, knowledge-, and rule-based memory systems depending on the learning experience of the animal. It is conceivable that the information represented in the knowledge-based memory system is constantly updated by the information represented in the event-based memory system, particularly, when novelty is encountered in the event that requires a significant alteration in the "knowledge" an animal possesses for the world. In return, it is possible that the knowledge-based memory system constantly influences the event-based memory system regarding how ambiguity that may be present in the current event needs to be reduced on the basis of one's knowledge-based memory in similar situations.

The interactive relationships between the knowledge-based memory system and event-based memory system for influencing each other may remind one of the concept of "schema," an organized set of information or knowledge, and how individual mnemonic experiences interact with schema. Some studies have suggested that schema play a significant role in memory consolidation. Morris et al. (Tse et al. 2007, 2011) have found, for example, that memory consolidation to neocortex can occur within 48 h after experience (which is significantly shorter compared to conventional thinking) when the experienced event fits or mildly deviate from the schema. Furthermore, learning can be obtained through a single trial if the event being experienced fits the schema. If one equates schema with "knowledge" obtained through many experiences of events in the knowledge-based memory system, the Kesner knowledge-based memory system and the Morris' schema-based memory system may refer to the same cognitive system in the brain.

Goal-Directed Task, Attribute Theory, and Behavioral Neurophysiology

Kesner's attribute theory is proposed largely to explain how disparate memory systems work together as various kinds of events need to be remembered in association with different time points. Another important, yet somewhat hidden, assumption of the model is that the theory is made for explaining how memory systems work in goal-directed tasks, and various predictions of the theory have been tested in goal-directed memory tasks since it was first proposed (Kesner et al. 1993; Kesner and Hardy 1983; Kesner and Williams 1995). Almost all of Kesner's experiments involve rats that encode a new event and "respond" to the event for the purpose of achieving reward (the famous Froot-Loops cereal!). Attributes such as affect and response might not be essential components of the theory if goal-directed responses were not required in behavioral paradigms (e.g., foraging or spontaneous object exploration).

I, Inah Lee, learned rodent behavioral testing methods in the Kesner laboratory as a graduate student. I remember discussing with Ray everyday how to motivate rats for optimal performance and to teach various rules of behavioral tasks using cereal reward. In other words, an experimenter in the laboratory should always have his/her goals in mind and practice the best ways to teach the rat to adopt those goals as their own and perform reliably on every trial to achieve goals. I later experienced other ways of testing behavior (especially for electrophysiological recording in freely moving rats) after graduating from the Kesner laboratory, but I should acknowledge that watching a tiny rat through all courses of cognitive development, from a naive and untrained state to a perfectly goal-oriented and "professional" performing stage, is the most exciting part of behavioral neuroscience to me. All of the behavioral paradigms I have designed in my own laboratory have been, thus, goal-directed tasks and their academic root can be found in Kesner's laboratory.

It is worth mentioning that those goal-directed memory tasks we used in the Kesner laboratory were useful for perturbation studies using lesioning and drug-inactivating techniques to examine behavioral alterations, but artful modifications should be made if one wants to use those paradigms for recording neural signals (e.g., single-unit activity or local field potential) at the same time in freely behaving rats. This is related to both technical and analytical issues. On the technical side, the current techniques for recording neural activity in freely moving rats require tethers or cables attached to the animal's head. With telemetry still at its infancy to fully replace the tether-based recording techniques, it is inevitable to restrict the animal's movement to comply with the capacity of the tethering system. As technology advances rapidly these days, however, it is expected that these technical barriers will eventually subside significantly in the foreseeable future. On the analytical side, the current analytical techniques rely heavily on averaging the number of spikes generated from a neuron over time or space. This naturally requires an adequate amount of sampling over a wide range of space and/or time for achieving reliability of interpretation. For example, the rat should not only pass a particular location many times if one wants to find a neuronal "place field" but should also pass other locations many times to make sure that the cell only fires in its place field but not in other locations. Therefore, it is difficult to use, for example, the so-called cheese-board maze, a round open field with many food wells forming grids within the field (Fig. 10.2), as easily as in lesion studies for a behavioral neurophysiological study because the rat may generate only a few samples of data per cell for each path.

Some modifications, however, would make most of the behavioral tasks available for behavioral neurophysiological experiments and I illustrate this by using one of the studies conducted in my laboratory. Kesner originally developed an object–place paired–associate task (Fig. 10.2) in which an object is rewarded when appearing at a particular location on the cheese-board maze, but not when it appears in another location (i.e., go/no-go paradigm). Another object may appear with the opposite object–place paired associative contingency. The latency to choose the object was measured and the average difference between rewarding (go) trials and nonrewarding (no-go) trials served as an index for learning. The task essentially requires the rat to associate an object with its rewarding place in the open arena and

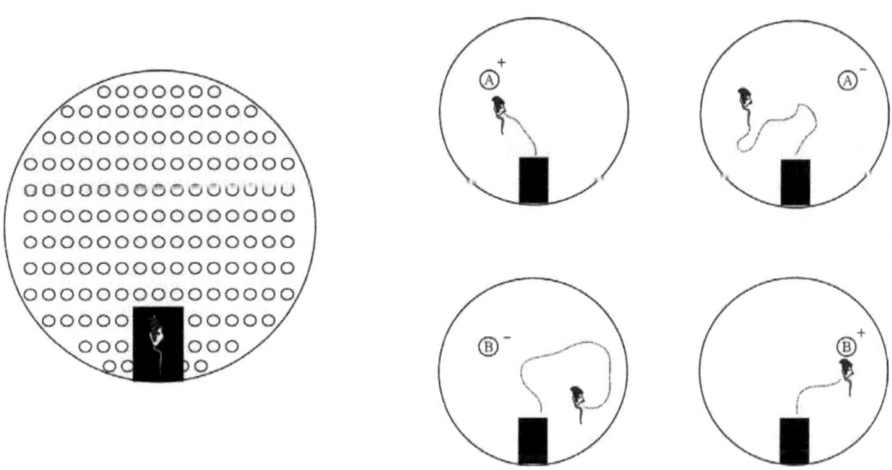

Fig. 10.2 Cheese-board maze (*left*) and representative behavioral response patterns of the rat in an object–place paired–associate task (Object-A is rewarded at the *left* position but not at the *right* position, and *vice versa* for Object-B). A well-trained rat would dash to the object positioned in its rewarded location, but not in the other conditions. The *black* rectangle is the start box

to naturally inhibit its response to the object when it appears at the nonrewarding location. It resembles a typical mnemonic experience of remembering an event with certain object–place associations being more significant for achieving goals than other object–place events. The hippocampus is critical for normal performance in this task presumably as CA3 serves as the center for associating object and place information (Kesner 2013; Kesner et al. 2008; Kesner and Rolls 2001; Rolls and Kesner 2006).

If one analyzes the task components, the object–place paired–associate task contains some aspects that may not be compatible with electrophysiological recording techniques. First, the rat leaves unrestricted position traces during the task that may be different from trial to trial. Second, on no-go trials, the rat may not even approach the object once well trained because it notices the nonrewarding situation from the start box as soon as the door is opened. These factors may cause the sampling problems mentioned above, which explains why most no-go response-involving tasks are not amenable to electrophysiological techniques. These problems of the task for electrophysiology, however, can be overcome by modifying the design as follows: First, instead of using the open field, a radial maze with linear tracks can be used for restricting the rat's trajectories confined within restricted zones in the environment. Second, instead of using a go/no-go paradigm, two objects, one being correct and the other incorrect, may always appear with the object–place paired associative rule still in place. That is, one of the objects within the object pair is rewarded when the pair appears in a particular arm, but not when the objects appear in another arm, and *vice versa,* for the other object. In this way (i.e., removing the no-go component), the rat can always choose an object on every trial.

By implementing the above components, we designed a new task that allowed testing object–place paired–associate memory with electrophysiological recording techniques (Fig. 10.3). As shown in Fig. 10.3, the rat was only allowed to enter a single arm (either arm 3 or arm 5) on a given trial and there were two objects at the end of the arm with the same object placed sometimes on the left side of the animal and other times on the right side. Regardless, the rat was required to choose one of the objects that was always rewarded in that particular arm. The object that was rewarded in one of the arms was not rewarded in the other arm (i.e., the other object was rewarded instead). The task essentially tests the place and sensory-perceptual (or object more precisely) attributes according to the attribute theory.

As the attribute model would predict, the above task was dependent on the hippocampus (Gilbert et al. 1998) presumably because place was an important factor in the task. Interestingly, when rats were first introduced to the task, the animals performed the task as if place and response, but not object, attributes were key features of the task. That is, they tended to adopt a response strategy by choosing an object located on a particular side of the rat as the animal approached the objects. Using such a response strategy made the rat obtain only half of the rewards available in a behavioral session and therefore is certainly not an optimal strategy. According to the attribute model, the animal's choice behavior at that stage of learning should be controlled by the striatum for processing memory involving the response attribute. It would be interesting to find out whether certain brain regions within the same

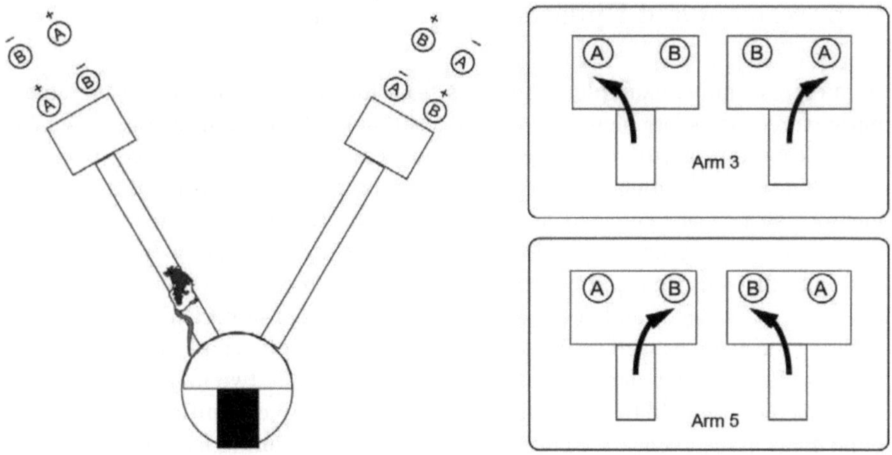

Fig. 10.3 Object–place paired–associate task using a modified radial-arm maze. Rats were allowed to enter one arm at a time. Two objects were presented on the choice platform (all possible configurations and rewarding objects are shown for illustration purposes on the *left* and more in detail on the *right; arrow*—correct turning direction for choosing the rewarded object) attached at the end of the arm. When arriving at the choice platform, the rat chose an object by displacing it

memory system (e.g., event-based memory system) engage at different stages during learning as the animal experiences certain events (Packard 1999; Packard and McGaugh 1996). In this case, the striatum for processing the response attribute appears to take priority over the sensory-perceptual attribute-processing system (in this case, probably the temporal association area (TE) and the perirhinal cortex, or PRC). As learning progressed, however, the rats learned to inhibit the response strategy-based choice behavior and adopted an object-in-place strategy by choosing the same object in a given arm regardless of the object locations within the arm. It took on average 7 days (32 trials per day) perhaps for the rats to realize that rule. Performance jumped suddenly as the rat acquired the rule (Kim et al. 2011; Lee and Solivan 2010).

Neural correlates of performance across learning in the above task revealed that the hippocampal and mPFC networks indeed underwent transitions in the mode of operation from the response-based choice mode to the object–place association-based choice mode during learning (Kim et al. 2011). Specifically, after the rat learned the task, firing patterns of a single unit in the hippocampus and mPFC became more similar between the trial conditions in which choices were made in compatible with the object-in-place rule of the task than between the trial conditions in which response-biased choices were made within the same session. Furthermore, the hippocampal-mPFC synchrony measured by the degree of phase-locking at theta rhythm between the two regions was higher after the rats learned the task as compared to the pre-learning stage. Although the PRC was not physiologically recorded in this task, inactivation of the PRC severely disrupted performance in the task presumably causing impairment in object-recognition memory (i.e., sensory-perceptual attribute in the attribute model; Jo and Lee 2010a).

Improvement and Updates for the Attribute Model

As a representative example, the results from the above studies using the object–place paired–associate memory task suggest that multiple brain regions may be recruited (possibly at different stages of learning) as events are experienced over time. According to the attribute memory model, different brain regions in the event-based memory system are engaged as their associated attributes need to be processed while experiencing an event. However, an event understandably is composed of multiple attributes (in the above example, sensory-perceptual or object, place, time, response, affect, etc.) and some attributes may be more important than others for defining the event as well as for achieving goals. In the object–place paired–associate task, for example, the place and object factors are the two major features to be remembered. It would be interesting to know how relative significance of different brain areas in the event-based memory system changes during learning. Our results suggest that a rule of the task that maximizes the amount of reward certainly changes the order of engagement or priority between the regions belonging to the event-based memory system. Although striatal neural responses were not recorded in the same task, if the attribute model is correct, the striatum-based response memory attribute should take priority over the hippocampus-based place memory attribute (and the PRC-based sensory-perceptual attribute) at the earlier stages of learning. Although the response strategy does not match the rule of the task, it guarantees at least 50 % of food reward if faithfully applied. Thus, it is in a sense the best strategy the rat may adopt while finding out the exact rule of the task. Such relative weights should change, however, as the rule is learned by trial and error while maximizing reward during acquisition and it appears that the mPFC plays a critical role during such processes.

Another clarification one may need to make for the application of the attribute model would be regarding the sensory-perceptual attribute, its definition and boundaries in particular. The sensory-perceptual attribute is indeed a broad category and may need to be processed with most attributes (place, time, language, affect) included in the model. If one may acknowledge the hierarchical, yet dynamically interacting nature of the sensory-perceptual systems (e.g., more primitive visual information processing in the striate cortex and higher perceptual information processing such as object recognition in the IT cortex), it may certainly be difficult to draw a definitive line between brain regions to indicate where information processing for the "memory" attribute starts and where "perception" ends. This issue is not trivial and its related debates have been taking place in the field of object recognition memory as people study the roles of the PRC and its related areas. The issue here is whether the PRC's main functions are perceptual or mnemonic. Detailed discussions on this topic appeared elsewhere repeatedly (Baxter 2009; Bussey and Saksida 2005; Murray et al. 2007; Suzuki 2009) and will not be repeated here. From our experiences with testing the roles of the PRC in various behavioral tasks, our laboratory has never seen the PRC playing critical roles in object recognition when simply a pair of objects needs to be visually discriminated. However, perturbations in the PRC resulted in severe deficits in performance when the rats were required to remember object identity with its associated place (Jo and Lee 2010a, b).

It would also be worthwhile to compare the event-based memory system proposed in the attribute memory model with the event-based memory system frequently appearing in the literature recently. Specifically, it has been suggested that the spatial information and nonspatial (or item) information merge in the hippocampus to form a memory representation of an event (Hargreaves et al. 2005). According to this model, spatial information is dominantly processed by the postrhinal cortex (or parahippocampal cortex in primates and humans) and the medial entorhinal cortex (MEC), and nonspatial information is mostly processed by the PRC and the lateral entorhinal cortex (LEC). These areas may well serve at the same time for representing the memories of spatial and nonspatial events. When items and their associated places need to be remembered, the item–place associative memory representation is formed in the hippocampus according to the theory. In comparison, the attribute memory theory does not clearly explain the functions of the postrhinal cortex or parahippocampal cortex in the event memory system. This again requires more sophisticated analysis on the sensory-perceptual attribute mainly because of the significant qualitative differences between spatial information and nonspatial information in addition to other issues mentioned earlier.

Conclusion

By providing a theoretical framework for investigating how different aspects of memory are processed by different areas of the brain, the attribute memory model of Kesner, along with other multiple memory theories, contributed significantly to our system-level understanding of memory. Kesner's research has also provided a gateway through which computational models for the hippocampal subfields can be experimentally tested. Kesner's research, most of all, highlighted how goal-oriented animals learn and remember information and how those cognitive processes can be tested behaviorally using ingenious tasks. As electrophysiological and imaging techniques are rapidly advancing, more emphasis is being placed on controlling the animal's behavior for accurate and detailed measurement of neural signals these days (Bartko et al. 2011; Dombeck et al. 2010; Ravassard et al. 2013). Despite the valuable information we learn from these highly controlled experiments with respect to the animal's behavior, controlling behavior too much inevitably entails the risk of inducing artificiality when testing behavior and its neural mechanisms. Therefore, the Kesner-style behavioral approaches will always hold some values.

References

Barnes, D. C., Hofacer, R. D., Zaman, A. R., Rennaker, R. L., & Wilson, D. A. (2008). Olfactory perceptual stability and discrimination. *Nature Neuroscience, 11*(12), 1378–1380. doi:10.1038/Nn.2217.

Bartko, S. J., Vendrell, I., Saksida, L. M., & Bussey, T. J. (2011). A computer-automated touch-screen paired-associates learning (PAL) task for mice: Impairments following administration of scopolamine or dicyclomine and improvements following donepezil. *Psychopharmacology (Berl), 214*(2), 537–548. doi:10.1007/s00213-010-2050-1.

Baxter, M. G. (2009). Involvement of medial temporal lobe structures in memory and perception. *Neuron, 61*(5), 667–677. doi:10.1016/j.neuron.2009.02.007.

Bussey, T. J., & Saksida, L. M. (2005). Object memory and perception in the medial temporal lobe: An alternative approach. *Current Opinion in Neurobiology, 15*(6), 730–737. doi:10.1016/j.conb.2005.10.014.

Dombeck, D. A., Harvey, C. D., Tian, L., Looger, L. L., & Tank, D. W. (2010). Functional imaging of hippocampal place cells at cellular resolution during virtual navigation. *Nature Neuroscience, 13*(11), 1433–1440. doi:10.1038/Nn.2648.

Eichenbaum, H., & Cohen, Neal J. (2001). *From conditioning to conscious recollection: Memory systems of the brain*. New York: Oxford University Press.

Gilbert, P. E., & Kesner, R. P. (2006). The role of the dorsal CA3 hippocampal subregion in spatial working memory and pattern separation. *Behavioural Brain Research, 169*(1), 142–149. doi:10.1016/j.bbr.2006.01.002.

Gilbert, P. E., Kesner, R. P., & DeCoteau, W. E. (1998). Memory for spatial location: Role of the hippocampus in mediating spatial pattern separation. *Journal of Neuroscience, 18*(2), 804–810.

Gluck, M. A., Meeter, M., & Myers, C. E. (2003). Computational models of the hippocampal region: Linking incremental learning and episodic memory. *Trends in Cognitive Sciences, 7*(6), 269–276.

Hargreaves, E. L., Rao, G., Lee, I., & Knierim, J. J. (2005). Major dissociation between medial and lateral entorhinal input to dorsal hippocampus. *Science, 308*(5729), 1792–1794. doi:10.1126/science.1110449.

Hasselmo, M. E., & Wyble, B. P. (1997). Free recall and recognition in a network model of the hippocampus: simulating effects of scopolamine on human memory function. *Behavioural Brain Research, 89*(1–2), 1–34. doi:10.1016/S0166-4328(97)00048-X.

Jo, Y. S., & Lee, I. (2010a). Disconnection of the hippocampal-perirhinal cortical circuits severely disrupts object-place paired associative memory. *Journal of Neuroscience, 30*(29), 9850–9858. doi:10.1523/JNEUROSCI.1580-10.2010.

Jo, Y. S., & Lee, I. (2010b). Perirhinal cortex is necessary for acquiring, but not for retrieving object-place paired association. *Learning & Memory, 17*(2), 97–103. doi:10.1101/Lm.1620410.

Kesner, R. P. (1980). An attribute analysis of memory—the role of the hippocampus. *Physiological Psychology, 8*(2), 189–197.

Kesner, R. P. (2009). Tapestry of memory. *Behavioral Neuroscience, 123*(1), 1–13. doi:10.1037/A0014004.

Kesner, R. P. (2013). A process analysis of the CA3 subregion of the hippocampus. *Front Cell Neuroscience, 7,* 78. doi:10.3389/fncel.2013.00078.

Kesner, R. P., & Hardy, J. D. (1983). Long-term-memory for contextual attributes—dissociation of amygdala and hippocampus. *Behavioural Brain Research, 8*(2), 139–149. doi:10.1016/0166-4328(83)90050-5.

Kesner, R. P., & Rogers, J. (2004). An analysis of independence and interactions of brain substrates that subserve multiple attributes, memory systems, and underlying processes. *Neurobiology of Learning and Memory, 82*(3), 199–215. doi 10.1016/j.nlm.2004.05.007.

Kesner, R. P., & Rolls, E. T. (2001). Role of long-term synaptic modification in short-term memory. *Hippocampus, 11*(3), 240–250. doi:10.1002/Hipo.1040.

Kesner, R. P., & Williams, J. M. (1995). Memory for magnitude of reinforcement—dissociation between the amygdala and hippocampus. *Neurobiology of Learning and Memory, 64*(3), 237–244.

Kesner, R. P., Bolland, B. L., & Dakis, M. (1993). Memory for spatial locations, motor-responses, and objects—triple dissociation among the hippocampus, caudate-nucleus, and extrastriate visual-cortex. *Experimental Brain Research, 93*(3), 462–470.

Kesner, R. P., Gilbert, P. E., & Wallenstein, G. V. (2000). Testing neural network models of memory with behavioral experiments. *Current Opinion in Neurobiology, 10*(2), 260–265. doi:10.1016/S0959-4388(00)00067-2.

Kesner, R. P., Hunsaker, M. R., & Warthen, M. W. (2008). The CA3 subregion of the hippocampus is critical for episodic memory processing by means of relational encoding in rats. *Behavioral Neuroscience, 122*(6), 1217–1225. doi:10.1037/A0013592.

Kim, J. J., Delcasso, S., & Lee, I. (2011). Neural correlates of object-in-place learning in hippocampus and prefrontal cortex. *Journal of Neuroscience, 31*(47), 16991–17006. doi:10.1523/Jneurosci.2859-11.2011.

Lee, I., & Kesner, R. P. (2002). Differential contribution of NMDA receptors in hippocampal subregions to spatial working memory. *Nature Neuroscience, 5*(2), 162–168. doi:10.1038/Nn790.

Lee, I., & Kesner, R. P. (2004). Encoding versus retrieval of spatial memory: Double dissociation between the dentate gyrus and the perforant path inputs into CA3 in the dorsal hippocampus. *Hippocampus, 14*(1), 66–76. doi:10.1002/Hipo.10167.

Lee, I., & Solivan, F. (2010). Dentate gyrus is necessary for disambiguating similar object-place representations. *Learning & Memory, 17*(5), 252–258.

Lee, I., Yoganarasimha, D., Rao, G., & Knierim, J. J. (2004). Comparison of population coherence of place cells in hippocampal subfields CA1 and CA3. *Nature, 430*(6998), 456–459. doi:10.1038/Nature02739.

Leutgeb, S., Leutgeb, J. K., Treves, A., Moser, M. B., & Moser, E. I. (2004). Distinct ensemble codes in hippocampal areas CA3 and CA1. *Science, 305*(5688), 1295–1298. doi:10.1126/science.1100265.

Leutgeb, J. K., Leutgeb, S., Moser, M. B., & Moser, E. I. (2007). Pattern separation in the dentate gyrus and CA3 of the hippocampus. *Science, 315*(5814), 961–966. doi:10.1126/science.1135801.

Marr, D. (1971). Simple memory—a theory for archicortex. *Philosophical Transactions of the Royal Society of London Series B-Biological Sciences, 262*(841), 23–81. doi:10.1098/rstb.1971.0078.

McDonald, R.J., & White, N.M. (1993). A triple dissociation of memory systems: hippocampus, amygdala, and dorsal striatum. *Behavioral Neuroscience, 107*, 3–22.

Murray, E. A., Bussey, T. J., & Saksida, L. M. (2007). Visual perception and memory: A new view of medial temporal lobe function in primates and rodents. *Annual Review of Neuroscience, 30*, 99–122. doi:10.1146/annurev.neuro.29.051605.113046.

Niessing, J., & Friedrich, R. W. (2010). Olfactory pattern classification by discrete neuronal network states. *Nature, 465*(7294), 47–52. doi:10.1038/Nature08961.

Oreilly, R. C., & Mcclelland, J. L. (1994). Hippocampal conjunctive encoding, storage, and recall—avoiding a trade-off. *Hippocampus, 4*(6), 661–682. doi:10.1002/hipo.450040605.

Packard, M. G. (1999). Glutamate infused posttraining into the hippocampus or caudate-putamen differentially strengthens place and response learning. *Proceedings of the National Academy of Sciences of the United States of America, 96*(22), 12881–12886. doi:10.1073/pnas.96.22.12881.

Packard, M. G., & McGaugh, J. L. (1996). Inactivation of hippocampus or caudate nucleus with lidocaine differentially affects expression of place and response learning. *Neurobiology of Learning and Memory, 65*(1), 65–72. doi:10.1006/nlme.1996.0007.

Ravassard, P., Kees, A., Willers, B., Ho, D., Aharoni, D., Cushman, J., et al. (2013). Multisensory control of hippocampal spatiotemporal selectivity. *Science, 340*(6138), 1342–1346. doi:10.1126/science.1232655.

Rolls, E. T. (1996). A theory of hippocampal function in memory. *Hippocampus, 6*(6), 601–620. doi:10.1002/(Sici)1098-1063(1996)6:63.0.Co;2–J.

Rolls, E. T., & Kesner, R. P. (2006). A computational theory of hippocampal function, and empirical tests of the theory. *Progress in Neurobiology, 79*(1), 1–48. doi:10.1016/j.pneurobio.2006.04.005.

Squire, L. R. (2004). Memory systems of the brain: A brief history and current perspective. *Neurobiology of Learning and Memory, 82*(3), 171–177. doi:10.1016/j.nlm.2004.06.005.

Squire, L. R., & Zola-Morgan, S. (1991). The medial temporal lobe memory system. *Science, 253*(5026), 1380–1386.

Suzuki, W. A. (2009). Perception and the medial temporal lobe: Evaluating the current evidence. *Neuron, 61*(5), 657–666. doi:10.1016/j.neuron.2009.02.008.

Treves, A., & Rolls, E. T. (1994). Computational analysis of the role of the hippocampus in memory. *Hippocampus, 4*(3), 374–391. doi:10.1002/hipo.450040319.

Tse, D., Langston, R. F., Kakeyama, M., Bethus, I., Spooner, P. A., Wood, E. R., et al. (2007). Schemas and memory consolidation. *Science, 316*(5821), 76–82. doi:10.1126/science.1135935.

Tse, D., Takeuchi, T., Kakeyama, M., Kajii, Y., Okuno, H., Tohyama, C., et al. (2011). Schema-dependent gene activation and memory encoding in neocortex. *Science, 333*(6044), 891–895. doi:10.1126/science.1205274.

Tulving, E. (1972). Episodic and semantic memory. In E. D. Tulving & W. Donaldson (Ed.), *Organization of memory* (pp. 382–402). New York: Academic.

Tulving, E. (1985). How many memory-systems are there. *American Psychologist, 40*(4), 385–398. doi:10.1037/0003-066x.40.4.385.

Wallenstein, G. V., Eichenbaum, H., & Hasselmo, M. E. (1998). The hippocampus as an associator of discontiguous events. *Trends in Neurosciences, 21*(8), 317–323. doi:10.1016/S0166-2236(97)01220-4.

Yassa, M. A., & Stark, C. E. (2011). Pattern separation in the hippocampus. *Trends in Neurosciences, 34*(10), 515–525. doi:10.1016/j.tins.2011.06.006.

Chapter 11
Prefrontal Cortex and Basal Ganglia Attributes Underlying Behavioral Flexibility

Michael E. Ragozzino and Phillip M. Baker

Introduction

The attribute model of memory was an early multiple memory systems conceptualization of how different brain areas encode and store specific types of memory (Kesner and DiMattia 1987). This neurobiologically based model proposes that the nature of memory can be explained by different attributes such as space, time, sensory-perception, response, and reward (affect), which are stored as memories in different forebrain areas. To test hypotheses based on the attribute idea, different behavioral paradigms have been developed for rats that emphasize the learning and memory of specific attribute information, for example, temporal order or egocentric response. After learning occurs, a specific brain area is lesioned, for example, hippocampus or striatum, and rats are tested on the retention of the originally learned information. Using this experimental procedure, several studies have demonstrated that lesioning a particular brain area produces a memory deficit for specific attribute information (Kesner 2009). Thus, the neurobiologically based attribute model of memory developed from evidence that certain brain areas store memories for particular attribute information.

Subsequent to this original formulation, the attribute model of memory was applied to investigate the structure of memory representation in the rodent prefrontal cortex (Kesner et al. 1996; Kesner 2000; Ragozzino and Kesner 1999, 2001; Ragozzino et al. 1998; Ragozzino et al. 2002). The prefrontal cortex is an interesting region of the brain to explore the attribute model as this area consists of several

M. E. Ragozzino (✉)
Biological Sciences and Department of Psychology, University of Illinois at Chicago, Chicago, IL 60607, USA
e-mail: mrago@uic.edu

P. M. Baker
Department of Psychology, University of Washington, Seattle, WA 98195, USA
e-mail: pbaker2@uw.edu

P. A. Jackson et al. (eds.), *The Neurobiological Basis of Memory*,
DOI 10.1007/978-3-319-15759-7_11

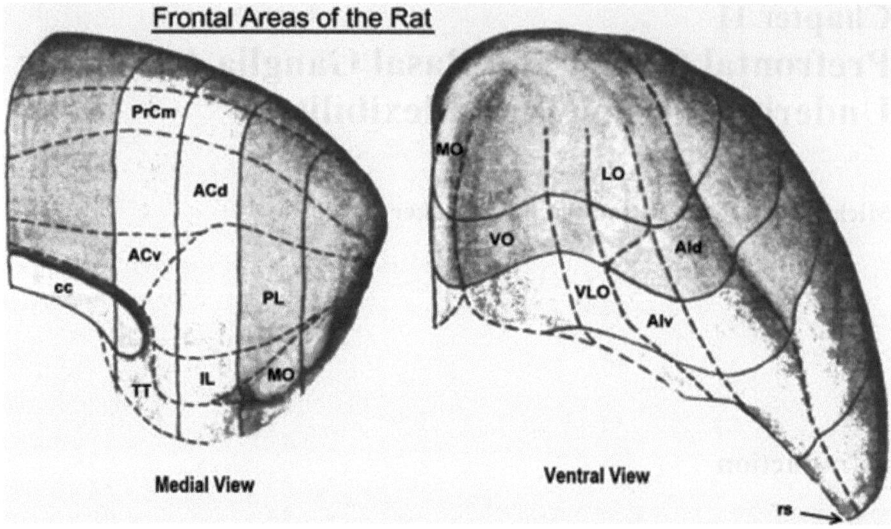

Fig. 11.1 Medial and ventral views of the rat frontal cortex. Abbreviations: *PrCm* precentral cortex, *AC* dorsal and ventral anterior cingulate, *PL–IL* prelimbic and infralimbic cortex, *MO* medial orbital cortex, *AI* dorsal and ventral agranular insular cortex, *LO* lateral orbital cortex, *VO* ventral orbital cortex, *VLO* ventrolateral orbital cortex. Reprinted with permission from Elsevier (Kesner and Churchwell 2011)

different subregions (see Fig. 11.1) that, to varying degrees, have distinct afferent and efferent connections (Uylings and van Eden 1990). The rodent prefrontal cortex can be subdivided based on structure and connectivity. One broad division of the prefrontal cortex is to separate it into a medial and a lateral sector. The medial area consists of the infralimbic cortex, the prelimbic cortex, the anterior cingulate cortex, and the medial precentral areas (Fig. 11.1). These delineations are based on the architectural makeup of the cortical layers as well as the thalamic projections that each area receives. Located centrally is the prelimbic area that is comparable to Brodmann's areas; area 24 and 32 (Uylings and van Eden 1990). The prelimbic area is densely interconnected with other areas of the prefrontal cortex (Eden et al. 1992; Heidbreder and Groenewegen 2003). It also sends projections to the dorsomedial striatum, as well as the subthalamic nucleus (Sesack et al. 1989; Gabbott et al. 2005). These areas represent the major inputs in the basal ganglia, a key area for motor actions. Additionally, the prelimbic area is one of the few areas of the brain that has reciprocal connections with the majority of the neuromodulatory neurotransmitter systems of the brain. Specifically, it has reciprocal projections with the ventral tegmental nucleus and substantia nigra pars compacta, the major dopaminergic neurons of the brain; the dorsal and median raphe nuclei, the serotonergic cells of the brain; the locus coeruleus, the primary source of noradrenergic innervation in the brain; and the nucleus basalis as well as the brainstem cholinergic nuclei, two major acetylcholine systems (Vertes 2004; Boix-Trelis et al. 2006; Hoover and

Vertes 2007). The connections of the prelimbic cortex with limbic and motor areas of the brain as well as its interconnections with the majority of the neuromodulatory systems of the brain suggest that it may play a critical role in the coordination of complex behavior such as those required for cognitive flexibility.

The lateral prefrontal cortex consists of the dorsal and ventral agranular insular along with the lateral orbital region. The primary afferent connections of these areas include the pyriform cortex and olfactory bulb, gustatory cortex and gustatory thalamus, parts of somatosensory I and II, visual association cortex, parietal cortex, perirhinal cortex, as well as the medial dorsal nucleus and central medial nucleus of the thalamus. The agranular region projects to the ventrolateral part of the striatum, whereas the lateral orbital region projects to the central part (Mailly et al. 2013). A ventral lateral region of the prefrontal cortex includes the ventral orbital and ventrolateral orbital cortices. These regions receive input from the parietal cortex, visual association cortex, medial dorsal nucleus, and central medial nucleus of the thalamus. These ventral orbital subregions project to dorsal and ventral striatum, posterior parietal cortex, secondary visual cortex, pyriform cortex, and olfactory bulb. Thus, several different prefrontal cortex subregions project to different areas of the basal ganglia, in particular the striatum, that suggests that the prefrontal cortex and basal ganglia may act in a cooperative manner to support various cognitive functions (Kesner and Churchwell 2011).

As the rat prefrontal cortex comprises various subregions, Kesner et al. investigated whether separate rat prefrontal cortex subregions facilitate working memory for specific attribute information (Kesner et al. 1996; Ragozzino and Kesner 1999; Ragozzino et al. 1998; 2002). For example, prelimbic and infralimbic cortex lesions impair working memory for spatial locations but not working memory for egocentric responses (Kesner et al. 1996; Ragozzino et al. 1998). In contrast, anterior cingulate and medial precentral lesions do not impair working memory for spatial locations, but do impair working memory for egocentric responses (Ragozzino and Kesner 2001; Ragozzino et al. 1998). Moreover, there is also evidence that the agranular insular cortex supports working memory for reward value (Ragozzino and Kesner 1999). Thus, at least for working memory, there is support for the idea that the rodent prefrontal cortex is organized such that separate subregions represent particular attribute information in memory.

Although the findings described above focus on the functional organization of the prefrontal cortex related to working memory, the prefrontal cortex is a brain area that has been proposed to support some of the most complex functions in mammals, including planning, temporal ordering, and behavioral flexibility (Kesner and Churchwell 2011). Many of these functions have been categorized into the singular, broader label of executive functioning (Kesner and Churchwell 2011). Thus, the prefrontal cortex offers an opportunity to examine whether the attribute model applies to the functional organization of the prefrontal cortex beyond learning and memory. My laboratory has been particularly interested in the neural basis of behavioral flexibility that developed during my postdoctoral training with Ray Kesner. We broadly define behavioral flexibility as the ability to adapt an individual's behavior when a change in the internal or external environment signals that an ongoing choice pattern is no longer optimal. This chapter focuses on three main themes related to the neural basis of behavioral flexibility: (1) the role of different prefrontal cortex areas in behavioral flexibility. These studies developed from earlier experiments investigating whether different prefrontal cortex areas support working memory for specific attribute information; (2) the role of the dorsal striatum in behavioral flexibility. The dorsal striatum receives inputs from specific areas of the medial prefrontal cortex and lateral prefrontal cortex (Berendse et al. 1992; Mailly et al. 2013). Thus, there was interest in determining whether this striatal area plays a similar or distinct role in behavioral flexibility as prefrontal cortex areas that project to the dorsal striatum; and (3) the chapter will describe interactions between prefrontal cortex and basal ganglia circuitry in supporting behavioral flexibility using conditional discrimination tests. Most neurobiological studies of behavioral flexibility have used paradigms in which a change in outcomes, for example, a choice is no longer reinforced, signals that a switch in choice patterns should occur. However, many situations demand behavioral adaptations to external cues which proactively signal that a behavioral switch should occur. Less is known about the role of the prefrontal cortex and basal ganglia areas in behavioral flexibility under these conditions. Recent findings from our laboratory are presented, which indicate that distinct prefrontal cortex and basal ganglia circuitry interact to enable rapid adaptations under these conditions.

Prefrontal Cortex, Attributes and Rules

In studying the role of the prefrontal cortex in behavioral flexibility, the attribute model was influential in shaping the design of early experiments. In particular, the attribute model would predict that separate prefrontal cortex subregions contribute to behavioral flexibility based on the type of attribute information needed to flexibly adapt. Following this idea, various studies have been carried out in which a rat had to learn one type of discrimination for specific attribute information and then, by changing the reinforcement contingencies, had to learn using different attribute information (set-shifting) or learn a different choice using the same attribute information (reversal learning). In a set-shift test, a subject must learn to make a choice based on one attribute, while inhibiting a choice based on different attribute information. For example, in one study rats learned to choose between two different sand cups that were filled with distinct odors, that is, cinnamon and nutmeg, while each cup was also in a distinct spatial location in a maze (Ragozzino et al. 2003). The scented sand cups are randomly switched between spatial locations across trials. A subject first learns to choose a sand cup based on spatial location to receive a reinforcement independent of odor. After learning to choose based on spatial location, reinforcement is changed so that it is always associated with a particular odor, that is, sand cup scented with cinnamon. Thus, the rat must shift to always choose the sand cup scented with cinnamon independent of spatial location. In a reversal-learning test, a subject must reverse what specific choice it employs to receive a reinforcement and learn to use a different choice based on the same attribute information. For example, a subject chooses between two different odors to receive a reinforcement. A cinnamon-scented sand cup is initially associated with reinforcement, while a nutmeg-scented cup is not associated with reinforcement. After initial learning, the contingencies are reversed such that the nutmeg-scented sand cup is associated with reinforcement. The prelimbic cortex and the orbitofrontal cortex are two prefrontal cortex subregions that have been most commonly studied to understand their contributions to behavioral flexibility.

Based on findings indicating that the prelimbic cortex supports working memory when spatial or visual object information must be used (Kesner et al. 1996; Ragozzino et al. 1998), one idea was that the prelimbic cortex would also support behavioral flexibility when conditions require the flexible use of spatial and/or visual object information. In set-shifting tests, prelimbic inactivation with a local anesthetic impaired performance when rats had to shift between using a spatial and visual object strategy (Ragozzino et al. 1999a). However, other studies indicated that lesions or targeted drug manipulations of the prelimbic cortex also impaired set-shifting performance even under conditions that did not involve the use of spatial and/or visual object information (Birrell and Brown 2000; Ragozzino 2002; Ragozzino et al. 2003; Ragozzino 2007; Stefani et al. 2003). Furthermore, prelimbic cortex inactivation or lesions did not impair visual object or place reversal learning or other types of reversal learning (Birrell and Brown 2000; Ragozzino et al. 1999b; Ragozzino et al. 2003). Important to note, these different discrimination tests do not have a salient working memory component although there is a delay between each trial. However, unlike the working memory tests used to examine the effects of prelimbic and infralimbic lesions where recently presented information had to be remembered and changes trial to trial, in these tests a subject must learn a particular rule that remains constant across a range of consecutive trials. Thus, the prelimbic cortex may contribute to different cognitive functions under some conditions that are dependent on specific attribute information and in other conditions independent of specific attribute information. Related to behavioral flexibility, the pattern of results suggests that the prelimbic area supports behavioral flexibility when conditions require a shift in strategies that can be determined by requiring rats to switch between different attribute information. In contrast, the prelimbic area is not involved in behavioral flexibility when conditions require a shift in choices using similar attribute information as is required in reversal learning.

Studies investigating the contribution of the orbitofrontal cortex to behavioral flexibility have yielded results that suggest that this prefrontal cortex subregion makes different contributions than the prelimbic cortex. Experiments involving orbitofrontal cortex inactivation or lesions found that these manipulations do not affect acquisition of different discrimination learning tests, for example, olfactory or visuospatial discrimination , but do impair reversal learning (Boulougouris et al. 2007; Churchwell et al. 2009; Ghods-Sharifi et al. 2008; Kim and Ragozzino 2005; McAlonan and Brown 2003; Riceberg and Shapiro 2012; Schoenbaum et al. 2002). This occurred in reversal-learning tests that involved the flexible use of odor, visual cue, tactile, or spatial information. In contrast, set-shifting is not impaired by lesions, local anesthetics, or gamma-aminobutyric acid (GABA) agonist infusions in the orbitofrontal cortex (Ghods-Sharifi et al. 2008; McAlonan and Brown 2003). Considered together, these findings from several investigations suggest that the orbitofrontal cortex does not support behavioral flexibility based on a particular type of attribute information, but more on the type of rule required for flexibly adapting an individual's behavior.

A comparison of results following prelimbic cortex versus orbitofrontal cortex manipulations on set-shifting and reversal-learning tests show a double dissociation in function between these areas. In particular, the prelimbic cortex supports behavioral flexibility when conditions require a set-shift, but not a reversal in choice patterns. Conversely, the orbitofrontal cortex supports behavioral flexibility when conditions require a reversal in choice patterns, but not a set-shift. Taken together, these results support a model proposed by Wise et al. (1996) to explain the functional organization of the primate frontal cortex in which different conditions require different types of rules to facilitate behavioral flexibility. These authors further proposed that these rules are mediated by separate primate prefrontal cortex areas. Specifically, the model proposes that there is a lower-order rule for the shifting of specific choices within a dimension. This rule allows the approach to and avoidance of a particular stimulus or scene as required in discrimination tasks that involve reversal learning. There is also a higher-order rule when conditions demand learning about stimulus attributes as opposed to within a stimulus. In these cases, learning must go beyond simply attaching a positive or negative valence to stimuli within a particular dimension and instead require attention to components of an object or scene or abstract rules about component objects or scenes. Thus, higher-order rules enable a subject to reconceptualize his or her approach to a task and attend to a new type of information. This model may be applicable to rodents such that the orbitofrontal cortex supports a lower-order rule to enable behavioral flexibility, while the prelimbic cortex supports a higher-order rule to facilitate behavioral flexibility.

Although there is considerable evidence that the prelimbic cortex and orbitofrontal cortex subregions support different rules to enable behavioral flexibility, these different prefrontal cortex subregions may support similar processes to enable various forms of behavioral flexibility. For example, a brain region may facilitate the ability to *initially* inhibit a previously relevant strategy and/or to generate a new strategy. In this case, inactivation of a prefrontal cortex subregion should produce a predominance of errors during the initial trials in a shift or reversal phase. These errors are commonly referred to as "perseverative errors." Another possibility is that a brain region supports a process that allows an individual to reliably execute or learn a new choice pattern once the new choice pattern is selected. This process would prevent or minimize regressions to the previously relevant choice pattern once the new, presently relevant choice pattern is selected. In this case, inactivation of a prefrontal cortex subregion should not produce a significant increase in errors during the initial trials of the shift or reversal phase, but rather should lead to a

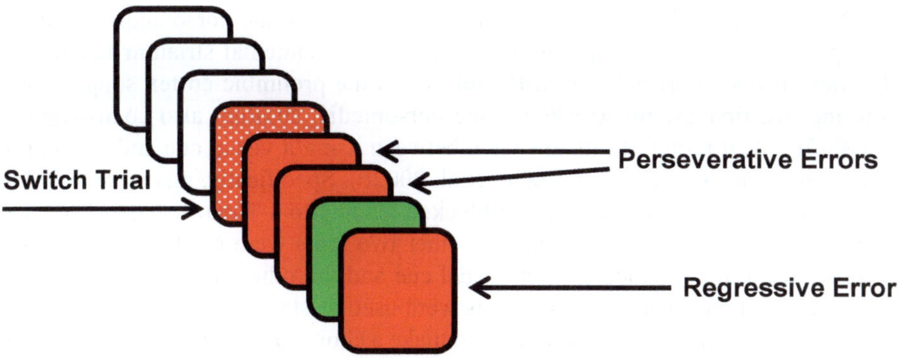

Fig. 11.2 Errors committed during set-shifting and reversal-learning tests. Errors are scored as either perseverative or regressive errors. *White blocks* represent trials from acquisition phase. *Patterned red block* represents the first switch trial from either a set-shifting or reversal-learning test. A *solid red block* represents an error trial. A *solid green block* represents a correct trial. Perseverative errors occur when errors immediately follow a switch trial error, until a correct response is made. Once a rat makes a correct response in a session, any errors following that correct response are considered regressive errors

greater number of errors once a rat has selected the new, presently relevant strategy. We have referred to these errors as "regressive errors" because a subject has chosen the new correct choice and has been reinforced for it, but regresses to the previous choice that is no longer reinforced (see Fig. 11.2). In multiple experiments where prelimbic cortex manipulations impaired set-shifting, a deficit resulted from an increase in perseverative errors but not regressive errors (Ghods-Sharifi et al. 2008; Ragozzino 2002 Ragozzino et al. 1999b, 2003).

Orbitofrontal cortex manipulations that have impaired reversal learning have also resulted from an increase in perseverative errors (Boulougouris et al. 2007; Kim and Ragozzino 2005). These findings suggest that despite the prelimbic cortex and orbitofrontal cortex supporting different forms of behavioral flexibility based on rules, both subregions facilitate the ability to *initially* inhibit a previously relevant choice pattern and/or to generate a new choice pattern.

Dorsomedial Striatum, Attributes and Rules

In view of the evidence that the prelimbic cortex and orbitofrontal cortex support behavioral flexibility related to the behavioral operation required to flexibly adapt, it is of interest that both regions project to the dorsomedial striatum (Berendse et al. 1992). Using a similar approach to investigate the contributions of the prefrontal cortex subregions to behavioral flexibility, a series of experiments have examined the effects of dorsomedial striatal inactivation on acquisition, reversal learning, and set-shifting tests.

Because the prelimbic cortex prominently projects to the dorsomedial striatum, one possibility is that the prelimbic cortex and dorsomedial striatum functionally interact to support behavioral flexibility. As the prelimbic cortex supports set-shifting, we first examined whether the dorsomedial striatum also contributed to set-shifting that required rats to switch between using a visual cue and egocentric spatial response strategy (Ragozzino et al. 2002b). Specifically, rats were tested in a cross-maze in which one arm was blocked leading to a T-maze shape. The stem arm was used as the start arm and the other two arms were used as choice arms. One choice arm contained a black visual cue and the other choice arm contained a white visual cue. The other two arms were used as start arms that were changed after every few trials. A rat could learn to make a choice based on a particular visual cue, for example, always enter the black arm, or due to an egocentric response, for example, always turn left. Dorsomedial striatal inactivation with a local anesthetic did not impair initial learning of a visual cue or egocentric response discrimination, but did impair set-shifting (Ragozzino et al. 2002b). More recent findings following neurotoxic lesions of the dorsomedial striatum also indicate that the dorsomedial striatum enables behavioral flexibility when a shift across attribute dimensions is required (Lindgren et al. 2013). However, dorsomedial striatal manipulations not only impair performance in set-shifting tests, but also produce deficits in reversal learning tests (Pisa and Cyr 1990; Ragozzino and Choi 2004). As N-methyl-D-aspartate (NMDA) receptors support synaptic plasticity (Spencer and Murphy 2000a; Boettiger and Doupe 2001; Akopian and Walsh 2002; Dang et al. 2006), the role of these receptors in the dorsomedial striatum related to behavioral flexibility has been examined. Comparable to dorsomedial striatal lesions or inactivation, infusion of the NMDA receptor antagonist (2R)-amino-5-phosphonovaleric acid (AP5) in this region impairs reversal learning but not acquisition (Palencia and Ragozzino 2004). Moreover, AP5 in the dorsolateral striatum does not impair reversal learning (Palencia and Ragozzino 2005). Thus, the dorsomedial striatum appears to play a broader role in behavioral flexibility then either the prelimbic cortex or orbitofrontal cortex alone.

While dorsomedial striatal manipulations lead to deficits in set-shifting or reversal learning, the behavioral flexibility deficit does not result from an increase in perseveration. Instead, dorsomedial striatal inactivation or NMDA receptor blockade in the dorsomedial striatum selectively increases regressive errors (Ragozzino et al. 2002b; Palencia and Ragozzino 2004; Ragozzino and Choi 2004). These results suggest that the dorsomedial striatum may dynamically interact with multiple prefrontal cortex subregions to facilitate behavioral flexibility in a distinct but complementary manner. More specifically, prefrontal cortex subregions may be critical for the generation of a new strategy or response pattern. This allows the initial inhibition of the previously relevant strategy. However, once a new strategy is generated, it must be executed into an appropriate response pattern. The striatum, in coordination with different prefrontal cortex areas, may facilitate the execution of an appropriate response pattern for a particular strategy that is generated. Thus, the striatum may link a particular response pattern with a specific strategy allowing for the reliable execution of a strategy once generated, as well as continual inhibition of previously relevant strategies.

Prefrontal Cortex—Basal Ganglia Interactions for Proactive Cue-Guided Behavioral Flexibility

The studies described above focused on understanding how distinct prefrontal cortex and striatal areas contribute to behavioral flexibility based on different discrimination tests in which a change in outcomes indicated that a behavioral switch should occur. However, these studies did not address more directly how different prefrontal cortex and basal ganglia regions may interact to facilitate behavioral flexibility. In addition, many of these investigations involved paradigms in which there was a single behavioral switch that had to occur over an extended time period, for example, a daily session as opposed to a few trials. Changes in environmental conditions often require rapid and repeated adaptations to achieve a goal. Moreover, previous paradigms to study behavioral flexibility have predominantly involved a change in outcomes to signal that a switch in a response pattern should occur. In many situations, cue information may be used proactively to switch actions to obtain a goal (Hikosaka and Isoda 2010).

To date, there has been significantly less examination of whether prefrontal cortex and basal ganglia areas support behavioral switching when cues can be used to proactively switch response patterns for an upcoming choice. Moreover, it is unknown whether the brain areas that support behavioral flexibility based on set-shifting and reversal learning support similar processes, for example, reduction in perseveration of a previously correct response pattern, under conditions that require cue-guided behavioral switching. Conditional discrimination tests offer a behavioral paradigm in which cues can be used to proactively switch behavior. In these paradigms, a cue, for example, 40 Hz tone, is associated with making a specific response, for example, press the right lever, to receive a reinforcement. On other trials a different cue is presented, for example, 200 Hz tone, that is associated with making a different response, for example, press the left lever, to receive a reinforcement. The cues are presented prior to making a response and are switched after a certain number of trials. Related to the prelimbic cortex, prelimbic lesions alone or prelimbic and infralimbic lesions do not impair acquisition of a conditional discrimination task (Chudasama et al. 2001; Delatour and Gisquet-Verrier 1999). More recently, a study trained rats on a conditional discrimination task in which one of the two different cue–response associations was presented for 5–10 consecutive trials before a switch to the other cue–response association (Leenaars et al. 2012). In this test, prelimbic inactivation selectively impairs performance for a switch trial. These findings suggest that the prelimbic cortex also supports behavioral flexibility when cue information must be used to proactively switch. However, it is unclear whether prelimbic inactivation also increases perseveration of the previous cue–response association and/or maintenance of the currently correct response pattern. Therefore, it is unknown whether the prelimbic cortex supports a similar process when a change in cues signals a switch, for example, inhibiting perseveration of a previously relevant response pattern, as when a change in outcomes can be used to switch a response pattern.

There is also limited understanding of how the prelimbic area may interact with other brain areas to support cue-guided behavioral switching. The prelimbic cortex has extensive projections to basal ganglia structures and together these areas may act in a cooperative manner to facilitate behavioral flexibility when a change in outcomes or a change in cues guides a behavioral switch (Afsharpour 1985; Chudasama and Robbins 2006; Jahfari et al. 2011; Kehagia et al. 2010; Mailly et al. 2013). The subthalamic nucleus and dorsomedial striatum are the two areas of basal ganglia that receive direct excitatory input from the prelimbic cortex that is mediated, at least in part, by NMDA receptors (Berendse et al. 1992; Conde et al. 1995; Gabbott et al. 2005; Magill et al. 2006; Maurice et al. 1998; Nambu et al. 2000; Sesack et al. 1989). Individual neurons in the nonhuman primate subthalamic nucleus show increased activity in response to a cue that signals when a switch from one response pattern to another will be rewarded suggesting that this area may be important for a proactive behavioral switch (Isoda and Hikosaka, 2008). In addition, dorsomedial striatal lesions or inactivation impair behavioral switching in conditional discrimination tests (Adams et al. 2001; Featherstone and McDonald 2005; Hallock et al. 2013). While the findings implicate the subthalamic nucleus and dorsomedial striatum in proactive behavioral switching, these paradigms typically involved cues for switching every 1 or 2 trials which may not be sufficient to establish a response set leading to switch costs as measured by switching errors or increased reaction time.

To begin addressing some of these issues, we recently completed a series of experiments to test conditional discrimination performance following a contralateral disconnection of the prelimbic cortex and subthalamic nucleus, as well as the prelimbic cortex and dorsomedial striatum. This involved infusions of the GABA agonists, baclofen and muscimol in the prelimbic cortex (Leenaars et al. 2012) and the NMDA receptor antagonist, AP-5 in the subthalamic nucleus (Baunez and Robbins 1999). The experiments further examined whether these pharmacological manipulations affected switch trial performance, initial perseveration of a previously relevant response pattern and/or maintenance of the currently relevant response pattern once selected.

To carry out these experiments, we developed a conditional discrimination test in a modified cross-maze (see Fig. 11.3). The stem arm served as the start arm and the other two arms served as choice arms. A white or black insert was placed in the start arm that covered the floor and side walls of the arm. Rats were trained to associate a start arm cue with choosing one particular choice arm, for example, a spatial location, to receive a food reward. Rats were tested for 57 trials per session. In the conditional cue–place association, the visual cue was changed in blocks of every 3–6 trials indicating that a behavioral switch should occur for the upcoming choice (see Baker and Ragozzino 2014a, b for details). The relatively short block length was chosen in order to emphasize the need to monitor task cues on every trial while also having a rat establish a response pattern prior to a switch. This is common in a proactive switch task in order to incur a switch cost such that performance is more difficult on a switch trial compared to that of non-switch trials (Hikosaka and Isoda 2010; Hyafil et al. 2009; Konishi et al. 2005). Consistent with the task

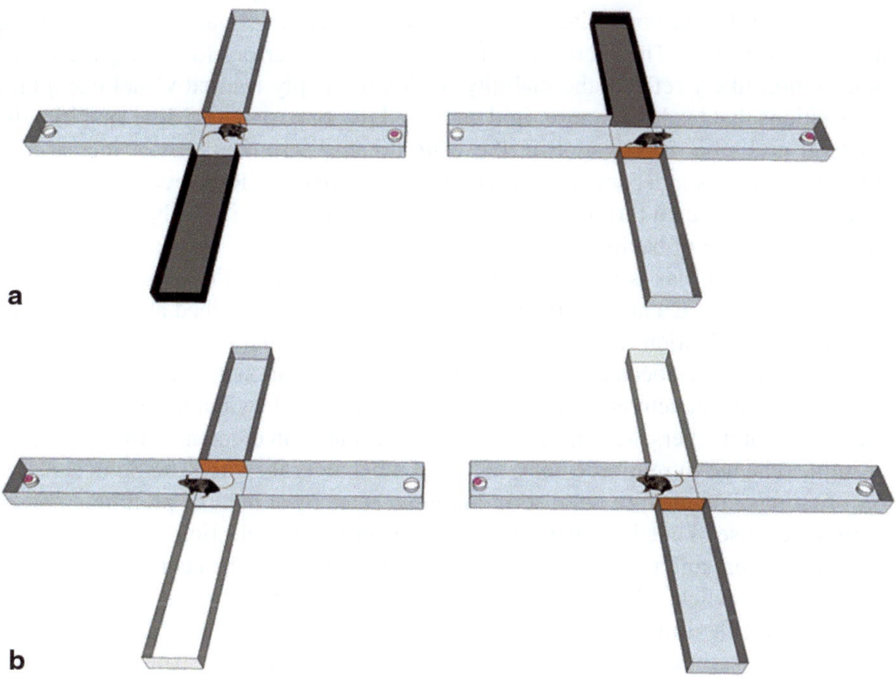

Fig. 11.3 Visual cue–place conditional discrimination. A visual cue was placed in the start arm. **a** In one condition, a *black* visual cue is placed in one of the two start arms and a rat must always enter the same maze arm to receive a cereal reinforcement. **b** In the other condition, a *white* visual cue is placed in one of the two start arms and a rat must enter the other maze arm to receive a cereal reinforcement. Rats learned to associate a start arm cue with entering a particular choice arm to receive a cereal reinforcement. Extra-maze visual cues surround the maze (not shown) that a rat can use to spatially guide their choice. The visual cues were randomly changed in blocks of every 3–6 trials within a 57 trial session. The *copper block* prevented entry into that arm on a trial. The *O-shaped* object in the foodwell represents a cereal piece reinforcement

having switch costs, we found that vehicle treated rats committed a significantly greater percentage of errors on switch trials compared to that of non-switch trials.

Studies using this task led to a unique and interesting set of results across brain areas. More specifically, bilateral prelimbic inactivation impaired conditional discrimination performance by significantly increasing switch, perseverative, and maintenance errors (Baker and Ragozzino 2014a, b). This contrasts with past studies using set-shifting tests in which a change in outcomes signaled a shift to occur such that prelimbic cortex inactivation selectively increased perseveration of the previously relevant response (Dias and Aggleton 2000; Ragozzino 2007; Ragozzino et al. 1999b). In this test, one possibility in the cue-guided behavioral switch test is that prelimbic cortex inactivation simply impairs discrimination performance independent of behavioral switching. To test this, Baker and Ragozzino (2014a, b) trained rats in a conditional discrimination test as before, but in a control test required rats to execute a single visual cue–place discrimination without any switches to other condition. Prelimbic cortex inactivation did not affect performance in a non-switch discrimination test. The increase in multiple types of errors following prelimbic inactivation likely reflects the inability to flexibly apply learned visual cue–place associations that leads to a more rigid and fixed response pattern. More specifically, bilateral prelimbic inactivation in the conditional discrimination test increased a turn bias that was independent of current cue information. Rats, even under saline treatment, exhibited a turn bias in the test, but this was significantly enhanced under the high dose of baclofen/muscimol injected in prelimbic cortex. However, the exaggerated turn bias is not a necessary consequence of prelimbic inactivation as this did not occur in a non-switch discrimination test. As described above, this conditional discrimination test is distinct from set-shifting and reversal-learning tests used in past studies because in a conditional discrimination test cues can be used on each trial to proactively determine when a behavioral switch must occur, while set-shifting and reversal-learning tests involve a change in outcome information, for example, change in reinforcement, to signal a behavioral switch should occur. Recent findings in a conditional discrimination test, suggest that the prelimbic cortex supports the use of cue information to allow the proactive selection of an alternative response pattern and maintenance of that response pattern when conditions require a behavioral switch. These results support the model of prefrontal cortex control of behavioral flexibility set forth by Wise et al. (1996). Specifically, although the conditional discrimination test requires a rat to reverse a response pattern based on a single attribute (spatial), as these reversals are determined by integrating visual cue and visuospatial information, a higher-order rule must be applied to successfully perform the task. As predicted by Wise et al. (1996) this higher-order rule processing requires the prelimbic cortex.

The role of the subthalamic nucleus in proactive behavioral switching was also examined using a conditional discrimination test. NMDA receptor blockade in the subthalamic nucleus also impaired performance in the proactive behavioral switch test (Baker and Ragozzino 2014a). However, in contrast to the effects of prelimbic cortex inactivation, NMDA receptor blockade in the subthalamic nucleus selectively increased switch and perseverative errors, but did not affect maintenance errors. Similarly, contralateral disconnection of the prelimbic cortex and subthalamic nucleus also increased switch errors in the conditional discrimination test. In addition, the contralateral disconnection increased perseverative errors leading a rat to repeatedly choose the previously relevant response pattern after the initial switch trial. In contrast, ipsilateral disconnection of the prelimbic cortex and subthalamic nucleus had no effect on performance. The findings following NMDA receptor blockade in the subthalamic nucleus are comparable to those in which subthalamic nucleus lesions impair inhibition of an initiated response in the stop-signal test (Eagle et al. 2008) and further suggest that the subthalamic nucleus is critical not only for inhibiting an initiated response, but also for inhibiting an ongoing response pattern when cues indicate an alternate response pattern should occur. Interestingly, this is true even after the initial switch as evidenced by the increase in perseveration if a switch error was committed.

To determine whether another basal ganglia region that receives prelimbic cortex input contributes to proactive behavioral switching, NMDA receptor blockade in the dorsomedial striatum, as well as contralateral disconnection of the prelimbic cortex and dorsomedial striatum was investigated (Baker and Ragozzino 2014b). Bilateral AP5 infusions in the dorsomedial striatum, as well as a contralateral disconnection of the prelimbic cortex and dorsomedial striatum impaired overall conditional discrimination performance. Similar to that observed with prelimbic cortex and subthalamic nucleus, ipsilateral disconnection of the prelimbic cortex and dorsomedial striatum did not impair performance. Besides increasing the number of switch errors, these manipulations significantly elevated the number of perseverative and maintenance errors. The significant increase in all error types following dorsomedial striatal NMDA receptor blockade emerged because this led a rat to commit errors across an entire block of trials 1–3 times in a session. This effect committing errors across an entire trial block was not due to the length of the previous block or the length of the block which was missed. Thus, the previous block of trials being short, for example, 3 trials or long, for example, 6 trials, nor the block in which errors were committed in all trials being short or long could explain the finding. One explanation for the failure to perform a given block is that the change in cue–reward contingencies fails to update the ongoing choice pattern resulting in the previous choice pattern being continually executed. In rats, the dorsomedial striatum has been implicated in relaying information about the expected value of an action based on recent task demands. In a recent study, rats were trained in a two-choice discrimination in which there were different probabilities for reward. The choices were reversed after 35 trials with multiple reversals in a session (Kim et al. 2013). Similar to the current experiments, rats were well-trained in the task in which multiple single units were recorded during the test. Although the activity of any single neuron only correlated weakly with a choice, there was an ensemble of activity in the dorsomedial striatum that preceded the actual choice and would change dynamically with a reversal in reward probabilities (Kim et al. 2013). This supports that the dorsomedial striatum is critical for the updating of expected value of an action or strategy.

Overall, while drug manipulations of all three brain areas impaired conditional discrimination performance, the pattern of errors that emerged were somewhat distinct and also differed from the same drug manipulations which also impaired performance on set-shifting and reversal learning tests. Moreover, the findings from these contralateral disconnection experiments suggest that the prelimbic cortex connections with specific basal ganglia areas dynamically interact to support proactive behavioral switching. The pattern of results raises the possibility that the prelimbic cortex is acting in a top–down fashion to control behavioral flexibility through two different basal ganglia pathways (see Fig. 11.4). Narayanan and Laubach (2006, 2009) have proposed that the dorsomedial frontal cortex encodes both prepotent responses and proactive inhibition such that when neurons encoding proactive inhibition predominate, a rat will be less likely to make a premature response. A similar top–down process may occur to allow proactive behavioral switching such that the prelimbic cortex encodes both inhibition of an ongoing strategy and

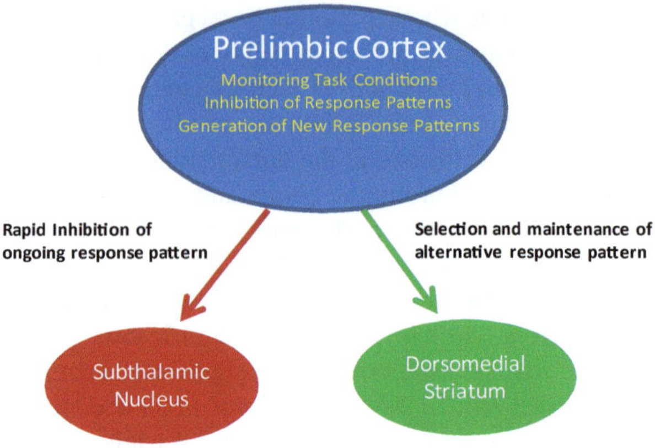

Fig. 11.4 Prelimbic cortex interactions with the dorsomedial striatum and subthalamic nucleus to facilitate proactive behavioral switching

generation of relevant strategies in response to specific cues. In this fashion, the prelimbic cortex would be critical for the monitoring of task cues to guide appropriate responses or rule applications on a trial-to-trial basis. When excitatory input from the prelimbic cortex to the subthalamic nucleus predominates, it allows an inhibition of the ongoing response pattern and selection of a different response pattern. In this manner, the prelimbic cortex and subthalamic nucleus together can rapidly terminate an ongoing or prepotent response when no longer relevant. Physiological evidence suggests that the prelimbic cortex–subthalamic nucleus circuit is ideally suited to this function. Prelimbic cortex stimulation is followed by a large burst of neuronal firing in the subthalamic nucleus after 4–8 ms (Maurice et al. 1998; Magill et al. 2006). Furthermore, recordings in the substantia nigra pars reticulata reveal that input from the subthalamic nucleus arrives before that from the direct pathway coming from the striatum (Fujimoto and Kita 1993; Ryan and Sanders 1994; Maurice et al. 1999). This is important for a proposed model of prelimbic cortex–subthalamic nucleus input in overriding a prepotent or ongoing behavior (Mathai and Smith 2011). The signal from this pathway arrives at basal ganglia output structures before that of the direct and indirect pathway allowing for modification of the output back to the motor cortex. In this way, the prelimbic cortex–subthalamic nucleus circuit represents an ideal mechanism for the top–down inhibition of an ongoing behavior or strategy when cues indicate the choice pattern should not be used.

Prelimbic cortex inactivation not only led to switch errors, but also increased maintenance errors. This would suggest that the prelimbic cortex interacts with other areas to support proactive switching. Results from prelimbic cortex–dorsomedial striatal areas suggest that these areas functionally interact differently than the prelimbic cortex and subthalamic nucleus to support behavioral switching. This is because contralateral disconnection of the prelimbic cortex and dorsomedial striatal areas selectively increased the likelihood of rats to miss an entire block of trials. One possibility is that the prelimbic cortex input to the dorsomedial striatum provides information about possible strategies or choice patterns in a context and the dorsomedial striatum facilitates the appropriate strategy selection (Kim et al. 2009; Tai et al. 2012). In fact, neuronal signals in the dorsomedial striatum have been shown to encode information about the expected reward value of a given behavioral response based on previous reward feedback from making that choice (Stalnaker et al. 2012; Kim et al. 2013). One possibility is that cue information also can be used proactively by the dorsomedial striatum to select a strategy. If input from the prelimbic cortex to the dorsomedial striatum is disrupted, this may decrease information about possible strategies and limit the accuracy of selecting a strategy (Ragozzino 2007), which could lead on occasion to making errors for an entire block of trials. Thus, in the conditional discrimination test rats may have been unable to generate a different choice pattern appropriate to the cues on a given trial and the previous choice pattern is repeatedly selected. Thus, we propose that when cue information should be used to proactively switch choice patterns that a neural system that includes the prelimbic cortex and subthalamic nucleus supports the rapid inhibition of an ongoing choice pattern while concomitantly a neural system that includes the prelimbic cortex and dorsomedial striatum enables selection of an alternative choice pattern. This latter system also continues to be critical for maintaining the alternative choice pattern after being initially selected.

Conclusions

The neurobiologically based attribute model of memory asserts that the nature of memory can be explained by different attributes such as space, time, sensory-perception, response, and reward (affect), which are stored as memories in different forebrain areas. Tests of this model led to the development of several novel learning paradigms that emphasized the learning and memory of a specific type of attribute, for example, visuospatial information. Our investigations of prefrontal cortex and basal ganglia structures in behavioral flexibility employed the attribute model of memory approach by whether the prefrontal cortex contributed to behavioral flexibility based on separate subregions supporting the flexible use of specific attribute information. The findings from numerous studies suggest that different prefrontal cortex subregions support different forms of behavioral flexibility based on the level of the operation required to flexibly adapt (Kesner and Churchwell 2011; Ragozzino 2007; Wise et al. 1996). Although different prefrontal cortex subregions may support different forms of behavioral flexibility when a change in outcomes signals a behavioral switch should occur, these different subregions appear particularly important for initially inhibiting perseveration of a previously relevant strategy.

The dorsomedial striatum is an area that receives input from both the orbitofrontal cortex and prelimbic cortex. There is considerable support for the idea that this striatal region plays a role in various types of behavioral flexibility when a change in outcomes occurs. This is consistent with the diverse prefrontal cortex input it receives. In set-shifting and reversal-learning tests, the dorsomedial striatum supports behavioral flexibility by maintaining the new choice pattern after it has been initially selected. Thus, the dorsomedial striatum likely plays a distinct, but complementary role from different prefrontal cortex subregions in facilitating set-shifting and reversal learning.

There is recent evidence that the prelimbic cortex and different basal ganglia areas interact to enhance proactive behavioral switching. Under conditions in which cues signal that an upcoming response should be switched, these brain areas act in a cooperative manner to facilitate behavioral flexibility. During proactive behavioral switching conditions, the prelimbic cortex and subthalamic nucleus are part of a neural system that enables the rapid inhibition of an ongoing choice pattern while concomitantly a neural system that includes the prelimbic cortex and dorsomedial striatum enables selection of an alternative choice pattern and maintenance of that selection. These results demonstrate that specific prefrontal–basal ganglia circuitry not only supports behavioral flexibility when there is a change in outcomes but also when cues can be used to proactively switch response patterns. Further, the effect of a general increase in errors with prelimbic cortex inactivation in a conditional discrimination test suggests that under certain conditions, the prefrontal cortex may be required for more than just the initial abandonment of the previous choice pattern, but plays a critical role in monitoring task conditions to concomitantly inhibit one choice pattern and facilitate the use of a different choice pattern. This is particularly the case when cue information must be monitored on a trial-by-trial basis to switch ongoing behavioral responses. Overall, there is accumulating evidence that prefrontal cortex and basal ganglia structures are crucial to allow rapid and repeated adaptations across a variety of stimulus attributes in which changes in reward feedback or proactive cue information signal a behavioral switch should occur.

Acknowledgments We are grateful to Dante A. Ragozzino for the maze illustration design in Fig. 11.3. We are particularly grateful to Ray Kesner who has had a deep influence on the research described throughout this chapter. He is a wonderful mentor with his constant positive and supportive approach. Ray Kesner has many magnificent attributes as a human being and a scientist, but surely one of his greatest attributes is his insatiable pursuit of scientific discovery that is expressed with pure joy and excitement which serves as an inspiration for current and future scientists.

References

Adams, S., Kesner, R. P., & Ragozzino, M. E. (2001). Role of the medial and lateral caudate-putamen in mediating an auditory conditional response association. *Neurobiology of Learning and Memory, 76,* 106–116.

Afsharpour, S. (1985). Topographical projections of the cerebral cortex to the subthalamic nucleus. *The Journal of Comparative Neurology, 236,* 14–28.

Akopian, G., & Walsh, J. P. (2002). Corticostriatal paired-pulse potentiation produced by voltage-dependent activation of NMDA receptors and L-type Ca(2+) channels. *Journal of Neurophysiology, 87,* 157–165.

Baker, P. M., & Ragozzino, M. E. (2014a). The prelimbic cortex and subthalamic nucleus contribute to cue-guided behavioral switching. *Neurobiology of Learning and Memory, 107,* 65–78.

Baker, P. M., & Ragozzino, M. E. (2014b). Contralateral disconnection of the rat prelimbic cortex and dorsomedial striatum impairs cue-guided behavioral switching. *Learning and Memory, 21,* 368–379.

Baunez, C., & Robbins, T. W. (1999). Effects of transient inactivation of the subthalamic nucleus by local muscimol and APV infusions on performance on the five-choice serial reaction time task in rats. *Psychopharmacology, 141*(1), 57–65.

Berendse, H. W., Galis-de Graaf, Y., & Groenewegen, H. J. (1992). Topographical organization and relationship with ventral striatal compartments of prefrontal corticostriatal projections in the rat. *The Journal of Comparative Neurology, 316,* 314–347.

Birrell, J. M., & Brown, V. J. (2000). Medial frontal cortex mediates perceptual attentional set shifting in the rat. *The Journal of Neuroscience: The Official Journal of the Society for Neuroscience, 20,* 4320–4324.

Boettiger, C. A., & Doupe, A. J. (2001). Developmentally restricted synaptic plasticity in a songbird nucleus required for song learning. *Neuron, 31,* 809–818.

Boix-Trelis, N., Vale-Martinez, A., Guillazo-Blanch, G., Costa-Miserachs, D., & Marti-Nicolovius, M. (2006). Effects of nucleus basalis magnocellularis stimulation on a socially transmitted food preference and c-Fos expression. *Learning & Memory (Cold Spring Harbor, N. Y.), 13,* 783–793.

Boulougouris, V., Dalley, J. W., & Robbins, T. W. (2007). Effects of orbitofrontal, infralimbic and prelimbic cortical lesions on serial spatial reversal learning in the rat. *Behavioural Brain Research, 179,* 219–228.

Chudasama, Y., & Robbins, T. W. (2006). Functions of frontostriatal systems in cognition: Comparative neuropsychopharmacological studies in rats, monkeys and humans. *Biological Psychology, 73*(1), 19–38.

Chudasama, Y., Bussey, T. J., & Muir, J. L. (2001). Effects of selective thalamic and prelimbic cortex lesions on two types of visual discrimination and reversal learning. *The European Journal of Neuroscience 0953-816X (P), 14,* 1009–1020.

Churchwell, J. C., Morris, A. M., Heurtelou, N. M., & Kesner, R. P. (2009). Interactions between the prefrontal cortex and amygdala during delay discounting and reversal. *Behavioral Neuroscience, 123,* 1185–1196.

Conde, F., Maire-Lepoivre, E., Audinat, E., & Crepel, F. (1995). Afferent connections of the medial frontal cortex of the rat. II. Cortical and subcortical afferents. *The Journal of Comparative Neurology, 352,* 567–593.

Dang, M. T., Yokoi, F., Yin, H. H., Lovinger, D. M., Wang, Y., & Li, Y. (2006). Disrupted motor learning and long-term synaptic plasticity in mice lacking NMDAR1 in the striatum. *Proceedings of the National Academy of Sciences of the United States of America, 103,* 15254–15259.

Delatour, B., & Gisquet-Verrier, P. (1999). Lesions of the prelimbic-infralimbic cortices in rats do not disrupt response selection processes but induce delay-dependent deficits: Evidence for a role in working memory? *Behavioral Neuroscience, 113,* 941–955.

Dias, R., & Aggleton, J. P. (2000). Effects of selective excitotoxic prefrontal lesions on acquisition of nonmatching- and matching-to-place in the T-maze in the rat: Differential involvement of the prelimbic infralimbic and anterior cingulate cortices in providing behavioural flexibility. *The European Journal of Neuroscience 0953-816X (P), 12,* 4457–4466.

Eagle, D. M., Baunez, C., Hutcheson, D. M., Lehmann, O., Shah, A. P., & Robbins, T. W. (2008). Stop-signal reaction-time task performance: Role of prefrontal cortex and subthalamic nucleus. *Cerebral Cortex (New York, N. Y.: 1991), 18*(1), 178–188.

Eden, C. G., Lamme, V., & Uylings, H. (1992). Heterotopic cortical afferents to the medial prefrontal cortex in the rat. A combined retrograde and anterograde tracer study. *European Journal of Neuroscience, 4,* 77–97.

Featherstone, R., McDonald, R. (2005). Lesions of the dorsolateral striatum impair the acquisition of a simplified stimulus-response dependent conditional discrimination task. *Neuroscience, 136,* 387–395.

Fujimoto, K., & Kita, H. (1993). Response characteristics of subthalamic neurons to the stimulation of the sensorimotor cortex in the rat. *Brain Research, 609,* 185–192.

Gabbott, P. L., Warner, T. A., Jays, P. R., Salway, P., & Busby, S. J. (2005). Prefrontal cortex in the rat: Projections to subcortical autonomic, motor, and limbic centers. *The Journal of Comparative Neurology, 492,* 145–177.

Ghods-Sharifi, S., Haluk, D. M., & Floresco, S. B. (2008). Differential effects of inactivation of the orbitofrontal cortex on strategy set-shifting and reversal learning. *Neurobiology of Learning and Memory, 89,* 567–573.

Hallock, H. L., Arreola, A. C., Shaw, C. L., & Griffin, A. L. (2013). Dissociable roles of the dorsal striatum and dorsal hippocampus in conditional discrimination and spatial alternation T-maze tasks. *Neurobiology of Learning and Memory, 100,* 108–116.

Heidbreder, C. A., & Groenewegen, H. J. (2003). The medial prefrontal cortex in the rat: Evidence for a dorso-ventral distinction based upon functional and anatomical characteristics. *Neuroscience & Biobehavioral Reviews, 27,* 555-579.

Hikosaka O., Isoda M. (2010). Switching from automatic to controlled behavior: Cortico-basal ganglia mechanisms. *Trends in Cognitive Sciences, 14,* 154–161.

Hoover, W. B., & Vertes, R. P. (2007). Anatomical analysis of afferent projections to the medial prefrontal cortex in the rat. *Brain Structure and Function, 212,* 149-179.

Hyafil, A., Summerfield, C., & Koechlin, E. (2009). Two mechanisms for task switching in the prefrontal cortex. *The Journal of Neuroscience: The Official Journal of the Society for Neuroscience, 29*(16), 5135–5142.

Isoda, M., & Hikosaka, O. (2008). Role for subthalamic nucleus neurons in switching from automatic to controlled eye movement. *The Journal of Neuroscience: The Official Journal of the Society for Neuroscience, 28,* 7209–7218.

Jahfari, S., Waldorp, L., van den Wildenberg, W. P. M., Scholte, H. S., Ridderinkhof, K. R., & Forstmann, B. U. (2011). Effective connectivity reveals important roles for both the hyperdirect (fronto-subthalamic) and the indirect (fronto-striatal-pallidal) fronto-basal ganglia pathways during response inhibition. *Journal of Neuroscience, 31*(18), 6891–6899.

Kehagia, A. A., Murray, G. K., & Robbins, T. W. (2010). Learning and cognitive flexibility: Frontostriatal function and monoaminergic modulation. *Current Opinion in Neurobiology, 20*(2), 199–204.

Kesner, R. P. (2000). Subregional analysis of mnemonic functions of the prefrontal cortex in the rat. *Psychobiology, 28,* 219–222.

Kesner, R. P. (2009). Tapestry of memory. *Behavioral Neuroscience, 123,* 1–13.

Kesner, R. P., & Churchwell, J. C. (2011). An analysis of rat prefrontal cortex in mediating executive function. *Neurobiology of Learning and Memory, 96,* 417–431.

Kesner, R. P., & DiMattia, B. V. (1987). Neurobiology of an attribute model of memory. In A. R. Morrison & A. N. Epstein (Eds.), *Progress in psychobiology and physiological psychology* (pp. 207–277). New York: Academic.

Kesner, R. P., Hunt, M. E., Williams, J. M., & Long, J. M. (1996). Prefrontal cortex and working memory for spatial response, spatial location and visual object information in the rat. *Cerebral Cortex, 6,* 311–318.

Kim, J., & Ragozzino, M. E. (2005). The involvement of the orbitofrontal cortex in learning under changing task contingencies. *Neurobiology of Learning and Memory, 83,* 125–133.

Kim, H., Sul, J. H., Huh, N., Lee, D., & Jung, M. W. (2009). Role of striatum in updating values of chosen actions. *The Journal of Neuroscience: The Official Journal of the Society for Neuroscience, 29,* 14701–14712.

Kim, H., Lee, D., & Jung, M. W. (2013). Signals for previous goal choice persist in the dorsomedial, but not dorsolateral striatum of rats. *The Journal of Neuroscience, 33,* 52–63.

Konishi, S., Chikazoe, J., Jimura, K., Asari, T., & Miyashita, Y. (2005). Neural mechanism in anterior prefrontal cortex for inhibition of prolonged set interference. *Proceedings of the National Academy of Sciences of the United States of America, 102*(35), 12584–12588.

Leenaars, C. H., Joosten, R. N., Zwart, A., Sandberg, H., Ruimschotel, E., Hanegraaf, M. A., Dematteis, M., Feenstra, M. G., & van Someren E. J. (2012). Switch-task performance in rats is disturbed by 12 h of sleep deprivation but not by 12 h of sleep fragmentation. *Sleep, 35,* 211.

Lindgren, H. S., Wickens, R., Tait, D. S., Brown, V. J., & Dunnett, S. B. (2013). Lesions of the dorsomedial striatum impair formation of attentional set in rats. *Neuropharmacology, 71,* 148–153.

Magill, P. J., Sharott, A., Bolam, J. P., & Brown, P. (2006). Delayed synchronization of activity in cortex and subthalamic nucleus following cortical stimulation in the rat. *The Journal of Physiology, 574*(3), 929–946.
Mailly, P., Aliane, V., Groenewegen, H. J., Haber, S. N., & Deniau, J. M. (2013). The rat prefronto-striatal system analyzed in 3D: Evidence for multiple interacting functional units. *The Journal of Neuroscience: the Official Journal of the Society for Neuroscience, 33,* 5718–5727.
Mathai, A., & Smith, Y. (2011). The corticostriatal and corticosubthalamic pathways: two entries, one target. So what? Frontiers in Systems Neuroscience 5.
Maurice, N., Deniau, J. M., Glowinski, J., & Thierry, A. M. (1998). Relationships between the prefrontal cortex and the basal ganglia in the rat: Physiology of the corticosubthalamic circuits. *The Journal of Neuroscience: The Official Journal of the Society for Neuroscience, 18*(22), 9539–9546.
Maurice, N., Deniau, J. M., Glowinski, J., & Thierry, A. M. (1999). Relationships between the prefrontal cortex and the basal ganglia in the rat: Physiology of the cortico-nigral circuits. *The Journal of Neuroscience: The Official Journal of the Society for Neuroscience, 19,* 4674–4681.
McAlonan, K., & Brown, V. J. (2003). Orbital prefrontal cortex mediates reversal learning and not attentional set shifting in the rat. *Behavioural Brain Research, 146,* 97–103.
Nambu, A., Tokuno, H., Hamada, I., Kita, H., Imanishi, M., Akazawa, T., & Hasegawa, N. (2000). Excitatory cortical inputs to pallidal neurons via the subthalamic nucleus in the monkey. *Journal of Neurophysiology, 84*(1), 289–300.
Narayanan, N. S., & Laubach, M. (2006). Top-down control of motor cortex ensembles by dorsomedial prefrontal cortex. *Neuron, 52*(5), 921–931.
Narayanan, N. S., & Laubach, M. (2009). Delay activity in rodent frontal cortex during a simple reaction time task. *Journal of Neurophysiology, 101*(6), 2859–2871.
Palencia, C. A., & Ragozzino, M. E. (2004). The influence of NMDA receptors in the dorsomedial striatum on response reversal learning. *Neurobiology of Learning and Memory, 82,* 81–89.
Palencia, C. A., & Ragozzino, M. E. (2005). The contribution of NMDA receptors in the dorsolateral striatum to egocentric response learning. *Behavioral Neuroscience, 119,* 953–960.
Pisa, M., & Cyr, J. (1990). Regionally selective roles of the rat's striatum in modality-specific discrimination learning and forelimb reaching. *Behavioural Brain Research, 37,* 281–292.
Ragozzino, M. E. (2002). The effects of dopamine D(1) receptor blockade in the prelimbic-infralimbic areas on behavioral flexibility. *Learning & Memory* (*Cold Spring Harbor, N. Y.*), *9,* 18–28.
Ragozzino, M. E. (2007). The contribution of the medial prefrontal cortex, orbitofrontal cortex, and dorsomedial striatum to behavioral flexibility. *Annals of the New York Academy of Sciences, 1121,* 355–375.
Ragozzino, M. E., & Choi, D. (2004). Dynamic changes in acetylcholine output in the medial striatum during place reversal learning. *Learning & Memory* (*Cold Spring Harbor, N. Y.*), *11,* 70–77
Ragozzino, M. E., & Kesner, R. P. (1999). The role of the agranular insular cortex in working memory for food reward value and allocentric space in rats. *Behavioural Brain Research, 1,* 103–112.
Ragozzino, M. E., & Kesner, R. P. (2001). The role of rat dorsomedial prefrontal cortex in working memory for egocentric responses. *Neuroscience Letters, 308,* 145–148.
Ragozzino, M. E., Adams, S., & Kesner, R. P. (1998). Differential involvement of the dorsal anterior cingulate and prelimbic/infralimbic areas of the rodent prefrontal cortex in spatial working memory. *Behavioral Neuroscience, 112,* 293–303.
Ragozzino, M. E., Wilcox, C., Raso, M., & Kesner, R. P. (1999a). Involvement of rodent prefrontal cortex subregions in strategy switching. *Behavioral Neuroscience, 113,* 32–41.
Ragozzino, M. E., Detrick, S., & Kesner, R. P. (1999b). Involvement of the prelimbic-infralimbic areas of the rodent prefrontal cortex in behavioral flexibility for place and response learning. *The Journal of Neuroscience, 19,* 4585–4594.

Ragozzino, M. E., Detrick, S., & Kesner, R. P. (2002). The effects of prelimbic and infralimbic lesions on working memory for visual objects in rats. *Neurobiology of Learning and Memory, 77,* 29–43.

Ragozzino, M. E., Ragozzino, K. E., Mizumori, S. J., & Kesner, R. P. (2002b). Role of the dorsomedial striatum in behavioral flexibility for response and visual cue discrimination learning. *Behavioral Neuroscience, 116,* 105–115.

Ragozzino, M. E., Kim, J., Hassert, D., Minniti, N., & Kiang, C. (2003). The contribution of the rat prelimbic-infralimbic areas to different forms of task switching. *Behavioral Neuroscience, 117,* 1054–1065.

Riceberg, J. S., & Shapiro, M. L. (2012). Reward stability determines the contribution of orbitofrontal cortex to adaptive behavior. *The Journal of Neuroscience: The Official Journal of the Society for Neuroscience, 32,* 16402–16409.

Ryan, L. J., & Sanders, D. J. (1994). Subthalamic nucleus and globus pallidus lesions alter activity in nigrothalamic neurons in rats. *Brain Research Bulletin, 34,* 19–26.

Schoenbaum, G., Nugent, S. L., Saddoris, M. P., & Setlow, B. (2002). Orbitofrontal lesions in rats impair reversal but not acquisition of go, no-go odor discriminations. *Neuroreport, 13,* 885–890.

Sesack, S. R., Deutch, A. Y., Roth, R. H., & Bunney, B. S. (1989). Topographical organization of the efferent projections of the medial prefrontal cortex in the rat: An anterograde tract-tracing study with *Phaseolus vulgaris* leucoagglutinin. *The Journal of Comparative Neurology, 290,* 213–242.

Spencer, J. P., & Murphy, K. P. (2000a). Bi-directional changes in synaptic plasticity induced at corticostriatal synapses in vitro. *Experimental Brain Research. Experimentelle Hirnforschung. Experimentation Cerebrale, 135,* 497–503.

Stalnaker, T. A., Calhoon, G. G., Ogawa, M., Roesch, M. R., & Schoenbaum, G. (2012). Reward prediction error signaling in posterior dorsomedial striatum is action specific. *The Journal of Neuroscience, 32,* 10296–10305.

Stefani, M. R., Groth, K., & Moghaddam, B. (2003). Glutamate receptors in the rat medial prefrontal cortex regulate set-shifting ability. *Behavioral Neuroscience, 117,* 728–737.

Tai, L. H., Lee, A. M., Benavidez, N., Bonci, A., & Wilbrecht, L. (2012). Transient stimulation of distinct subpopulations of striatal neurons mimics changes in action value. *Nature Neuroscience, 15,* 1281–1289.

Uylings, H. B., & van Eden C. G. (1990). Qualitative and quantitative comparison of the prefrontal cortex in rat and in primates, including humans. *Progress in Brain Research, 85,* 31–62.

Vertes, R. P. (2004). Differential projections of the infralimbic and prelimbic cortex in the rat. *Synapse, 51,* 32–58.

Wise, S. P., Murray, E. A., & Gerfen, C. R. (1996). The frontal cortex-basal ganglia system in primates. *Critical Reviews in Neurobiology, 10,* 317–356.

Chapter 12
Balancing the Contributions of Multiple Neural Systems During Learning and Memory

Paul E. Gold

Research described here was supported by NIA R01 AG07648, NIDA DA024129, NSF IOS 08-43175, and 13-18490, and by a grant from Alzheimer's Association.

My Early Days with Ray Kesner

I met Ray Kesner when he was an organizer of the first Winter Memory Conference over 40 years—40 years!—ago. The years have not diminished his unbounded enthusiasm for research on the neurobiology of memory.

That first memory conference was small, with perhaps 15 or so people in attendance. I still enjoy reminiscing about that meeting. I was a postdoc in Jim McGaugh's lab, flattered to have been invited to the meeting. Ray was, like me, just shy of 30 year old, youngsters in a crowd of old men—no women as I recall—"old men" who were mostly in their 40s at the time. That first meeting was held in Salt Lake City, from where we commuted to the mountains for skiing during the day. Very soon, the meeting organizers recognized the utility of staying up in the mountains to skip the commute, an outstanding decision in my view.

In thinking about Ray's contributions to the field, his research defining the concept of multiple memory systems has impacted the deep rationales behind the work of many researchers, including me, about brain mechanisms of learning and memory. In this regard, it may be instructive to consider that the first meeting—again, viewed through my own recollections—dealt with issues like the precise time necessary for memories to be formed, using retrograde amnesia gradients as the basis for temporal properties of memory consolidation. There were temporal

P. E. Gold (✉)
Department of Biology, Syracuse University,
Life Sciences Complex 107 College Place, Syracuse, NY 13244, USA
e-mail: pegold@syr.edu

P. A. Jackson et al. (eds.), *The Neurobiological Basis of Memory*,
DOI 10.1007/978-3-319-15759-7_12

gradients presented that ranged from 0.5 s, to 30 s, to 1 h, to several days (e.g., Chorover and Schiller 1965; Paolino et al. 1966; Quartermain et al. 1965; Zornetzer and Gold 1976), and later findings extended the temporal retrograde amnesia gradient to weeks in mice (Squire and Spanis 1984) as in humans (Squire et al. 1975). At the early Winter Memory Conference meeting, Cherkin (1969) showed that retrograde amnesia gradients in chicks varied directly from seconds to days with the duration and dose of the analeptic drug flurothyl, results analogous to those obtained in rodents (cf.: McGaugh 1966; Gold et al. 1973; Gold and Zornetzer 1983). The variable time "constants" presented at that meeting influenced the development of memory modulation as a concept distinct from consolidation as a way to look at memory formation (Gold and McGaugh 1975).

Some of the treatments of that era included protein synthesis inhibitors. For these drugs too the time courses varied widely across experimental conditions. The variable retrograde and anterograde amnesia gradients were consistent with the developing view of memory modulation, which could accommodate multiple amnesia gradients rather easily, but were (and are) more difficult to reconcile with the use of memory consolidation as the underlying construct. The multiple time courses of amnesia produced by protein synthesis inhibitors were then, and remain now, more readily compatible with the view that amnesia produced by protein synthesis inhibitors reflects alterations in systems that modulate memory—some hormones and some neurotransmitters—rather than providing insights into the molecular bases of memory consolidation (Gold 2008). With additional recent evidence that anisomycin blocks local neurophysiological activity acutely when injected into the hippocampus (Sharma et al. 2012), that neurophysiological network properties of motor cortex are dysfunctional for days after treatment (Kleim et al. 2003), and that long-term potentiation can be seen even when protein synthesis is blocked (Abbas et al. 2009; Villers et al. 2012; cf. Nichol and Roche 2013), it may be time to move beyond the view that the formation of long-term memory and long-term potentiation are dependent on training-initiated protein synthesis and instead turn toward other considerations regarding molecular bases of memory formation (e.g., Routtenberg and Rekart 2005; Routtenberg 2013).

Recalling that first Winter Memory Meeting, the considerable attention to properties of memory consolidation overshadowed discussion of memory systems. I freely admit that the overshadowing of my memory of that time likely reflects my own primary interest in memory consolidation. But I think there is a simpler explanation of the little attention given to discussing memory systems in the early meetings compared to the prominence of the topic in more recent years. In the early 1970s, the data did not yet exist that would permit full and deep discussions about memory systems. Instead, people were perplexed about how to match the impact of patient H.M. in understanding memory systems in humans (Milner et al. 1968) with the rather weak evidence that hippocampal damage in rodents produced impairments of memory that approached those so evident in the patient H.M. (a few examples: Jarrard et al. 1964; Thomas 1971; Walker et al. 1972; Means et al. 1970). In rodents, the clear association of hippocampal functions with memory awaited the development of new behavioral tasks (Squire 1992; White and McDonald 2002), a research agenda to which Ray Kesner has contributed so importantly (Kesner 1985, 2009). Ray had and continues to have a remarkable ability to devise novel tasks that define specific cognitive attributes and then to relate these to the functions of specific brain areas.

In the mid-1990s, Ray published a triple dissociation of task *x* brain lesion (Kesner et al. 1993). This report, together with the triple-dissociation published at about the same time by McDonald and White (1993), led directly to our tests of pharmacological manipulations and neurochemical measures that might reflect and participate in memory processing of different attributes in different brain areas. Some of these findings are presented below.

Epinephrine and Glucose Modulation of Memory

Many treatments that enhance learning and memory appear to do so for a wide array of tasks (Gold 1995). These results seem somewhat surprising when contrasted with the specific effects of brain damage on learning and memory. There are two important differences between drug and lesion effects on memory that may explain the generality vs. specificity of the results across tasks. The simplest is that peripheral administration of drugs influences multiple memory systems at once and therefore can impact performance across many cognitive domains. Therefore, a drug that impairs or enhances learning and memory for one attribute may also regulate multiple neural systems to impair or enhance memory for another attribute.

It may be useful to apply this line of thought to a classic task used in studies of memory consolidation, the inhibitory avoidance task. The task is widely used in research on the molecular bases of memory. Inhibitory avoidance tasks are often referred to as hippocampus-dependent or -sensitive tasks and a myriad set of studies show that manipulations of the hippocampus influence learning and memory for inhibitory avoidance tasks. However, inhibitory avoidance tasks are not "pure" tasks, in the sense that the hippocampus is specifically associated with learning and memory in the task. First, the effects on inhibitory avoidance learning and memory of manipulations of the hippocampus are not always evident, but require that the rats have more time in the start and/or shock compartments task than is necessary to support later memory (Rudy et al. 2002; Huff et al. 2005; Rudy and Matus-Amat 2005; McHugh and Tonegawa 2007; Qi and Gold 2009). Second, lesions of many brain regions other than the hippocampus impair inhibitory avoidance learning and memory. In addition to the hippocampus, inhibitory avoidance learning and memory are also impaired by damage to or pharmacological down-regulation of the amygdala, striatum, insular cortex, ventral tegmental area, and olfactory bulbs (e.g., Gold et al. 1975; Salado-Castillo et al. 1996; Ghanbarian and Motamedi 2013; Archer et al. 1984; Bermudez-Rattoni et al. 1991) as well as other brain areas. Perhaps it is the broad integration of inhibitory avoidance memory across neural systems that underlies the task generality for treatments identified as memory enhancing and impairing agents when tested on additional tasks. Third, the behavioral measure of inhibitory avoidance memory itself can have multiple bases. The operational definition of memory in this task is of course the latency to cross from a start location to a shock location. The increased latencies may, for example, reflect association of the start compartment with fear, leading to freezing behavior and long latencies to enter the shock compartment, or associating the shock compartment with the aversive stimulus leading to avoiding the shock compartment more actively.

Looking at the behaviors of many rats during an inhibitory avoidance memory trial is illustrative here. Some rats turn to the corner farthest from the shock compartment and remain there until the end of the trial. Other rats actively explore the safe compartment while avoiding the shock compartment. Others stop at the end of the safe compartment nearest to the shock compartment, staring into that space. All of these behaviors result in high latencies, i.e., operationally good avoidance scores, even as it seems likely that different rats are achieving high latencies through different means—or possibly strategies—to solve the task. The multiple memory systems involved in inhibitory avoidance learning and the multiple behavioral responses that converge on the same operational definition of memory—high latencies—may explain in part why the effects of many treatments shown to enhance and impair of memory generalize readily to many other tasks and species (Gold 1992).

The findings that pharmacological treatments could enhance memory formation (McGaugh 1966) led to the idea that endogenous factors such as neuroendocrine responses to experience might modulate memory processing by regulating mechanisms of brain plasticity (cf.: Gold and McGaugh 1975; McGaugh 1983; McGaugh and Roozendaal 2002; Korol and Gold 2007; Gold and Korol 2010). One hormone that is particularly effective at enhancing memory is epinephrine. Epinephrine enhances memory assessed not only for inhibitory avoidance training in both rats and mice (Gold and van Buskirk 1975, 1976; Gold et al. 1977), but also for memory of other tasks, such as appetitive (Sternberg et al. 1985), object recognition (Dornelles et al. 2007), and spatial working memory (Talley et al. 2000) tasks, as well as long-term potentiation in rats (Korol and Gold 2007) and memory in humans (Cahill and Akire 2003). Epinephrine appears to act in large part through adrenergic receptors on hepatocytes, where the hormone action is to break down glycogen to glucose, with subsequent increases in blood glucose levels (Gold and Korol 2010). Consistent with this putative mechanism, peripheral injections of epinephrine do not enhance memory in food-restricted rats, in which the hormone is unable to increase blood glucose levels in the absence of liver glycogen availability (Talley et al. 2000).

Like epinephrine, glucose enhances memory for a wide range of tasks in laboratory rodents (Gold and Korol 2010, 2012) and also enhances memory in many human subject populations, from healthy college students to healthy elderly, and in individuals with Alzheimer's disease, Down syndrome, and schizophrenia (Hall et al. 1989; Manning et al. 1990, 1992, 1993, 1997, 1998; Parsons and Gold 1992; Korol 2002; Gold 2001; Stone et al. 2003; Newcomer et al. 1999; Stone and Seidman 2008). Of particular interest, glucose effects on memory are evident not only with systemic injections but also with direct central injections, where the actions are generally consistent with canonical memory functions of the target brain areas.

For example, intrahippocampal injections of glucose enhance working memory assessed with spontaneous alternation testing (Ragozzino et al. 1998; Stefani and Gold 1998) and also memory after training in inhibitory avoidance tasks (Krebs and Parent 2005; Morris and Gold 2013). Intra-amygdala infusions of glucose enhance memory for conditioned place preference training (Schroeder and Packard 2003), a task impaired by amygdala lesions (McDonald and White 1993; Naeem and White 2011). Together with evidence that extracellular glucose levels are depleted in some brain areas during training and that exogenous systemic glucose repletes those levels (McNay et al. 2000, 2001; McNay and Gold 2001), the findings provide strong evidence that glucose availability is a key regulator of learning and memory processing.

When administered near the time rats are engaged in learning and memory tasks, glucose augments the release of acetylcholine in the hippocampus (Ragozzino et al. 1996, 1998). Acetylcholine release across neural systems, in turn, reveals dynamic shifts in the balance of the contributions of multiple neural systems activated by learning (Pych et al. 2005a, b). In addition, recent findings suggest that an important contributor to glucose effects on memory is as a substrate for lactate production, with subsequent increases in lactate provision from astrocytes to neurons (Newman et al. 2011). Mediation of glucose effects on memory by acetylcholine release and by increasing energy availability layer well on multiple memory system approaches to brain processing and will be considered in the next two major sections of this review.

ACh Modulation of Multiple Memory Systems

We have conducted many studies using spontaneous alternation tasks to assess spatial working memory. Although multiple brain areas participate in memory for this task, we have focused largely on the hippocampus as a region in which to examine glucose effects on acetylcholine release. As shown in Fig. 12.1 (left panel), systemic injections of glucose enhanced alternation scores on this task in an inverted-U dose–response manner (Ragozzino et al. 1996). Figure 12.1 (right panel) also shows the effects of glucose on acetylcholine release in the hippocampus of rats while they were engaged in alternation testing. In concert with the enhancement of memory, the findings revealed that systemic injections of glucose also augmented the behavior-induced increases in acetylcholine release in the hippocampus, following the inverted-U dose–response curve seen with memory scores. Similarly, microinjections of glucose directly into the hippocampus also enhanced memory and augmented the release of acetylcholine in the hippocampus during spontaneous alternation testing (Ragozzino et al. 1998) (Fig. 12.2). Of interest, increases in acetylcholine release were not evident when the glucose injections were administered to rats at rest in holding cages. Therefore, glucose-initiated increases in acetylcholine release appear to require activation of the neural system.

To examine the role of acetylcholine release in multiple memory systems, further experiments have looked at acetylcholine release using tasks other than spontaneous alternation. These tasks have more specific relationships to particular memory

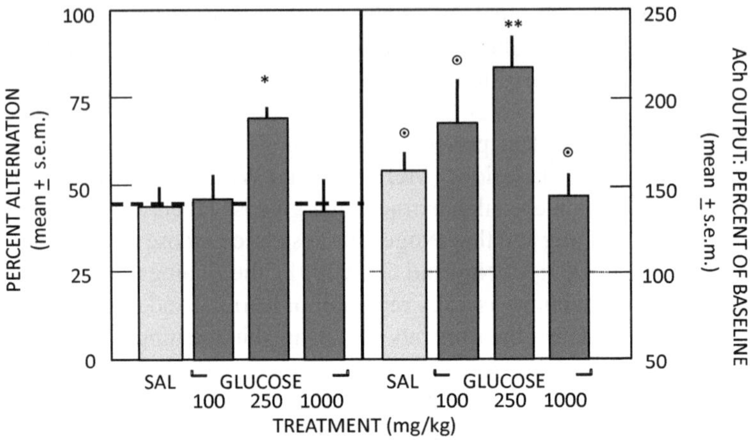

Fig. 12.1 Enhancement of spontaneous alternation memory scores and of acetylcholine (*ACh*) release in the hippocampus by glucose (I.P.). Chance alternation scores and baseline glucose levels are indicated by the horizontal lines. ★P < 0.05 vs. SAL (saline). ⊙P < 0.05 vs. baseline. ★P < 0.02 vs., saline. (Adapted from Ragozzino et al. 1996)

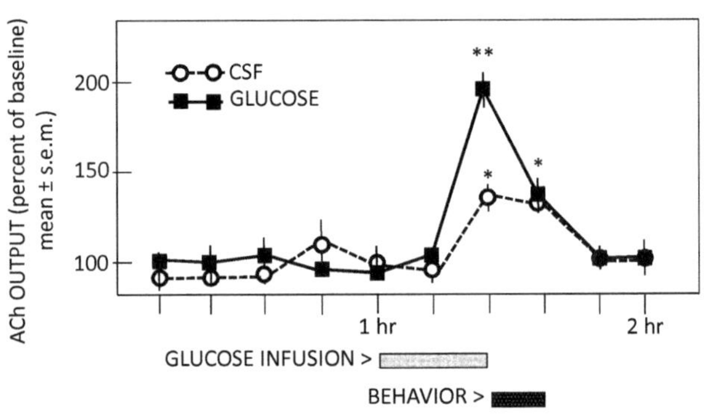

Fig. 12.2 Enhancement of acetylcholine (*ACh*) output in the hippocampus by contralateral infusions of glucose (24 min), compared to artificial cerebrospinal fluid, directly into the hippocampus during spontaneous alternation testing. The times of glucose infusion and behavioral testing on the alternation maze are indicated by bars below the figure. ACh was measured by in vivo microdialysis. ★P < 0.01 vs. baseline. ★★P < 0.05 vs. artificial cerebrospinal fluid. (Adapted from Ragozzino et al. 1998)

systems. In one set of experiments (Pych et al. 2006), rats were trained on either a place or response version of a 4-arm plus-shaped maze (Fig. 12.3). Throughout training, microdialysis samples were collected from the hippocampus and striatum in different sets of rats. Acetylcholine release increased in both brain areas during training. The pattern of increase in release in the hippocampus was similar whether rats were trained on the place or response task, suggesting that the hippocampus and perhaps spatial cues were used in both versions of the maze (Fig. 12.4, top). As in the hippocampus, acetylcholine release in the striatum also increased when rats were

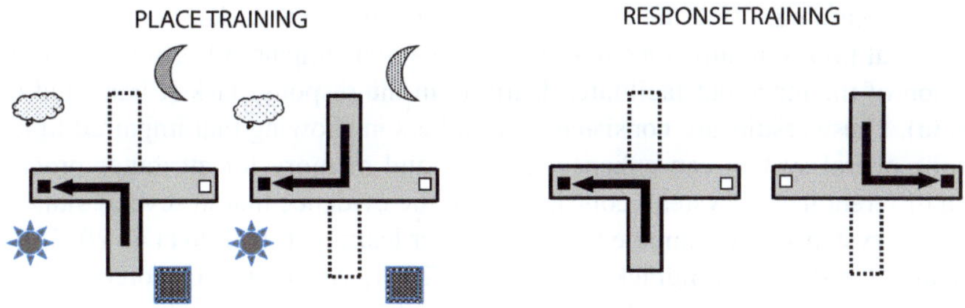

Fig. 12.3 Illustration of place and response versions of the T-shaped maze. Start arms in both versions were from either the north or south and goals were in either the east or west arms. The arm directly across from the start arm was blocked during all training trials. In the place version of the maze, rats were trained to find food in a particular location in the room (between the cloud and star in this figure). In the response version, rats were trained to find food by making the same egocentric turn on each trial (to the *left* in this figure)

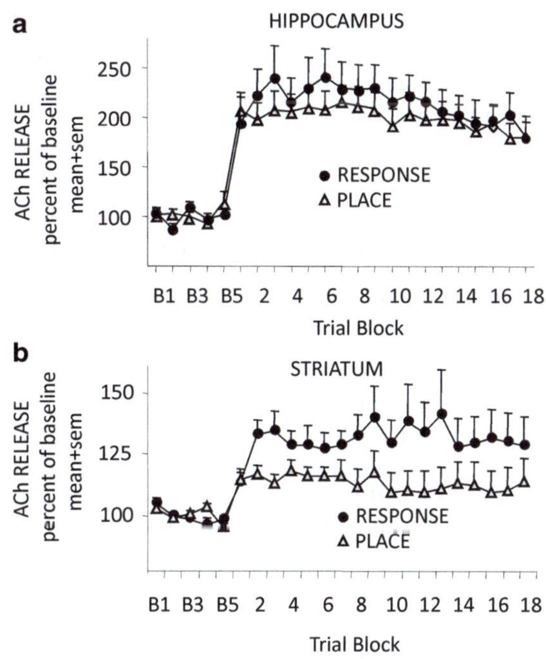

Fig. 12.4 Acetylcholine (*ACh*) output in the hippocampus (**a**) and striatum (**b**) in microdialysis samples obtained simultaneously from these brain areas while rats were trained on either the place or response version of the maze. Acetylcholine release increased in both brain areas during training on both tasks. However, output of acetylcholine in the striatum was greater in those rats trained on the response version of the task than those trained on the place version. (Adapted from Pych et al. 2005b)

trained on either the place or response task (Fig. 12.4, bottom). However, response training resulted in significantly greater magnitude of release of acetylcholine in the striatum than was seen during place training. Thus, there was a clear task-related difference in acetylcholine release in the striatum.

These findings suggest that the balance between the hippocampus and striatum contributions to learning is mainly driven by increases in acetylcholine release in the striatum in these tasks. Perhaps the hippocampus participates in both place and response learning, using the cues to attempt to solve both tasks. In prior work, we found that lidocaine infusions into the hippocampus impaired learning in the place version of the maze but facilitated learning in the response task (Chang and Gold 2003a). These results are consistent with others in showing that impaired function of one neural system can enhance learning and memory for attributes processed by a different neural system, contributing to the evidence that in several situations, memory systems may compete for control over learning (cf. Gold et al. 2013; Packard and Cahill 2001; Poldrack and Packard 2003; White 2008). Moreover, inactivation of the striatum with lidocaine impaired learning more when training was conducted in a cue-poor room than when conducted in a cue-rich room (Chang and Gold 2004), suggesting that the hippocampus might contribute to learning a "response" rule that is actually based on conditional discrimination, e.g., if starting from the north, turn left. With this thinking in mind, we examined acetylcholine release in the hippocampus when rats were trained on the response task in a room with rich extramaze cues and with reduced extramaze cues (Chang and Gold 2004). Release of acetylcholine in the hippocampus increased about equally upon the start of training in either the cue-rich or cue-poor condition. However, hippocampal acetylcholine release decreased across trials in the cue-poor condition, suggesting that the balance of control over learning was now in the purview of other brain areas, including the striatum.

Contributions of the dynamics of acetylcholine release to learning have also been seen in other settings. For example, when trained on a rewarded spontaneous alternation task, rats gradually shift from typical alternation scores to a clear response strategy (always turn right [or left]) (Pych et al. 2005a). A similar shift in strategy to solve a maze is also evident in a dual-solution T-maze (Tolman 1948; Restle 1957). In this case, rats often shift from a place solution early in training to a response solution later in training. Pharmacological evidence suggests that the change in expressed solution is a function of hippocampal control over the learned response early in training and striatal control over the learned response later in training (Packard and McGaugh 1996; Packard 1999).

In our experiments using the dual-solution T-maze (McIntyre et al. 2003a; Chang and Gold 2003b), rats were trained within a single session while acetylcholine release was measured simultaneously in both the dorsal striatum and the hippocampus. Rats were trained to a criterion of 9/10 correct, at which time a probe test was given to determine whether the rat used a place or response strategy to solve the maze. The results indicated that individual differences in the ratio of acetylcholine release in the hippocampus and striatum predicted which solution a rat would use on the probe trial (McIntyre et al. 2003a) (Fig. 12.5). Those rats with high hippocampus : striatum ratios of acetylcholine release during training were the rats likely to exhibit a place solution in the T-maze. Interestingly, although acetylcholine release increased in both brain areas regardless of the solution expressed on the probe trial, the difference in ratios was evident even prior to the start of training, i.e., before the rats had received any experience on a maze. Thus, the rats' baseline acetylcholine levels in hippocampus and striatum predicted the later solution to training on the T-maze. The factors controlling this difference are unknown and may reflect either

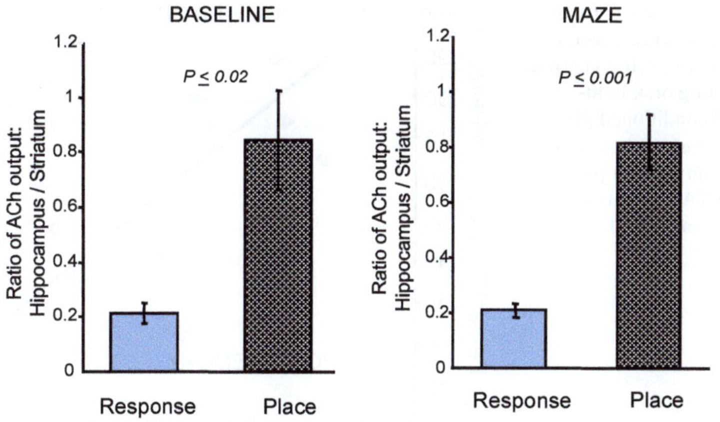

Fig. 12.5 Ratios of acetylcholine (*ACh*) output in hippocampus/striatum in rats trained on a T-maze task that can be solved effectively using either response or place strategies. The strategy used by each rat was determined on a probe trial administered after the rat had reached a criterion of 9/10 correct. The ratios of acetylcholine output in hippocampus/striatum in samples collected before (baseline) training paralleled individual differences in the strategy that would be employed at the end of training. These ratios were maintained during training (maze). Thus, the individual neurochemical balance across brain regions before rats were exposed to this (or any) training, as well as the balance during training, were associated with differences in the learning strategy used to solve the maze (Adapted from McIntyre et al. 2003a)

differences in acetylcholine release across long times or perhaps day-to-day differences based on undefined variables. Of interest, acetylcholine release interacts with estrogen treatment in a manner associated with estrogen enhancement of place learning in female rats (Marriott and Korol 2003), suggesting that acetylcholine levels, and the difference in these levels across brain areas, may fluctuate with hormonal and other state changes that alter the balance of the contributions of different neural systems to learning.

Release of acetylcholine measured in different neural systems during learning also provides opportunities to see examples of both competition and cooperation across those systems. On the basis of lesion work, conditioned cue preference tasks have been identified as amygdala sensitive, i.e., lesions of the amygdala interfere with learning in this task (e.g., McDonald and White 1993; Naeem and White 2011). In contrast, hippocampal lesions *enhance* learning in this task, an effect interpreted to be a release from hippocampal "attempts" to learn about spatial features of the task in a manner that interferes with acquisition of a reward-heavy conditioned place preference (White and McDonald 1993; McDonald and White 1995). We examined hippocampal acetylcholine release while rats learned this amygdala-dependent task. The results, shown in Fig. 12.6, indicate that high levels of acetylcholine release in the hippocampus are correlated with poor acquisition of the conditioned cue preference (McIntyre et al. 2002). Thus, acetylcholine release in the hippocampus is a negative predictor of learning in the amygdala-based task.

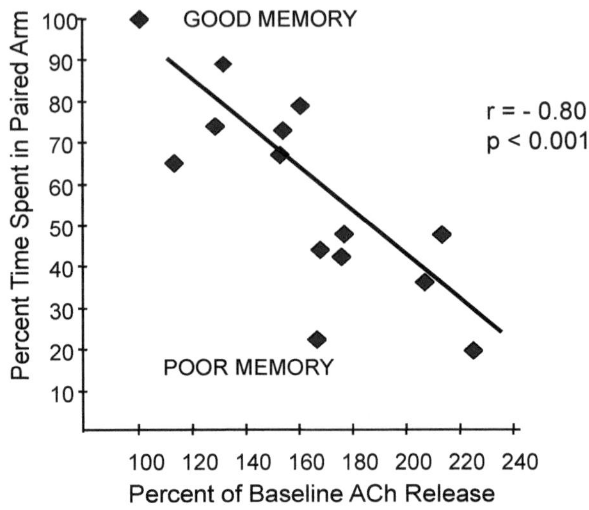

Fig. 12.6 Negative correlation between percent increase in acetylcholine (*ACh*) output during testing on a food-motivated conditioned place preference task and percent time spent in the arm paired with food. (Adapted from McIntyre et al. 2002)

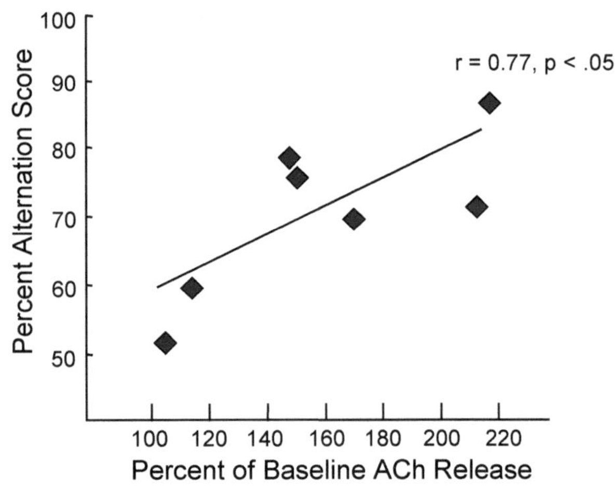

Fig. 12.7 Positive correlation between percent increase in acetylcholine (*ACh*) output during testing on a spontaneous alternation task and memory scores on that task. (Adapted from McIntyre et al. 2003b)

Cooperation, as well as competition, across memory systems is readily evident using correlational studies of the same type. Figure 12.7 shows the results of an experiment that measured release of acetylcholine in the amygdala while rats participated in a non-amygdala, and in part hippocampal, spontaneous alternation task (McIntyre et al. 2003b). In this case, the relationship is a positive one, indicating that activation of the amygdala, defined here by acetylcholine release, accompanies increased working memory scores in this task. These findings, together with others (e.g., Gaskin and White 2006), suggest that multiple memory systems can interact collaboratively with each contributing to learning an experience, or can interact competitively such that inactivation or lower activation of one brain area can enhance learning of a task associated with another brain area.

The findings described above indicate that glucose enhances learning and memory processing, and also augments release of acetylcholine during training. Still unknown is the extent to which the glucose and acetylcholine findings will converge. Together with the relationship between regional differences in acetylcholine levels and multiple memory systems, there may be important functions for glucose at a local level to regulate memory. Recent progress has begun to address both, mechanisms by which glucose enhances memory and the importance of glucose across multiple memory systems.

Glucose Control of Energy Needs for Memory Processing

The mechanisms by which glucose can modulate memory processing are not fully identified. However, it now appears that glucose levels are not always adequate to support optimal memory functions, with evidence that these levels are depleted during learning and memory testing. Thus, cognitive functions may therefore place a metabolic demand on neurons, one that glucose itself cannot meet. However, there is an additional energy substrate available to neurons. Astrocytes produce lactate from glucose and store glycogen, which can be quickly metabolized to lactate, an energy substrate delivered to neurons when needed as an energy reserve. This general view forms the basis of the research reviewed next.

While rats are tested for memory on a spontaneous alternation task, extracellular glucose levels decrease in the hippocampus (McNay et al. 2000, 2001; McNay and Gold 2001; Newman et al. 2011); systemic injections of glucose blunt this depletion while enhancing memory. In young adult rats, extracellular glucose levels recover after the initial decrease, even as the rats continue to be engaged in the behavioral task. The recovery coincides with an endogenous increase in blood glucose levels, apparently triggered by the arousal of being on a novel maze (McNay et al. 2001). The decrease in extracellular glucose levels during alternation testing is greatly exaggerated in senescent rats as compared to young adult rats; the depletion of extracellular glucose in the hippocampus of aged rats accompanies age-related memory impairments (McNay and Gold 2001). In these aged rats, glucose administration enhances memory assessed in a spontaneous alternation task, restoring memory scores to those of young rats. Glucose, given either systemically or directly into the hippocampus, also ameliorates age-related memory impairments for inhibitory avoidance training (Morris and Gold 2013). Importantly, the increase in blood glucose levels in response to arousal from foot shock, immersion in water, or maze testing is evident in young rats but not in aged rats, which have severely diminished ability to release glucose in response to arousal or to epinephrine injections (Mabry et al. 1995; Morris et al. 2010). Thus, systemic and central injections of glucose enhance memory in both young and old rats while reversing the depletion of glucose (Morris et al. 2013), providing possible neuroendocrine bases for age-related memory impairments in rats and opening future investigations that apply these findings to multiple neural systems responsible for learning and remembering different task attributes.

The findings that extracellular glucose decreases during memory processing and that glucose administration enhances memory processing suggest that glucose may, in a sense, be a limiting substrate for optimal mnemonic functions. This possibility led to recent studies that examine glucose as an energy substrate that can regulate learning and memory processes. The results have identified a potentially important role for astrocytes in mediating the regulation of learning and memory processing by glucose. Astrocytes release lactate into extracellular space for provision of an energy substrate to supplement glucose when its levels flag during training-related consumption. The astrocytes contain glycogen stores that can be metabolized quickly to lactate, with release of lactate making it available for neuronal uptake. Glycogenolysis and release of lactate likely occurs in response to receptors on astrocytes that are activated by several neurotransmitters. A schematic of our working hypothesis is shown in Fig. 12.8.

There are several results supporting the importance of astrocyte-derived lactate as an important regulator of learning and memory and as an important mediator of glucose effects on learning and memory. One key piece of information is

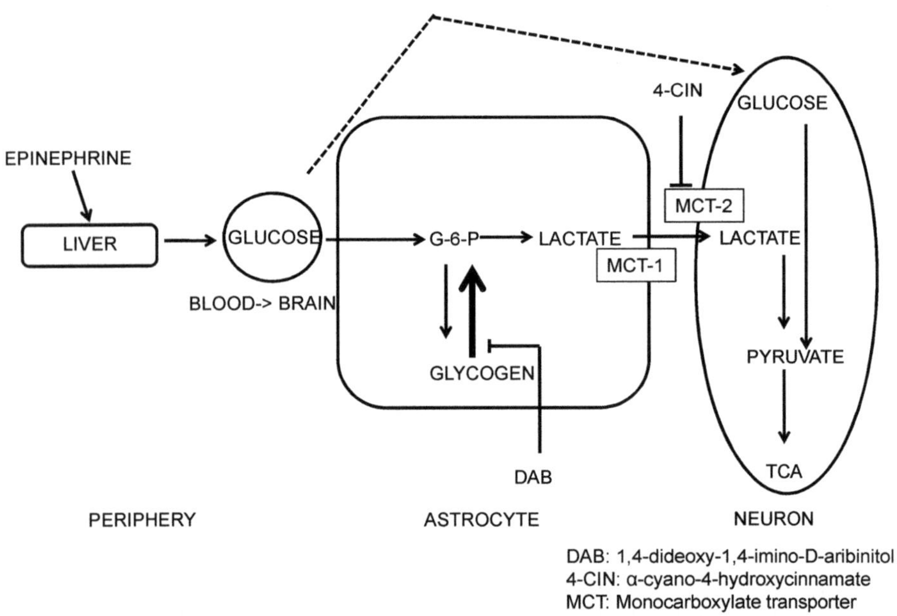

Fig. 12.8 Model of a role of astrocytes and glycogenolysis in the production of lactate to support neuronal functions during memory processing. In this model, training-related arousal results in epinephrine release from the adrenal medulla. Epinephrine activates hepatic adrenergic receptors to initiate glycogen breakdown in the liver with subsequent increases in blood glucose levels. Glucose from blood enters neurons and astrocytes. Neuronal uptake of glucose appears to be important for baseline neuronal metabolism but is inadequate during times of demand, as shown by decreases in hippocampus glucose levels in extracellular fluid during memory testing. Astrocytic uptake of glucose produces glycogen as an energy reserve that can be quickly metabolized to lactate, delivered to neurons during times of high demand for energy substrates. (Adapted from Newman et al. 2011)

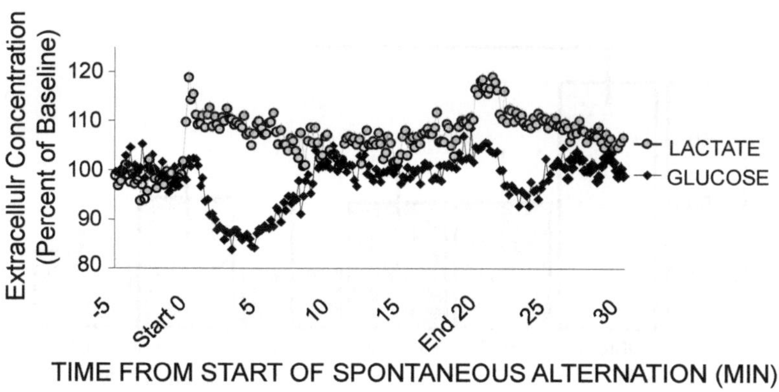

Fig. 12.9 Extracellular lactate and glucose levels in the hippocampus, expressed as percent of baseline, measured before, during, and after spontaneous alternation testing. Measurements were made with lactate- and glucose-specific biosensors using 1-s sampling times. Lactate concentrations significantly increased at the beginning of behavioral testing generally mirroring a decrease in glucose concentrations. The increase in glucose levels at 5–10 into the testing session corresponds to an increase in blood glucose levels. After the rat was removed from the maze there was a significant increase in lactate compared to baseline levels, most likely due to handling. (Adapted from Newman et al. 2011)

that extracellular lactate levels in the hippocampus increase around the time of training (Suzuki et al. 2011; McNay and Sherwin 2004; Newman et al. 2011). During spontaneous alternation testing, increases in lactate levels largely mirror the decreases seen in glucose levels (Newman et al. 2011; Fig. 12.9). Intrahippocampal infusions of lactate enhance memory in the spontaneous alternation task (Fig. 12.10). In addition, pharmacological (Newman et al. 2011) (Fig. 12.11) and genetic (Suzuki et al. 2011) blockade of the neuronal monocarboxylate transporter 2 (MCT2) responsible for lactate uptake interfere with enhancement of memory by either glucose or lactate. The finding that glucose administration, which can enhance memory in the absence of blockade of MCT2, is ineffective when lactate

Fig. 12.10 Memory enhancement produced by intrahippocampal injections of lactate (50 nmol), administered 5 min before testing on a 4-arm delayed spontaneous alternation task. Higher and lower doses of lactate did not significantly improve alternation scores. (Adapted from Newman et al. 2011)

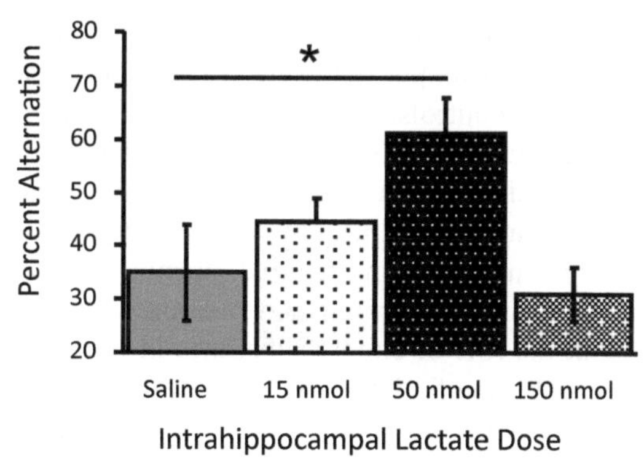

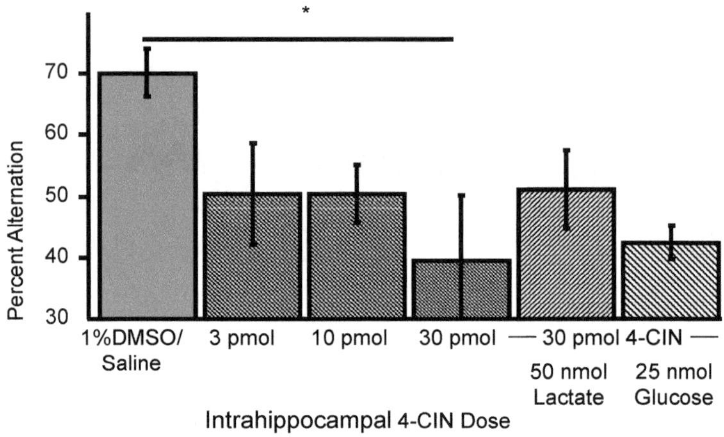

Fig. 12.11 Impairment of memory by intrahippocampal injections of α-cyano-4- hydroxycinnamate (4-CIN), a drug that blocks the neuronal monocarboxylate transporter (MCT2) responsible for uptake of lactate into neurons. The impairment was not reversed by either lactate or glucose. (Adapted from Newman et al. 2011)

uptake is blocked suggests that glucose must pass through astrocytic conversion to lactate before it enhances memory. This finding is consistent with evidence that both glucose and lactate are taken up by neurons in barrel cortex under conditions of rest, but lactate and not glucose uptake by neurons increases upon activation (Chuquet et al. 2010).

These findings indicate that lactate is a major contributor to modulation of memory by glucose, and perhaps by neurotransmitters that can act on astrocytes to produce lactate. We recently have begun to examine the role of astrocytic lactate in different brain areas for different tasks, using the place vs. response mazes and hippocampus vs. striatum brain areas as the initial subjects of interest (Newman et al. in preparation). Some early results (Fig. 12.12) illustrate the changes in extracellular levels of lactate in the hippocampus, while rats are trained on either a place or response version of the 4-arm maze. The controls received food reward on a schedule similar to that obtained by rats during training. These initial findings show that training on the place version of the maze resulted in increases in lactate in the hippocampus that were greater than those seen in rats trained on the response task or in controls. Early indications are that similar increases in hippocampal and striatal lactate levels are seen in rats trained for water reward, a condition in which controls do not have large increases in brain lactate levels. These increases in extracellular lactate are likely to contribute to the metabolic needs of neurons processing the information being learned, opening new investigations into the role of energy metabolism in learning and memory. In addition to providing new information about the role of energy substrates in learning and memory processing for appetitive tasks, the findings also suggest that these methods and results will offer a new perspective with which to examine participation of, and interactions of, multiple memory systems in conditions assessing different attributes of training experiences

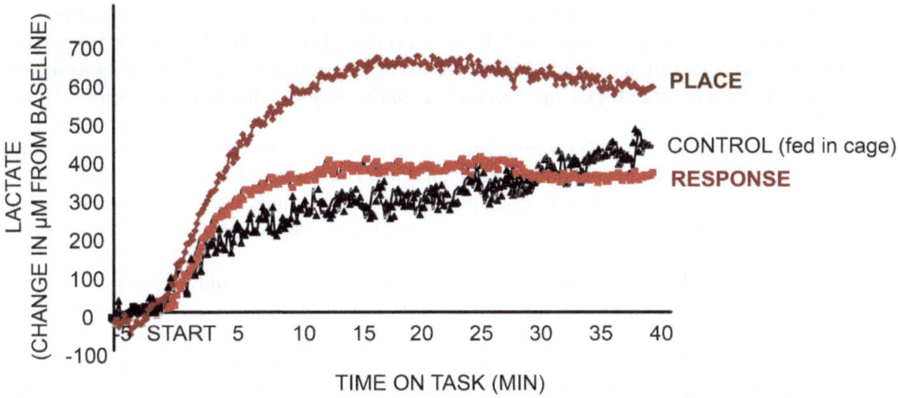

Fig. 12.12 Changes in extracellular lactate concentrations in the hippocampus sampled with biosensors while rats were trained on the place or response versions of mazes as in Fig. 12.3, and in rats that received food reward in a holding cage on a schedule like that of the trained groups. Extracellular lactate levels increased significantly more in the place-trained rats than in either response-trained or control-fed rats. (Adapted from Newman et al. 2011)

as affected by many short- and long-term modulators of learning and memory (Gold et al. 2013; Korol et al. 2013).

Conclusions

The identification of neural systems that process different attributes of experiences to be learned and remembered is a hallmark of the work that Ray Kesner has provided, including the triple-dissociation of task *x* brain lesion study (Kesner et al. 1993) that pushed his Attribute Model of Memory to new limits. The findings described here owe their genesis to this model and to other formulations of multiple memory systems. The integration of neurochemical measures and pharmacological manipulations to explore the interactions of these systems is still only available for a small set of brain regions thus far, and does not include tests of additional brain areas and subdivisions of those already tested that are implicated in mediating different attributes of learning and memory, attributes so clearly delineated in past research (Kesner 2009).

References

Abbas, A. K., Dozmorov, M., Li, R., Huang, F. S., Hellberg, F., Danielson, J., Tian, Y., Ekstrom, J., Sandberg, M., & Wigstrom, H. (2009). Persistent LTP without triggered protein synthesis. *Neuroscience Research, 63,* 59–65.

Archer, T., Söderberg, U., Ross, S. B., & Jonsson, G. (1984). Role of olfactory bulbectomy and DSP4 treatment in avoidance learning in the rat. *Behavioral Neuroscience, 98,* 496–505.

Bermudez-Rattoni, F., Introini-Collison, I. B., & McGaugh, J. L. (1991). Reversible inactivation of the insular cortex by tetrodotoxin produces retrograde and anterograde amnesia for inhibitory avoidance and spatial learning. *Proceedings of the National Academy of Science USA, 88,* 5379–5382.

Cahill, L., & Alkire, M. T. (2003). Epinephrine enhancement of human memory consolidation: Interaction with arousal at encoding. *Neurobiology of Learning and Memory, 79,* 194–198.

Chang, Q., & Gold, P. E. (2003a). Intra-hippocampal lidocaine injections impair acquisition of a place task and facilitate acquisition of a response task in rats. *Behavioural Brain Research, 144,* 19–24.

Chang, Q., & Gold, P. E. (2003b). Switching memory systems during learning: Changes in patterns of brain acetylcholine release in the hippocampus and striatum in rats. *Journal of Neuroscience, 23,* 3001–3005.

Chang, Q., & Gold, P. E. (2004). Inactivation of dorsolateral striatum impairs acquisition of response learning in cue-deficient but not cue-available conditions. *Behavioral Neuroscience, 118,* 383–388.

Cherkin, A. (1969). Kinetics of memory consolidation: Role of amnesic treatment parameters. *Proceedings of the National Academy of Science USA, 63,* 1094–1101.

Chorover, S. L., & Schiller, P. H. (1965). Short-term retrograde amnesia in rats. *Journal of Comparative and Physiological Psychology, 59,* 73–78.

Chuquet, J., Quilichini, P., Nimchinsky, E. A., & Buzsáki, G. (2010). Predominant enhancement of glucose uptake in astrocytes versus neurons during activation of the somatosensory cortex. *Journal of Neuroscience, 30,* 15298–15303.

Dornelles, A., de Lima, M. N. M., Grazziotin. M., Presti-Torres, J., Garcia, V. A., Scalco, F. S., Roesler, R., & Schröder, N. (2007). Adrenergic enhancement of consolidation of object recognition memory. *Neurobiology of Learning and Memory, 88,* 137–142.

Gaskin, S., & White, N. M. (2006). Cooperation and competition between the dorsal hippocampus and lateral amygdala in spatial discrimination learning. *Hippocampus, 16,* 577–585.

Ghanbarian, E., & Motamedi, F. (2013). Ventral tegmental area inactivation suppresses the expression of CA1 long term potentiation in anesthetized rat. *PloS One, 8,* e58844.

Gold, P. E. (1992). Modulation of memory processing: Enhancement of memory in rodents and humans. In L. R. Squire & N. Butters (Eds.), Neuropsychology of memory, (2nd Ed., pp. 402–414). New York: Guilford.

Gold, P. E. (1995). Modulation of emotional and non-emotional memories: Same pharmacological systems, different neuroanatomical systems. In J. L. McGaugh, N. M. Weinberger, & G. S. Lynch (Eds.), Brain and memory: Modulation and mediation of neural plasticity, (pp. 41–74). New York: Oxford University Press.

Gold, P. E. (2001). Drug enhancement of memory in aged rodents and humans. In: Carroll, M. E. & Overmier, J. B. (Eds.), Animal research and human health: Advancing human welfare through behavioral science, (pp. 293–304). Washington DC: American Psychological Association.

Gold, P. E. (2008). Protein synthesis inhibition: Memory formation vs. amnesia. *Neurobiology of Learning and Memory, 89,* 201–211.

Gold, P. E., & Korol, D.L. (2010). Hormones and memory. In G. Koob, M. Le Moal, & R. F. Thompson (Eds.), Encyclopedia of behavioral neuroscience (Vol. 2, pp. 57–64). Oxford: Academic Press.

Gold, P. E., & Korol, D. L. (2012). Making memories matter. Special issue: The impact of emotion on cognition—dissociating between enhancing and impairing effects. F. Dolcos, L. Wang, and M. Mather, hosts. *Frontiers in Integrative Neuroscience, 6,* 116. doi:10.3389/fnint.2012.00116.

Gold, P. E., & McGaugh, J. L. (1975). A single trace, two process view of memory storage processes. In: D. Deutsch & J. A. Deutsch (Eds.), Short term memory (pp. 355–390). New York: Academic Press.

Gold, P. E., & Zornetzer, S. F. (1983). The mnemon and its juices: Neuromodulation of memory processes. *Behavioral and Neural Biology, 38,* 151–189,

Gold, P. E., Macri, J., & McGaugh, J. L. (1973). Retrograde amnesia gradients: Effects of direct cortical stimulation. *Science, 197,* 1343–1345.

Gold, P. E., Hankins, L., Edwards, R. M., Chester, J., & McGaugh, J. L. (1975). Memory interference and facilitation with posttrial amygdala stimulation: Effect on memory varies with footshock level. *Brain Research, 86,* 509–513.

Gold, P. E., Newman, L. A., Scavuzzo, C. J., & Korol, D. L. (2013). Modulation of multiple memory systems: From neurotransmitters to metabolic substrates. *Hippocampus, 23,* 1053–1065.

Gold, P. E., & van Buskirk, R. B. (1975). Facilitation of time dependent memory processes with posttrial epinephrine injections. *Behavioral Biology, 13,* 145–153.

Gold, P. E., & van Buskirk, R. B. (1976). Effects of posttrial hormone injections on memory processes. *Hormones and Behavior, 7,* 509–517.

Gold, P. E., van Buskirk, R. B., & Haycock, J. W. (1977). Effects of post training epinephrine injections on retention of avoidance training in mice. *Behavioral Biology, 20,* 197–204.

Hall, J. L., Gonder-Frederick, L. A., Chewning, W. W., Silveira, J., & Gold, P. E. (1989). Glucose enhancement of performance on memory tests in young and aged humans. *Neuropsychologia, 27,* 1129–1138.

Huff, N. C., Wright-Hardesty, K. J., Higgins, E. A., Matus-Amat, P., & Rudy, J. W. (2005). Context pre-exposure obscures amygdala modulation of contextual-fear conditioning. *Learning and Memory, 12,* 456–460.

Jarrard, L. E., Isaacson, R. L., & Wickelgren, W. O. (1964). Effects of hippocampal ablation and intertrial interval on runway acquisition and extinction. *Journal of Comparative and Physiological Psychology, 57,* 442–444.

Kesner, R. P. (1985). Correspondence between humans and animals in coding of temporal attributes: Role of hippocampus and prefrontal cortex. *Annals of the New York Academy of Science, 44,* 122–136.

Kesner, R. P. (2009). Tapestry of memory. *Behavioral Neuroscience, 123,* 1–13.

Kesner, R. P., Bolland, B. L., & Dakis, M. (1993). Memory for spatial locations, motor responses, and objects: Triple dissociation among the hippocampus, caudate nucleus, and extrastriate visual cortex. *Experimental Brain Research, 93,* 462–470.

Kleim, J. A., Bruneau, R., Calder, K., Pocock, D., VandenBerg, P. M., MacDonald, E., Monfils, M. H., Sutherland, R. J., & Nader, K. (2003). Functional organization of adult motor cortex is dependent upon continued protein synthesis. *Neuron, 40,* 167–176.

Korol, D. L. (2002). Enhancing cognitive function across the life span. *Annals of the New York Academy of Science, 959,* 167–179.

Korol, D. L., & Gold, P. E. (2007). Modulation of learning and memory by adrenal and ovarian hormones. In R. P. Kesner & J. L. Martinez (Eds.), Neurobiology of learning and memory (pp. 243–268). New York: Elsevier Science.

Korol, D. L., Gold, P. E., & Scavuzzo, C. J. (2013). Use it and boost it with physical and mental activity. *Hippocampus, 23,* 1125–1135

Krebs, D. L., & Parent, M. B. (2005). The enhancing effects of hippocampal infusions of glucose are not restricted to spatial working memory. *Neurobiology of Learning and Memory, 83,* 168–172.

Mabry, T. R., Gold, P. E., & McCarty, R. (1995). Age-related changes in plasma catecholamine and glucose responses of F-344 rats to footshock as in inhibitory avoidance training. *Neurobiology of Learning and Memory, 64,* 146–155.

Manning, C. A., Hall, J. L., & Gold, P. E. (1990). Glucose effects on memory and other neuropsychological tests in elderly humans. *Psychological Sciences, 1,* 307–311.

Manning, C. A., Parsons, M. W., & Gold, P. E. (1992). Anterograde and retrograde enhancement of 24-hour memory by glucose in elderly humans. *Behavioral and Neural Biology, 58,* 125–130.

Manning, C. A., Ragozzino, M., & Gold, P. E. (1993). Glucose enhancement of memory in patients with Alzheimer's disease. *Neurobiology of Aging, 14,* 523–528.

Manning, C. A., Parsons, M. W., Cotter, E. M., & Gold, P. E. (1997). Glucose effects on declarative and nondeclarative memory in healthy elderly and young adults. *Psychobiology, 25,* 103–108.

Manning, C. A., Honn, V. S., Stone, W. S., Jane, J. S., & Gold, P. E. (1998). Glucose effects on cognition in adults with Down's Syndrome. *Neuropsychology, 12,* 479–484.

Marriott, L. K., & Korol, D. L. (2003). Short-term estrogen treatment in ovariectomized rats augments hippocampal acetylcholine release during place learning. *Neurobiology of Learning and Memory, 80,* 315–322.

McDonald, R. J., & White, N. M. (1993). A triple dissociation of memory systems: Hippocampus, amygdala, and dorsal striatum. *Behavioral Neuroscience, 107,* 3–22.

McDonald, R. J., & White, N. M. (1995). Information acquired by the hippocampus interferes with acquisition of the amygdala-based conditioned-cue preference in the rat. *Hippocampus, 5,* 189–197.

McGaugh, J. L. (1966). Time-dependent processes in memory storage. *Science, 153,* 1351–1358.

McGaugh, J. L. (1983). Hormonal influences on memory. *Annual Review Psychology, 34,* 297–323.

McGaugh, J. L., & Roozendaal, B. (2002). Role of adrenal stress hormones in forming lasting memories in the brain. *Current Opinion Neurobiology, 12,* 205–210.

McHugh, T. J., & Tonegawa, S. (2007). Spatial exploration is required for the formation of contextual fear memory. *Behavioral Neuroscience, 121,* 335–339.

McIntyre, C. K., Marriott, L. K., & Gold, P. E. (2003a). Patterns of brain acetylcholine release predict individual differences in preferred learning strategies in rats. *Neurobiology of Learning and Memory, 79,* 177–183

McIntyre, C. K., Marriott, L. K., & Gold, P. E. (2003b). Cooperation between memory systems: Acetylcholine release in the amygdala correlates positively with good performance on a hippocampus-dependent task. *Behavioral Neuroscience, 117,* 320–326.

McIntyre, C. K., Pal, S. N., Marriott, L. K., & Gold, P. E. (2002). Competition between memory systems: Acetylcholine release in the hippocampus correlates negatively with good performance on an amygdala-dependent task. *Journal of Neuroscience, 22,* 1171–1176.

McNay, E. C., & Gold, P. E. (2001). Age-related differences in hippocampal extracellular fluid glucose concentration during behavioral testing and following systemic glucose administration. *Journal of Gerontology: Biological Sciences, 56A,* B66–B71.

McNay, E. C., & Sherwin, R. S. (2004). Effect of recurrent hypoglycemia on spatial cognition and cognitive metabolism in normal and diabetic rats. *Diabetes, 53,* 418–425.

McNay, E. C., Fries, T. M., & Gold, P. E. (2000). Decreases in rat extracellular hippocampal glucose concentration associated with cognitive demand during a spatial task. *Proceedings of the National Academy of Science USA, 97,* 2881–2885.

McNay, E. C., McCarty, R. M., & Gold, P. E. (2001). Fluctuations in glucose concentration during behavioral testing: Dissociations both between brain areas and between brain and blood. *Neurobiology of Learning and Memory, 75,* 325–337.

Means, L. W., Walker, D. W., & Isaacson, R. L. (1970). Facilitated single alternation go, no- go performance following hippocampectomy in the rat. *Journal of Comparative and Physiological Psychology, 72,* 278–285

Milner, B., Corkin, S., & Teuber, H. L. (1968). Further analysis of the hippocampal amnesic syndrome: 14-year follow-up study of HM. *Neuropsychologia, 6,* 215–234.

Morris, K. A., & Gold, P. E. (2013). Epinephrine and glucose modulate training-related CREB phosphorylation in old rats: Relationships to age-related memory impairments. *Experimental Gerontology, 48,* 115–127.

Morris, K. A., Chang, Q., Mohler, E. G., & Gold, P. E. (2010). Age-related memory impairments due to reduced blood glucose responses to epinephrine. *Neurobiology of Aging, 31,* 2136–2145.

Naeem, M., & White, N. M. (2011). Lesions of basolateral and central amygdala differentiate conditioned cue preference learning with and without unreinforced preexposure. *Behavioral Neuroscience, 125,* 84–92.

Newcomer, J. W., Craft, S., Fucetola, R., Moldin, S. O., Selke, G., Paras, L., & Miller, R. (1999). Glucose-induced increase in memory performance in patients with schizophrenia. *Schizophrenia Bulletin, 25,* 321–335.

Newman, L. A., Korol, D. L., & Gold, P. E. (2011). Lactate produced by glycogenolysis in astrocytes regulates memory. *PLoS One, 6,* e28427.

Nicoll, R. A., & Roche, K. W. (2013). Long-term potentiation: Peeling the onion. *Neuropharmacology, 74,* 18–22.

Packard, M. G. (1999). Glutamate infused posttraining into the hippocampus or caudate-putamen differentially strengthens place and response learning. *Proceedings of the National Academy of Science USA, 96,* 12881–12886.

Packard, M. G., & Cahill, L. (2001). Affective modulation of multiple memory systems. *Current Opinion in Neurobiology, 11,* 752–756.

Packard, M. G., & McGaugh, J. L. (1996). Inactivation of hippocampus or caudate nucleus with lidocaine differentially affects expression of place and response learning. *Neurobiology of Learning and Memory, 65,* 65–72.

Paolino, R. M., Quartermain, D., & Miller, N. E. (1966). Different temporal gradients of retrograde amnesia produced by carbon dioxide anesthesia and electroconvulsive shock. *Journal of Comparative and Physiological Psychology, 62,* 270–274.

Parsons, M., & Gold, P. E. (1992). Glucose enhancement of memory in elderly humans: An inverted-U dose-response curve. *Neurobiology of Aging, 13,* 401–404.

Poldrack, R. A., & Packard, M. G. (2003). Competition among multiple memory systems: Converging evidence from animal and human brain studies. *Neuropsychologia, 41,* 245–251.

Pych, J. C., Chang, Q., Colon-Rivera, C., & Gold, P. E. (2005a). Acetylcholine release in hippocampus and striatum during training on a rewarded spontaneous alternation task. *Neurobiology of Learning and Memory, 84,* 93–101.

Pych, J. C., Chang, Q., Colon-Rivera, C., Haag, R., & Gold, P. E. (2005b). Acetylcholine release in the hippocampus and striatum during place and response training. *Learning and Memory, 12,* 564–572.

Pych, J. C., Kim, M., & Gold, P. E. (2006). Effects of injections of glucose into the dorsal striatum on learning of place and response mazes. *Behavioural Brain Research, 167,* 373–378.

Qi, Z., & Gold, P. E. (2009). Intrahippocampal infusions of anisomycin produce amnesia: Contributions of increased release of norepinephrine, dopamine and acetylcholine. *Learning and Memory, 16,* 308–314.

Quartermain, D., Paolino, R. M., & Miller, N. E. (1965). A brief temporal gradient of retrograde amnesia independent of situational change. *Science, 149,* 1116–1118.

Ragozzino, M. E., Unick, K. E., & Gold, P. E. (1996). Hippocampal acetylcholine release during memory testing in rats: Augmentation by glucose. *Proceedings of the National Academy of Science USA, 93,* 4693–4698.

Ragozzino, M. E., Pal, S. N., Unick, K., Stefani, M. R., & Gold, P. E. (1998). Modulation of hippocampal acetylcholine release and of memory by intrahippocampal glucose injections. *Journal of Neuroscience, 18,* 1595–1601.

Restle, F. (1957). Discrimination of cues in mazes: A resolution of the place-vs.- response question. *Psychological Review, 64,* 217–228.

Routtenberg, A. (2013). Lifetime memories from persistently supple synapses. *Hippocampus, 2,* 202–206.

Routtenberg, A., & Rekart, J. L. (2005). Post-translational protein modification as the substrate for long-lasting memory. *Trends in Neuroscience, 28,* 12–19.

Rudy, J. W., & Matus-Amat, P. (2005). The ventral hippocampus supports a memory representation of context and contextual fear conditioning: Implications for a unitary function of the hippocampus. *Behavioral Neuroscience, 119,* 154–163

Rudy, J. W., Barrientos, R. M., & O'Reilly, R. C. (2002). Hippocampal formation supports conditioning to memory of a context. *Behavioral Neuroscience, 116,* 530–538.

Salado-Castillo, R., Guante, M. A., Alvarado, R., Quirarte, G. L., & Prado-Alcalá, R. A. (1996). Effects of regional GABAergic blockade of the striatum on memory consolidation. *Neurobiology of Learning and Memory, 66,* 102–108.

Schroeder, J. P., & Packard, M. G. (2003). Systemic or intra-amygdala injections of glucose facilitate memory consolidation for extinction of drug-induced conditioned reward. *European Journal of Neuroscience, 17,* 1482–1488.

Sharma, A. V., Nargang, F. E., & Dickson, C. T. (2012). Neurosilence: Profound suppression of neural activity following intracerebral administration of the protein synthesis inhibitor anisomycin. *Journal of Neuroscience, 32,* 2377–2387.
Squire, L. R. (1992). Memory and the hippocampus: A synthesis from findings with rats, monkeys, and humans. *Psychological Review, 99,* 195–231.
Squire, L. R., & Spanis, C. W. (1984). Long gradient of retrograde amnesia in mice: Continuity with the findings in humans. *Behavioral Neuroscience, 98,* 345–348.
Squire, L. R., Slater, P. C., & Chace, P. M. (1975). Retrograde amnesia: Temporal gradient in very long term memory following electroconvulsive therapy. *Science, 187,* 77–79.
Stefani, M. R., & Gold, P. E. (1998). Intra-septal injections of glucose and glibenclamide attenuate galanin-induced spontaneous alternation performance deficits in the rat. *Brain Research, 813,* 50–56.
Sternberg, D. B., Isaacs, K., Gold, P. E., & McGaugh, J. L. (1985). Epinephrine facilitation of appetitive learning: Attenuation with adrenergic receptor antagonists. *Behavioral and Neural Biology, 44,* 447–453.
Stone, W. S., & Seidma, L. J. (2008). Toward a model of memory enhancement in schizophrenia: Glucose administration and hippocampal function. *Schizophrenia Bulletin, 34,* 93–108.
Stone, W. S., Seidman, L. J., Wojcik, J. D., & Green, A. I. (2003). Glucose effects on cognition in schizophrenia. *Schizophrenia Research, 62,* 93–103.
Suzuki, A., Stern, S. A., Bozdagi, O., Huntley, G. W., Walker, R. H., Magistretti, P. J., & Alberini, C. M. (2011). Astrocyte-neuron lactate transport is required for long-term memory formation. *Cell, 144,* 810–823.
Talley, C. E. P., Kahn, S., Alexander, L., & Gold, P.E. (2000). Epinephrine fails to enhance performance of food-deprived rats on a delayed spontaneous alternation task. *Neurobiology of Learning and Memory, 73,* 79–86.
Thomas, G. J. (1971). Maze retention by rats with hippocampal lesions and with fornicotomies. *Journal of Comparative and Physiological Psychology, 75,* 41–49.
Tolman, E. C. (1948). Cognitive maps in rats and men. *Psychological Review, 55,* 189–208.
Villers, A., Godaux, E., & Ris, L. (2012). Long-lasting LTP requires neither repeated trains for its induction nor protein synthesis for its development. *PLoS One, 7,* e40823.
Walker, D. W., Messer, L. G., Freund, G., & Means, L. W. (1972). Effect of hippocampal lesions and intertrial interval on single-alternation performance in the rat. *Journal of Comparative and Physiological Psychology, 80,* 469–477.
White, N. M. (2008). Multiple memory systems in the brain: Cooperation and competition. In H. B. Eichenbaum (Ed.), Memory systems of Byrne, J., editor, Learning and memory: A comprehensive reference (Vol 3, pp. 9–46). Oxford: Elsevier.
White, N. M., & McDonald, R. J. (1993). Acquisition of a spatial conditioned place preference is impaired by amygdala lesions and improved by fornix lesions. *Behavioural Brain Research, 55,* 269–281.
White, N. M., & McDonald, R. J. (2002). Multiple parallel memory systems in the brain of the rat. *Neurobiology of Learning and Memory, 77,* 125–184.
Zornetzer, S. F., & Gold, M. S. (1976). The locus coeruleus: Its possible role in memory consolidation. *Physiology & Behavior, 16,* 331–336.

Part III
Attribute Theory of Memory Applied to Models of Neurological Disorders

Chapter 13
Memory Disruption Following Traumatic Brain Injury

Robert F. Berman, Bruce G. Lyeth, Kiarash Shahlaie and Gene G. Gurkoff

Introduction

This chapter is focused on the consequences of traumatic brain injury (TBI) for memory and learning. We briefly review human studies that have examined memory impairments after TBI. This is followed by a review of studies in animal models of brain injury and how they have contributed to understanding the types of memory loss, as well as the specific memory structures and systems that appear to be particularly vulnerable to TBI. Recent developments in the field of memory research have proposed multiple types of memory as well as multiple memory systems in the brain, and this conceptualization is well represented by the Kesner attribute model of memory (Kesner 2009b). Such models have generated novel research strategies and new ways to understand how memory systems and memory processes are affected by TBI. However, as discussed in this review, TBI research has generally been slow to embrace some of these recent developments, and we argue that this situation should be remedied in future preclinical and clinical studies of TBI.

It is important to note at the outset that the damage following TBI is uncontrolled, typically asymmetrical and not complete within, or limited to a specific brain structure. Furthermore, the memory impairments are often graded in magnitude, dependent on the size of the lesions produced by brain injury. Cognitive loss following TBI also depends on the physical and mechanical forces that are applied to the brain. For example, blunt head injury can produce localized damage to specific

---

R. F. Berman (✉) · B. G. Lyeth · G. G. Gurkoff
Department of Neurological Surgery, University of California Davis, 1515 Newton Ct., Davis, CA 95616, USA
e-mail: rfberman@ucdavis.edu

K. Shahlaie
Department of Neurological Surgery, U.C Davis School of Medicine, Sacramento, CA 95817, USA

P. A. Jackson et al. (eds.), *The Neurobiological Basis of Memory*,
DOI 10.1007/978-3-319-15759-7_13

structures associated with the brain's memory systems, including temporal lobe structures (e.g., hippocampus, amygdala) and cortex (e.g., prefrontal cortex). At the other end of the spectrum are injuries that result from rapid acceleration, deceleration, or rotational forces that produce shearing forces, that produce diffuse axonal damage in major white matter pathways, that can disconnect memory systems in the brain (Gennarelli and Graham 1998; Smith et al. 2003). This is true whether one is considering clinical TBI or brain injury produced using the animal models described in this chapter. This situation makes studies of TBI and memory systems particularly challenging. This is in contrast to the majority of memory studies in laboratory animals (e.g., rodents, nonhuman primates) where the extent of the lesion is typically well controlled and usually bilaterally symmetrical in order to simplify linkage between brain structures damaged and memory impairment. However, even when lesion size is relatively well controlled in experimental animal studies, the relationship between the extent of damage and memory loss can be complicated (Baxter and Murray 2001; Zola and Squire 2001).

In view of these considerations, the Kesner attribute model provides a comprehensive and systematic framework in which to incorporate the effects of damage to multiple memory systems, as well as to understand how graded insult to the brain may affect memory following TBI (Kesner 2009b). The attribute model was developed to capture the complexity of memory and memory processes by recognizing that individual memories are constructed of multiple attributes, including space, time, affect, sensory-perceptual, response and language in humans, and that this experiential information is processed by multiple memory systems. The advantage of conceptualizing memory within such a framework is that it encompasses the interactions of many brain regions, systems and memory processes involved in the storage and recall of memory, rather than focusing a limited set of brain regions (e.g., hippocampus, prefrontal cortex) or processes. Recent studies in animal models of TBI using concepts from the attribute model are described in this review that illustrate the value of such a theoretical framework in providing a better understanding and new insights into the nature of memory loss following TBI.

Epidemiology of Traumatic Brain Injury

TBI is defined as damage to the brain resulting from external mechanical forces, such as rapid acceleration or deceleration, impact, blast waves, or penetration by a projectile (Maas et al. 2008). With more than 5.3 million people requiring long-term care as a result of injury, TBI is a major public health concern in the USA (Thurman et al. 1999). Each year approximately 1.7 million additional people suffer head trauma, and, as over 97% of those individuals survive, the population of TBI patients with chronic deficits is rapidly growing (Faul et al. 2010). TBI is also the leading cause of disability and death in children and young adults in the USA. Currently, there is little that can be done to effectively reduce the neurological consequences of TBI, and there is a critical need for improved understanding into how TBI damages specific brain regions in order to develop new treatment strategies that can improve neurological outcome and substantially improve patient lives following brain injury.

Memory Impairment Following TBI

The most common complaint of patients after TBI is cognitive impairment, including problems with memory and executive functions (McAllister et al. 2004; Vanderploeg et al. 2001). Although some recovery after TBI occurs during the first 6 months after injury, recovery often plateaus or declines thereafter and many patients are left with lifelong memory problems (Christensen et al. 2008; Esbjornsson et al. 2013). The most widely reported memory impairment following TBI is for episodic memory (Tulving 1983). Episodic memory is memory for facts and events within a specific spatial or temporal context, and consists of information that can be explicitly stored and retrieved. Medial temporal lobe (MTL) structures, including the hippocampus, perirhinal, entorhinal, and parahippocampal cortices are important for episodic memory (Eichenbaum et al. 2012; Hunsaker et al. 2008a; Squire and Zola-Morgan 1991). In contrast, semantic memory is memory for facts and knowledge about the external world that is independent of the spatial and temporal context in which it was acquired (Manns et al. 2003; Tulving 1983). Most studies have found that the semantic memory system generally remains intact following TBI, and that semantic deficits that are found reflect problems accessing and efficiently using semantic knowledge (McWilliams and Schmitter-Edgecombe 2008). Procedural memory, sometimes called implicit memory, is unconscious memory of skills and procedures, such as how to use objects or ride a bike (Baddeley 1995). Procedural memory has rarely been tested in patients following TBI, but when examined is typically reported to be relatively unaffected (Ewert et al. 1989). However, in a recent study of children who had sustained severe TBI at least 1 year prior to testing, injured individuals performed worse than controls in a repetition primary task that tests implicit memory (Lah et al. 2011).

Focal Injury to the Medial Temporal Lobe (MTL) Memory System

The MTL memory system consists of the hippocampus, perirhinal, entorhinal and parahippocampal cortices, and amygdala (Squire and Zola-Morgan 1991). Damage to structures within the MTL is frequently seen with TBI (Kotapka et al.1992). Such injuries can also lead to progressive atrophy across brain regions and of the hippocampus in particular following moderate to severe TBI. Based on MRI studies, significant MTL damage is thought to be the cause of enduring or permanent cognitive impairment (Bigler et al. 1997; Bigler et al. 1996). A study of 14 adult patients with moderate to severe TBI who were examined at 4.5 months and 2.5 years post-injury found significant bilateral decreases in hippocampal volumes when compared with published normative data (Ng et al. 2008). In a recent study in children 10 years after TBI, smaller hippocampal volumes and a general reduction in both gray and white brain matter were found. Reductions in hippocampal volumes, lateral ventricular enlargement, and associated memory impairment have been reported as long as 30 years after initial injury (Himanen et al. 2005). In many studies, brain volume loss was predictive of long-term cognitive disability and a poor prognosis for memory rehabilitation (Strangman et al. 2010; Warner et al. 2010). In contrast, amygdala volume was increased in severely injured patients compared to mild and moderate injury, indicating that childhood TBI can affect brain development even a decade after injury (Beauchamp et al. 2011). Findings from diffusion tensor imaging and T1 weighted structural MRI imaging show reduced volume and increased mean diffusion in the mediodorsal thalamus and anterior hippocampus, showing that TBI results in structural damage in both the MTL and related diencephalic memory systems (Avants et al. 2008; Warner et al. 2010). Bigler et al. investigated changes in the perforant pathway (PP) in TBI patients using diffusion tensor imaging and structural volumetric analysis. They found that patients with severe TBI had decreased fractional anisotrophy (FA) and higher apparent diffusion coefficients (ADCs) for the PP zone and higher ADCs bilaterally in the hippocampus. Volumetric analysis showed significantly decreased volumes bilaterally in hippocampi and temporal gray matter. Nonverbal memory (immediate and delayed recall) was significantly associated with right and left zone PP ADC, left hippocampal volume, and gray and white matter temporal volumes. These studies demonstrate that morphological changes in temporal lobe structures in patients with severe TBI are highly associated with memory impairment. Such studies are also important because memory improvement during rehabilitation following TBI is positively associated with the volume of the hippocampus (Strangman et al. 2010).

Although the importance of the hippocampus for spatial learning and memory is well established (Nadel 1991), only a few clinical studies have examined the effects of TBI on spatial processing. When 12 individuals with moderate to severe TBI were compared to normal controls in a virtual Morris water maze (MWM), brain injured patients were impaired in learning the location of a hidden "platform" and showed poor memory for platform location during probe trials (Skelton et al. 2000). However, no MRI or other structural information concerning the location or extent of TBI pathology was presented, including whether or not there was damage or atrophy of the hippocampus or other MTL structures. Children tested 4 years following TBI in an open-field apparatus similar to the radial arm maze (Olton and Samuelson 1976) had a significant memory deficit for spatial location during a memory probe trial (Lehnung et al. 2003). Brain injury in this study was mainly contusional from motor vehicle accidents and falls, but no MRI findings were available to indicate the specific site or extent of brain damage. The results of these studies demonstrate that spatial learning and memory deficits occur following TBI, and are consistent with the possibility that hippocampal damage may have been present in the test subjects. However, it is difficult to draw firm conclusions about the relationship of damage to spatial deficits without the requisite anatomical findings (e.g., magnetic resonance imaging MRI).

Diffuse Axonal Injury (DAI) and Memory Impairment

Another common type of TBI, diffuse axonal injury (DAI) or traumatic axonal injury, results from shearing forces induced by rapid acceleration or deceleration of the brain that damage axons. This white matter damage to axons results in disconnections between brain regions (Buki et al. 2000), and is considered a major cause of memory problems, even after otherwise mild head injury (Scheid et al. 2006). Diffusion tensor imaging (DTI) has been used to visualize and quantify DAI following TBI (Bigler et al. 2010; Kraus et al. 2007; Spitz et al. 2013). This technique is based on extraction of fractional anisotropy (FA) from DTI, where loss of structural integrity of white matter tracts indicated by decreased FA in fiber pathways. Decreased FA can be seen in most major fiber bundles in patients with moderate to severe TBI, and reduced FA is associated with poor cognitive function in several cognitive domains, including verbal learning and visual spatial memory (Kraus et al. 2007; Spitz et al. 2013). Loss of functional connectivity studies by functional MRI (fMRI) can also occur in patients with brain injuries consistent with DAI. A recent study of 25 patients with DAI found significantly lower interhemispheric functional connectivity for the hippocampus and anterior cingulate cortex compared to healthy controls, and a less focused recruitment of the default mode network for the dorsolateral prefrontal cortex (Marquez de la Plata et al. 2011). Evidence that damage restricted to axonal pathways can result in severe memory impairment can also be seen in a case report of a gunshot patient who suffered bilateral transection of the fornix without evidence of collateral damage to the hippocampus, anterior or dorsomedial thalamus or amygdala (D'Esposito et al. 1995). This patient showed severe anterograde episodic memory impairment for daily events since the injury, a deficit consistent with damage to the MTL memory system. One of the difficulties in understanding memory impairment following DAI is the complexity of injury that may involve many brain regions, as well as the fact that some reorganization of axonal connections and remyelination likely occurs following injury (Levin 2003). Recent advances in quantitative neuroimaging has made it possible to carry out morphometric studies across multiple brain regions, and the potential to combine DTI with fMRI data should provide a powerful approach to associate structural changes in the brain following TBI with specific types of memory impairments that are seen in patients.

Repeated Traumatic Head Injury

Repeated head injury such as those observed in professional and amateur athletes may also result in cumulative damage to the brain with cognitive loss and emotional problems emerging years after trauma (Hylin et al. 2013; Matser et al. 1998; McKee et al. 2009; Prins et al. 2013). Omalu et al. (2005, 2011) described histopathological findings in the postmortem brain of a retired professional football player who had a medical history of cognitive impairment, mood disorder, and Parkinsonian symptoms (McKee et al. 2009; Omalu et al. 2005). They found neuropathological changes now labeled “chronic traumatic encephalopathy” (CTE) consistent with repetitive concussive brain injury. This included mild neuronal loss in the frontal, parietal, and temporal neocortex, as well as widespread distribution of amyloid plaques, sparse neurofibrillary tangles and tau-positive neuritic threads throughout the neocortex. In a later study 10 of 14 (71 %) athletes studied showed evidence of CTE including small and large globose neurofibrillary tangles and neurotic threads in hippocampus, cerebral cortex, basal ganglia, and brainstem nuclei (Omalu et al. 2011). Further studies confirm a progressive tauopathy in professional athletes exposed to repetitive head trauma (McKee et al. 2009) and in blast-exposed veterans (Goldstein et al. 2012). Repeated head trauma in amateur soccer players was studied in the Netherlands and was found to be associated with memory problems assessed using the Weschler memory scale and deficits in planning on the Wisconsin card sorting test, as well as deficits in fine motor skills (Matser et al. 1998). Theriault et al. studied event-related potentials in athletes with a history of concussions and found that individuals with a history of three or more concussions had attenuated posterior contralateral negativity that was associated with poor retention of visual information and poor visual short-term memory (Theriault et al. 2011). In a recent study of aging former professional football players who had sustained a mean of four concussions during their life span, 35 % showed impairments in visual (Rey-Osterrieth Complex Figure Test, Delayed Recall Subtest) and verbal episodic memory (California Verbal Learning Test), as well as in name and word finding. White matter abnormalities were evident in frontal and parietal cortex, corpus callosum, and left temporal lobe, along with reduced cerebral blood flow compared with controls in the left temporal pole and right occipital region (Hart et al. 2013). Not only do these studies highlight some of the consequences of repeated TBI on learning systems, but also the importance of understanding how damage to multiple brain systems affects memory following injury.

Can New Approaches Better Define How TBI Affects the Brain’s Memory Systems?

It is clear that TBI can result in brain atrophy that can progress over decades, and that brain areas associated with memory and learning, including the MTL structures (i.e., hippocampus, entorhinal cortex) and the diencephalon (e.g., mediodorsal thalamus) are vulnerable to injury. Deficits in memory and learning, attention and executive function following TBI are well documented. However, the majority of research on memory impairment following TBI has focused on tests of executive function including immediate memory, working memory, verbal learning and memory, and visuo-spatial memory using standard neuropsychological tests (e.g.,

Wechsler Memory Scale-Revised, California verbal learning scale) (Anderson and Catroppa 2007; Catroppa and Anderson 2002; Di Stefano et al. 2000; Farmer et al. 2002; Levin et al. 1996). Very little memory research in the TBI field has incorporated more recent views on the organization of multiple memory systems in the brain (Eichenbaum et al. 2007; Kesner 2009b; Kesner and Goodrich-Hunsaker 2010; Squire and Wixted 2011) or included behavioral testing strategies that assess how TBI affects processes associated with these systems. For example, few studies have directly assessed retrograde amnesia in TBI patients using tests of remote autobiographical memory, such as those that have been used in patients with damage to the MTL (Squire and Wixted 2011). Similarly, declarative memory is thought to include two memory processes, recollection and familiarity (Eichenbaum et al. 2007). Recollection memory involves remembering specific details about an event whereas familiarity memory is a feeling that an event was previously experienced. Studies in hypoxic patients who have restricted hippocampal damage and show reduced hippocampal volume have provided strong evidence that MTL structures are important for both types of memory. However, a current debate centers around whether the hippocampus is specifically involved in recognition memory and the perirhinal cortex in familiarity (Eichenbaum et al. 2007), or whether the hippocampus and the perirhinal cortex contribute to both types of memory (Dede et al. 2013; Squire and Wixted 2011). This question has important implications for understanding the cognitive processes and neuroanatomical substrates of multiple memory systems, but has not yet been addressed by the TBI field.

A critical role for the hippocampus in spatial cognition is now well established (Burgess 2008; Nadel 1991). For example, amnestic patients with well-defined damage restricted to the hippocampus due to a hypoxic insult are impaired in spatial learning assessed in a virtual water maze (Goodrich-Hunsaker et al. 2010). Such findings have important implications for understanding the role of the hippocampus in spatial navigation and spatial memory. As another example, hypoxic subjects with circumscribed hippocampal damage show impairments in object–place associative memory as well as in odor–place associative memory. Specifically when amnesic patients are asked to remember the position of an object in space (Cave and Squire 1991) or a location where an odor was presented they exhibit significant impairments (Goodrich-Hunsaker et al. 2009). Unfortunately, only a limited number of spatial memory and spatial mapping studies have been carried out in TBI patients.

Perhaps it is not surprising that so little research has focused on describing the precise nature of the memory and cognitive impairments seen in TBI patients, because the injuries sustained by the brain following trauma are rarely discrete or localized to a single region of the brain. Furthermore, neither the specific injury nor the pathophysiology resulting from TBI is exactly the same in any two patients. As a result, past research has focused on relating prehospital admission injury severity with the degree of cognitive loss, the long-term consequences of TBI on memory and cognition, risks associated with repeated head injuries, and the differences between focal versus DAI on memory. However, the tools and concepts from research on patients with more selective damage to memory systems (e.g., hypoxia subjects) could be applied to understanding cognitive loss and memory impairments in TBI patients. Similarly, preclinical animal studies of memory have developed new behavioral testing procedures and findings from these studies have contributed substantially to understanding the neurobiological substrates of memory. The following sections briefly review preclinical TBI models and studies that have examined memory functions following TBI. Finally, we will conclude this chapter with studies that have begun to incorporate more contemporary concepts of memory into their design and interpretation.

Preclinical Research in Animal Models of TBI

Research in animal models has provided new insights into the nature of memory and how memory systems are organized in the brain. Much of this research has come from the development of new behavioral approaches in combination with the use of discrete lesions to cortical and MTL systems to dissect their contributions to memory processes (Kesner 2009b; Kesner and Goodrich-Hunsaker 2010). In contrast to human research in TBI, researchers using animal models of TBI have begun to embrace many of these newer behavioral testing approaches, concepts and experimental tools, and such research is providing new insights into memory loss following TBI. However, as evident in the following sections, with the exception of the radial arm maze and the MWM, both traditional spatial learning paradigms, there is a paucity of preclinical research into the effects of TBI on processing of spatial information, including spatial relationships between objects, principles of pattern separation, or the inclusion of spatiotemporal information as embodied in the Kesner memory attribute theory (Kesner 2009b). The following sections briefly review the three major preclinical models of TBI and how they have been used to study cognitive loss and memory impairment associated with TBI.

Models of Traumatic Brain Injury

As previously described, TBI has been defined as damage to the brain resulting from external mechanical force (Maas et al. 2008). One of the difficulties in studying and treating TBI is the heterogeneity of the clinical condition (Saatman et al. 2008). Over the years several animal models have been developed that reproduce key aspects of clinical TBI, including deficits in learning and memory (Xiong et al. 2013). In the following section the three most commonly used rodent models of TBI are briefly described, including the fluid percussion ("contrecoup" concussive-like injury), the cortical contusion injury (contusion), and the impact acceleration or Marmarou model of DAI.

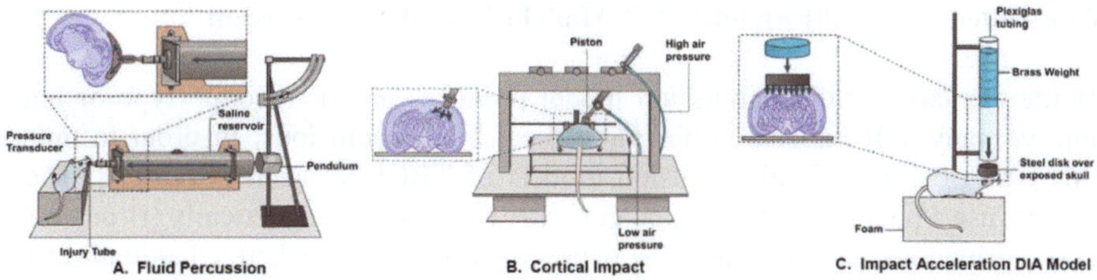

Fig. 13.1 Experimental models of traumatic brain injury in anesthetized rats. **a** Lateral fluid percussion (*LFP*) model which uses a rapid injection of a small volume of saline into the epidural space to produce brain injury. **b** Controlled cortical impact (*CCI*) model of contusional injury in which a pneumatic or electromagnetically driven piston penetrates the brain at a predetermine velocity, depth, and impact area. **c** Impact acceleration model of DAI. A metal disk is glued to the surface of the exposed skull to prevent fracture. A weight is then dropped from a calibrated height onto the disk, rapidly accelerating the brain into a foam mattress. Modified from Xiong et al. (Xiong et al. 2013) and reprinted with permission Macmillan Publishers Limited, 2013

The Fluid Percussion Model

The fluid percussion injury is the most widely used model of TBI in rodents. It was initially developed to produce a controlled mechanical injury to the brain of rabbits (Gurdjian et al. 1954; Lindgren and Rinder 1965, 1966; Rinder 1969), and was later modified for use in cats (Hayes et al. 1987; Stalhammar et al. 1987; Sullivan et al. 1976), rats (Dixon et al. 1988), and mice (Carbonell et al. 1998). The model produces brain injury by rapidly injecting a small volume of fluid (< 100 μl) against the exposed dural surface of the brain of an anesthetized animal, resulting in damage to the underlying cortex and proximal subcortical regions (Fig. 13.1a). The magnitude of injury (e.g., mild, moderate, or severe) is controlled by the force at which fluid is injected into the cranium. A pressure transducer is used to quantify the force of the fluid pulse, and high-speed X-ray cineradiography has revealed that the fluid pulse spreads out radially, resulting in a deformation of the dura and underlying brain over a much larger area than the diameter of the craniectomy (Dixon et al. 1988; Lighthall et al. 1989). Typically, the fluid pulse is delivered lateral to the saggital suture, termed lateral fluid percussion (LFP), and generates a combination of focal injury to the ipsilateral hemisphere as well as diffuse injury to the contralateral hemisphere (i.e., contrecoup injury) (Hallam et al. 2004; Katayama et al. 1990; McIntosh et al. 1989). Specifically, the LFP model produces widespread ipsilateral cell death in the somatosensory and primary motor cortices, focal cell death in the CA3 and hilus of the hippocampus, and scattered cell death in the thalamus (Conti et al. 1998; Hallam et al. 2004; Hicks et al. 1993; Smith et al. 1997). There is also intraparenchymal hemorrhage at grey/white interfaces and in the brainstem (McIntosh et al. 1989). The fluid pulse can also be delivered centrally directly over the saggital suture resulting in diffuse bilateral injury even in the absence of significant cell loss (Lifshitz et al. 2007).

Controlled Cortical Impact (CCI) Model of Cortical Contusion

While the lateral fluid percussion model produces a "contrecoup" type of brain injury, many TBI patients suffer more focal brain contusions. In order to model this type of injury, focal contusion models of TBI have been developed. These models use a pneumatically (Dixon et al. 1991) or electromagnetically (Brody et al. 2007) driven piston that strikes the exposed dura of an anesthetized animal resulting in damage restricted to the footprint area of the stainless steel tip of the piston (Fig. 13.1b). Injury severity is manipulated by adjusting the acceleration, depth and duration of the piston's contact with the dura (Dixon et al. 1991). Mild injury typically does not produce gross tissue damage (Hamm et al. 1992a), while more moderate and severe CCI can lead to significant cell death in the underlying cortex and subcortical structures including the hippocampus (Dixon et al. 1991; Goodman et al. 1994). With severe injury there may be only sparse cell death in the CA1, the region closest to the impact site, but considerable cell death and damage in the highly vulnerable dentate gyrus and CA3 regions of the hippocampus (Anderson et al. 2005; Colicos et al. 1996).

Impact Acceleration Model of Diffuse Axonal Injury (DAI)

DAI is characterized by widespread shearing and subsequent degeneration of axons. This type of injury is quite common, particularly following injuries that result in rapid acceleration, deceleration, or rotation of the brain such as in motor vehicle accidents (Adams et al. 1989). The most severe cases of DAI result in a high rate of coma and mortality (Adams et al. 1989), but even mild and moderate DAI can cause persistent neurological deficits, often including impairments in memory. Modeling DAI has been a technical challenge, and none of the existing models adequately models this type of injury. However, the most widely used procedure is the weight-drop impact-acceleration model of DAI (Marmarou et al. 1994). In this procedure, a small stainless steel disk (10-mm diameter and 3-mm thick) is lightly cemented to the exposed skull of an anesthetized animal. The animal's head is placed on a foam support, and a calibrated weight is dropped directly onto the disk (Fig. 13.1c). This results in a rapid acceleration of the animal's head into the foam support, resulting in shearing forces that damage axonal connections in the brain with very little evidence of a more focal contusion. Damage includes bilateral diffuse axonal injury in the corpus callosum, internal capsule, and optic tracts as well as in the cerebral and cerebellar peduncles (Buki et al. 2000; Greer et al. 2013; Hallam et al. 2004; Lifshitz et al. 2007), neurofilament compaction (Marmarou et al. 2005; Zakaria et al. 2012), impaired axonal transport (Marmarou et al. 2005) and a breakdown in dendritic structure (Folkerts et al. 1998). In addition, axonal injury extends into the long tracts entering the brain stem (Foda and Marmarou 1994). However, significant cell death is not a major feature of this impact-acceleration injury model.

Mechanisms of Cell Injury Following TBI

The mechanism(s) of cell death following focal TBI are complex, and involve an initial depolarization of injured cells, calcium influx, and release of toxic levels of glutamate (Gurkoff et al. 2013a; Katayama et al. 1990). Depolarization activates voltage-gated calcium channels (VGCCs), and the influx of calcium can activate secondary cell injury mechanisms, including calcium-dependent proteolytic enzymes (e.g., calpains), generation of toxic free radicals and additional glutamate release (Gurkoff et al. 2013a; Shahlaie et al. 2013; Shahlaie et al. 2010). Depolarization coincides with a dramatic increase in glucose utilization (i.e., hyperglycolysis), presumably to generate the ATP necessary to activate the ion pumps needed to restore ionic homeostasis (Hovda et al. 1992; Kawamata et al. 1992; Yoshino et al. 1991). Hyperglycolysis in the acute phase after injury is followed by reduced glucose uptake and utilization that is thought to contribute to cell injury and death (Bergsneider et al. 2001; Bergsneider et al. 1997; Yoshino et al. 1991). These neurochemical and metabolic changes are associated with failures in generation and maintenance of hippocampal long-term potentiation that undoubtedly contributes to cognitive loss following brain injury (Kelley et al. 2007; Sanders et al. 2001; Sanders et al. 2000). Similar to models with some form of contusion, DAI also leads to metabolic deficits (Tavazzi et al. 2005) and abnormal cerebral blood flow that contribute to cell injury and cell death (Engelborghs et al. 2000).

Inflammation following TBI also contributes to the overall pathology. The inflammatory response following brain injury involves vascular damage, alterations in the blood–brain barrier (BBB), and activation of the immune system, including microglial activation. Prolonged microglial activation can result in continued release of cytotoxic proinflammatory substances (e.g., cytokines, chemokines, proteolytic enzymes, reactive oxygen species) that can damage neurons and lead to progressive central nervous system (CNS) injury. Astrocytes can also be injured following TBI and can release proinflammatory substances, as well as form glial scars that can inhibit axonal regeneration and disrupt connections. Finally, microvascular damage to the BBB can allow macrophages to enter the brain and contribute to the informatory process (Mayer et al. 2013).

Summary of Experimental Models of TBI

One of the complexities of studying TBI and its effect on memory and learning is the inherent heterogeneity of injury experienced by patient populations. As described above, head injury can result in focal or multifocal contusions, as well as more diffuse axonal injuries, or a combination of injury types. In addition, the area of brain that is injured is not uniform across patients, nor is the severity of injury. As a result, all three experimental approaches described above, including fluid percussion, cortical confusion, and impact acceleration, have been used to

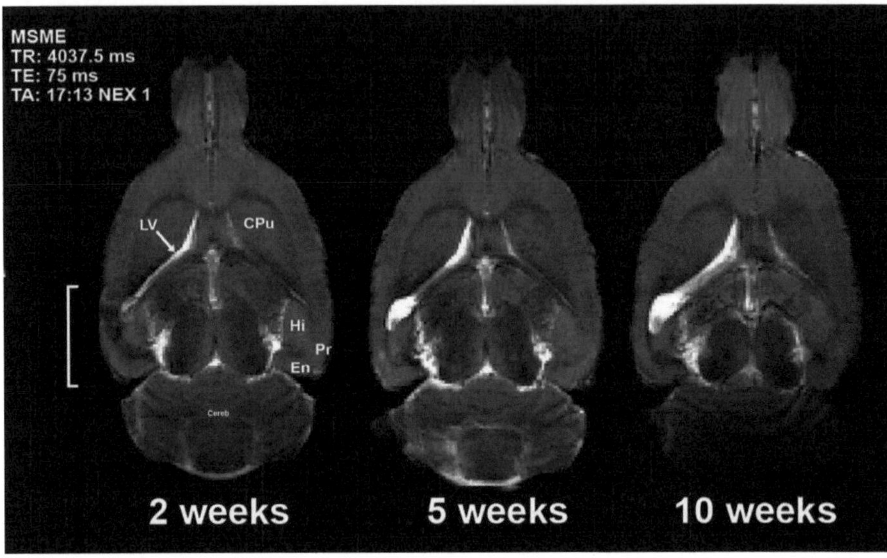

Fig. 13.2 T2-weighted MRI showing atrophy of hippocampus and overlying cortex following lateral fluid percussion TBI in an adult rat. Imaging was repeated at 2, 5, and 10 weeks following TBI. Bracket indicates location of fluid percussion injury in the right hemisphere. Note progressive atrophy of the hippocampus and overlying cortex and enlargement of the lateral ventricle in the injured right hemisphere compared to the opposite uninjured hemisphere. *LV* lateral ventricle, *CPu* caudate/putamen, *Hi* hippocampus, *Pr* perirhinal cortex, *En* entorhinal cortex

investigate the effects of TBI on memory and learning, as well as to evaluate therapeutic strategies to improve neurological function after brain damage to memory systems. The results from such studies in animal models are reviewed in the following sections.

TBI and Spatial Maze Learning

Evidence of damage to MTL structures including the hippocampus, overlying cortices, and amygdala is frequently observed in patients following TBI. Damage to these same structures can also be produced with each of the TBI models described above. Figure 13.2 is an MRI showing progressive atrophy of the hippocampus, overlying cortical areas, and amygdala following lateral fluid percussion brain injury in the rat. Because of the focus on MTL structures, the vast majority of behavioral experiments on cognitive loss following TBI have used tests that are sensitive to hippocampal damage. These include tests of spatial maze learning and memory, including the radial arm maze (Olton and Samuelson 1976), the MWM (Morris et al. 1990), and the Barnes maze (Barnes 1979). In each of these maze-learning paradigms animals are required to use distal visual cues to identify target locations within the maze. These tasks are often described as "hippocampally dependent" because performance in each is impaired by damage to the hippocampus (Morris et al. 1982).

Radial Arm Maze

The radial arm maze (Olton and Samuelson 1976) consists of eight equidistantly spaced arms that radiate out from a central platform. Each of the arms can be baited with food and the animal's task is to enter baited arms of the maze to consume the food reward without reentering a previously visited arm. When each of the arms is baited the task is predominantly one of working memory, where an animal has to remember which arms it has traveled down and which it has not for an individual training day, in order to most rapidly consume the reward. Good performance is characterized by a rapid progression through the arms retrieving all of the food pellets without a repeat visit to an arm. In another version of the radial arm maze not all of the arms are baited (e.g., 4 baited/4 unbaited). In this task an animal must establish a reference memory, remembering the location of arm that are baited and arms that are not, using external spatial visual cues. There is also a working memory component to this task because animals must recall which of the baited arms they have already visited to complete the task in the least amount of time and with the fewest errors. In pretrained animals, damage to the hippocampal circuitry (e.g., fornix transection) impairs working memory, but leaves reference memory intact (Olton and Papas 1979). In the TBI literature there is a fairly even split between studies using the working memory 8-baited arm task and those using the combined working memory/reference memory paradigms. In addition there is a relatively even split between experiments in which animals were pretrained on the radial arm maze (therefore testing the effect of TBI on retention) as compared to initiating training (acquisition) in the days following injury. A summary of these studies is presented below.

Fluid Percussion Injury

Fluid percussion injury consistently impairs memory performance in the radial arm maze. In the first of such studies, rats were trained in an 8-arm maze, where 6 arms were baited and the same 2 arms were never baited (Lyeth et al. 1990). Both mild- and moderate-injury severity impaired working memory, assessed by counting the number of times animals returned to a previously visited arm, while reference memory, evaluated by counting the number of times the animals visited arms that were not baited by food, was unaffected (Lyeth et al. 1990). Similar working memory impairment was seen after lateral fluid percussion injury in a radial arm configuration where only 4 of the 8 arms were baited (Enomoto et al. 2005). Even with all arms baited, injured animals made a significant number of working memory errors, returning to previously visited arms, as compared to sham animals (Hallam et al. 2004; Lyeth et al. 2001b), and these deficits were accompanied by damage to the hippocampus, although damage is not exclusively to the hippocampus.

Controlled Cortical Impact (CCI)

The radial arm maze has also been used to evaluate outcome following CCI. In an initial study animals were pretrained in the radial maze with 4 of the 8 arms baited. Unlike what was seen in fluid percussion, animals with CCI displayed deficits in reference memory as opposed to working memory (Soblosky et al. 1996). Brain injury in this study was mainly restricted to the cortex, including the parietal cortex, and the authors suggested that the effects on reference memory were more likely due to parietal rather than hippocampal damage. This is consistent with previous studies linking the posterior parietal cortex with long-term spatial reference memory (Kesner 2009a; Kesner et al. 1987). Following a lateral CCI that injured both fimbria/fornix, corpus callosum, and hippocampus, as well as parietal cortex, injured rats had significant deficits in reacquisition of the radial arm maze task 8 months after the injury (Lindner et al. 1998). These results are consistent with spatial memory deficits following contusional injury that damages the parietal cortex and hippocampus.

Impact Acceleration Model of DAI

Fairly limited data are available on the effects of traumatic DAI on radial-arm maze performance. Following DAI, animals that had been pretrained to visit each of the 8 arms only once had significant increases in the number of errors made (i.e., repeated entry into an arm) as well as increased latency to complete the maze 4 weeks following injury indicating a working memory deficit (Berman et al. 2000). Initial spatial learning in the radial arm maze has also been reported to be impaired after DAI, although the data were not shown in these studies and the authors did not indicate whether errors were made to baited or unbaited arms so it is not possible to determine whether the impairment was for working or reference memory (Kreipke et al. 2011; Shenaq et al. 2012; Wang et al. 2013a).

Morris Water Maze

The most commonly used paradigm to evaluate the effect of TBI on cognition has been, by a large margin, the MWM. In the MWM, rats are trained to utilize distal extra-maze cues to swim to a hidden escape platform submerged in a large tank of water (Morris 1984). Because the starting location varies from trial to trial and the platform is not visible, animals must develop a spatial map to find the location of the escape platform. Several experiments have shown that learning and performance in this task are severely disrupted by hippocampal lesions (Morris et al. 1982; Sutherland et al. 1982). A review of the TBI literature shows that there is considerable variability between laboratories in how the water maze is run, with key differences including whether the water is clear or opaque, whether the temperature of the water is cold or warm, the total number of trials and days that an animal is tested (e.g., from as few as 2 to as many as 10 days), the number and size of extra-maze cues, and size of the tank. Interestingly, several paradigms have been adapted for the maze including the standard reference memory with a probe trial, pretraining prior to injury, reversal learning and a working memory version of the task. While the majority of literature focuses on injury and water maze performance in adult rats, there are a limited number of studies for each of the three experimental models of injury in both pediatric rat models and adult mice. Regardless of differences in setup, paradigm, age, or species, the findings consistently demonstrate that TBI causes deficits in MWM performance.

Fluid Percussion Injury

The effect of TBI on MWM performance was first assessed using the LFP model. Rats pretrained in the MWM and then retested 2 days after injury were impaired in recalling platform location (Hicks et al. 1993; Smith et al. 1991). Rats trained and tested 11–15 days after LFP injury also show poor learning indicating a working memory impairment (Hamm et al. 1992b) and such deficits can persist for as long as 8 weeks following injury (Sanders et al. 1999). Using a reference memory design where rats were started from a fixed position and required to swim to a fixed escape location, injured rats learned the platform location and performed similar to uninjured controls (Hamm et al. 1993). A working memory version of the water maze task has also been used where rats received four pairs of trials per day over a 5-day period. Each pair of trials had a different start position and unique platform location. Injured animals displayed profound deficits in working memory as measured by increased latency to find the platform on the second trial of each pair as compared to controls (Hamm et al. 1996). Specific location of the fluid percussion injury (lateral versus medial, rostral versus caudal) affect the degree of impairment in MWM acquisition, with greater impairment with injury placements that damaged the CA2/3 hippocampal and dentate gyrus (Adelson et al. 2013; Vink et al. 2001). Cognitive deficits have been observed following mild or moderate fluid percussion even in the absence of gross hippocampal damage (Gurkoff et al. 2006; Lyeth et al. 1990). These findings indicate that TBI can result in cognitive impairment in the absence of gross tissue destruction, and that other mechanisms, including alterations in cholinergic (Lyeth et al. 1988) or glutamatergic (Lyeth et al. 2001a; Zwienenberg et al. 2001) transmission, metabolic disruption (Barkhoudarian et al. 2011) or interference with intrinsic electrical activity (e.g., theta rhythms) in the hippocampus (Fedor et al. 2010; Lee et al. 2013) may underlie such impairments. Impairments in MWM acquisition have also been reported in studies of aged rats (Hamm et al. 1992b) as well as in young rats used to model pediatric TBI (Gurkoff et al. 2006; Prins and Hovda 1998). Finally, mice also have significant deficits in acquisition as well as retention of MWM training following LFP injury (Carbonell et al. 1998).

Controlled Cortical Impact (CCI)

CCI also impairs spatial learning in the MWM. Rats receiving CCI were slower to learn to locate the escape platform location compared to the uninjured controls. However, some learning across trials was evident in the injured animals, and an analysis of swim patterns showed that they developed a nonspatial strategy of swimming to the center of the maze which served to facilitate finding the escape platform (Hamm et al. 1992a). The authors speculated that the primary deficit was, therefore, one of the spatial learning, while reference memory was intact. Surprisingly, no evidence of tissue destruction or cell loss was evident when brains were examined by light microscope 35 days after injury. Therefore, similar to fluid percussion, mild or moderate CCI can cause deficits in MWM in the absence of gross histological cell loss. Unilateral CCI of sufficient severity to reduce the number of morphologically intact CA1 pyramidal neurons in the injured hemisphere impairs both initial spatial learning in the MWM as well as spatial working member when rats are required to remember several escape locations each day (Kline et al. 2002). Such deficits are seen in older rats (Swan et al. 2011) as well as rats injured as young as postnatal day 7 (Adelson et al. 2013). A few studies have assessed spatial learning in the MWM in mice after CCI and the results consistently show impairments in MWM acquisition (Hannay et al. 1999; Smith et al. 1995; Whalen et al. 1999).

Impact Acceleration Model of DAI

The impact acceleration model of DAI also impairs acquisition and performance in the MWM (Beaumont et al. 1999), with older rats showing greater impairment than younger rats (Maughan et al. 2000). The DAI model has been frequently used in studies of pediatric TBI, perhaps because it is more easily scalable for smaller animals. In pediatric head injury, rats injured as young as 7 days show impaired MWM acquisition, and poor recall during a probe trial (Adelson et al. 2013). Similar results are seen in rats undergoing injury at 17 (Adelson et al. 2000) or 22 days (Adelson et al. 1997) of age. With severe DAI, deficits can persist for over 90 days (Zohar et al. 2011). Even when injured animals learn the location of the platform, they never became as proficient as sham-injured control animals (Zohar et al. 2011; Zohar et al. 2003). It has also been reported that repeated mild impact acceleration in mice also caused significant deficits in water maze performance (DeFord et al. 2002).

Barnes Maze

The Barnes maze is a "dry land" spatial maze task conceptually similar to the MWM (Barnes 1979). In this, maze animals are placed in the center of a brightly illuminated circular platform with up to 18 holes located around the periphery, one of which leads to a darkened escape box. The animal must use extra-maze spatial cues to locate the hole and enter the escape box. An advantage of the Barnes maze over the other spatial maze tasks is that animals can be scored for one of the three search strategies: spatial strategy relying on extra-maze cues, a peripheral response strategy where an animal sequentially searches holes around the maze, and random searches that lack a systematic search strategy (Barnes 1979; Fedor et al. 2010). Similar to the radial arm maze, the number of incorrect holes visited (errors) is quantified.

Lateral Fluid Percussion Injury

Following lateral fluid percussion injury, rats have significantly increased escape latencies and make more errors (i.e., repeat visit to incorrect hole) as compared to uninjured control rats (Doll et al. 2009; Maegele et al. 2005), a deficit that can be observed for as many as 3 months following injury (Lima et al. 2008). Injured animals do show improvement in Barnes maze performance with repeated testing, but many of the animals rely on the use of a peripheral response strategy rather than a spatial strategy (Fedor et al. 2010; Lee et al. 2013). Poor performance in the Barnes maze following lateral fluid percussion injury has been related to damage to the hippocampus and a decrease in hippocampal theta electroencephalography (EEG) power (Fedor et al. 2010; Lee et al. 2013).

Controlled Cortical Impact (CCI)

Rats receiving either a single (Peruzzaro et al. 2013) or repeated (Vonder Haar et al. 2013) cortical contusion with the CCI model show significant increases in latency to find the escape box in the Barnes maze. However, no data related to errors or search strategy has been reported in these studies. Similarly, mice receiving a CCI (Fox et al. 1998) failed to develop spatial search strategies as compared to sham-injured animals resulting in significant increases in latency to find the goal box (Fox et al. 1998; Fox et al. 1999). Therefore contusion injury, like fluid percussion, leads to deficits in Barnes maze performance.

Impact Acceleration Model of DAI

To date, only one group has investigated the effect of the impact acceleration model of DAI on Barnes maze performance in rats (Cernak et al. 2001). In this study, DAI resulted in increased latencies to find the escape hole in the Barnes maze. Interestingly, levels of cyclo-oxygenase-2 (COX-2) were dramatically elevated in the injured hippocampus, indicating post-injury brain inflammation and the generation of reactive oxygen species due to increased prostanoid synthesis, ultimately resulting in the disruption of hippocampal function.

Effects of TBI with Second Insults on Spatial Maze Performance

As previously described, many TBI patients experience a second insult (i.e., hypoxia–hypotension, seizures, etc.) sometime in their acute posttraumatic care. Additionally, repeat TBI, and particularly repeat mild TBI has become a key focus of clinical research. A second hypoxic insult in the initial hours following TBI ($F_iO_2 = 11\,\%$) is characterized by increased damage to the hippocampus (Clark et al. 1997; Feng et al. 2012) and diminished performance in the MWM task (Beaumont et al. 1999; Bramlett et al. 1999; Gurkoff et al. 2013b) as well as the radial arm maze (Hallam et al. 2004). This effect was also observed following a second hypoxic insult in pediatric rats (Adelson et al. 1997). Similarly, seizures in the acute hours following TBI have been associated with increased hippocampal cell death (Bao et al. 2011; Zanier et al. 2003). However, there is very little, if any, direct evidence that posttraumatic seizures per se worsen memory outcome following TBI in humans (Mazzini et al. 2003) or in animal models (Gurkoff et al. 2009; Hamm et al. 1995). This important question should be addressed in future studies.

While repetitive TBI is a very relevant clinical problem, surprisingly few studies have evaluated cognitive function following repetitive injuries in animal models of TBI. Studies in adult mice have found that repeated TBI using the CCI model (Laurer et al. 2001; Longhi et al. 2005), DAI model (DeFord et al. 2002), or lateral fluid percussion model (Wang et al. 2013b) results in greater deficits in the MWM than a single TBI. Repeated mild TBI in a rat model of pediatric TBI generates lasting metabolic deficits (Prins et al. 2013) as well as evidence of DAI (Huh et al. 2007; Prins et al. 2010) in the absence of significant hippocampal cell death. Available evidence from animal models also indicates a temporal window of vulnerability to a second TBI of approximately 1–3 days, after which the second injury becomes less damaging (Longhi et al. 2005; Prins et al. 2013). While most investigators have reported repeated injury results in greater impairments in the MWM, (Prins et al. 2010), other groups have not (Huh et al. 2007).

Memory Loss Following TBI within the Attribute Theory of Memory

As described in this review, the majority of animal studies of memory loss following TBI have relied on general tests of spatial memory using the radial arm maze and MWM, tests of novel object visual recognition memory (Donkin et al. 2011a), and more recently tests of motor learning (Chen et al. 2013). These studies have tended to focus attention on the behavioral consequences of damage to relatively large brain regions, including the hippocampus, and sensory and motor cortices. However, the patterns of brain injury following various types of TBI (e.g., focal versus diffuse), and their consequences for memory are complex and call for an encompassing model of memory systems in the brain, as exemplified by the attribute theory of memory proposed by Kesner (Kesner 2007, 2009b). The model puts forth a comprehensive view of how memory systems are organized in the brain. Under this model memories are composed of multiple attributes, including information about space, time, affect, sensory perceptual, response, and language in humans, to name a few. Importantly, individual attributes can be further subdivided, so that, for example, spatial attributes can include information about precise distances and angles between objects in space (i.e., egocentric metric spatial distance) and about relative object location in space (i.e., allocentric topological spatial location), while the time attribute can include information about duration and order of temporally separated events or stimuli. These attributes are then processed by three memory systems; the event-based, knowledge-based, and rule-based systems. The event-based system provides temporary representations of incoming information about specific memory attributes, the knowledge-based system provides for more permanent representations, and the rule-based system includes processes for maintaining and manipulating information in the three memory systems for subsequent action (e.g., attention, rehearsal, retrieval, long-term storage). This view of memory has led to novel and testable predictions about memory systems, including the discovery that lesions to subregions of the hippocampus result in different types of spatial memory deficits. For example, lesions of the dentate gyrus and CA3 subregions of the hippocampus selectively impair processing of egocentric metric spatial information, while lesions of the CA1 impair processing of information about the temporal order of events (Goodrich-Hunsaker et al. 2008a; Kesner 2007). Similarly, the process of pattern separation which allows for discrimination among similar spatial, temporal, or sensory inputs is carried out at several levels in the nervous systems, with the dorsal dentate gyrus involved in spatial pattern separation, and the CA3 involved in spatial pattern completion (Rolls and Kesner 2006). Although poorly studied at present, spatial and temporal memory deficits can occur following TBI and such deficits can be understood at a neurobiological level under the attribute model. Specifically, deficits in metric and topological memory, as well as deficits in memory for temporal order have been recently studied in rats following lateral fluid percussion TBI as described below (Gurkoff et al. 2013c).

One Trial Object and Odor Based Tasks

The use of standard maze learning tasks has allowed investigators to evaluate several aspects of cognition that are impaired following TBI, such as reference, working and long-term memory as they pertain to spatial learning. However, while many of these tasks require intact hippocampal function, they are often time consuming to run and were not designed to evaluate how damage to particular subregions of the hippocampus contribute to impaired memory and learning. Furthermore, many of these tasks use food deprivation or require an escape response and are therefore stressful, and the addition of stress to behavioral testing may confound interpretation of results. Therefore, visual object recognition and odor-based tasks have been developed that take advantage of rodents natural exploratory behavior and can be designed to dissociate the functions of specific subregions of the hippocampus and cortex as they pertain to several clinically relevant aspects of learning and memory (Goodrich-Hunsaker et al. 2005; Kesner 2009b; Kesner and Goodrich-Hunsaker 2010). Finally, many of these tasks have the potential to increase our understanding of the spectrum of memory and learning deficits in TBI patients (Goodrich-Hunsaker et al. 2008a). To date, these tasks have been underutilized in the field of TBI.

Novel Object Recognition

The novel object recognition task tests whether animals can recognize a novel from a familiar object. In this test, animals are allowed to freely explore two identical objects during a brief (e.g., 15 min) study phase, and then are removed from the test apparatus. They are returned to the apparatus after a delay for the test phase where one copy of the initial object and a second, novel object are available for exploration. Animals are scored based on how much time they spend with the novel as compared to the familiar object (Ennaceur and Delacour 1988). The test was designed to be similar to visual recognition tests used in nonhuman primates, does not involve primary reinforcement such as food or electric shocks, and allows for interspecies comparisons. Also, unlike maze learning, the test takes advantage of the spontaneous exploratory behavior typical of rodents. The novel object task is a test of working memory, and increasing the interval between the initial presentation of the objects and the test trial makes the task more difficult. The task can be easily scaled, reducing the size of both the environment and the objects, so that both rat pups and mice can be evaluated (Dodart et al. 1997; Reger et al. 2009). In an adaptation of the novel object task, rats can be tested sequentially for recognition using several object pairs. Recollection memory can then be tested by presenting object pairs presented earlier or later in a test series, and determining whether rats reexplore the older familiar object pair (i.e., primacy) or the more recently presented familiar object pair (i.e., recency) (Albasser et al. 2012).

Lesions of either the perirhinal cortex (Aggleton et al. 1997) or combined hippocampal + amygdala lesions (Aggleton et al. 1989), disrupt rodent's performance in visual nonspatial novel object recognition tests, whereas lesions restricted to the hippocampus, amygdala, cingulate cortex, cingulum bundle, fornix, or medial prefrontal cortex do not (Ennaceur et al. 1997). Following TBI, regardless of injury model or species, there are deficits in novel object recognition. For example, DAI in adult rats results in widespread axonal injury to major fiber pathways and also impairs visual novel object recognition, as well as MWM acquisition (Donkin et al. 2011b). In a pediatric model of brain contusion using CCI, young rats that received injury to the left parietal cortex between postnatal days 21–22 showed impaired novel object recognition when tested 1 week after injury (Scafidi et al. 2010). Novel object recognition has also been evaluated using in mice. Both cortical contusion (Han et al. 2011; Wakade et al. 2010) and impact acceleration (Siopi et al. 2012) TBI result in deficits in novel object recognition in mice. Impairments in recognition memory suggest that the resulting damage from TBI is not restricted to the hippocampus. However, there has been no further investigation to determine whether hippocampal damage combined with additional damage (e.g., amygdala, perirhinal cortex or DAI) is responsible for the observed deficits.

One of the potential reasons that the novel object test has not been extensively used to evaluate memory loss following TBI is that performance differences between sham and traumatic brain injured rats are relatively small. When testing potential therapeutic interventions investigators tend to focus on outcome measures that result in the largest separation between groups, such as the MWM. However, there are ways to optimize the novel object task such as increasing/decreasing the duration of the sample/study phase to adjust the difficulty of the task (Dodart et al. 1997; Reger et al. 2009).

The Metric Spatial Processing Task

The metric spatial processing task (Fig. 13.3a), described by Kesner and colleagues (Goodrich-Hunsaker et al. 2005), is a test of spatial pattern separation that investigates a rodent's ability to discriminate absolute positions of two distinct objects based on the precise angles and exact distances that separate the objects in the environment (Gallistel and Cramer 1996). Briefly, two novel objects are positioned near the center of a platform at a fixed distance (e.g., 68 cm) from each other. Animals are allowed to freely explore the environment including the two objects for 15 min. Rats (Goodrich-Hunsaker et al. 2005; Gurkoff et al. 2013c) and mice (Hunsaker et al. 2009) quickly habituate to the objects over time, showing more exploration of the objects in the first 5 min of the trial as compared to the last 5 min. Animals are then returned to their home cage for 5 min and the distance between the two objects is reduced (e.g., 34 cm). Control rats and mice reexplore the repositioned objects indicating that they have processed metric information about object location (Goodrich-Hunsaker et al. 2005; Gurkoff et al. 2013c; Hunsaker et al. 2009). Alternatively, objects can be started closer together and then moved farther apart. The difficulty of the task can easily be adjusted by making the change in the distance either more subtle (smaller change) or more obvious (larger change).

Experimental lesions of the dorsal hippocampus, and specifically the dentate gyrus, hilus, and CA3, significantly disrupt object re-exploration during the test phase on the metric task (Goodrich-Hunsaker et al. 2005; Goodrich-Hunsaker et al. 2008b) while lesions of the parietal cortex do not (Goodrich-Hunsaker et al. 2008b). These findings led the authors of these studies to conclude that the dentate gyrus and CA3 of the hippocampus are critical to the formation of memory for the loca-

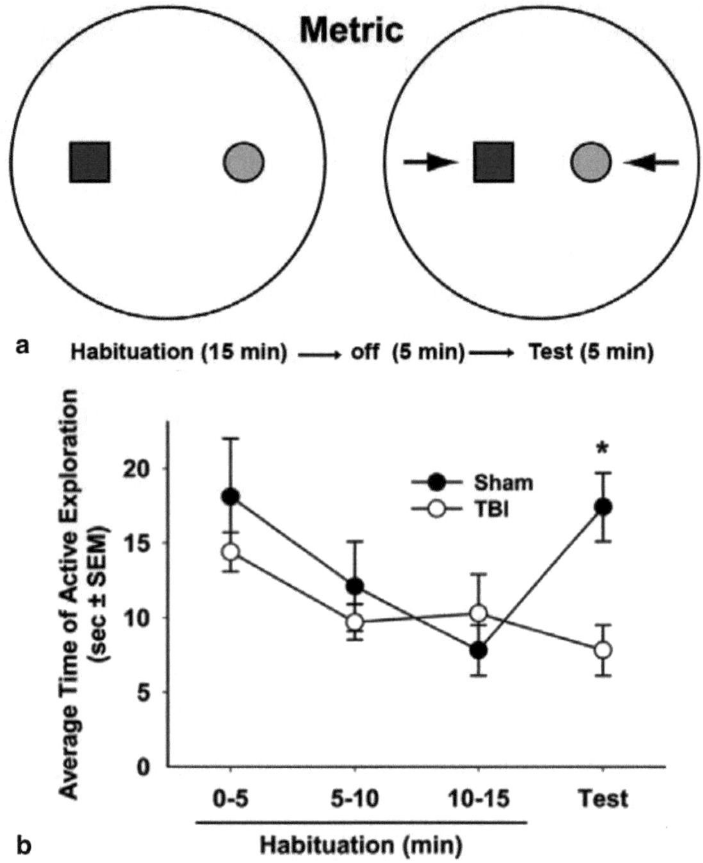

Fig. 13.3 **a** The metric task tests an animal's discrimination of a change in distance between two objects. **b** Both uninjured control ($n=9$) and TBI ($n=10$) animals habituated to the two objects over a 15 min period, but only Sham control animals reexplored the two objects after the distance between the two objects was reduced. $^*p<0.05$

tion of objects based on the relationships of distance and angle between objects (Gilbert and Kesner 2004). This process is referred to as pattern separation and data are accumulating that pattern separation is a fundamental computation process of the brain used to maximize the ability to discriminate between similar stimuli (Rolls and Kesner 2006). Moderate lateral fluid percussion injury results in specific and significant hippocampal neuronal loss in the CA2/CA3, the dentate gyrus and the hilus (Conti et al. 1998; Grady et al. 2003; Hallam et al. 2004; Hicks et al. 1996; Lowenstein et al. 1992). Similar to that observed in animals with lesions of the dorsal hippocampus, moderate lateral fluid percussion did not alter initial exploration or habituation of the two objects, but significantly impaired the injured rat's ability to detect a change in distance between two objects at the test phase (Fig. 13.3b) (Gurkoff et al. 2013c). These data suggest that rats with TBI to the hippocampus,

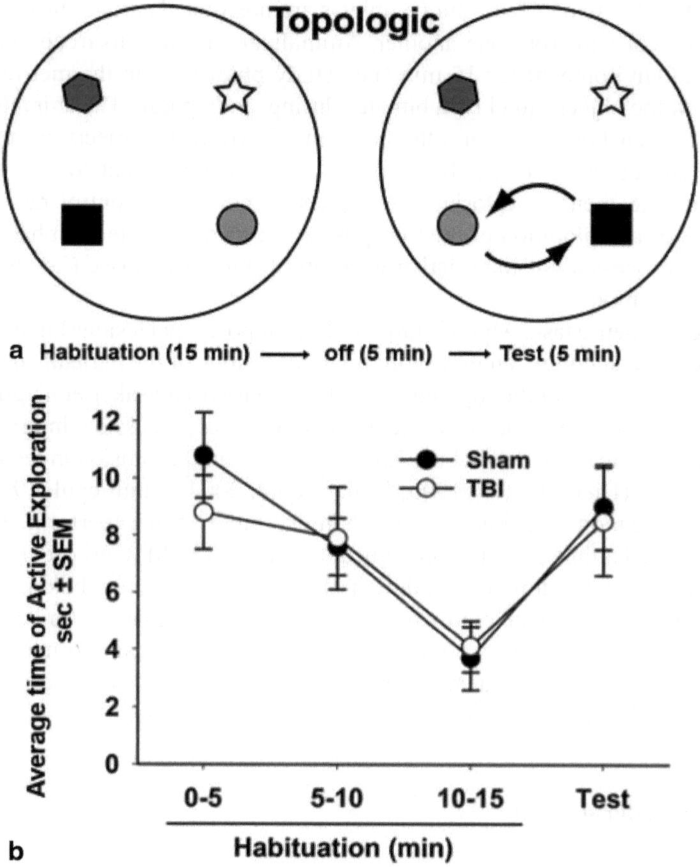

Fig. 13.4 **a** The topological task tests the ability to detect changes in spatial relationships between objects. Animals explore 4 different objects for 15 min (Habituation). The position of two objects is switched and animals reexplore for 5 min (Test). **b** Exploration declines for TBI ($n = 12$) and Sham Control ($n = 11$) groups during habituation, and both groups show equal reexploration of objects during the 5 min test

and CA2/CA3 and dentate gyrus in particular, are impaired in their ability to discriminate the spatial relationships between objects based on angles and spatial cues (i.e., metric relationships), and this may reflect deficits in the more general process of pattern separation (Rolls and Kesner 2006).

The Topological Spatial Processing Task

The topological spatial information task (Fig. 13.4a), described by Goodrich-Hunsaker et al. (2005), examines the ability of an animal to discriminate differences in the placement of two unique objects relative to each other using only spatial cues.

In this test, either two or four unique objects are positioned around the center of the platform equidistant from one another. Animals are allowed to freely explore the objects and environment for 15 min (i.e., study phase). As in the metric task, exploration of the objects quickly habituates during study phase. The animals are then returned to their home cage and the position of two of the objects are transposed in space, and the animals are then returned to the environment for the test phase where the time spent reexploring the objects is measured. Control rats and mice show increased exploration of the transposed objects during the test phase, indicating spatial processing of the relative location of objects in space (i.e., topological spatial information).

Unlike the metric task, which is impaired by hippocampal lesions but not by damage to the parietal cortex, animals with bilateral lesions of the parietal cortex fail to reexplore the objects in the topological task (Goodrich-Hunsaker et al. 2008b; Goodrich-Hunsaker et al. 2005). Moderate lateral fluid percussion injury not only causes hippocampal cell death, but also generates a small contusion in the ipsilateral parietal cortex (Dietrich et al. 1996; Hallam et al. 2004; Smith et al. 1997). However, following moderate fluid percussion injury, these animals performed similarly to sham-injured animals on the topological task (Fig. 13.4b) (Gurkoff et al. 2013c). The failure to detect a topological deficit in injured animals may be due to the fact that the majority of parietal cortex remains intact following lateral fluid percussion. The importance of the parietal cortex for processing topological information following TBI may be better evaluated in cortical contusions models (e.g., CCI) designed to selectively damage the parietal cortex unilaterally or bilaterally.

Temporal Ordering Task

Temporal ordering, the ability to recall the sequence of events, is also dependent on hippocampal function (Fortin et al. 2002; Hunsaker et al. 2008b; Hunsaker and Kesner 2009; Hunsaker et al. 2008b). The temporal ordering memory task (Fig. 13.5a) was designed to investigate a rodents' ability to remember the temporal sequence of odor (Hunsaker et al. 2008b), object (Hunsaker et al. 2008b), or spatial presentations (Hunsaker et al. 2008b). For temporal order of odor presentation, rats are exposed to three different pairs of identical odors (e.g., mace–mace, ginger-ginger, clove-clove) with a short delay between the three odor pairs. During the test phase, the animal is presented the first and the third presented odors (e.g., mace versus clove), and exploration (e.g., sniffing) of the odors is recorded. Uninjured control rats and mice preferentially spend more time exploring the first presented odor compared to the last presented odor (Gurkoff et al. 2013c; Hunsaker et al. 2008b; Hunsaker et al. 2010; Hunsaker and Kesner 2008, 2009). The same experimental design logic can also be used to examine temporal processing of novel objects or even of spatial location.

Rats with lesions of the CA1, but not CA3, dentate gyrus or fornix, have impairments in temporal ordering tasks. While uninjured controls spent significantly more time exploring conditions from the first sample phase, CA1 lesioned rats spent more

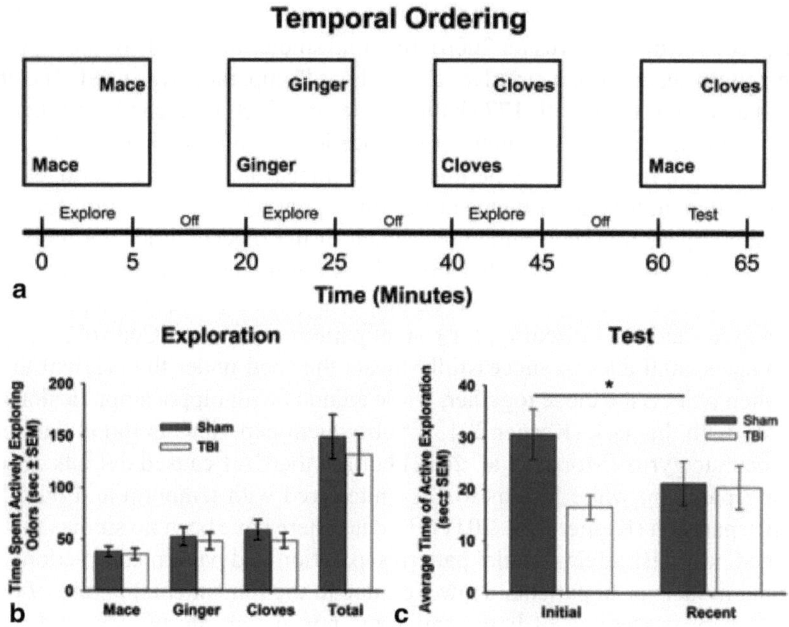

Fig. 13.5 **a** The temporal ordering task investigated rats' ability to remember a sequence of three odors. Rats were presented with two cups filled with the same odor for 5 min for each of three odors (i.e., *mace, ginger, cloves*), with 5 min between odors. Cups containing the first and third odors were then presented for 5 min, and active investigation cups (i.e., sniffing, physical contact) were recorded. **b** There was no difference in exploration of the three odors (*mace, ginger, cloves*) between uninjured control and TBI animals. **c** However, Sham control animals ($n=12$) spent significantly more time (s) exploring the first odor in the test phase, while TBI animals ($n=11$) did not differ in the time investigating the two odors. *$p<0.05$

time exploring the conditions from the third sample phase (Hunsaker et al. 2008b; Hunsaker and Kesner 2008, 2009). Two weeks following a fluid percussion or sham-injury rats were tested for their ability to discriminate the temporal order of presentation of three odors. While uninjured control rats spent significantly more time with the first odor presented during the sample/study phase, rats with a moderate lateral fluid percussion demonstrated no preference for either odor (Fig. 13.5b, c). Although in this study, the magnitude of lateral fluid percussion TBI was not associated with significant CA1 cell death, the authors argued that TBI may damage circuitry or signaling in the CA1, hippocampal, or even MTL that could disrupt behavior even in the absence of significant cell loss (Gurkoff et al. 2013c).

Pattern Separation and Pattern Completion

Pattern separation is defined as a computational process that reduces redundant information from similar inputs allowing them to be separated from each other, thereby reducing interference and generating a more categorized set of information (Kesner 2013; Rolls and Kesner 2006). In initial studies of the role of the hippocampus in pattern separation, a circular "cheese board" apparatus was used. The cheese board is a circular table with 177 shallow holes in which food can be hidden under objects. The rat's task is to remember the location of objects where food is hidden (Kesner 2009b). In a typical experiment, two objects are placed close together and food is hidden under one of them. In the study phase the rat must find food under one of the objects and then remember which of the objects hides the food during the test phase. The closer together the objects are in space the more difficult the task, and the greater reliance there is on hippocampal function, and in particular the dentate gyrus and CA1 circuitry, for spatial pattern separation. Control animals are able to use spatial cues to successfully locate the food under the original location even when objects are close together, while animals with hippocampal lesions have difficulty with this task (Kesner 2013). Subsequent experiments found that lesions of the dentate gyrus (Morris et al. 2012) but not the CA1 caused deficits in spatial pattern separation, while lesions to CA1 interfered with temporal and odor-based pattern separation (Kesner et al. 2011). To date, there have been no studies that have examined how TBI affects spatial pattern separation and pattern completion, either in animal models or in patients. However, due to the fundamental nature of pattern separation for processing of both spatial and nonspatial information, and the fact that the hippocampus and its circuitry are often targets of damage in TBI, future research should include such memory tests in order to better characterize the cognitive deficits associated with various types of brain injury.

Summary

This chapter reviews data from more than 30 years of research on the effects of TBI on memory and learning in both patient populations and in animal models of TBI. The existing data are reviewed within the context of more recent experimental approaches to understanding the underlying neurobiological substrates of memory. As should be evident in the review, the field of memory and learning has made significant advances in understanding the nature and anatomical systems supporting multiple memory systems in the brain. However, it is also apparent that the field of TBI has yet to embrace many of these more contemporary principles or begun to use recently developed behavioral paradigms (e.g., temporal order, pattern separation) to study memory. This should be viewed as a current limitation of the TBI field that should be addressed in the future. Such a revised approach could help to better understand the nature of cognitive loss in brain injured patients and how damage to specific brain regions would be expected to alter cognition, including memory. For example, instead of comparing patients based on the standard categories of mild, moderate, or severe injury, a better strategy may be to determine how specific injuries are correlated with specific outcome patterns (Saatman et al. 2008). Are there deficits in global cognition following TBI, or are certain types of memory impaired while others remain intact? Such information could also provide important insights into the types of rehabilitation strategies that might be designed to target specific types of cognitive loss. In order to answer these questions it will be critical to integrate the findings from research on patients with well-defined damage to memory systems (e.g., patient H.M., hypoxic patients) as well as from animal models used to study memory systems in the brain. There is also a need to critically evaluate whether current brain injury models in rodents accurately model the patterns of cognitive loss found in human TBI patients. If not, strategies must be found to adapt the rodent models to make them more clinically relevant. Gaining a better understanding of the relationship between TBI and cognitive dysfunction offers the best hope for developing future therapies that could help the millions of individuals who suffer chronic cognitive impairments following TBI.

References

Adams, J. H., Doyle, D., Ford, I., Gennarelli, T. A., Graham, D. I., & McLellan, D. R. (1989). Diffuse axonal injury in head injury: Definition, diagnosis and grading. *Histopathology, 15*(1), 49–59.

Adelson, P. D., Dixon, C. E., Robichaud, P., & Kochanek, P. M. (1997). Motor and cognitive functional deficits following diffuse traumatic brain injury in the immature rat. *Journal of Neurotrauma, 14*(2), 99–108.

Adelson, P. D., Dixon, C. E., & Kochanek, P. M. (2000). Long-term dysfunction following diffuse traumatic brain injury in the immature rat. *Journal of Neurotrauma, 17*(4), 273–282.

Adelson, P. D., Fellows-Mayle, W., Kochanek, P. M., & Dixon, C. E. (2013). Morris water maze function and histologic characterization of two age-at-injury experimental models of controlled cortical impact in the immature rat. *Child's Nervous System, 29*(1), 43–53.

Aggleton, J. P., Blindt, H. S., & Rawlins, J. N. (1989). Effects of amygdaloid and amygdaloid-hippocampal lesions on object recognition and spatial working memory in rats. *Behavioral Neuroscience, 103*(5), 962–974.

Aggleton, J. P., Keen, S., Warburton, E. C., & Bussey, T. J. (1997). Extensive cytotoxic lesions involving both the rhinal cortices and area TE impair recognition but spare spatial alternation in the rat. *Brain Research Bulletin, 43*(3), 279–287.

Albasser, M. M., Amin, E., Lin, T. C., Iordanova, M. D., & Aggleton, J. P. (2012). Evidence that the rat hippocampus has contrasting roles in object recognition memory and object recency memory. *Behavioral Neuroscience, 126*(5), 659–669.

Anderson, V., & Catroppa, C. (2007). Memory outcome at 5 years post-childhood traumatic brain injury. *Brain Injury, 21*(13–14), 1399–1409.

Anderson, K. J., Miller, K. M., Fugaccia, I., & Scheff, S. W. (2005). Regional distribution of Fluoro-Jade B staining in the hippocampus following traumatic brain injury. *Experimental Neurology, 193*(1), 125–130.

Avants, B., Duda, J. T., Kim, J., Zhang, H., Pluta, J., Gee, J. C., et al. (2008). Multivariate analysis of structural and diffusion imaging in traumatic brain injury. *Academic Radiology, 15*(11), 1360–1375.

Baddeley, A. D. (1995). The psychology of memory. In A. D. Baddeley, B.A. Wilson, & F.N. Watts (Ed.), *Handbook of memory disorders* (pp. 3–25). Chichester: Wiley.

Bao, Y. H., Bramlett, H. M., Atkins, C. M., Truettner, J. S., Lotocki, G., Alonso, O. F., et al. (2011). Post-traumatic seizures exacerbate histopathological damage after fluid-percussion brain injury. *Journal of Neurotrauma, 28*(1), 35–42.

Barkhoudarian, G., Hovda, D. A., & Giza, C. C. (2011). The molecular pathophysiology of concussive brain injury. *Clinics in Sports Medicine, 30*(1), 33–48, vii–iii.

Barnes, C. A. (1979). Memory deficits associated with senescence: A neurophysiological and behavioral study in the rat. *Journal of Comparative Physiological Psychology, 93*(1), 74–104.

Baxter, M. G., & Murray, E. A. (2001). Opposite relationship of hippocampal and rhinal cortex damage to delayed nonmatching-to-sample deficits in monkeys. *Hippocampus, 11*(1), 61–71.

Beauchamp, M. H., Ditchfield, M., Maller, J. J., Catroppa, C., Godfrey, C., Rosenfeld, J. V.., et al. (2011). Hippocampus, amygdala and global brain changes 10 years after childhood traumatic brain injury. *International Journal of Developmental Neuroscience, 29*(2), 137–143.

Beaumont, A., Marmarou, A., Czigner, A., Yamamoto, M., Demetriadou, K., Shirotani, T.., et al. (1999). The impact-acceleration model of head injury: Injury severity predicts motor and cognitive performance after trauma. *Neurological Research, 21*(8), 742–754.

Bergsneider, M., Hovda, D. A., Shalmon, E., Kelly, D. F., Vespa, P. M., Martin, N. A., et al. (1997). Cerebral hyperglycolysis following severe traumatic brain injury in humans: A positron emission tomography study. *Journal of NeuroSurgery, 86*(2), 241–251.

Bergsneider, M., Hovda, D. A., McArthur, D. L., Etchepare, M., Huang, S. C., Sehati, N., et al. (2001). Metabolic recovery following human traumatic brain injury based on FDG-PET: Time course and relationship to neurological disability. *The Journal of Head Trauma Rehabilitation, 16*(2), 135–148.

Berman, R. F., Verweij, B. H., & Muizelaar, J. P. (2000). Neurobehavioral protection by the neuronal calcium channel blocker ziconotide in a model of traumatic diffuse brain injury in rats. *Journal of NeuroSurgery, 93*(5), 821–828.

Bigler, E. D., Blatter, D. D., Johnson, S. C., Anderson, C. V., Russo, A. A., Gale, S. D., et al. (1996). Traumatic brain injury, alcohol and quantitative neuroimaging: Preliminary findings. *Brain Injury, 10*(3), 197–206.

Bigler, E. D., Blatter, D. D., Anderson, C. V., Johnson, S. C., Gale, S. D., Hopkins, R. O., et al. (1997). Hippocampal volume in normal aging and traumatic brain injury. *AJNR. American Journal of Neuroradiology, 18*(1), 11–23.

Bigler, E. D., McCauley, S. R., Wu, T. C., Yallampalli, R., Shah, S., MacLeod, M., et al. (2010). The temporal stem in traumatic brain injury: Preliminary findings. *Brain Imaging and Behavior, 4*(3–4), 270–282.

Bramlett, H. M., Dietrich, W. D., & Green, E. J. (1999). Secondary hypoxia following moderate fluid percussion brain injury in rats exacerbates sensorimotor and cognitive deficits. *Journal of Neurotrauma, 16*(11), 1035–1047.

Brody, D. L., Mac Donald, C., Kessens, C. C., Yuede, C., Parsadanian, M., Spinner, M., et al. (2007). Electromagnetic controlled cortical impact device for precise, graded experimental traumatic brain injury. *Journal of Neurotrauma, 24*(4), 657–673.

Buki, A., Okonkwo, D. O., Wang, K. K., & Povlishock, J. T. (2000). Cytochrome c release and caspase activation in traumatic axonal injury. *The Journal of Neuroscience, 20*(8), 2825–2834.

Burgess, N. (2008). Spatial cognition and the brain. *Annals of the New York Academy of Sciences, 1124,* 77–97.

Carbonell, W. S., Maris, D. O., McCall, T., & Grady, M. S. (1998). Adaptation of the fluid percussion injury model to the mouse. *Journal of Neurotrauma, 15*(3), 217–229.

Catroppa, C., & Anderson, V. (2002). Recovery in memory function in the first year following TBI in children. *Brain Injury, 16*(5), 369–384.

Cave, C. B., & Squire, L. R. (1991). Equivalent impairment of spatial and nonspatial memory following damage to the human hippocampus. *Hippocampus, 1*(3), 329–340.

Cernak, I., Wang, Z., Jiang, J., Bian, X., & Savic, J. (2001). Cognitive deficits following blast injury-induced neurotrauma: Possible involvement of nitric oxide. *Brain Injury, 15*(7), 593–612.

Chen, C. Y., Noble-Haeusslein, L. J., Ferriero, D., & Semple, B. D. (2013). Traumatic injury to the immature frontal lobe: A new murine model of long-term motor impairment in the absence of psychosocial or cognitive deficits. *Developmental Neuroscience, 35,* 474–490.

Christensen, B. K., Colella, B., Inness, E., Hebert, D., Monette, G., Bayley, M., et al. (2008). Recovery of cognitive function after traumatic brain injury: A multilevel modeling analysis of Canadian outcomes. *Archives of Physical Medicine and Rehabilitation, 89*(12 Suppl), S3–15.

Clark, R. S., Kochanek, P. M., Dixon, C. E., Chen, M., Marion, D. W., Heineman, S.., et al. (1997). Early neuropathologic effects of mild or moderate hypoxemia after controlled cortical impact injury in rats. *Journal of Neurotrauma, 14*(4), 179–189.

Colicos, M. A., Dixon, C. E., & Dash, P. K. (1996). Delayed, selective neuronal death following experimental cortical impact injury in rats: Possible role in memory deficits. *Brain Research, 739*(1–2), 111–119.

Conti, A. C., Raghupathi, R., Trojanowski, J. Q., & McIntosh, T. K. (1998). Experimental brain injury induces regionally distinct apoptosis during the acute and delayed post-traumatic period. *The Journal of Neuroscience, 18*(15), 5663–5672.

D'Esposito, M., Verfaellie, M., Alexander, M. P., & Katz, D. I. (1995). Amnesia following traumatic bilateral fornix transection. *Neurology, 45*(8), 1546–1550.

Dede, A. J., Wixted, J. T., Hopkins, R. O., & Squire, L. R. (2013). Hippocampal damage impairs recognition memory broadly, affecting both parameters in two prominent models of memory. *Proceedings of the National Academy of Sciences of the United States of America, 110*(16), 6577–6582.

DeFord, S. M., Wilson, M. S., Rice, A. C., Clausen, T., Rice, L. K., Barabnova, A., et al. (2002). Repeated mild brain injuries result in cognitive impairment in B6C3F1 mice. *Journal of Neurotrauma, 19*(4), 427–438.

Di Stefano, G., Bachevalier, J., Levin, H. S., Song, J. X., Scheibel, R. S., & Fletcher, J. M. (2000). Volume of focal brain lesions and hippocampal formation in relation to memory function after closed head injury in children. *Journal of Neurology, Neurosurgery, and Psychiatry, 69*(2), 210–216.

Dietrich, W. D., Alonso, O., Busto, R., & Finklestein, S. P. (1996). Posttreatment with intravenous basic fibroblast growth factor reduces histopathological damage following fluid-percussion brain injury in rats. *Journal of Neurotrauma, 13*(6), 309–316.

Dixon, C. E., Lighthall, J. W., & Anderson, T. E. (1988). Physiologic, histopathologic, and cineradiographic characterization of a new fluid-percussion model of experimental brain injury in the rat. *Journal of Neurotrauma, 5*(2), 91–104.

Dixon, C. E., Clifton, G. L., Lighthall, J. W., Yaghmai, A. A., & Hayes, R. L. (1991). A controlled cortical impact model of traumatic brain injury in the rat. *Journal of Neuroscience Methods, 39*(3), 253–262.

Dodart, J. C., Mathis, C., & Ungerer, A. (1997). Scopolamine-induced deficits in a two-trial object recognition task in mice. *Neuroreport, 8*(5), 1173–1178.

Doll, H., Truebel, H., Kipfmueller, F., Schaefer, U., Neugebauer, E. A., Wirth, S., et al. (2009). Pharyngeal selective brain cooling improves neurofunctional and neurocognitive outcome after fluid percussion brain injury in rats. *Journal of Neurotrauma, 26*(2), 235–242.

Donkin, J. J., Cernak, I., Blumbergs, P. C., & Vink, R. (2011a). A substance P antagonist reduces axonal injury and improves neurologic outcome when administered up to 12 hours after traumatic brain injury. *Journal of Neurotrauma, 28*(2), 217–224.

Donkin, J. J., Cernak, I., Blumbergs, P. C., & Vink, R. (2011b). A substance P antagonist reduces axonal injury and improves neurologic outcome when administered up to 12 hours after traumatic brain injury. *Journal of Neurotrauma, 28*(2), 217–224.

Eichenbaum, H., Yonelinas, A. P., & Ranganath, C. (2007). The medial temporal lobe and recognition memory. *Annual Review of Neuroscience, 30,* 123–152.

Eichenbaum, H., Sauvage, M., Fortin, N., Komorowski, R., & Lipton, P. (2012). Towards a functional organization of episodic memory in the medial temporal lobe. *Neuroscience and Biobehavioral Reviews, 36*(7), 1597–1608.

Engelborghs, K., Haseldonckx, M., Van Reempts, J., Van Rossem, K., Wouters, L., Borgers, M., et al. (2000). Impaired autoregulation of cerebral blood flow in an experimental model of traumatic brain injury. *Journal of Neurotrauma, 17*(8), 667–677.

Ennaceur, A., & Delacour, J. (1988). A new one-trial test for neurobiological studies of memory in rats. 1: Behavioral data. *Behavioural Brain Research, 31*(1), 47–59.

Ennaceur, A., Neave, N., & Aggleton, J. P. (1997). Spontaneous object recognition and object location memory in rats: The effects of lesions in the cingulate cortices, the medial prefrontal cortex, the cingulum bundle and the fornix. *Experimental Brain Research, 113*(3), 509–519.

Enomoto, T., Osugi, T., Satoh, H., McIntosh, T. K., & Nabeshima, T. (2005). Pre-Injury magnesium treatment prevents traumatic brain injury-induced hippocampal ERK activation, neuronal loss, and cognitive dysfunction in the radial-arm maze test. *Journal of Neurotrauma, 22*(7), 783–792.

Esbjornsson, E., Skoglund, T., Mitsis, M. K., Hofgren, C., Larsson, J., & Sunnerhagen, K. S. (2013). Cognitive impact of traumatic axonal injury (TAI) and return to work. *Brain Injury, 27*(5), 521–528.

Ewert, J., Levin, H. S., Watson, M. G., & Kalisky, Z. (1989). Procedural memory during post-traumatic amnesia in survivors of severe closed head injury. Implications for rehabilitation. *Archives of Neurology, 46*(8), 911–916.

Farmer, J. E., Kanne, S. M., Haut, J. S., Williams, J., Johnstone, B., & Kirk, K. (2002). Memory functioning following traumatic brain injury in children with premorbid learning problems. *Developmental Neuropsychology, 22*(2), 455–469.

Faul, M., Xu, L., Wald, M. M., & Coronado, V. G. (Eds.). (2010). *Traumatic brain injury in the United States: Emergency department visits, hospitalizations and deaths 2002–2006*. Atlanta: U.S. Department of Health and Human Services.

Fedor, M., Berman, R. F., Muizelaar, J. P., & Lyeth, B. G. (2010). Hippocampal theta dysfunction after lateral fluid percussion injury. *Journal of Neurotrauma, 27*(9), 1605–1615.

Feng, J. F., Zhao, X., Gurkoff, G. G., Van, K. C., Shahlaie, K., & Lyeth, B. G. (2012). Post-traumatic hypoxia exacerbates neuronal cell death in the hippocampus. *Journal of Neurotrauma, 29*(6), 1167–1179.

Foda, M. A., & Marmarou, A. (1994). A new model of diffuse brain injury in rats. Part II: Morphological characterization. *Journal of NeuroSurgery, 80*(2), 301–313.

Folkerts, M. M., Berman, R. F., Muizelaar, J. P., & Rafols, J. A. (1998). Disruption of MAP-2 immunostaining in rat hippocampus after traumatic brain injury. *Journal of Neurotrauma, 15*(5), 349–363.

Fortin, N. J., Agster, K. L., & Eichenbaum, H. B. (2002). Critical role of the hippocampus in memory for sequences of events. *Nature Neuroscience, 5*(5), 458–462.

Fox, G. B., Fan, L., LeVasseur, R. A., & Faden, A. I. (1998). Effect of traumatic brain injury on mouse spatial and nonspatial learning in the Barnes circular maze. *Journal of Neurotrauma, 15*(12), 1037–1046.

Fox, G. B., LeVasseur, R. A., & Faden, A. I. (1999). Behavioral responses of C57BL/6, FVB/N, and 129/SvEMS mouse strains to traumatic brain injury: Implications for gene targeting approaches to neurotrauma. *Journal of Neurotrauma, 16*(5), 377–389.

Gallistel, C. R., & Cramer, A. E. (1996). Computations on metric maps in mammals: Getting oriented and choosing a multi-destination route. *The Journal of Experimental Biology, 199* (1), 211–217.

Gennarelli, T. A., & Graham, D. I. (1998). Neuropathology of the head injuries. *Semininars in Clinical Neuropsychiatry, 3*(3), 160–175..

Gilbert, P. E., & Kesner, R. P. (2004). Memory for objects and their locations: The role of the hippocampus in retention of object-place associations. *Neurobiology of Learning and Memory, 81*(1), 39–45..

Goldstein, L. E., Fisher, A. M., Tagge, C. A., Zhang, X. L., Velisek, L., Sullivan, J. A., et al. (2012). Chronic traumatic encephalopathy in blast-exposed military veterans and a blast neurotrauma mouse model. *Science Translational Medicine, 4*(134), 134ra160.

Goodman, J. C., Cherian, L., Bryan, R. M. Jr., & Robertson, C. S. (1994). Lateral cortical impact injury in rats: Pathologic effects of varying cortical compression and impact velocity. *Journal of Neurotrauma, 11*(5), 587–597.

Goodrich-Hunsaker, N. J., Hunsaker, M. R., & Kesner, R. P. (2005). Dissociating the role of the parietal cortex and dorsal hippocampus for spatial information processing. *Behavioral Neuroscience, 119*(5), 1307–1315.

Goodrich-Hunsaker, N. J., Howard, B. P., Hunsaker, M. R., & Kesner, R. P. (2008a). Human topological task adapted for rats: Spatial information processes of the parietal cortex. *Neurobiology of Learning and Memory, 90*(2), 389–394.

Goodrich-Hunsaker, N. J., Hunsaker, M. R., & Kesner, R. P. (2008b). The interactions and dissociations of the dorsal hippocampus subregions: How the dentate gyrus, CA3, and CA1 process spatial information. *Behavioral Neuroscience, 122*(1), 16–26.

Goodrich-Hunsaker, N. J., Gilbert, P. E., & Hopkins, R. O. (2009). The role of the human hippocampus in odor-place associative memory. *Chemical Senses, 34*(6), 513–521.

Goodrich-Hunsaker, N. J., Livingstone, S. A., Skelton, R. W., & Hopkins, R. O. (2010). Spatial deficits in a virtual water maze in amnesic participants with hippocampal damage. *Hippocampus, 20*(4), 481–491.

Grady, M. S., Charleston, J. S., Maris, D., Witgen, B. M., & Lifshitz, J. (2003). Neuronal and glial cell number in the hippocampus after experimental traumatic brain injury: Analysis by stereological estimation. *Journal of Neurotrauma, 20*(10), 929–941.

Greer, J. E., Hanell, A., McGinn, M. J., & Povlishock, J. T. (2013). Mild traumatic brain injury in the mouse induces axotomy primarily within the axon initial segment. *Acta Neuropathologica, 126*(1), 59–74.

Gurdjian, E. S., Lissner, H. R., Webster, J. E., Latimer, F. R., & Haddad, B. F. (1954). Studies on experimental concussion: Relation of physiologic effect to time duration of intracranial pressure increase at impact. *Neurology, 4*(9), 674–681.

Gurkoff, G. G., Giza, C. C., & Hovda, D. A. (2006). Lateral fluid percussion injury in the developing rat causes an acute, mild behavioral dysfunction in the absence of significant cell death. *Brain Research, 1077*(1), 24–36.

Gurkoff, G. G., Giza, C. C., Shin, D., Auvin, S., Sankar, R., & Hovda, D. A. (2009). Acute neuroprotection to pilocarpine-induced seizures is not sustained after traumatic brain injury in the developing rat. *Neuroscience, 164*(2), 862–876.

Gurkoff, G., Shahlaie, K., Lyeth, B. G., & Berman, R. F. (2013a). Voltage-gated calcium channel antagonists and traumatic brain injury. *Pharmaceuticals, 6*(7), 788–812.

Gurkoff, G. G., Feng, J. F., Van, K. C., Izadi, A., Ghiasvand, R., Shahlaie, K., et al. (2013b). NAAG peptidase inhibitor improves motor function and reduces cognitive dysfunction in a model of TBI with secondary hypoxia. *Brain Research, 1515*, 98–107.

Gurkoff, G. G., Gahan, J. D., Ghiasvand, R. T., Hunsaker, M. R., Van, K., Feng, J. F., et al. (2013c). Evaluation of metric, topological, and temporal ordering memory tasks after lateral fluid percussion injury. *Journal of Neurotrauma, 30*(4), 292–300.

Hallam, T. M., Floyd, C. L., Folkerts, M. M., Lee, L. L., Gong, Q. Z., Lyeth, B. G., et al. (2004). Comparison of behavioral deficits and acute neuronal degeneration in rat lateral fluid percussion and weight-drop brain injury models. *Journal of Neurotrauma, 21*(5), 521–539.

Hamm, R. J., Dixon, C. E., Gbadebo, D. M., Singha, A. K., Jenkins, L. W., Lyeth, B. G., et al. (1992a). Cognitive deficits following traumatic brain injury produced by controlled cortical impact. *Journal of Neurotrauma, 9*(1), 11–20.

Hamm, R. J., White-Gbadebo, D. M., Lyeth, B. G., Jenkins, L. W., & Hayes, R. L. (1992b). The effect of age on motor and cognitive deficits after traumatic brain injury in rats. *Neurosurgery, 31*(6), 1072–1077. discussion 1078.

Hamm, R. J., Lyeth, B. G., Jenkins, L. W., O'Dell, D. M., & Pike, B. R. (1993). Selective cognitive impairment following traumatic brain injury in rats. *Behavioural Brain Research, 59*(1–2), 169–173.

Hamm, R. J., Pike, B. R., Temple, M. D., O'Dell, D. M., & Lyeth, B. G. (1995). The effect of post-injury kindled seizures on cognitive performance of traumatically brain-injured rats. *Experimental Neurology, 136*(2), 143–148.

Hamm, R. J., Temple, M. D., Pike, B. R., O'Dell, D. M., Buck, D. L., & Lyeth, B. G. (1996). Working memory deficits following traumatic brain injury in the rat. *Journal of Neurotrauma, 13*(6), 317–323.

Han, X., Tong, J., Zhang, J., Farahvar, A., Wang, E., Yang, J., et al (2011). Imipramine treatment improves cognitive outcome associated with enhanced hippocampal neurogenesis after traumatic brain injury in mice. *Journal of Neurotrauma, 28*(6), 995–1007.

Hannay, H. J., Feldman, Z., Phan, P., Keyani, A., Panwar, N., Goodman, J. C., et al. (1999). Validation of a controlled cortical impact model of head injury in mice. *Journal of Neurotrauma, 16*(11), 1103–1114.

Hart, J., Kraut, M. A., Womack, K. B., Strain, J., Didehbani, N., Bartz, E., et al. (2013). Neuroimaging of cognitive dysfunction and depression in aging retired National Football League players: A cross-sectional study. *JAMA Neurology, 70*(3), 326–335.

Hayes, R. L., Stalhammar, D., Povlishock, J. T., Allen, A. M., Galinat, B. J., Becker, D. P., et al. (1987). A new model of concussive brain injury in the cat produced by extradural fluid volume loading: II. Physiological and neuropathological observations. *Brain Injury, 1*(1), 93–112.

Hicks, R. R., Smith, D. H., Lowenstein, D. H., Saint Marie, R., & McIntosh, T. K. (1993). Mild experimental brain injury in the rat induces cognitive deficits associated with regional neuronal loss in the hippocampus. *Journal of Neurotrauma, 10*(4), 405–414.

Hicks, R., Soares, H., Smith, D., & McIntosh, T. (1996). Temporal and spatial characterization of neuronal injury following lateral fluid-percussion brain injury in the rat. *Acta Neuropathologica, 91*(3), 236–246.

Himanen, L., Portin, R., Isoniemi, H., Helenius, H., Kurki, T., & Tenovuo, O. (2005). Cognitive functions in relation to MRI findings 30 years after traumatic brain injury. *Brain Injury, 19*(2), 93–100.

Hovda, D. A., Becker, D. P., & Katayama, Y. (1992). Secondary injury and acidosis. *Journal of Neurotrauma, 9*(Suppl 1), S47–60.

Huh, J. W., Widing, A. G., & Raghupathi, R. (2007). Basic science; repetitive mild non-contusive brain trauma in immature rats exacerbates traumatic axonal injury and axonal calpain activation: A preliminary report. *Journal of Neurotrauma, 24*(1), 15–27.

Hunsaker, M. R., & Kesner, R. P. (2008). Evaluating the differential roles of the dorsal dentate gyrus, dorsal CA3, and dorsal CA1 during a temporal ordering for spatial locations task. *Hippocampus, 18*(9), 955–964.

Hunsaker, M. R., & Kesner, R. P. (2009). Transecting the dorsal fornix results in novelty detection but not temporal ordering deficits in rats. *Behavioural Brain Research, 201*(1), 192–197.

Hunsaker, M. R., Lee, B., & Kesner, R. P. (2008a). Evaluating the temporal context of episodic memory: The role of CA3 and CA1. *Behavioural Brain Research, 188*(2), 310–315.

Hunsaker, M. R., Fieldsted, P. M., Rosenberg, J. S., & Kesner, R. P. (2008b). Dissociating the roles of dorsal and ventral CA1 for the temporal processing of spatial locations, visual objects, and odors. *Behavioral Neuroscience, 122*(3), 643–650.

Hunsaker, M. R., Wenzel, H. J., Willemsen, R., & Berman, R. F. (2009). Progressive spatial processing deficits in a mouse model of the fragile X premutation. *Behavioral Neuroscience, 123*(6), 1315–1324.

Hunsaker, M. R., Goodrich-Hunsaker, N. J., Willemsen, R., & Berman, R. F. (2010). Temporal ordering deficits in female CGG KI mice heterozygous for the fragile X premutation. *Behavioural Brain Research, 213*(2), 263–268.

Hylin, M. J., Orsi, S. A., Rozas, N. S., Hill, J. L., Zhao, J., Redell, J. B., et al. (2013). Repeated mild closed head injury impairs short-term visuospatial memory and complex learning. *Journal of Neurotrauma, 30*(9), 716–726..

Katayama, Y., Becker, D. P., Tamura, T., & Hovda, D. A. (1990). Massive increases in extracellular potassium and the indiscriminate release of glutamate following concussive brain injury. *Journal of NeuroSurgery, 73*(6), 889–900.

Kawamata, T., Katayama, Y., Hovda, D. A., Yoshino, A., & Becker, D. P. (1992). Administration of excitatory amino acid antagonists via microdialysis attenuates the increase in glucose utilization seen following concussive brain injury. *Journal of Cerebral Blood Flow and Metabolism, 12*(1), 12–24.

Kelley, B. J., Lifshitz, J., & Povlishock, J. T. (2007). Neuroinflammatory responses after experimental diffuse traumatic brain injury. *Journal of Neuropathology and Experimental Neurology, 66*(11), 989–1001.

Kesner, R. P. (2007). Neurological views of memory. In R. P. K. J. L. Martinez (Ed.), *The neurobiology of learning and memory* (2nd edn, pp. 271–304). San Diego: Academic.

Kesner, R. P. (2009a). The posterior parietal cortex and long-term memory representation of spatial information. *Neurobiology of Learning and Memory, 91*(2), 197–206.

Kesner, R. P. (2009b). Tapestry of memory. *Behavioral Neuroscience, 123*(1), 1–13.

Kesner, R. P. (2013). Role of the hippocampus in mediating interference as measured by pattern separation processes. *Behavioural Processes, 93,* 148–154.

Kesner, R. P., & Goodrich-Hunsaker, N. J. (2010). Developing an animal model of human amnesia: The role of the hippocampus. *Neuropsychologia, 48*(8), 2290–2302.

Kesner, R. P., DiMattia, B. V., & Crutcher, K. A. (1987). Evidence for neocortical involvement in reference memory. *Behavioral and Neural Biology, 47*(1), 40–53.

Kesner, R. P., Hunsaker, M. R., & Ziegler, W. (2011). The role of the dorsal and ventral hippocampus in olfactory working memory. *Neurobiology of Learning and Memory, 96*(2), 361–366.

Kline, A. E., Massucci, J. L., Marion, D. W., & Dixon, C. E. (2002). Attenuation of working memory and spatial acquisition deficits after a delayed and chronic bromocriptine treatment regimen in rats subjected to traumatic brain injury by controlled cortical impact. *Journal of Neurotrauma, 19*(4), 415–425.

Kotapka, M. J., Graham, D. I., Adams, J. H., & Gennarelli, T. A. (1992). Hippocampal pathology in fatal non-missile human head injury. *Acta Neuropathologica, 83*(5), 530–534.

Kraus, M. F., Susmaras, T., Caughlin, B. P., Walker, C. J., Sweeney, J. A., & Little, D. M. (2007). White matter integrity and cognition in chronic traumatic brain injury: A diffusion tensor imaging study. *Brain: A Journal of Neurology, 130*(10), 2508–2519.

Kreipke, C. W., Rafols, J. A., Reynolds, C. A., Schafer, S., Marinica, A., Bedford, C., et al. (2011). Clazosentan, a novel endothelin A antagonist, improves cerebral blood flow and behavior after traumatic brain injury. *Neurological Research, 33*(2), 208–213.

Lah, S., Epps, A., Levick, W., & Parry, L. (2011). Implicit and explicit memory outcome in children who have sustained severe traumatic brain injury: Impact of age at injury (preliminary findings). *Brain Injury: [BI], 25*(1), 44–52..

Laurer, H. L., Bareyre, F. M., Lee, V. M., Trojanowski, J. Q., Longhi, L., Hoover, R., et al. (2001). Mild head injury increasing the brain's vulnerability to a second concussive impact. *Journal of NeuroSurgery, 95*(5), 859–870.

Lee, D. J., Gurkoff, G. G., Izadi, A., Berman, R. F., Ekstrom, A. D., Muizelaar, J. P., et al. (2013). Medial septal nucleus theta frequency deep brain stimulation improves spatial working memory after traumatic brain injury. *Journal of Neurotrauma, 30*(2), 131–139.

Lehnung, M., Leplow, B., Ekroll, V., Benz, B., Ritz, A., Mehdorn, M., et al. (2003). Recovery of spatial memory and persistence of spatial orientation deficits after traumatic brain injury during childhood. *Brain Injury, 17*(10), 855–869.

Levin, H. S. (2003). Neuroplasticity following non-penetrating traumatic brain injury. *Brain Injury, 17*(8), 665–674.

Levin, H. S., Fletcher, J. M., Kusnerik, L., Kufera, J. A., Lilly, M. A., Duffy, F. F., et al. (1996). Semantic memory following pediatric head injury: Relationship to age, severity of injury, and MRI. *Cortex, 32*(3), 461–478.

Lifshitz, J., Kelley, B. J., & Povlishock, J. T. (2007). Perisomatic thalamic axotomy after diffuse traumatic brain injury is associated with atrophy rather than cell death. *Journal of Neuropathology and Experimental Neurology, 66*(3), 218–229.

Lighthall, J. W., Dixon, C. E., & Anderson, T. E. (1989). Experimental models of brain injury. *Journal of Neurotrauma, 6*(2), 83–97.

Lima, F. D., Souza, M. A., Furian, A. F., Rambo, L. M., Ribeiro, L. R., Martignoni, F. V., et al. (2008). Na^+, K^+-ATPase activity impairment after experimental traumatic brain injury: Relationship to spatial learning deficits and oxidative stress. *Behavioural Brain Research, 193*(2), 306–310.

Lindgren, S., & Rinder, L. (1965). Experimental studies in head injury. I. Some factors influencing results of model experiments. *Biophysik, 2*(5), 320–329.

Lindgren, S., & Rinder, L. (1966). Experimental studies in head injury. II. Pressure propagation in "percussion concussion". *Biophysik, 3*(2), 174–180.

Lindner, M. D., Plone, M. A., Cain, C. K., Frydel, B., Francis, J. M., Emerich, D. F., et al. (1998). Dissociable long-term cognitive deficits after frontal versus sensorimotor cortical contusions. *Journal of Neurotrauma, 15*(3), 199–216..

Longhi, L., Saatman, K. E., Fujimoto, S., Raghupathi, R., Meaney, D. F., Davis, J., et al. (2005). Temporal window of vulnerability to repetitive experimental concussive brain injury. *Neurosurgery, 56*(2), 364–374; discussion 364–374.

Lowenstein, D. H., Thomas, M. J., Smith, D. H., & McIntosh, T. K. (1992). Selective vulnerability of dentate hilar neurons following traumatic brain injury: A potential mechanistic link between head trauma and disorders of the hippocampus. *The Journal of Neuroscience, 12*(12), 4846–4853.

Lyeth, B. G., Dixon, C. E., Jenkins, L. W., Hamm, R. J., Alberico, A., Young, H. F., et al. (1988). Effects of scopolamine treatment on long-term behavioral deficits following concussive brain injury to the rat. *Brain Research, 452*(1–2), 39–48.

Lyeth, B. G., Jenkins, L. W., Hamm, R. J., Dixon, C. E., Phillips, L. L., Clifton, G. L., et al. (1990). Prolonged memory impairment in the absence of hippocampal cell death following traumatic brain injury in the rat. *Brain Research, 526*(2), 249–258.

Lyeth, B. G., Gong, Q. Z., Shields, S., Muizelaar, J. P., & Berman, R. F. (2001). Group I metabotropic glutamate antagonist reduces acute neuronal degeneration and behavioral deficits after traumatic brain injury in rats. *Experimental Neurology, 169*(1), 191–199.

Maas, A. I., Stocchetti, N., & Bullock, R. (2008). Moderate and severe traumatic brain injury in adults. *Lancet Neurology, 7*(8), 728–741.

Maegele, M., Lippert-Gruener, M., Ester-Bode, T., Sauerland, S., Schafer, U., Molcany, M., et al. (2005). Reversal of neuromotor and cognitive dysfunction in an enriched environment combined with multimodal early onset stimulation after traumatic brain injury in rats. *Journal of Neurotrauma, 22*(7), 772–782.

Manns, J. R., Hopkins, R. O., & Squire, L. R. (2003). Semantic memory and the human hippocampus. *Neuron, 38*(1), 127–133.

Marmarou, A., Foda, M. A., van den Brink, W., Campbell, J., Kita, H., & Demetriadou, K. (1994). A new model of diffuse brain injury in rats. Part I: Pathophysiology and biomechanics. *Journal of NeuroSurgery, 80*(2), 291–300.

Marmarou, C. R., Walker, S. A., Davis, C. L., & Povlishock, J. T. (2005). Quantitative analysis of the relationship between intra- axonal neurofilament compaction and impaired axonal transport following diffuse traumatic brain injury. *Journal of Neurotrauma, 22*(10), 1066–1080.

Marquez de la Plata, C. D., Garces, J., Shokri Kojori, E., Grinnan, J., Krishnan, K., Pidikiti, R., et al. (2011). Deficits in functional connectivity of hippocampal and frontal lobe circuits after traumatic axonal injury. *Archives of Neurology, 68*(1), 74–84.

Matser, J. T., Kessels, A. G., Jordan, B. D., Lezak, M. D., & Troost, J. (1998). Chronic traumatic brain injury in professional soccer players. *Neurology, 51*(3), 791–796.

Maughan, P. H., Scholten, K. J., & Schmidt, R. H. (2000). Recovery of water maze performance in aged versus young rats after brain injury with the impact acceleration model. *Journal of Neurotrauma, 17*(12), 1141–1153.

Mayer, C. L., Huber, B. R., & Peskind, E. (2013). Traumatic brain injury, neuroinflammation, and post-traumatic headaches. *Headache, 53*(9), 1523–1530.

Mazzini, L., Cossa, F. M., Angelino, E., Campini, R., Pastore, I., & Monaco, F. (2003). Posttraumatic epilepsy: Neuroradiologic and neuropsychological assessment of long-term outcome. *Epilepsia, 44*(4), 569–574.

McAllister, T. W., Flashman, L. A., Sparling, M. B., & Saykin, A. J. (2004). Working memory deficits after traumatic brain injury: Catecholaminergic mechanisms and prospects for treatment—a review. *Brain Injury, 18*(4), 331–350.

McIntosh, T. K., Vink, R., Noble, L., Yamakami, I., Fernyak, S., Soares, H., et al. (1989). Traumatic brain injury in the rat: Characterization of a lateral fluid-percussion model. *Neuroscience, 28*(1), 233–244.

McKee, A. C., Cantu, R. C., Nowinski, C. J., Hedley-Whyte, E. T., Gavett, B. E., Budson, A. E., et al. (2009). Chronic traumatic encephalopathy in athletes: Progressive tauopathy after repetitive head injury. *Journal of Neuropathology and Experimental Neurology, 68*(7), 709–735.

McWilliams, J., & Schmitter-Edgecombe, M. (2008). Semantic memory organization during the early stage of recovery from traumatic brain injury. *Brain Injury, 22*(3), 243–253.

Morris, R. (1984). Developments of a water-maze procedure for studying spatial learning in the rat. *Journal of Neuroscience Methods, 11*(1), 47–60.

Morris, R. G., Garrud, P., Rawlins, J. N., & O'Keefe, J. (1982). Place navigation impaired in rats with hippocampal lesions. *Nature, 297*(5868), 681–683.

Morris, R. G., Schenk, F., Tweedie, F., & Jarrard, L. E. (1990). Ibotenate lesions of hippocampus and/or subiculum: Dissociating components of allocentric spatial learning. *The European Journal of Neuroscience, 2*(12), 1016–1028.

Morris, A. M., Churchwell, J. C., Kesner, R. P., & Gilbert, P. E. (2012). Selective lesions of the dentate gyrus produce disruptions in place learning for adjacent spatial locations. *Neurobiology of Learning and Memory, 97*(3), 326–331.

Nadel, L. (1991). The hippocampus and space revisited. *Hippocampus, 1*(3), 221–229.

Ng, K., Mikulis, D. J., Glazer, J., Kabani, N., Till, C., Greenberg, G., et al. (2008). Magnetic resonance imaging evidence of progression of subacute brain atrophy in moderate to severe traumatic brain injury. *Archives of Physical Medicine and Rehabilitation, 89*(12 Suppl), S35–44.

Olton, D. S., & Papas, B. C. (1979). Spatial memory and hippocampal function. *Neuropsychologia, 17*(6), 669–682.

Olton, D. S., & Samuelson, R. J. (1976). Remembrance of places passed: Spatial memory in rats. *Journal of Experimental Psychology: Animal Behavior Processes, 2,* 97–116.

Omalu, B. I., DeKosky, S. T., Minster, R. L., Kamboh, M. I., Hamilton, R. L., & Wecht, C. H. (2005). Chronic traumatic encephalopathy in a National Football League player. *Neurosurgery, 57*(1), 128–134; discussion 128–134.

Omalu, B., Bailes, J., Hamilton, R. L., Kamboh, M. I., Hammers, J., Case, M., et al. (2011). Emerging histomorphologic phenotypes of chronic traumatic encephalopathy in American athletes. *Neurosurgery, 69*(1), 173–183. Discussion 183.

Peruzzaro, S. T., Gallagher, J., Dunkerson, J., Fluharty, S., Mudd, D., Hoane, M. R., et al. (2013). The impact of enriched environment and transplantation of murine cortical embryonic stem cells on recovery from controlled cortical contusion injury. *Restorative Neurology and Neuroscience, 31*(4), 431–450.

Prins, M. L., & Hovda, D. A. (1998). Traumatic brain injury in the developing rat: Effects of maturation on Morris water maze acquisition. *Journal of Neurotrauma, 15*(10), 799–811.

Prins, M. L., Hales, A., Reger, M., Giza, C. C., & Hovda, D. A. (2010). Repeat traumatic brain injury in the juvenile rat is associated with increased axonal injury and cognitive impairments. *Developmental Neuroscience, 32*(5–6), 510–518.

Prins, M. L., Alexander, D., Giza, C. C., & Hovda, D. A. (2013). Repeated mild traumatic brain injury: Mechanisms of cerebral vulnerability. *Journal of Neurotrauma, 30*(1), 30–38.

Reger, M. L., Hovda, D. A., & Giza, C. C. (2009). Ontogeny of Rat Recognition Memory measured by the novel object recognition task. *Developmental Psychobiology, 51*(8), 672–678.

Rinder, L. (1969). "Concussive response" and intracranial pressure changes at sudden extradural fluid volume input in rabbits. *Acta Physiologica Scandinavica, 76*(3), 352–360.

Rolls, E. T., & Kesner, R. P. (2006). A computational theory of hippocampal function, and empirical tests of the theory. *Progress in Neurobiology, 79*(1), 1–48.

Saatman, K. E., Duhaime, A. C., Bullock, R., Maas, A. I., Valadka, A., Manley, G. T., et al. (2008). Classification of traumatic brain injury for targeted therapies. *Journal of Neurotrauma, 25*(7), 719–738.

Sanders, M. J., Dietrich, W. D., & Green, E. J. (1999). Cognitive function following traumatic brain injury: Effects of injury severity and recovery period in a parasagittal fluid-percussive injury model. *Journal of Neurotrauma, 16*(10), 915–925.

Sanders, M. J., Sick, T. J., Perez-Pinzon, M. A., Dietrich, W. D., & Green, E. J. (2000). Chronic failure in the maintenance of long-term potentiation following fluid percussion injury in the rat. *Brain Research, 861*(1), 69–76.

Sanders, M. J., Dietrich, W. D., & Green, E. J. (2001). Behavioral, electrophysiological, and histopathological consequences of mild fluid-percussion injury in the rat. *Brain Research, 904*(1), 141–144.

Scafidi, S., Racz, J., Hazelton, J., McKenna, M. C., & Fiskum, G. (2010). Neuroprotection by acetyl-L-carnitine after traumatic injury to the immature rat brain. *Developmental Neuroscience, 32*(5–6), 480–487.

Scheid, R., Walther, K., Guthke, T., Preul, C., & von Cramon, D. Y. (2006). Cognitive sequelae of diffuse axonal injury. *Archives of Neurology, 63*(3), 418–424.

Shahlaie, K., Lyeth, B. G., Gurkoff, G. G., Muizelaar, J. P., & Berman, R. F. (2010). Neuroprotective effects of selective N-type VGCC blockade on stretch-injury-induced calcium dynamics in cortical neurons. *Journal of Neurotrauma, 27*(1), 175–187.

Shahlaie, K., Gurkoff, G. G., Lyeth, B. G., Muizelaar, J. P., & Berman, R. F. (2013). Neuroprotective effects of SNX-185 in an in vitro model of TBI with a second insult. *Restorative Neurology and Neuroscience, 31*(2), 141–153.

Shenaq, M., Kassem, H., Peng, C., Schafer, S., Ding, J. Y., Fredrickson, V., et al. (2012). Neuronal damage and functional deficits are ameliorated by inhibition of aquaporin and HIF1alpha after traumatic brain injury (TBI). *Journal of the Neurological Sciences, 323*(1–2), 134–140.

Siopi, E., Llufriu-Daben, G., Fanucchi, F., Plotkine, M., Marchand-Leroux, C., & Jafarian-Tehrani, M. (2012). Evaluation of late cognitive impairment and anxiety states following traumatic brain injury in mice: The effect of minocycline. *Neuroscience Letters, 511*(2), 110–115.

Skelton, R. W., Bukach, C. M., Laurance, H. E., Thomas, K. G., & Jacobs, J. W. (2000). Humans with traumatic brain injuries show place-learning deficits in computer-generated virtual space. *Journal of Clinical and Experimental Neuropsychology, 22*(2), 157–175.

Smith, D. H., Okiyama, K., Thomas, M. J., Claussen, B., & McIntosh, T. K. (1991). Evaluation of memory dysfunction following experimental brain injury using the Morris water maze. *Journal of Neurotrauma, 8*(4), 259–269..

Smith, D. H., Soares, H. D., Pierce, J. S., Perlman, K. G., Saatman, K. E., Meaney, D. F., et al. (1995). A model of parasagittal controlled cortical impact in the mouse: Cognitive and histopathologic effects. *Journal of Neurotrauma, 12*(2), 169–178.

Smith, D. H., Chen, X. H., Pierce, J. E., Wolf, J. A., Trojanowski, J. Q., Graham, D. I., et al. (1997). Progressive atrophy and neuron death for one year following brain trauma in the rat. *Journal of Neurotrauma, 14*(10), 715–727..

Smith, D. H., Meaney, D. F., & Shull, W. H. (2003). Diffuse axonal injury in head trauma. *The Journal of Head Trauma Rehabilitation, 18*(4), 307–316.

Soblosky, J. S., Tabor, S. L., Matthews, M. A., Davidson, J. F., Chorney, D. A., & Carey, M. E. (1996). Reference memory and allocentric spatial localization deficits after unilateral cortical brain injury in the rat. *Behavioural Brain Research, 80*(1–2), 185–194.

Spitz, G., Maller, J. J., O'Sullivan, R., & Ponsford, J. L. (2013). White matter integrity following traumatic brain injury: The association with severity of injury and cognitive functioning. *Brain Topography, 26*(4), 648–660.

Squire, L. R., & Wixted, J. T. (2011). The cognitive neuroscience of human memory since H.M. *Annual Review of Neuroscience, 34,* 259–288.

Squire, L. R., & Zola-Morgan, S. (1991). The medial temporal lobe memory system. *Science, 253*(5026), 1380–1386.

Stalhammar, D., Galinat, B. J., Allen, A. M., Becker, D. P., Stonnington, H. H., & Hayes, R. L. (1987). A new model of concussive brain injury in the cat produced by extradural fluid volume loading: I. Biomechanical properties. *Brain Injury, 1*(1), 73–91.

Strangman, G. E., O'Neil-Pirozzi, T. M., Supelana, C., Goldstein, R., Katz, D. I., & Glenn, M. B. (2010). Regional brain morphometry predicts memory rehabilitation outcome after traumatic brain injury. *Frontiers in Human Neuroscience, 4,* 182.

Sullivan, H. G., Martinez, J., Becker, D. P., Miller, J. D., Griffith, R., & Wist, A. O. (1976). Fluid-percussion model of mechanical brain injury in the cat. *Journal of NeuroSurgery, 45*(5), 521–534.

Sutherland, R. J., Kolb, B., & Whishaw, I. Q. (1982). Spatial mapping: Definitive disruption by hippocampal or medial frontal cortical damage in the rat. *Neuroscience Letters, 31*(3), 271–276.

Swan, A. A., Chandrashekar, R., Beare, J., & Hoane, M. R. (2011). Preclinical efficacy testing in middle-aged rats: Nicotinamide, a novel neuroprotectant, demonstrates diminished preclinical efficacy after controlled cortical impact. *Journal of Neurotrauma, 28*(3), 431–440.

Tavazzi, B., Signoretti, S., Lazzarino, G., Amorini, A. M., Delfini, R., Cimatti, M., et al. (2005). Cerebral oxidative stress and depression of energy metabolism correlate with severity of diffuse brain injury in rats. *Neurosurgery, 56*(3), 582–589; discussion 582–589.

Theriault, M., De Beaumont, L., Tremblay, S., Lassonde, M., & Jolicoeur, P. (2011). Cumulative effects of concussions in athletes revealed by electrophysiological abnormalities on visual working memory. *Journal of Clinical and Experimental Neuropsychology, 33*(1), 30–41.

Thurman, D. J., Alverson, C., Dunn, K. A., Guerrero, J., & Sniezek, J. E. (1999). Traumatic brain injury in the United States: A public health perspective. *The Journal of Head Trauma Rehabilitation, 14*(6), 602–615..

Tulving, E. (1983). *Elements of episodic memory*. New York: Oxford University Press.

Vanderploeg, R. D., Crowell, T. A., & Curtiss, G. (2001). Verbal learning and memory deficits in traumatic brain injury: Encoding, consolidation, and retrieval. *Journal of Clinical and Experimental Neuropsychology, 23*(2), 185–195.

Vink, R., Mullins, P. G., Temple, M. D., Bao, W., & Faden, A. I. (2001). Small shifts in craniotomy position in the lateral fluid percussion injury model are associated with differential lesion development. *Journal of Neurotrauma, 18*(8), 839–847.

Vonder Haar, C., Friend, D. M., Mudd, D. B., & Smith, J. S. (2013). Successive bilateral frontal controlled cortical impact injuries show behavioral savings. *Behavioural Brain Research, 240,* 153–159.

Wakade, C., Sukumari-Ramesh, S., Laird, M. D., Dhandapani, K. M., & Vender, J. R. (2010). Delayed reduction in hippocampal postsynaptic density protein-95 expression temporally correlates with cognitive dysfunction following controlled cortical impact in mice. *Journal of NeuroSurgery, 113*(6), 1195–1201.

Wang, T., Chou, D. Y., Ding, J. Y., Fredrickson, V., Peng, C., Schafer, S., et al. (2013a). Reduction of brain edema and expression of aquaporins with acute ethanol treatment after traumatic brain injury. *Journal of NeuroSurgery, 118*(2), 390–396.

Wang, T., Van, K. C., Gavitt, B. J., Grayson, J. K., Lu, Y. C., Lyeth, B. G., et al. (2013b). Effect of fish oil supplementation in a rat model of multiple mild traumatic brain injuries. *Restorative Neurology Neuroscience, 31*(5), 647–659.

Warner, M. A., Youn, T. S., Davis, T., Chandra, A., Marquez de la, Plata, C., Moore, C., et al. (2010). Regionally selective atrophy after traumatic axonal injury. *Archives of neurology, 67*(11), 1336–1344..

Whalen, M. J., Carlos, T. M., Dixon, C. E., Schiding, J. K., Clark, R. S., Baum, E., et al. (1999). Effect of traumatic brain injury in mice deficient in intercellular adhesion molecule-1: Assessment of histopathologic and functional outcome. *Journal of Neurotrauma, 16*(4), 299–309..

Xiong, Y., Mahmood, A., & Chopp, M. (2013). Animal models of traumatic brain injury. *Nature Reviews. Neuroscience, 14*(2), 128–142.

Yoshino, A., Hovda, D. A., Kawamata, T., Katayama, Y., & Becker, D. P. (1991). Dynamic changes in local cerebral glucose utilization following cerebral conclusion in rats: Evidence of a hyper- and subsequent hypometabolic state. *Brain Research, 561*(1), 106–119.

Zakaria, N., Kallakuri, S., Bandaru, S., & Cavanaugh, J. M. (2012). Temporal assessment of traumatic axonal injury in the rat corpus callosum and optic chiasm. *Brain Research, 1467,* 81–90.

Zanier, E. R., Lee, S. M., Vespa, P. M., Giza, C. C., & Hovda, D. A. (2003). Increased hippocampal CA3 vulnerability to low-level kainic acid following lateral fluid percussion injury. *Journal of Neurotrauma, 20*(5), 409–420.
Zohar, O., Schreiber, S., Getslev, V., Schwartz, J. P., Mullins, P. G., & Pick, C. G. (2003). Closed-head minimal traumatic brain injury produces long-term cognitive deficits in mice. *Neuroscience, 118*(4), 949–955.
Zohar, O., Rubovitch, V., Milman, A., Schreiber, S., & Pick, C. G. (2011). Behavioral consequences of minimal traumatic brain injury in mice. *Acta Neurobiologiae Experimentalis, 71*(1), 36–45.
Zola, S. M., & Squire, L. R. (2001). Relationship between magnitude of damage to the hippocampus and impaired recognition memory in monkeys. *Hippocampus, 11*(2), 92–98.
Zwienenberg, M., Gong, Q. Z., Berman, R. F., Muizelaar, J. P., & Lyeth, B. G. (2001). The effect of groups II and III metabotropic glutamate receptor activation on neuronal injury in a rodent model of traumatic brain injury. *Neurosurgery, 48*(5), 1119–1126. Discussion 1126–1117.

Chapter 14
Altered Neural Synchronies Underlying Cognitive Deficits in a Transgenic Mouse Model of Huntington's Disease

Yoon H. Cho and Yannick Jeantet

Huntington's Disease (HD) and Mouse Genetics

Memory disturbances due to brain damage or neuropathology have greatly contributed to our understanding of how memories are organized in normal brain. Many neurological diseases are associated with severe cognitive impairments, and some of them are caused by the mutation of one or multiple genes encoding for proteins or enzymes important for nerve cell functions. Powerful mouse genetics have been used to dissect the pathogenesis of these diseases, which provide an interesting setting for the investigation of the neurobiological basis of learning and memory, and behaviors in pathological conditions. This is the case, for example, for Alzheimer's disease, Fragile X syndrome, and HD to cite only a few.

HD is particularly interesting because the disease is caused by the mutation of a single gene, which was cloned 20 years ago by the collaborative HD research group (1993). The mutation is transmitted by inheritance in an autosomal dominant manner. The mutated HD gene is located in chromosome 4 and encodes for Huntingtin protein whose functional roles are not well established. The mutation is characterized by an expansion of a trinucletide CAG repeat coding for glutamine, which results in toxic protein fragments that accumulate in the cell in the form of aggregates.

While the (mutated) protein is widely expressed (DiFiglia et al. 1995; Gutekunst et al. 1999), this inherited disease is associated with selective neurodegeneration of the basal ganglia and more specifically the striatum (Vonsattel et al. 1985). Accordingly, motor disturbances and chorea are cardinal clinical features. However, cognitive disturbances expressed in the form of cognitive inflexibility and/or working memory deficits as well as psychiatric problems often precede the motor im-

Y. H. Cho (✉) · Y. Jeantet
Institut de Neurosciences Cognitives et Intégratives d'Aquitaine, University of Bordeaux, Avenue des Facultes, 33405 Talence, Cedex, France
e-mail: yoon.cho@u-bordeaux.fr

Y. Jeantet
e-mail: mice.recorder@yahoo.fr

P. A. Jackson et al. (eds.), *The Neurobiological Basis of Memory*,
DOI 10.1007/978-3-319-15759-7_14

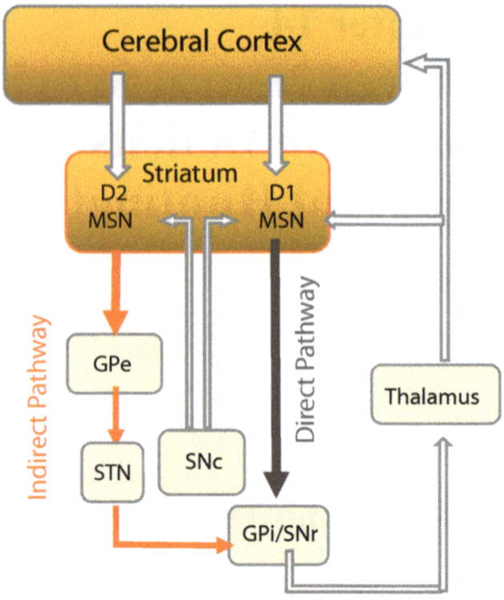

Fig. 14.1 Simplified basal ganglion loop with indirect and direct pathways. *MSN* medium spiny neuron, *GPe* external globus pallidus, *STN* subthalamic nuclei, *SNc* substance nigra compacta, *GPi* internal globus pallidus, *SNr* substance nigra reticulata

pairments by several years (Josiassen et al. 1983; Lawrence et al. 1996; Lawrence et al. 2000; Lemiere et al. 2004; Vonsattel et al. 1985). Other cognitive impairments are reminiscent of deficits observed with striatal (and cortex) pathology, which is known to support stimulus–response (S–R) associations, habit formation, procedural, nondeclarative, or implicit memory (Graybiel et al. 1994; Squire 1992; Yin and Knowlton 2006); incremental processes taking place with repeated trials and errors. These findings are consistent with the neurobiological based attribute theory of memory in which the striatum subserves a response attribute which involves memory representations based on feedback from motor responses (often based on proprioceptive and vestibular cues) that occur in specific situations (Kesner and Rogers 2004).

The vulnerable striatum constitutes the main output structure of the basal ganglia, an ensemble of regions formed by the cortex, globus pallidus, subthalaminc nuclei, thalamus, and substance nigra (Fig. 14.1). The striatum contains heterogeneous cell populations. Principal cells, also called medium spiny neurons (MSNs), send axons to downstream regions in the basal ganglia loop, and several classes of local interneurons (IN) form intricate and complex regulatory and interactive networks. One subclass of MSN that forms so called "indirect pathway" is affected earlier than "direct pathway" MSN neurons in HD. The indirect pathway MSNs project via internal segment of pallidum and subthalamic nuclei to substantia nigra, and the direct pathway MSNs project directly to the substantia nigra (Reiner et al. 1988). While reasons for this differential vulnerability are not well understood, a new hypothesis suggests that the imbalance of activities between the two pathways, caused by dysfunctional indirect pathway and maybe compensatory activity of direct pathway, may constitute a core pathological process in HD (Andre et al. 2011b; Andre et al. 2011a).

As mentioned previously, the hereditary nature of the disease transmission has made the transgenic mouse the prime choice to model HD in animals. Upon cloning of the HD gene in 1993 (The Huntington's Disease Collaborative Research Group 1993), several transgenic mouse lines (i.e., R6 lines, YAC128 mice, and knock-in mice) have been produced, with which the effects of the gene mutation could be explored from molecule to behavior. As such, a plethora of symptoms described in HD patients has been reproduced in HD mouse models, even though no one model reproduces all the symptoms seen in patients. Phenotypic changes in these transgenic mice concerned changes of biochemical and molecular properties in vulnerable cell populations (Cepeda et al. 2007; Cha 2007; Cha et al. 1999; Ferrante 2009), as well as behavioral and cognitive abnormalities. The latter domain is associated with deficits in visuospatial working memory, cognitive flexibility, emotionality, social interaction, and circadian rhythm among others (Brooks et al. 2006; Dumas et al. 2013; Lione et al. 1999; Morton et al. 2005; Naver et al. 2003; Nithianantharajah et al. 2008; Pietropaolo et al. 2011). Impaired synaptic plasticity and functional changes in the vulnerable corticostriatal pathway (Cummings et al. 2007; Cummings et al. 2006) have been associated with these symptoms.

Striatal Activity Changes in HD Transgenic Mice

While striatal degeneration is an incontestable hallmark of HD, some postmortem studies have shown only limited signs of cell loss in the brain including the striatum despite substantial clinical evidence of HD (Vonsattel et al. 1985). This suggests that neuronal dysfunction, rather than cell death, may contribute to behavioral manifestations, especially at early stage of HD. In line with this idea, most symptoms of HD have been reproduced in mouse models in which neuronal loss is not present at all. The nature of neural dysfunction in HD, in particular information coding process mediating cognitive and behavioral disturbances is not known, and mouse models with strong construct validity present a unique opportunity to explore this issue, which is not possible in human patients.

We thus recorded single-unit activity and local field potentials (LFPs) in the dorsal striatum of motor presymptomatic HD transgenic mice (i.e., R6/1 mice) and their nontransgenic littermates while they acquired and performed an operant conditioning task. In this task, mice learned to associate a nose-poke action with the arrival of a reward at the food port. Because it has been shown that striatal activity undergoes plasticity and displays activity changes in normal rodents throughout learning (Barnes et al. 2005; Jog et al. 1999), our recordings were initiated with behaviorally naïve mice and continued until they reached asymptotic performance levels (Cayzac et al. 2011). Since these HD mice displayed impaired synaptic plasticity in the cortico-striatal pathway, we hypothesized that learning related changes in cell firing may be absent in the transgenic mice (Cepeda et al. 2003; Cepeda et al. 2007; Cummings et al. 2007; Cummings et al. 2006; Cybulska-Klosowicz et al. 2004).

We chose to study the R6/1 line, which displayed delayed onset and progressive disease phenotypes, reminiscent of adult onset HD (Mangiarini et al. 1996). These mice exhibit impaired ability for spatial short-term (Grote et al. 2005; Naver et al. 2003; Nithianantharajah et al. 2008; Pang et al. 2009; Lebreton et al. 2015) and long-term memory (Nithianantharajah et al. 2008) at 10–12 weeks of age, abnormal social behaviors at 13 weeks (Pietropaolo et al. 2011), and impaired motor coordination and muscle strength at 18–20 weeks of age (Naver et al. 2003; Lebreton et al. 2015). The mice have a 6–7 month lifespan. Although mutant protein aggregates begin to appear at 8 weeks of age, no cell loss is observed even at the time of death (Naver et al. 2003; Nicniocaill et al. 2001). This progressive and sequential onset of phenotypic events is well adapted for studying early changes in neural processing associated with cognitive disturbances due to the HD mutation.

The activity of striatal cells recorded from wild-type (WT) littermates was tuned to the trajectory between nose-poke area and the food port, action preparation and initiation, and reaching the goal (food well) rather than the operant nose-poke behavior per se (Fig. 14.2). A majority of cells responded to these behavioral/task stimuli and action by increasing their discharge rates, and only a small (<5 %) proportion of striatal cells decreased discharge rates especially during reward consumption period (Fig. 14.2). This was the case for not only phasically active MSN, but also IN, most of which are suspected to be the ones expressing parvalbumin because of their well-known tonic discharge characteristics (Berke et al. 2004; Mallet et al. 2005). Our data confirm the recent controversial findings that striatal IN do display firing patterns that are sensitive to behaviors (Berke 2008; Gage et al. 2010; Stalnaker et al. 2010). In addition, the proportion of cells responding to task elements increased as mice acquired the task from early (60–75 % correct) to late (75–100 % correct) stages of learning (Fig. 14.2f). Surprisingly, striatal cells recorded from R6/1 mice also displayed similar task sensitivity as well as plastic changes of firings; the proportion of task sensitive neurons increased as mice learned the task (Fig. 14.2e-f). This suggests somewhat normal integrity of striatal function despite the HD pathology in these mice.

One intriguing observation was that the recorded neurons in the HD transgenics were mostly IN, while they represent only a half in proportion among the recorded neurons in WT mice (Fig. 14.3a). Even though MSNs represent 95–97 % of the striatal cell population, they are known to be silent in most physiological conditions, explaining such a low proportion (i.e., about 50 %) of MSN recorded in WT mice. As no cell loss has been reported in R6/1 mice, we suspected that most of MSN in transgenic mice did not reach spike firing threshold, and thus were dysfunctional. This activity pattern was associated with severe learning impairments. R6/1 mice required 4–5 times more sessions to attain an asymptote performance level, which

---

placed beneath the fourth wall. **b–d** Rastor plots color coded (with *white to black i*ndicating high to low activity levels) and average firing rates (circular peri-event time histogram) for three representative striatal cells across three (*A, B, and C*) sequential periods repeated during the nose-poke task. The three periods were bordered by nose-poke, animal's arrival at the food port marked by first lick, exit of the reward port and nose-poke again. The period A was repeated to illustrate the circular nature of the behavioral sequences. **e** Percent modulated cells during the three periods in both genotypes. **f** Percent modulated projection cells (*MSN*) and interneurons (*IN*) throughout four different stages of the operant conditioning in both genotypes

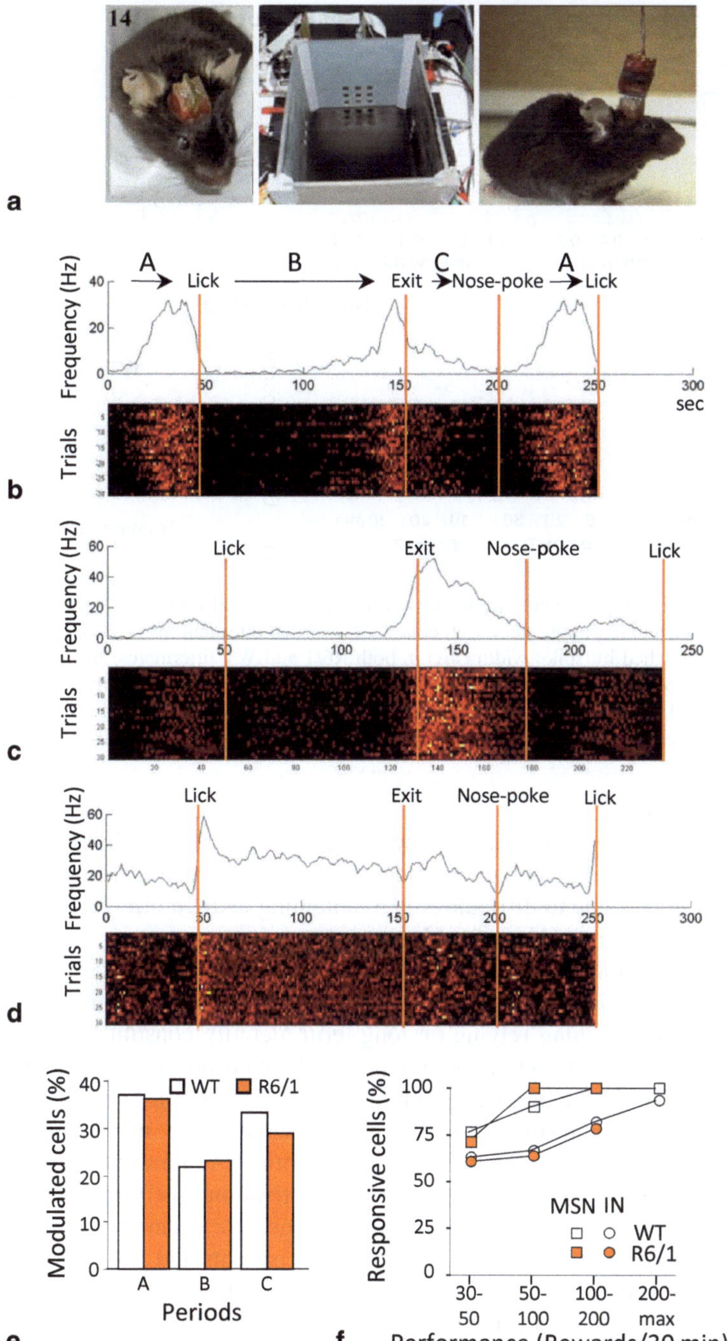

Fig. 14.2 Task-sensitive activity in the striatum. **a** Photographs of a mouse with an electrode implant and operant chamber with nose-poke holes placed in its three inner walls. A food port was

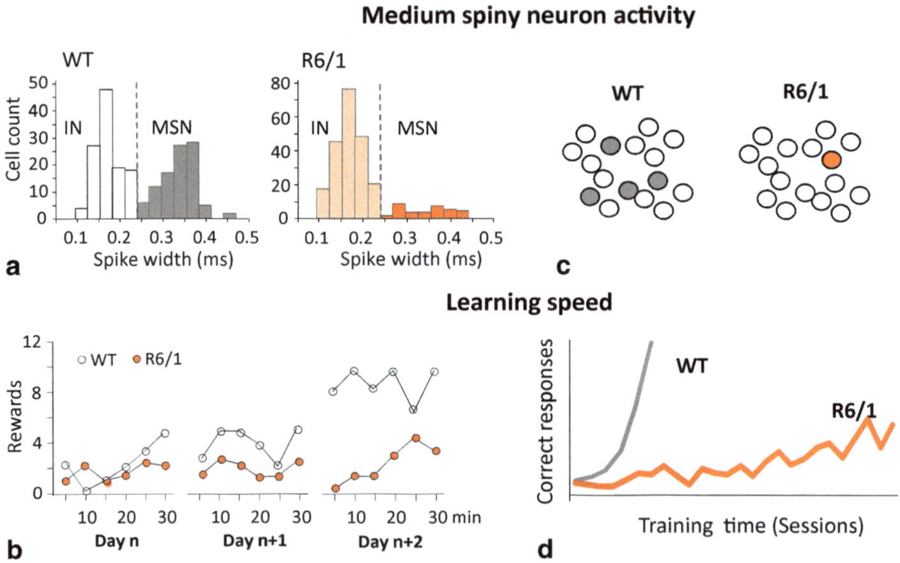

Fig. 14.3 Relationship between the scarce recruitment of striatal projection cells (*MSN*) in R6/1 transgenic mice and learning rate. **a** Cell counts of recorded interneurons (*IN*) and projection cells (*MSN*) distinguished by spike width (*ms*) in both R6/1 and WT littermates. **b** Example of learning curb for representative WT and R6/1 mice across three consecutive daily learning sessions (30 min). **c** and **d** Cartoons showing a diminished recruitment of MSNs and a retarded learning rate in R6/1 mice. Filled circles in (**c**) symbolize activated MSN and empty circles represent inactivated MSN

in addition, remained significantly lower than the level reached by WT littermates. When examining behavioral performances closely, retarded learning of the transgenics was in part due to difficulties in maintaining learned information over days, such that they seem to learn freshly at each session even at an asymptote stage (Fig. 14.3b). We noticed that interrupting training for one or two daily sessions was enough to disrupt significantly the learned operant behaviors. This might suggest that incremental learning relying on long-term memory consolidation requires an intact recruitment of striatal projection cells, which ensures the proper information transfer and flow in the basal ganglion circuit (Fig. 14.3c, d). The local interneuronal activity, which appeared normal in our transgenic mice, may not be sufficient to support the normal information flow and learning. Abnormalities in other brain regions or other physiological changes (e.g., motivation level) may also contribute to the behavioral changes seen in our mice, because, for example, the hippocampus also undergoes the cellular abnormalities in HD patients and R6/1 mice. However, because hippocampal lesioned mice were able to perform this task perfectly (Cho and Jeantet 2010), a hippocampal contribution to the learning abnormalities seems unlikely.

In addition, when striatal cell firing patterns in HD transgenics were closely examined, their discharges were abnormally regular and presented rhythmic firings at high frequency (60–80 Hz) referred to as high γ oscillation (Fig. 14.4a, c). Both MSN and IN displayed this abnormal firing regime in HD transgenics, while no such phenomenon was present in WT littermates. Therefore, not only the scarce recruitment of the principal projection cells during task learning, but also excessive γ synchrony in intrinsic firings characterizes "abnormal" striatal activity in HD mice. Pairs of neurons recorded either from the same tetrodes or tetrodes placed bilaterally also showed common synchronous discharges at this γ range (Fig. 14.4b, d). This suggests a more global synchrony within (or maybe beyond) the cortico-striatal loop. This was also confirmed by an additional observation that motor cortical cells above the striatum also resonate in the γ frequency.

When we looked at what precise moments during behavioral testing the γ oscillation in the population activity (reflected in the LFPs) becomes abnormally strong, we found that this activity was prominent when mice were immobile during reward consumption. It is known that dopamine (DA) is released while animals receive reward, and DA transmission is altered in HD (Andre et al. 2010; Cha et al. 1999; Chen et al. 2013; Jakel and Maragos 2000; Petersen et al. 2002). Therefore, it is tempting to speculate that the aberrant γ synchrony is associated with transient and abnormal DA transmission. Whether the γ oscillation is unique to our mice or this rhythm exists in HD patients remains yet unexplored. A few single-unit recording studies in humans limited to the pallidum, demonstrated increased or unchanged pallidal activity (Starr et al. 2008; Tang et al. 2005).

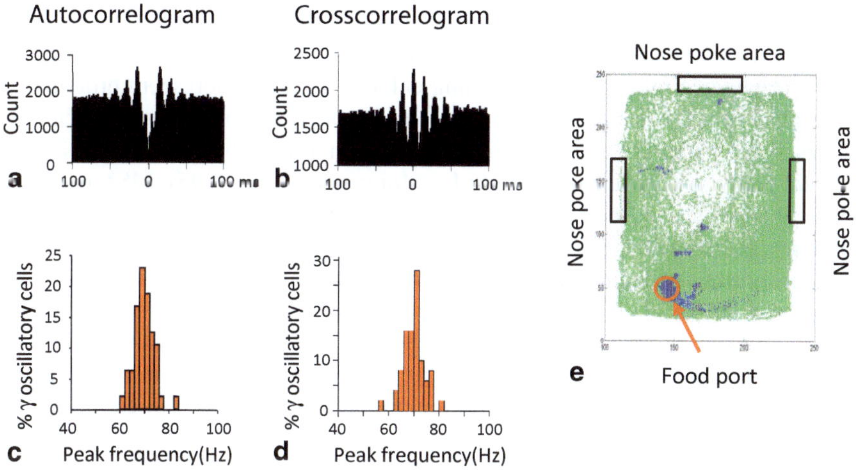

Fig. 14.4 High γ-oscillation in the striatum in R6/1 **a** and **b** Autocorrelogram and crosscorrelogram of striatal cells displaying γ synchrony. **c** and **d** Relative peak γ frequency distributions of single (**c**) and paired (**d**) cells. **e** Spatial positions of an R6/1 mouse in the operant chamber when his striatal LFP expressed high γ synchrony (blue dots, green dots show visited pixels)

Basal Ganglia Network Activity Across the Disease Progression in HD Mice

We previously formulated the hypothesis that dysfunctional striatal MSNs, which did not meet threshold for action potentials during recording, might be part of the indirect pathway vulnerable in HD. As a first step for testing this long standing but untested hypothesis, we employed immunocytochemistry for labeling the neural activity marker, c-Fos, to determine age-dependent changes in the basal ganglia activity.

Our data were surprising for two reasons. First, contrary to our expectation, in R6/1 mice, there was no diminution of striatal activity revealed by Fos-immunoreactivity (IR) subsequent to behavioral stimulation at any ages studied (Bassil et al. 2012). This suggests that the cellular machinery necessary for c-Fos transcriptional activity remains relatively intact in the transgenic mice even at symptomatic ages. Second, contrary to the expected diminution, when compared with WT counterpart, the level of Fos-IR in the dorsomedial striatum was unexpectedly increased at an early asymptomatic age well before the appearance of motor symptoms. These early cellular and possibly compensatory changes by the intact direct pathway were correlated with significant degradation of cognitive and social behaviors (Pietropaolo et al. 2011; Lebreton et al. 2015) in these young R6/1 mice. The increase of Fos-IR disappeared at older (4 and 6 months) ages in R6/1 mice, annulling differences between the genotypes. The pattern of Fos activation of infra- and prelimbic prefrontal cortex that project directly to the dorsomedial striatum (Voorn et al. 2004) studied here, followed closely the activation pattern of the neurons in the striatum. This confirms that changes in the integral cortico-striatal pathway are an early event in HD (Andre et al. 2011a). The exact mechanisms responsible for these unique and paradoxical changes at early ages remain to be investigated.

To dissociate the MSN activity of the indirect from that of the direct pathway we used a transgenic mouse line expressing enhanced green fluorescent protein under the control of a D1 receptor promoter. D1 receptors are expressed mainly in MSNs of the direct pathway, while D2 receptors are mainly expressed in MSNs of the indirect pathway (Gerfen et al. 1990; Le Moine and Bloch 1995). MSNs constitute about 95 % of striatal cell population as mentioned above, and MSNs of the two pathways are equivalent in numbers (Matamales et al. 2009), even though a negligible proportion of them coexpresses both D1 and D2 receptors. Using colabeling of Fos and DARP32 (labeling all MSNs) in crossbred mice (R6/1 x D1-GFP), we quantified the proportions (and difference) of D1 labeled MSNs among Fos-IR neurons. Preliminary results of this ongoing experiment confirmed an increased Fos-expression in the striatum at presymptomatic age in HD transgenics. In addition, the percent of MSN expressing D1 receptors among Fos-IR cells was significantly higher in the R6/1 mice as compared to WT mice, suggesting a hyperactivity of the direct D1 pathway in these asymptomatic transgenic mice.

β Synchrony and Its Paradoxical Relationship with Sleep–Wake Vigilance States

When analyzing striatal activity during the previous procedural/habit learning, we have noticed subtle changes in striatal LFPs in the frequencies of 25–35 Hz, which we referred to as β oscillations. Albeit different in the frequency, this oscillatory activity present in both normal and transgenic mice was associated with specific behaviors, for example, at the mouse's arrival at the food port. Previous recording studies performed in healthy rats also reported such oscillatory activity in the cortico-striatal circuit (Howes et al. 2011; Leventhal et al. 2000). Because the β bursts arising at discrete moments during habit learning increased significantly when animals acquired the task, this transient synchrony has been proposed to be a marker of the habit memory formation (Howes et al. 2011).

The procedural/habit memory formation requires over night sleep, and slow wave sleep (SWS) or rapid eye movement (REM) sleep deprivation perturbs the retention of the learned procedural memory and retards the task acquisition (Marshall and Born 2007; Walker and Stickgold 2004). Since HD patients and HD transgenic mice display sleep and circadian cycle disturbances (Wiegand et al. 1991; Petersen et al. 2005; Arnulf et al. 2008; Goodman et al. 2011; Morton et al. 2005), the examination of brain rhythmic activities during sleep might be informative for explaining cognitive and behavioral changes seen in HD.

We, therefore, performed electrophysiological recording of LFPs of different regions of the basal ganglia circuitry throughout disease onset and progression according to a longitudinal plan. The analysis of different LFPs during sleep revealed an abnormal β synchrony (20–40 Hz) in all R6/1 mice studied, but not in any of WT littermates across ages. This β oscillation thus constituted a hallmark brainwave of HD in R6/1 mice. Ages at which the β rhythm was first detected preceded neurological hind limb clasping by 1.5 months in most of the transgenics, announcing the later appearance of motor impairments (Jeantet et al. 2013).

In addition, an intriguing finding was that the β synchrony owned a unique and unexpected relationship with sleep–wake vigilance states. More precisely, the β rhythm was mainly associated with sleep state because it was present barely during waking state. However, when entering REM sleep, which is usually associated with intense desynchronized cortical activity similar to the waking state, the β oscillation continued to intensify its amplitude instead of disintegrating as major cortical synchrony (Fig. 14.5a-b). Therefore, characteristics of the β oscillation in our HD mice enabled a clear dissociation between REM sleep and waking states.

Finally, this pattern of β oscillation variation across vigilance levels became more pronounced with the disease progression. The β peak shifted with age toward lower frequency but higher amplitude within a given brain state (Fig. 14.5c-d). The phenomenon thus became more exaggerated with the disease progression. However, the frequency and amplitude of 7–10 Hz θ rhythm also strongly present in REM sleep spectra in both R6/1 and WT mice, remained constant over ages even in

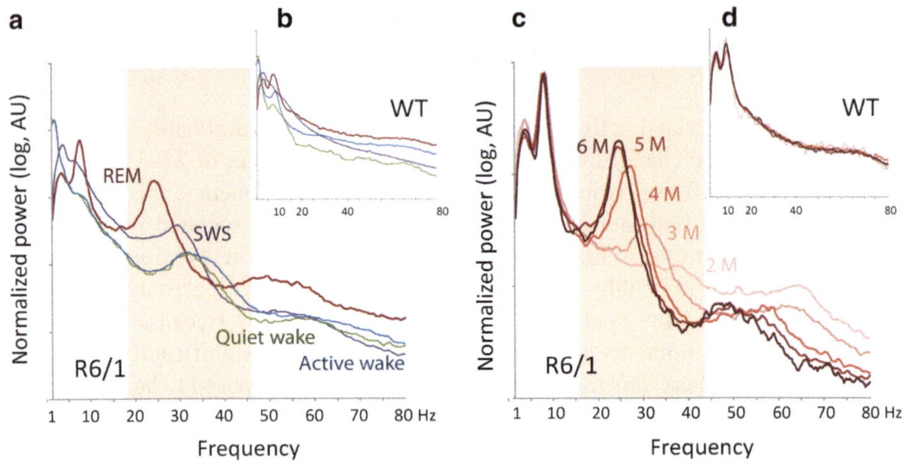

Fig. 14.5 β (20–40 Hz) oscillation in R6/1 mice **a** and **b** Power variation in the β frequency range across four vigilance (*active wake, quiet wake, slow wave sleep (SWS), and rapid eye movement sleep (REM)*) states in a representative symptomatic R6/1 mouse (**a**) and an age-matched littermate (**b**). *Blue, green, purple, and brown lines* in **a** and **b** represent active wake, quiet wake, SWS, and REM sleep, respectively. **c** and **d** REM sleep power spectral variations across disease progression (*2–6 months (M), symptomatic to symptomatic, respectively*) in the same R6/1 (**c**) and WT littermate (**d**) shown in **a** and **b**

R6/1 mice. However, its frequency remained 0.5–0.75 Hz lower in R6/1 mice than in WT mice. These data identify the β, but not θ synchrony, as a dynamic in vivo neurophysiological marker of HD, which accompanies sleep abnormalities (i.e., fragmentation and diurnal activity changes) and cognitive impairments in these mice (Lebreton et al. 2015). The fact that the β oscillation is present during SWS and REM sleep, periods critical for offline information processing and consolidation, and the aberrant oscillations may disturb these critical neural processes and information transfer among neural regions involved in learning, providing network "correlates" of the observed learning deficits mentioned earlier (Cayzac et al. 2011). This also highlights the interrelationship among the cortico-striatal and basal ganglia network activity, sleep and cognitive functions in general and especially in HD.

During typical sleep–wake cycles, two types of phenomena are known to announce REM sleep onset following SWS: (1) the return of certain brain (cortex and hippocampus) activities to awake-like state and (2) a further intensification of certain SWS characteristics such as increased threshold for awakening by sensory stimuli (during eye movement) (Ermis 2010) and loss of muscular tone (atonia). Since, in R6/1 mice, the β synchrony that appears during SWS and grows further during REM sleep, the cortico-striato–thalamo–cortical network at the origin of this synchrony may be subject to the second type of phenomenon, i.e., the further intensification of SWS characteristics. Neither the exact constituents of the β generating circuit (i.e., striatum, subthalamic nuclei, thalamus, cortex, etc.), nor the way in which this circuitry is modulated during REM sleep has been identified. However, our data suggest that the machinery for β, but not for θ synchrony, that is distinctively active during sleep, is dysfunctional in HD mice. This specificity may point to defective brain activation systems (e.g., DA, Acetylcholine, Serotonin, Orexin, etc.) or other mechanisms within the restricted circuit affected in HD, which when dissected, may offer new targets for therapeutic interventions.

Conclusion

Basal ganglia form a complex neural network involved in the selection and execution of response/action through interactions with multiple brain areas that process sensorimotor, emotional, and cognitive information. Here, we summarized how the functional network activity in the basal ganglion is compromised in a mouse model of HD, which expresses the HD mutation. Growing evidence suggests that synchronization of neuronal activity within and across different brain regions is a fundamental property of cortical and subcortical networks and serves a variety of functions in cognitive processes (Fries 2005; Singer 1999). Because remarkable and unique synchronies in the β and γ frequencies were associated with cognitive and behavioral perturbations in HD mice, it is tempting to speculate that these aberrant neural synchronies play critical roles in cognitive and motor dysfunctions associated with HD. Further work should aim at understanding of how these peculiar synchronies involving large populations of neurons perturb the proper information coding and flow in the basal ganglia loop and related limbic structures. These studies may ultimately bring a new insight into not only the pathophysiology of HD, but also the nature of the cognitive operations performed in/by the intact striatum, the main component of habit and procedural learning system.

Acknowledgment The works summarized here were supported by the Hereditary Disease Foundation, the University of Bordeaux 1, the HD Society of America, and the Agence Nationale de la Recherche (ANR-08-MNPS-019-01).

The authors wish to thank Vietminh Paz, Sebastien Delcasso, Sebastien Cayzac, Susanna Pietropaolo, Pauline Delage, Fanny Lebreton, Michele Pignatelli, Xavier Leinekugel, Fares Bassil, Nicole Mons, Huowel Du, Maurice Garret, Magali Cabanas for their fruitful collaborations.

References

Andre, V. M., Cepeda, C., & Levine, M. S. (2010). Dopamine and glutamate in Huntington's disease: A balancing act. *CNS Neuroscience & Therapeutics, 16*(3), 163–178.

Andre, V. M., Fisher, Y. E., & Levine, M. S. (2011a). Altered balance of activity in the striatal direct and indirect pathways in mouse models of Huntington's disease. *Frontiers in Systems Neuroscience, 5,* 46.

Andre, V. M., Cepeda, C., Fisher, Y. E., Huynh, M., Bardakjian, N., & Singh, S. (2011b). Differential electrophysiological changes in striatal output neurons in Huntington's disease. *The Journal of Neuroscience, 31*(4), 1170–1182.

Arnulf, I., Nielsen, J., Lohmann, E., Schiefer, J., Wild, E., Jennum, P., Konofal, E., Walker, M., Oudiette, D., Tabrizi, S., Durr, A. (2008). Rapid eye movement sleep disturbances in Huntington disease. *Archives of Neurololy, 65*(4), 482–488.

Barnes, T. D., Kubota, Y., Hu, D., Jin, D. Z., & Graybiel, A. M. (2005). Activity of striatal neurons reflects dynamic encoding and recoding of procedural memories. *Nature, 437*(7062), 1158–1161.

Bassil, F., Du, Z., Garret, M., Mons, N., & Cho, Y. H. (2012). Altered basal ganglion activity in the R6/1 mice. *Society for Neuroscience Abstract.*

Berke, J. D. (2008). Uncoordinated firing rate changes of striatal fast-spiking interneurons during behavioral task performance. *The Journal of Neuroscience, 28*(40), 10075–10080.

Berke, J. D., Okatan, M., Skurski, J., & Eichenbaum, H. B. (2004). Oscillatory entrainment of striatal neurons in freely moving rats. *Neuron, 43*(6), 883–896.

Brooks, S. P., Betteridge, H., Trueman, R. C., Jones, L., & Dunnett, S. B. (2006). Selective extra-dimensional set shifting deficit in a knock-in mouse model of Huntington's disease. *Brain Research Bulletin, 69*(4), 452–457.

Cayzac, S., Delcasso, S., Paz, V., Jeantet, Y., & Cho, Y. H. (2011). Changes in striatal procedural memory coding correlate with learning deficits in a mouse model of Huntington disease. *Proceedings of the National Academy of Sciences of the United States of America, 108*(22), 9280–9285.

Cepeda, C., Hurst, R. S., Calvert, C. R., Hernandez-Echeagaray, E., Nguyen, O. K., & Jocoy, E. (2003). Transient and progressive electrophysiological alterations in the corticostriatal pathway in a mouse model of Huntington's disease. *The Journal of Neuroscience, 23*(3), 961–969.

Cepeda, C., Wu, N., Andre, V. M., Cummings, D. M., & Levine, M. S. (2007). The corticostriatal pathway in Huntington's disease. *Progress in Neurobiology, 81*(5–6), 253–271.

Cha, J. H. (2007). Transcriptional signatures in Huntington's disease. *Progress in Neurobiology, 83*(4), 228–248.

Cha, J. H., Frey, A. S., Alsdorf, S. A., Kerner, J. A., Kosinski, C. M., & Mangiarini, L. (1999). Altered neurotransmitter receptor expression in transgenic mouse models of Huntington's disease. *Philosophical Transactions of the Royal Society of London. Series B, Biological Sciences, 354*(1386), 981–989.

Chen, J. Y., Wang, E. A., Cepeda, C., & Levine, M. S. (2013). Dopamine imbalance in Huntington's disease: A mechanism for the lack of behavioral flexibility. *Frontiers in Neurosciences, 7,* 114.

Cho, Y. H., & Jeantet, Y. (2010). Differential involvement of prefrontal cortex, striatum, and hippocampus in DRL performance in mice. *Neurobiology of Learning and Memory, 93*(1), 85–91.

Cummings, D. M., Milnerwood, A. J., Dallerac, G. M., Waights, V., Brown, J. Y., & Vatsavayai, S. C. (2006). Aberrant cortical synaptic plasticity and dopaminergic dysfunction in a mouse model of Huntington's disease. *Human Molecular Genetics, 15*(19), 2856–2868.

Cummings, D. M., Milnerwood, A. J., Dallerac, G. M., Vatsavayai, S. C., Hirst, M. C., & Murphy, K. P. (2007). Abnormal cortical synaptic plasticity in a mouse model of Huntington's disease. *Brain Research Bulletin, 72*(2/3), 103–107.

Cybulska-Klosowicz, A., Mazarakis, N. K., Van Dellen, A., Blakemore, C., Hannan, A. J., & Kossut, M. (2004). Impaired learning-dependent cortical plasticity in Huntington's disease transgenic mice. *Neurobiology of Diseases, 17*(3), 427–434.

DiFiglia, M., Sapp, E., Chase, K., Schwarz, C., Meloni, A., & Young, C. (1995). Huntingtin is a cytoplasmic protein associated with vesicles in human and rat brain neurons. *Neuron, 14*(5), 1075–1081.

Dumas, E. M., van den Bogaard, S. J., Middelkoop, H. A., & Roos, R. A. (2013). A review of cognition in Huntington's disease. *Frontiers in Bioscience (Scholar Edition), 5,* 1–18.

Ermis, U., Krakow, K., Voss, U. (2010). Arousal thresholds during human tonic and phasic REM sleep. *Jounal of Sleep Research, 19*(3), 400–406.

Ferrante, R. J. (2009). Mouse models of Huntington's disease and methodological considerations for therapeutic trials. *Biochimica et Biophysica Acta, 1792*(6), 506–520.

Fries, P. (2005). A mechanism for cognitive dynamics: Neuronal communication through neuronal coherence. *Trends in Cognitive Sciences, 9*(10), 474–480.

Gage, G. J., Stoetzner, C. R., Wiltschko, A. B., & Berke, J. D. (2010). Selective activation of striatal fast-spiking interneurons during choice execution. *Neuron, 67*(3), 466–479.

Gerfen, C. R., Engber, T. M., Mahan, L. C., Susel, Z., Chase, T. N., & Monsma, F. J. Jr. (1990). D1 and D2 dopamine receptor-regulated gene expression of striatonigral and striatopallidal neurons. *Science, 250*(4986), 1429–1432.

Graybiel, A. M., Aosaki, T., Flaherty, A. W., & Kimura, M. (1994). The basal ganglia and adaptive motor control. *Science, 265*(5180), 1826–1831.

Grote, H. E., Bull, N. D., Howard, M. L., van Dellen, A., Blakemore, C., & Bartlett, P. F. (2005). Cognitive disorders and neurogenesis deficits in Huntington's disease mice are rescued by fluoxetine. *The European Journal of Neuroscience, 22*(8), 2081–2088.

Gutekunst, C. A., Li, S. H., Yi, H., Mulroy, J. S., Kuemmerle, S., & Jones, R. (1999). Nuclear and neuropil aggregates in Huntington's disease: Relationship to neuropathology. *The Journal of Neuroscience, 19*(7), 2522–2534.

Howes, O., Bose, S., Turkheimer, F., Valli, I., Egerton, A., & Stahl, D. (2011). Progressive increase in striatal dopamine synthesis capacity as patients develop psychosis: A PET study. *Molecular Psychiatry, 16*(9), 885–886.

Jakel, R. J., & Maragos, W. F. (2000). Neuronal cell death in Huntington's disease: A potential role for dopamine. *Trends in Neurosciences, 23*(6), 239–245.

Jeantet, Y., Cayzac, S., & Cho, Y. H. (2013). Beta oscillation during slow wave sleep and rapid eye movement sleep in the electroencephalogram of a transgenic mouse model of Huntington's disease. *PLoS One, 8*(11), e79509.

Jog, M. S., Kubota, Y., Connolly, C. I., Hillegaart, V., & Graybiel, A. M. (1999). Building neural representations of habits. *Science, 286*(5445), 1745–1749.

Josiassen, R. C., Curry, L. M., & Mancall, E. L. (1983). Development of neuropsychological deficits in Huntington's disease. *Archives of Neurology, 40*(13), 791–796.

Kesner, R. P., & Rogers, J. (2004). An analysis of independence and interactions of brain substrates that subserve multiple attributes, memory systems, and underlying processes. *Neurobiology of Learning and Memory, 82*(3), 199–215.

Lawrence, A. D., Sahakian, B. J., Hodges, J. R., Rosser, A. E., Lange, K. W., & Robbins, T. W. (1996). Executive and mnemonic functions in early Huntington's disease. *Brain, 119*(Pt 5), 1633–1645.

Lawrence, A. D., Watkins, L. H., Sahakian, B. J., Hodges, J. R., & Robbins, T. W. (2000). Visual object and visuospatial cognition in Huntington's disease: Implications for information processing in corticostriatal circuits. *Brain, 123*(Pt 7), 1349–1364.

Le Moine, C., & Bloch, B. (1995). D1 and D2 dopamine receptor gene expression in the rat striatum: Sensitive cRNA probes demonstrate prominent segregation of D1 and D2 mRNAs in distinct neuronal populations of the dorsal and ventral striatum. *The Journal of Comparative Neurology, 355*(3), 418–426.

Lebreton, F., Cayzac, S., Pietropaolo, S., Jeantet, Y., Cho, Y. H. (2015). Sleep physiology alterations precede plethoric phenotypic changes in R6/1 Huntington's disease mice, *Plos One, i0(5), e0126972.*

Lemiere, J., Decruyenaere, M., Evers-Kiebooms, G., Vandenbussche, E., & Dom, R. (2004). Cognitive changes in patients with Huntington's disease (HD) and asymptomatic carriers of the HD mutation—a longitudinal follow-up study. *Journal of Neurology, 251*(8), 935–942.

Leventhal, L., Sortwell, C. E., Hanbury, R., Collier, T. J., Kordower, J. H., & Palfi, S. (2000). Cyclosporin A protects striatal neurons in vitro and in vivo from 3-nitropropionic acid toxicity. *The Journal of Comparative Neurology, 425*(4), 471–478.

Lione, L. A., Carter, R. J., Hunt, M. J., Bates, G. P., Morton, A. J., & Dunnett, S. B. (1999). Selective discrimination learning impairments in mice expressing the human Huntington's disease mutation. *The Journal of Neuroscience, 19*(23), 10428–10437.

Mallet, N., Le Moine, C., Charpier, S., & Gonon, F. (2005). Feedforward inhibition of projection neurons by fast-spiking GABA interneurons in the rat striatum in vivo. *The Journal of Neuroscience, 25*(15), 3857–3869.

Mangiarini, L., Sathasivam, K., Seller, M., Cozens, B., Harper, A., & Hetherington, C. (1996). Exon 1 of the HD gene with an expanded CAG repeat is sufficient to cause a progressive neurological phenotype in transgenic mice. *Cell, 87*(3), 493–506.

Marshall, L., & Born, J. (2007). The contribution of sleep to hippocampus-dependent memory consolidation. *Trends in Cognitive Sciences, 11*(10), 442–450.

Matamales, M., Bertran-Gonzalez, J., Salomon, L., Degos, B., Deniau, J. M., & Valjent, E. (2009). Striatal medium-sized spiny neurons: Identification by nuclear staining and study of neuronal subpopulations in BAC transgenic mice. *PLoS One, 4*(3), e4770.

Morton, A. J., Wood, N. I., Hastings, M. H., Hurelbrink, C., Barker, R. A., & Maywood, E. S. (2005). Disintegration of the sleep-wake cycle and circadian timing in Huntington's disease. *The Journal of Neuroscience, 25*(1), 157–163.

Naver, B., Stub, C., Moller, M., Fenger, K., Hansen, A. K., & Hasholt, L. (2003). Molecular and behavioral analysis of the R6/1 Huntington's disease transgenic mouse. *Neuroscience, 122*(4), 1049–1057.

Nicniocaill, B., Haraldsson, B., Hansson, O., O'Connor, W. T., & Brundin, P. (2001). Altered striatal amino acid neurotransmitter release monitored using microdialysis in R6/1 Huntington transgenic mice. *The European Journal of Neuroscience, 13*(1), 206–210.

Nithianantharajah, J., Barkus, C., Murphy, M., & Hannan, A. J. (2008). Gene-environment interactions modulating cognitive function and molecular correlates of synaptic plasticity in Huntington's disease transgenic mice. *Neurobiology of Diseases, 29*(3), 490–504.

Pang, T. Y., Du, X., Zajac, M. S., Howard, M. L., & Hannan, A. J. (2009). Altered serotonin receptor expression is associated with depression-related behavior in the R6/1 transgenic mouse model of Huntington's disease. *Human Molecular Genetics, 18*(4), 753–766.

Petersen, A., Puschban, Z., Lotharius, J., NicNiocaill, B., Wiekop, P., & O'Connor, W. T. (2002). Evidence for dysfunction of the nigrostriatal pathway in the R6/1 line of transgenic Huntington's disease mice. *Neurobiology of Diseases, 11*(1), 134–146.

Petersen, A., Gil, J., Maat-Schieman, M. L., Bjorkqvist, M., Tanila, H., Araújo, I. M., Smith, R., Popovic, N., Wierup, N., Norlén, P., Li, J. Y., Roos, R. A., Sundler, F., Mulder, H., Brundin, P. (2005). Orexin loss in Huntington's disease. *Human Molecular Genetics, 14*(1), 39–47.

Pietropaolo, S., Delage, P., Cayzac, S., Crusio, W. E., & Cho, Y. H. (2011). Sex-dependent changes in social behaviors in motor pre-symptomatic R6/1 mice. *PLoS One, 6*(5), e19965.

Reiner, A., Albin, R. L., Anderson, K. D., D'Amato, C. J., Penney, J. B., & Young, A. B. (1988). Differential loss of striatal projection neurons in Huntington disease. *Proceedings of the National Academy of Sciences of the United States of America, 85*(15), 5733–5737.

Singer, W. (1999). Neuronal synchrony: A versatile code for the definition of relations? *Neuron, 24*(1), 49–65, 111–125.

Squire, L. R. (1992). Memory and the hippocampus: A synthesis from findings with rats, monkeys, and humans. *Psychological Review, 99*(2), 195–231.

Stalnaker, T. A., Calhoon, G. G., Ogawa, M., Roesch, M. R., & Schoenbaum, G. (2010). Neural correlates of stimulus-response and response-outcome associations in dorsolateral versus dorsomedial striatum. *Frontiers in Integrative Neuroscience, 4,* 12.

Starr, P. A., Kang, G. A., Heath, S., Shimamoto, S., & Turner, R. S. (2008). Pallidal neuronal discharge in Huntington's disease: Support for selective loss of striatal cells originating the indirect pathway. *Experimental Neurology, 211*(1), 227–233.

Tang, J. K., Moro, E., Lozano, A. M., Lang, A. E., Hutchison, W. D., & Mahant, N. (2005). Firing rates of pallidal neurons are similar in Huntington's and Parkinson's disease patients. *Experimental Brain Research, 166*(2), 230–236.

The Huntington's Disease Collaborative Research Group (1993). A novel gene containing a trinucleotide repeat that is expanded and unstable on Huntington's disease chromosomes. The Huntington's disease collaborative research group. *Cell, 72*(6), 971–983.

Vonsattel, J. P., Myers, R. H., Stevens, T. J., Ferrante, R. J., Bird, E. D., & Richardson, E. P. Jr. (1985). Neuropathological classification of Huntington's disease. *Journal of Neuropathology and Experimental Neurology, 44*(6), 559–577.

Voorn, P., Vanderschuren, L. J., Groenewegen, H. J., Robbins, T. W., & Pennartz, C. M. (2004). Putting a spin on the dorsal-ventral divide of the striatum. *Trends Neuroscience, 27*(8), 468–474.

Walker, M. P., & Stickgold, R. (2004). Sleep-dependent learning and memory consolidation. *Neuron, 44*(1), 121–133.

Wiegand, M., Moller, A. A., Lauer, C. J., Stolz, S., Schreiber, W., Dose, M., Krieg, J. C. (1991). Nocturnal sleep in Huntington's disease. *Journal of Neurology, 238*(4), 203–208.

Yin, H. H., & Knowlton, B. J. (2006). The role of the basal ganglia in habit formation. *Nature Reviews. Neuroscience, 7*(6), 464–476.

Chapter 15
Applying the Attribute Model to Develop Behavioral Tasks that Phenocopy Human Clinical Phenotypes Using Mouse Disease Models: An Endophenotyping Approach

Michael R. Hunsaker

Introduction

This chapter is designed to demonstrate exactly where the attribute model as proposed by Ray Kesner is being extended. Specifically, the logic proposed in the attribute model is being used to guide advanced task selection and development in genetic models of human disease. This is important since there have been problems in translating mouse disease research, particularly as related to therapeutic studies, into humans (cf., Hunsaker 2012a, b).

I propose that this is due to poor task selection and not taking into account precisely what is being tested in the mice or rodents and what is being tested in the human population. Taking these factors into account is already providing novel information that can guide research into mouse models of human genetic disease as well as traumatic brain research in rodents. The introduction of the NIH Toolbox in 2012 has made it abundantly clear that behavioral research into mouse models need to develop a parallel assessment toolkit. The attribute model is a perfect match with the tasks that lie ahead: namely the necessity to specifically identify what cognitive domains are actually being tested in the human patient population and which are not, deciding what are the fundamental cognitive processes underlying the domain being tested in the patient population, and subsequently designing tests in the rat or

M. R. Hunsaker (✉)
Center for Integrative Neuroscience and Human Behavior, University of Utah,
26 South Wasatch Drive, Salt Lake City, UT 84112, USA
e-mail: ryanhunsaker@icloud.com

Department of Neurological Surgery, University of California, Davis, 1515 Newton Ct., Davis, CA 95616, USA

Granite School District, 2500 S. State Street, Salt Lake City, UT, USA
e-mail: mrhunsaker@graniteschools.org

P. A. Jackson et al. (eds.), *The Neurobiological Basis of Memory*,
DOI 10.1007/978-3-319-15759-7_15

mouse model that specifically serve as valid, *homologous,* tests when compared to the results from human patient populations.

General Introduction

With the increasing sophistication of the genetic techniques used to develop mouse models of genetic disorders, it is imperative that the techniques used to elucidate the behavioral phenotype of these models evolve just as rapidly. Although there is a movement toward adopting standardized behavioral phenotyping protocols, to a large degree neuroscientists evaluating mouse models of genetic disorders still lack sensitive behavioral assays needed to evaluate the core cognitive deficits present in genetic disorders. At present, mouse models, particularly those developed to study neurodevelopmental or other genetic disorders, demonstrate inconsistent phenotypes or entirely lack behavioral phenotypes when tested using the most common behavioral tasks: including the water maze, Barnes maze, active/passive avoidance, rotarod, or fear conditioning (Baker et al. 2010; Bohlen et al. 2009; Cannon and Keller 2006; Hasler et al. 2006; Kendler and Neale 2010; Long et al. 2006; Manji et al. 2003; Paylor and Lindsay 2006; Rustay et al. 2003; Spencer et al. 2011; Weiser et al. 2005; Yan et al. 2004). Furthermore, it has been shown that even extremely subtle differences between individual laboratory protocols for these common tasks changes observed phenotypes (e.g., differences in care taken while fixing a rough surface to a rotarod has been shown to dissociate performance across collaborating laboratories studying the same mouse strain; cf., Crabbe et al. 1999; Crabbe and Wahlsten 2003; Wahlsten 1972, 2001; Wahlsten et al. 2003a, b, c; 2006).

Additionally, mouse models often demonstrate phenotypes that are not specifically associated with any genetic disorder in particular, but are more aptly described as shared clinical phenotypes that similarly present across a wide array of disorders (e.g., gross learning and memory deficits, anxiety, depression). The interpretation of such inconclusive findings is often that the mouse model fails to recapitulate the phenotypes observed in patients (cf., Gottesman and Gould 2003; Gould and Gottesman 2006; Weiser et al. 2005). Unfortunately, these types of findings are analogous to inconsistent findings in clinical populations when standardized neuropsychological tests are administered—many different populations show very similar deficits despite nonoverlapping genetic or developmental disorders (cf., Fig. 15.1a wherein the final result is "shared clinical phenotype"). Such inconsistencies often render behavioral research into developmental or psychiatric disorders frustrating and such anomalous findings mask the differences that do exist. I propose that inconsistent behavioral results observed in clinical populations as well as mouse models do not infer the lack of cognitive impairments, but rather these "null" data reflect the often startling insensitivity of the behavioral tasks commonly employed.

In situations where, based on standardized behavioral tasks, mouse models do not appear to specifically model clinical phenotypes observed in patient populations, one strategy is to evaluate intermediate- or endophenotypes associated specifically with the genetic mutation and subserved by neuroanatomical structures

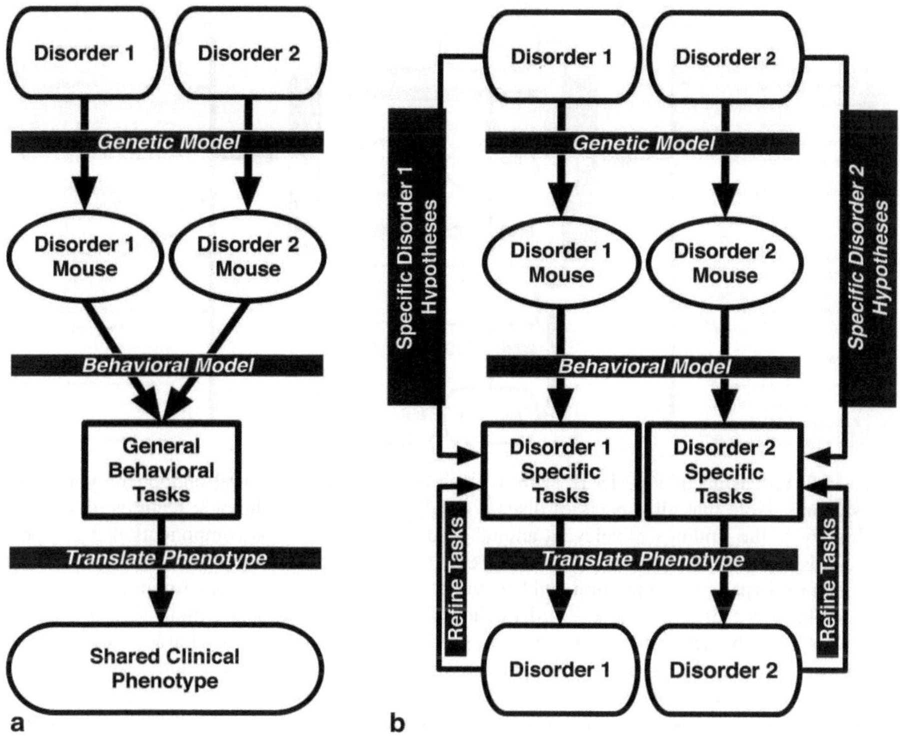

Fig. 15.1 a Diagram of standard behavioral phenotyping process in which different mouse models are given the same battery of tasks to define a behavioral phenotype. The outcome of the behavioral tasks is compared to the full clinical phenotype of the genetic disorders being modeled. This approach lacks the specificity and selectivity to identify phenotypes unique to a single disorder. This is a representation of the standard mouse behavioral phenotyping methods (e.g., SHIRPA) as well as the NIH Toolbox. b Diagram of behavioral endophenotyping process in which disorder-specific hypotheses are used to develop unique batteries of behavioral tasks that directly translate to the phenotype of the clinical disorder. This approach does not model the general deficits seen across genetic disorders, but rather specifically identifies phenotypes known to be unique to the genetic disorder being modeled. Figure 15.2 expands upon Fig. 15.1b. From Hunsaker (2012a)

disrupted by the mutation (cf., Figs. 15.1b, 15.2). A similar process applies to studies of human clinical populations when standardized tests fail to uncover phenotypes that are present, but only manifest at a subclinical level (cf., Gottesman and Gould 2003; Hunsaker 2012a, b; Simon 2008, 2011).

Endophenotypes are collections of quantitative traits hypothesized to represent risk for genetic disorders at more biologically (and empirically) tractable levels than the full clinical phenotype, which often contains little more than profound deficits shared across various genetic disorders (Gould and Einat 2007). A behavioral endophenotyping approach facilitates the identification of behavioral deficits that are clearly associated with both the specific genetic mutation and the pathological features observed in the clinical populations being modeled—and more importantly with the pathological/clinical features *unique* to the population being modeled (cf.,

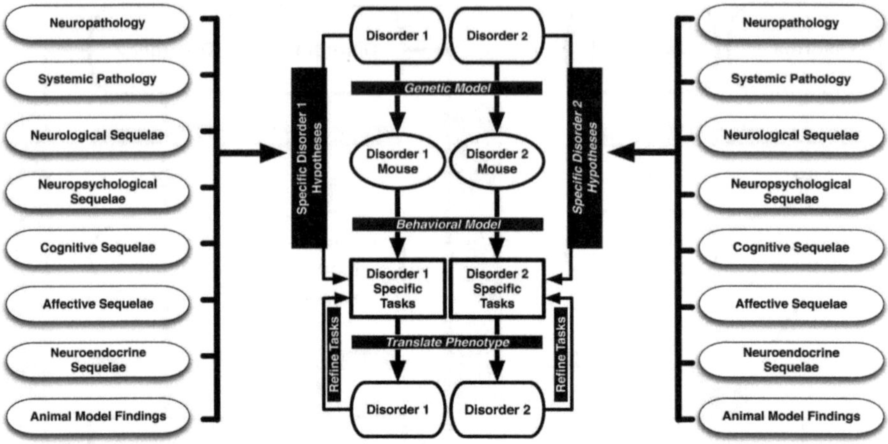

Fig. 15.2 Extension of Fig. 15.1b with a focus on what the behavioral endophenotyping process entails. Note that all aspects of disease ranging from neuropathologic features to systemic pathology to the findings of relevant animal models are included as components that comprise the disease specific hypotheses. Differential effects of disease states on these factors provide the raw data for hypothesis generation and behavioral dissociations of diseases. Importantly, this figure emphasizes that unless two disorders show identical sequelae and pathologies, one cannot be adequately characterized using the same paradigms as those used to characterize the other. Also emphasized is the use of the full spectrum of results from the mouse model being applied to inform the hypotheses concerning the genetic/clinical disorder. This emphasis is the goal for any translational study—that the clinic and the mouse model mutually inform each other's hypotheses

Fig. 15.2). When designed to evaluate such disease-specific hypotheses, behavioral endophenotypes model quantitative patterns of behavioral deficits that scale with the size and/or severity of the genetic mutation (Gottesman and Gould 2003; Gould and Gottesman 2006; Hasler et al. 2006; Hunsaker 2012a; Hunsaker et al. 2012; Weiser et al. 2005).

The behavioral endophenotyping process deviates from the currently accepted method for determining behavioral phenotypes. The present method is to determine phenotypes in clinical populations and mouse models—that of using behavioral tasks that were designed without prior consideration of the pathology and anecdotal features present in the population. Far too often an approach such as this is not sufficiently sensitive to characterize gene–brain–behavior interactions that underlie disease pathogenesis (Amann et al. 2010; Gur et al. 2007; Hunsaker 2012b; Karayiorgou et al. 2010; Simon 2007, 2008, 2011). In contrast with the currently utilized approach, behavioral endophenotyping emphasizes the use of behavioral paradigms that were developed to specifically evaluate a priori hypotheses concerning the alterations to nominal gene–brain–behavior interactions identified or proposed to exist in a given population using carefully selected tasks to identify unique phenotypes within each model; and thus are more capable of characterizing the neurocognitive consequences of the specific gene mutations underlying the genetic disorder (Gould and Gottesman 2006; cf., Fig. 15.1b wherein the final step results in separate phenotypes directly related to a disease or mutation; Fig. 15.2).

In this chapter, I will evaluate advances in the methods associated with neurobehavioral endophenotyping, and will propose a clear strategy to efficiently and comprehensively characterize neurobehavioral deficits in mouse models of genetic disorders. This approach uses neurocognitive theory to design and select behavioral tasks that test specific hypotheses concerning the genetic disorder being studied. I propose this novel approach will extend the utility of mouse models by integrating the expertise of clinical neurology and cognitive neuroscience into the mouse behavioral laboratory.

Additionally, I will discuss a new collection of psychological assessments called the NIH Toolbox. The NIH Toolbox was designed with the intent of extending the traditional behavioral phenotyping approach commonly used in animals into human clinical populations (cf., web resources at http://nihtoolbox.org; Gershon 2007; Gershon et al. 2010). Although limited in application, the development of the NIH Toolbox assessments is important for research into genetic disease models since having a standardized set of experiments in human clinical research gives behavioral neuroscientists a baseline against which to develop murine behavioral paradigms.

I propose that directly emphasizing the reciprocal translation of research between human disease states and the associated mouse models is essential for both groups to mutually inform each other's research to more efficiently generate hypotheses and elucidate treatment strategies, as has been the primary emphasis as late of the National Institutes of Health (NIH; cf., National Center for Advancing Translational Science (NCATS); http://www.ncats.nih.gov/). This type of translational science requires not only including the NIH Toolbox in studies of human clinical populations, but also extending beyond the NIH Toolbox to evaluate specific disease-related hypotheses that can be used as outcome measures or biomarkers for future studies into disease related risk. Similarly, researchers studying the behavior of mouse disease models also must extend beyond their comfort zones to make the scientific advancements that are required to inform progress taking place in the clinic.

Factors to Consider When Designing Behavioral (Endo) Phenotyping Batteries

Any discussion concerning the behavioral phenotyping of mouse models of genetic disorders must necessarily begin with a description of what a behavioral phenotype is and what assumptions underlie tasks used to evaluate them. In short, behavioral phenotyping quantifies performance of mutant mice across behavioral experiments; and the behavioral performance is related to the clinical population to identify parallels that may exist. The analogy between the phenotype of human genetic disorder and the behavioral phenotype of the mouse model can be expressed as a combination of three factors: face validity, construct (or content) validity, and predictive validity (cf., Crawley 2004; Guion 1977; Hunsaker 2012a).

Face validity is the surface similarity between the behavior of the mouse model and the patient on analogous tasks (i.e., does the performance of the mouse and human resemble each other at face value). In other words, if a mouse has to perform a similar response during a task as the patient makes during performance of a similar task, the task shows face validity. Similarly, if the mouse and human behavioral tasks can be intuitively interpreted as being similar, the task shows face validity.

Construct (or content) validity, so far as behavioral paradigms are concerned, refers to the similarity between the behavioral or cognitive domains being tested by a given task in the mouse model and the human patient. This means that for tests to show construct validity, the tasks must be designed to directly model specific aspects of the genetic disorder *and* that performance be subserved by similar neural substrates and/or cognitive process across species. More specifically, the tasks need to be developed to explicitly model the human disorder, not solely rely on creative post hoc interpretations of behavioral performance on general behavioral tasks (cf., Figs. 15.1, 15.2; Hunsaker 2012a, b). One necessity of construct validity is that a basic understanding of the disorder being modeled is required, such that the research translates a behavioral phenotype across species, not providing the primary elucidation of any phenotype at all in the mouse model—although this approach can provide useful data in limited situations such as in cases of rare genetic disorders.

Predictive validity refers to the utility of a mouse model as a proxy for the patient in studies of disease progression or therapeutic intervention—this can refer to either the endpoints of a behavioral study or the physiology of the model. Although predictive validity is commonly thought of as a characteristic of phenotyping approaches, it is more accurate to state that predictive validity is the quantified endpoint of an adequately designed behavioral phenotyping experiment—that is, to define some behavior or set of behaviors that serve as valid outcome measures for later studies (Berge 2011; Greene-Schloesser et al. 2011; Hunsaker 2012a). In other words, predictive validity is only present when behavioral performance of the model during a given experiment proves useful for inferring or correlating dosage of a given mutation, disease progression, or treatment outcomes in not only the model, but also the clinical population.

Approaches to Endophenotyping—Applying the Attribute Model

Since the tacit acceptance of the water maze, Barnes maze, passive/active avoidance, and contextual fear conditioning as the standard memory tasks for mouse models of disease (cf., Hunsaker 2012a, b; Llano Lopez et al. 2010; Whishaw and Tomie 1996), the development of behavioral tasks to dissect the role of brain regions affected by the mutation for memory processing has stalled—at least in mice. In contrast, during this same period the research into the neural systems underlying learning and memory processes has reached a boon in rats. An effort has begun to translate these more sophisticated paradigms developed for rats into mouse disease research, with a relatively high levels of success (Hunsaker 2012a, b; Hunsaker et al. 2009, 2010, 2012; Nakazawa et al. 2004; Rondi-Reig et al. 2006).

What has remained elusive in the field of behavioral genetics is a clear theoretical rationale underlying the choice of experiments performed on each given model (i.e., water maze and Barnes maze do not test all types of spatial memory, let alone all types of "learning and memory"). To facilitate comprehensively evaluating learning and memory processes across all mouse models, it is helpful to step back and separate learning and memory into component or attributes or domains that can be evaluated in turn (cf., Hunsaker 2012b; Hunsaker and Kesner 2012; Kesner and Hunsaker 2010; Kesner and Rogers 2004; White and McDonald 2002). In practice, this approach allows the murine researcher to evaluate brain function in a more sophisticated manner than previously possible using the standard behavioral tasks that were not developed to test any attribute or hypothesis (cf., Hunsaker 2012a, b).

Before moving into a closer analysis of the proposed approach, it is important to mention the pitfalls with the common memory tasks used widely in mice: the water maze, Barnes maze, passive/active avoidance, and contextual fear conditioning. All of these tasks can be useful as components of a phenotyping approach, but in themselves they do not allow researchers to specifically determine the nature of impaired memory in mouse models. For all these tasks there are uncontrolled factors relating to anxiety and, more importantly, motivational confounds involving the use of negative reinforcement as the primary motivation for task performance (cf., Barkus et al. 2010; Hunsaker 2012a, b). Furthermore, when negative reinforcement is used for motivation—especially when using assays such as contextual fear conditioning to evaluate spatial memory—models demonstrating disorders in affect (i.e., depression or anxiety disorders) may demonstrate "spatial" memory deficits for reasons other than de facto impairments to spatial processing (cf., Banik and Anand 2011).

Additionally, it has been suggested on numerous occasions that the water maze may not be an appropriate task for use in mice, as mouse performance is poorer than would be predicted when compared to performance on non-water-based paradigms compared to rat performance on similar tasks (Frick et al. 2000; cf., Whishaw and Tomie 1996)—and dry land alternatives for the water maze have been frustratingly slow to be adopted, despite being introduced to the field as early as the 1980s (cf., Kesner et al. 1989; Llano Lopez et al. 2010).

Attributes of Memory Processing

Table 15.1 outlines the first consideration in developing or choosing behavioral experiments to test mouse disease models, which is to consider what type of memory needs to be tested in the mouse. Briefly, one has to consider if the disorder being studied primarily results in an episodic (event-based) memory deficit, knowledge-based memory deficits, or executive function (rule-based) deficit (Hunsaker 2012b; Kesner and Hunsaker 2010; Kesner and Rogers 2004). Once the memory system being tested is determined, then the component memory domains can be identified and tested using experiments designed with each disorder and model in mind (Hunsaker 2012a, b; Kesner and Rogers 2004; Simon 2007, 2008, 2011; Fig. 15.2).

Table 15.1 Memory systems used in the attribute theory as applicable to research into mouse models of disease

	Event based	Knowledge based	Rule based
Encoding	Pattern separation	Selective attention	Strategy selection
	Transient representations	Associated with permanent memory representations	
	Short-term memory	Perceptual memory	Rule maintenance
	Intermediate-term memory		
Retrieval	Consolidation	Long-term memory	Short-term working memory
	Pattern completion	Retrieval based on flexibility and action	

Event-based memory refers to a memory system wherein information is processed online from active sensory/perceptual data and representations formed by each memory system using those sensory/perceptual data. This is the memory system that allows for trial unique responses and behavioral performance. Knowledge-based memory is often referred to as semantic memory in the human episodic memory literature. This chapter will use the term knowledge-based memory, because semantic memory has an implicit language component that cannot be directly modeled in rodents. Knowledge-based memory is most analogous to the reference memory system proposed by Olton et al. (1979) than to any other taxonomic system. The rule-based memory system spans both the event- and knowledge-based memory systems by providing a framework to guide behavioral performance. That is, the rule-based memory system is the memory system that allows an individual to generate rules and motivational contexts that guide behavioral performance across all timescales and allows for generalization across paradigms and situations.

Table 15.2 outlines neuroanatomical substrates underlying each attribute in mice that can be consulted to guide the development or application of behavioral tasks for mouse models of disease. Importantly, although these anatomical structures have been shown to underlie the attributes as mentioned in Table 15.2, this description is more of a blueprint of structures that are critically involved with these processes. Stated another way, when one brain region is shown to underlie or be involved in a process, it is more likely than not that a larger network including the candidate neuroanatomical structure actually subserves the process, and that the contributions of the larger network is more poorly understood than the role for any single structure. An example of this common over-simplification is the hippocampus: hippocampus ablations result in profound deficits for spatial and temporal processing (cf., Jerman et al. 2006; Hunsaker et al. 2008a), but removal of inputs/outputs from the entorhinal cortex and septal nuclei result often in qualitatively similar, if not identical, behavioral deficits (cf., Hunsaker et al. 2008b). As such, it is more correct to state that neural networks that include the hippocampus subserve spatial and temporal processing, rather than the hippocampus in isolation subserves these memory processes.

Table 15.2 Neuroanatomical correlates underlying each attribute

Attribute	Event based	Knowledge based	Rule based
Spatial	Hippocampus	Parietal cortex	IL/PL[b]
	Pulvinar		Retrosplenial cortex
Temporal	Hippocampus	Anterior cingulate	Anterior cingulate
	Basal ganglia	IL/PL[b]	IL/PL[b]
	Anterior thalamus	Cerebellum	
Sensory/perceptual	Primary sensory cortices	TE2 cortex[a]	IL/PL[b]
		Perirhinal cortex	
		Piriform cortex	
Response	Caudoputamen	Precentral cortex	Precentral cortex
		Cerebellum	Cerebellum
Affect	Amygdala	Agranular insula[c]	Agranular insula[c]
	VTA	Amygdala	IL/PL[b]
	Nucleus accumbens	Anterior cingulate	
Executive function	Basal ganglia	IL/PL[b]	IL/PL[b]
	IL/PL[b]	Parietal cortex	Parietal cortex
	Anterior cingulate		
Social proto-language	Underlying neural networks still being elucidated		

Murine homologs of human
[a] Inferior temporal cortex
[b] Medial prefrontal cortex
[c] Orbitofrontal cortex

Selection and Design of Cognitive Tasks for Humans: The NIH Toolbox

The NIH Toolbox, an NIH Blueprint for Neuroscience Initiative (http://neuroscienceblueprint.nih.gov) that began in fall 2006, set out to develop a set of brief, validated instruments to assess cognitive, emotional, motor, and sensory function across a wide age range (3–85 years of age; cf., Gershon 2007; Gershon et al. 2010). As developed, the NIH Toolbox is intended for use in epidemiological and longitudinal studies to identify those aspects of cognition that are associated with optimal function and health. Similarly, the NIH Toolbox is also intended for use in large-scale intervention and prevention trials.

At present, there are many ways in which data on neurocognitive function are collected. Unfortunately, the tools currently in common use are not standardized sufficiently to allow comparison among laboratories and protocols. This results in issues similar to the situation observed in murine research as described above (cf., Wahlsten et al. 2003a). By adopting a standard set of publicly available tools, the NIH Toolbox will be able to enable the efficient aggregation of data from multiple studies and perhaps even facilitate comparison across studies. Such features as these greatly enhance the value of information collected during the course any single research project. Importantly, by providing access to the exact tools and clear instructions for usage, the site to site variability in the results from NIH Toolbox assessments can be somewhat mitigated, such that the data can be compared across sites.

The NIH Toolbox currently includes 108 primary and supplemental instruments assessing the following constructs: cognition, emotional health, motor function, and sensory function. A full listing of paradigms is available at the NIH Toolbox web site (Gershon 2007; Gershon et al. 2010; cf., Hoffman et al. 2009; McClelland and Cameron 2011; Pilkonis et al. 2012; Quatrano and Cruz 2011; Wang et al. 2011) and will not be explicitly listed here, other than to point out that under each parent domain there are a number of subdomains that cover a much broader spectrum of function than the four parent domains would initially suggest. For example, the behavioral assays under the domain of cognition include tests of attention, executive function, processing speed, working memory, episodic memory, and language subdomains. The surveys that assess the emotional health domain evaluate positive affect, negative affect, social relationships, and stress and coping. To evaluate motor function, the NIH Toolbox includes tests of locomotion, strength, nonvestibular balance, endurance, and dexterity. To evaluate sensory function, the NIH Toolbox uses paradigms evaluating vision, audition, vestibular balance, taste, olfaction, and somatosensation.

The NIH Toolbox consists of these four parent domain batteries, each requiring an average of 30 min to administer (and approximately 20 min for children aged 3–5 years), with a total of 2 h administration time for the entire Toolbox (80–90 min for children). Importantly for studies involving children, each test has an upper time limit of 5–7 min, so the tests do not provide the complication of taxing the attentional capacity of children 3–5 years of age, although it is presumed adults would also appreciate the rapid nature of the tasks.

Utility of the NIH Toolbox in Clinical Populations

Previous to the introduction of the NIH Toolbox, behavioral and cognitive research using clinical population was hampered by similar problems as murine research: that is, different labs use similar paradigms in their research, but these tasks lack standardization. The effect of this lacking standardization is that labs are likely finding differences among their populations not because of actual differences in disease severity or observed phenotype, but rather due to small design or methodological differences in the behavioral tasks (i.e., different measures of executive function resulting in different neurocognitive phenotypes—cf., Wisconsin Card Sort and Color/Word Stroop vs. Behavioral Dyscontrol Scale-II (BDS-II); Allen et al. 2011; Grigsby et al. 2008). This has long been an issue with mouse research and has led to the development of standardized batteries of behavioral tasks such as SHIRPA (cf., Hatcher et al. 2001; Rogers et al. 1997, 1999, 2001).

Unfortunately for the traditional behavioral phenotyping strategy in mice as well as the NIH Toolbox, these batteries or collections of tools do not take into account the disease state of the mouse model or the clinical population. Using a standard set of tools rather than tools selected based on specific hypotheses does provide data that can compared against clear, accepted norms, but any interpretation of observed findings is often tortured (cf., Figs. 15.1 and 15.2).

Neglecting to take into account the specific skills and weaknesses of the population being studied results in not only misidentifying or mischaracterizing a population, but also missing behavioral or cognitive phenotypes entirely by omitting/neglecting necessary tests that would have uncovered a potential trove of information (Hunsaker 2012a; Fig. 15.2). To provide a hypothetical example, consider a situation in which an individual with type I bipolar disorder and an individual with schizophrenia show similar patterns of performance on the NIH Toolbox collection of assessments. In this case, the researcher is forced to either accept the NIH Toolbox is insufficient to characterize these populations and dissociate them from each other, or else to accept that type I bipolar disorder and schizophrenia have the same clinical presentation—at least on the assessments being used. Clearly, type I bipolar disorder and schizophrenia do not share an overall neurocognitive phenotype, but there is no clear way to tease them apart unless clear hypotheses concerning these two disease states are used to develop tasks to characterize them at a resolution sufficient to distinguish the two disorders (process diagrammed in Fig. 15.2). This logic can be extended to any set of disorders whose performance on the NIH Toolbox may be compared. This is where an endophenotyping strategy becomes a necessity.

That being said, it is by no means necessary or even advisable to reject the NIH Toolbox or a standardized behavioral battery outright, but it is necessary to accept the weaknesses inherent in using collections of experiments selected without regard to the population being studied. Figure 15.1 provides a diagrammatic representation of this argument with the NIH Toolbox/standard behavioral task battery represented by the left plate (Fig. 15.1a) and an endophenotype/hypothesis driven approach on the right (Fig. 15.1b). An analysis of these two approaches suggests that at best only large-scale cognitive/psychological disturbances can be identified with the NIH Toolbox that must be more carefully followed up on to provide information that can be acted upon in the development of outcome measures or therapeutic strategies. Looking at Fig. 15.2 best illustrates how different hypotheses for different disorders emerge, namely through a consideration of all disease related sequelae and how they may affect psychological and cognitive function. Comparing the sequelae for different disorders illustrates an important point: no two disorders present with identical systemic, pathologic, endocrine dysfunction, etc., so it follows that separate disorders should be not be tested using identical tools if the goal is to uncover the differential phenotypes of unique pathological states. Additionally, it is also essential that it be understood that null fundings using the NIH Toolbox tools do not mean that no deficits are present in the population, just that the NIH Toolbox assessments were insufficiently sensitive to uncover any behavioral deficits. Because these tools were developed and adopted without regard to populations being studied, these tools are ill-suited to characterize specific deficits or strengths in any population, but rather emphasize global deficits known to interfere with daily function.

Most importantly, the NIH Toolbox is intended to be brief (~2 h for the entire collection of assessments), minimally burdensome to respondents and administrators, relatively low in cost, psychometrically sound, free of intellectual property issues, and appropriate for use across a wide age range and with diverse populations (e.g., English and Spanish speakers). All these qualities are expected to make its use attractive to investigators.

The appeal of the NIH Toolbox is clear for those already planning to assess cognitive, emotional, motor, or sensory function tested by the NIH Toolbox and, perhaps more importantly, to investigators who might not assess areas of function were it not for the availability of the NIH Toolbox. This should lead to a greater number of studies collecting standard data on more areas of function, which, since these data can be aggregated and directly compared, significantly increases the likelihood of making new discoveries, and identifying currently unknown relationships between function and health, and function and disease. This could lead to new prevention strategies as well as additional treatment targets. Similarly, using NIH Toolbox instruments when evaluating treatments could reveal a broader range of treatment effects then it is typically possible to do in a single study or a few studies. This kind of finding, in turn, may lead to development of or an adjustment to the existing therapeutic medications in clinical practice.

Critically, when the NIH Toolbox is used across laboratories, it becomes possible to evaluate intervention strategies using the same outcome measures across multiple sites simultaneously, a process not straightforward at present. Although this may be problematical in the long term as similar outcome measures may not be appropriate to compare across populations, standardized testing makes finding solutions to any complications easier than trying to harmonize data collected using different protocols across disparate studies.

Reshuffling of the NIH Toolbox by Attribute

The NIH Toolbox is organized, as mentioned above, into four component domains that can be administered in blocks so as to minimize interference among the tasks contained within each domain. An alternate method of categorizing the tests contained within the NIH Toolbox is by the component attribute tested by each task. Table 15.3 contains the NIH Toolbox tools, reshuffled by attribute or domain evaluated.

What becomes apparent when these tasks are categorized by attribute is that there are attribute domains that are not evaluated by the NIH Toolbox, and moreover, a number of attributes are only cursorily tested. Additionally, the design of the NIH Toolbox requires that only certain types of behavioral tasks be chosen, particularly those that may be administered in a very short timeframe, thus limiting the efficacy of the NIH Toolbox as a whole by disqualifying tests that may be more dense so far as data collection is concerned but require more time to administer. Unfortunately,

Table 15.3 NIH Toolbox reshuffled and organized by attribute

Attribute	Event based	Knowledge based	Rule based
Spatial	Picture sequence memory	*Not addressed*	*Not addressed*
Temporal	Picture sequence memory	List sorting working memory	List sorting working memory
	List sorting working memory		
Sensory perceptual	All tests included in the sensory domain of the NIH Toolbox		*Not addressed*
Response	Grip strength[a]	Nine hole pegboard dexterity[a]	Flanker inhibitory control and attention
	Knee extension strength[a]		Dimensional change card sort
	Standing balance[a]		
	4 m walk gait speed[a]		
	2 min walk endurance[a]		
Affect	*Not addressed*	Surveys in the emotion domain of the NIH Toolbox	
		Subsections: psychological well being, negative affect, stress and self efficacy, social relationships	
Specific cross-domain tasks important for research into neurodevelopmental disorders			
Executive function	Flanker inhibitory control and attention		
	Dimensional change card sort		
	List sorting working memory		
	Pattern comparison speed		
	Oral symbol digit		
Social	All surveys included in the social relationships subsection of emotion domain of the NIH Toolbox		
Proto-language	Ray auditory verbal learning		
	Picture vocabulary		
	Oral reading recognition		

Action-outcome and stimulus-response tasks more classically test the response attribute in both the event- and knowledge-based memory systems

[a] These motor tasks are related to the response attribute

it appears that these requirements have led to the selection of behavioral tasks that are not always an accepted standard for evaluating function within a given domain.

For example, to evaluate executive function and attention the NIH Toolbox uses a flanker inhibitory control task and a dimensional card sorting task. These tasks have a rich history in attentional research, but it can be argued that a go/no go task, a stop signal task, or a self-ordered pointing task would dissect the executive function and attentional processes being evaluated in a more process-pure manner. In other words, using the NIH Toolbox there is no way to dissect a dysexecutive syndrome from focal attentional deficits or cognitive control processes, all of which affect behaviors differentially (cf., Grigsby et al. 2008).

Additionally, the separation of processing speed from attention seems an odd choice given that attentional processes are thought to underlie performance on processing speed tasks. Additionally, the emotion domain lacks any true behavioral assays that can evaluate anhedonia, hyper vigilance, fear, anxiety, panic, or other mood state that may be well relevant to the study, but rather limits the analysis to a battery of surveys. These surveys are important, but often experimental studies into fear-, anxiety-, and panic-related behaviors may reveal more to the experimenter than the data collected in a survey, particularly responses to acute stress as well as the efficacy of an individual's coping mechanisms to stressors.

Intriguingly, the executive function domain lacks any analysis of stimulus-response or action-outcome processes that are encompassed by the response attribute. The closest the NIH Toolbox comes to evaluating these functions is the dimensional change card sort, which assesses the ability to form and switch cognitive sets. This is an important test, but more fundamental processes that may underlie cognitive set shifting remain unevaluated. Serial reaction time tasks or other similar sensory-response tasks would be able to quantify intact rule or stimulus-response processing at a more tangible level than serial set switching or other higher level cognitive processes.

The motor domain can be construed to tax the response attribute as most tests of response-based memory use motor output as a dependent measure, but this is something of a procrustean solution as general motor function is at best only tangentially related to the response attribute. Additionally, as is the case with the tests of executive function and attention, the motor domain contains some tasks that are substandard to the questions commonly asked: For example, the nine hole pegboard was selected rather than the grooved pegboard as a test of dexterity (cf., Wang et al. 2011), resulting in the NIH Toolbox preferring a task that takes less time to administer, but at the cost of selecting a task that was not as data rich as the alternative; an alternative which takes only 2–3 more minutes to administer.

An additional, albeit necessary, weakness of the NIH Toolbox assessments is the relatively sparse amount of data collected by any given task. Whereas the common memory and executive function tests, for example, often take 15–30 min each to administer, the episodic memory test in the NIH Toolbox takes 7 min and has a limited number of trials. Similarly, the nine hole pegboard uses only a single trial from each hand for data collection, rather than multiple trials as is common practice. In a similar manner, inclusion of tasks only available for children as young as 3 for adults and aged individuals (i.e., 3–85 years of age; Gershon et al. 2010) limits the utility of the NIH Toolbox by introducing the potential for ceiling performance during adulthood with nonceiling performance at young and elderly ages, providing an inverted U function that may make analysis of longitudinal data difficult if not intractable. To remedy this potential issue, the NIH Toolbox utilizes computerized adaptive testing to accommodate performance differences across ages whenever possible, but it remains to be seen how reliable this method remains over time once truly longitudinal studies begin covering months to years within individual study participants.

Mouse Variant of the NIH Toolbox Organized by Attribute

Currently implemented behavioral screens have the benefit of clear face validity as the implications of behavioral deficits on a task or collection of tasks are intuitively applicable in the context of the clinical phenotype, but often these tasks lack construct validity (cf., Hunsaker 2012a, b). The behavioral endophenotyping process I am proposing emphasizes clearly defined construct validity across paradigms designed to test specific disease or mutation-related hypotheses. A starting point for the development of this test battery is the NIH Toolbox.

An optimal, comprehensive behavioral phenotyping strategy integrates common behavioral tasks as well as endophenotyping approaches performed across the lifespan. Such an approach is important because a number of genetic disorders show distinct early and late manifestations of disease that bear independent scrutiny. Often times, carriers of genetic mutations show few or at most subtle characteristics of later clinical disease early in life, but with increasing age these symptomatology emerge and the individuals receive a clinical diagnosis (Chonchaiya et al. 2009a, b; Pirogovsky et al. 2009; Rupp et al. 2009). This does not infer, however, that early in life these individuals are unaffected by the mutation; more likely the consequences of the mutation are present early in life, but require more sophisticated analyses to identify patterns of behavioral abnormalities (cf., Goodrich-Hunsaker et al. 2011a, b; Wong et al. 2012).

In cases of genetic disorders, it is useful to evaluate the cognitive domains that underlie later clinical phenotypes early in life to determine if there are markers that can quantify or predict disease progression (Devanand et al. 2000; Pirogovsky et al. 2009; Salomonczyk et al. 2010; Yong-Kee et al. 2010). Research into a number of neurodegenerative disorders have been able to characterize subclinical endophenotypes early in the disease process that seem to predict the severity of the disease or rate of disease progression (Gilbert and Murphy 2004a, b; Karayiorgou et al. 2010; Salomonczyk et al. 2010; Xu et al. 2010; Yong-Kee et al. 2010).

The NIH Toolbox is an important tool that facilitates translational research across human and murine research. As demonstrated in Table 15.3, the NIH Toolbox can be organized into the attribute model to facilitate the development of analogous behavioral tasks for mice that show face and construct validity with the end goal of predictive validity. A critical aside is that the mouse behavioral paradigms do not have to be exact copies of the paradigms used by the NIH Toolbox, but rather need to assess the same fundamental cognitive processes. This assumption suggests that for a mouse model of behavioral deficits to model the human disorder only a similar pattern of behavioral deficits across tasks used in the NIH Toolbox is important, not the exact structure of any given task. In other words, it is nearly always better to err on the side of construct (content) rather than face validity when presented with the choice (cf., Hunsaker 2012a, b).

Table 15.4 outlines a collection of simple tasks based on each component attribute that can be used to test cognitive dysfunction in mouse disease models and provide a functional analog to the NIH Toolbox. Aside from spatial attributes

Table 15.4 Murine options for an NIH Toolbox analog. Italicized tasks are proposed for the mouse NIH Toolbox and the rest are recommended for in depth follow-up studies

Attribute	Event based	Knowledge based	Rule based
Spatial	*Metric processing*	*Biconditional discrimination*	Covert attention tasks
	Topological processing	Delay match to place with variable cues	*Self ordered nonmatch to sample*
	Magnitude estimation	Declarative sequence learning	
	Delay match to place with variable interference	Cheeseboard	
	Biconditional discrimination for trial unique associations		
Temporal	Trace conditioning	Sequence completion	*Five choice serial reaction time*
	Temporal ordering	Duration discrimination	Peak interval timing
	Sequence learning		Time left task
Sensory	*Delay match to sample with variable interference*	*Biconditional discrimination*	
Perceptual	Acoustic startle		
	Prepulse inhibition		
	Psychonomic threshold		
Response	*Ladder walking task*	Delay match to direction	Reversal learning
	Acquisition of skilled reaching	Direction discrimination	*Probabilistic reversal learning*
	Working memory for motor movements	Nondeclarative sequence learning	Operant conditioning
	Capellini handling task		Stop signal task
	Seed shelling tasks		Serial reversal learning
Affect	*Reward contrast with variable reward value*	Classical conditioning	Operant conditioning
		Trace conditioning	Gambling Task
		Conditioned preference	Latent inhibition
		Anticipatory contrast	
Specific cross-domain tasks important for murine research into neurodevelopmental disorders			
Executive function	Contextually cued biconditional discrimination		
	Five choice serial reaction time		
	Operant conditioning		
	Covert attention tasks		
	Intra-extra dimensional set shifting		
	Reversal learning		

Table 15.4 (continued)

Attribute	Event based	Knowledge based	Rule based
	Probabilistic (80/20) reversal learning		
	Serial reversal learning		
	Stop signal task		
	Gambling task		
	Latent inhibition		
Social	Social transmission of food preference		
	Social novelty detection		
Proto-language	*Spectrographic analysis of ultrasonic vocalizations*		

commonly tested, along with the temporal, response, social, and sensory/perceptual attributes, it is also critical to evaluate the role of affect, proto-language, and executive functioning attributes in mouse models of neurodevelopmental disorders, because these domains are often profoundly affected in these clinical populations (Hunsaker 2012a, b; Simon 2007, 2008, 2011).

An often overlooked, but critical, consideration in choosing behavioral assays is that of the neuropathology associated with any disorder being modeled (Fig. 15.2). It seems an obvious point that one would choose behavioral paradigms that emphasize spatial (and temporal) processing to evaluate disorders with known hippocampal pathology (e.g., Alzheimer disease and Down syndrome) and tasks emphasizing response learning in tasks showing clear basal ganglia pathology (e.g., Parkinson disease and Huntington disease), but unfortunately this is not consistently taken into consideration in experiments using mouse models of genetic disorders (cf., Taylor et al. 2010; Wesson et al. 2011). Similarly, the NIH Toolbox appears to omit an explicit consideration of specific application of tasks that relate to common neuropathological features observed in aging and neurodegenerative disease. This is seen in the limited coverage of the NIH Toolbox for selectively testing the different attributes of memory (cf., Table 15.3).

In the case of the present analysis, the choice of behavioral paradigm used depends largely upon which behavioral tasks are being used in the NIH Toolbox in a particular disorder. Additionally, the mouse researcher can take the spirit or rationale behind the selection of a given behavioral paradigm and test the same specific process in the mouse model. The list of tasks provided in Table 15.4 is somewhat over-encompassing to be a direct one to one match with the NIH Toolbox, but the tasks presented test the same processes in a manner that is faithful to the intentions of the NIH Toolbox. More critically, all of the tasks listed in Table 15.2 have been either previously used or pharmacologically validated in murine and rodent models and thus only require a pilot project be undertaken for each lab, rather than a laborious development period for a novel task, prior to data collection.

Mouse Model of the NIH Toolbox: Behavioral Endophenotyping

Once the researcher has developed a mouse model for the NIH Toolbox, they may then extend beyond the NIH Toolbox assessments and apply an endophenotyping approach to select additional tasks that test disease and domain specific hypotheses. In other words, an explicit murine behavioral model of the NIH Toolbox can serve as a core service for all models, with follow up experiments that are unique to each disorder or model being tested (cf., Figs. 15.1 and 15.2).

In cases of mouse models that have never been behaviorally assessed, having an explicit model of the NIH Toolbox allows for an easy translation of research findings across the human disorder (provided they use the NIH Toolbox to develop a clinical phenotype in the population) and mouse model. Once the initial phenotype is elucidated using a mouse model of the NIH Toolbox, then a more careful selection of behavioral assays can be selected based upon the behavioral findings and any analyses of pathology in the clinical population or mouse model.

The proposed analog of the NIH Toolbox is included in Table 15.4 as the italicized elements under each attribute and memory system. These selections have been made to be as simple as possible as well as relatively high throughput tasks. Other tasks included in each section tend to be more complicated tasks and are candidate tasks for follow up studies based upon the results of the initial screen. Also, these tasks that are intended for follow-up research rather than an initial screen are based directly upon paradigms used in cognitive research with clinical populations, so as to directly parallel the mouse model with human clinical populations (cf., Hunsaker 2012a, b).

Examples of Behavioral Endophenotyping

Rodent Traumatic Brain Injury (TBI)

Although not in mice, an example in parallel with the NIH Toolbox is the recent finding that using the metric and topological tasks, as well as the temporal ordering tasks listed in Table 15.4 in lieu of the water maze and/or Barnes maze demonstrated clear specificity in characterizing the behavioral deficits of experimental TBI caused by lateral fluid percussion in rats (Gurkoff et al. 2012). In this task, rats with TBI showed deficits for hippocampus-dependent behavioral performance requiring spatial and temporal processing, but spared parietal cortex function as well as spared cortical function related to identifying sensory/perceptual stimuli. This was a model for episodic memory deficits often demonstrated in TBI populations.

Gurkoff et al. (2012) asserted that the primary strengths of this approach in their hands was twofold: (1) the spatial and temporal ordering tasks they used had been used in human clinical populations previously, albeit under different task names (e.g., categorical and coordinate; episodic sequence learning). They emphasized this fact as a strength as there are no clear analogs to the water maze or Barnes maze in the human TBI literature: rather episodic/general memory deficits uncovered by list learning tasks are not comparable to the rodent research, whereas research into the spatial and temporal processing have been previously done in TBI and non-TBI clinical populations. (2) The tasks Gurkoff et al. (2012) used did not require extensive training, as any deficits for executive function could result in behavioral deficits on tasks that require training due to not learning the rules of the task rather than any deficits for spatial and temporal processing per se. This is a very important consideration as TBI reliable results in altered executive function that interferes with adaptive function.

In selecting/designing their experiments, Gurkoff et al. (2012) applied the attribute model to select tasks that were most applicable to the population being studied (TBI) and developed hypotheses based upon the neuropathological features and clinical manifestation of TBI cases seen by their collaborators in the clinic, as well as taking into consideration the results of previous studies into their model (cf., procedure outlined in Fig. 15.2; Table 15.4). This is analogous to using the NIH Toolbox to specifically assess episodic memory and executive function in individuals with TBI using more sensitive and standard measures than simple memory tests and clinical neuropsychological tools.

Mouse Model of Fragile X Premutation

For an example of this behavioral endophenotyping process in mice, research into the mouse model of the fragile X premutation, a polymorphic CGG repeat expansion on the *FMR1* gene will be discussed. The fragile X premutation is associated with a late onset neurodegenerative disorder called fragile X-associated tremor ataxia/syndrome (FXTAS). FXTAS occurs in ~40% of male premutation carriers and ~16% of female premutation carriers and is associated with an intention tremor and cerebellar gait ataxia, as well as cognitive decline and executive function impairments (Hagerman and Hagerman 2004; Jacquemont et al. 2004). Unfortunately, there are no agreed upon cognitive effects of the premutation on carriers not showing FXTAS motor signs (Allen et al. 2011; Hunter et al. 2011; Goodrich-Hunsaker et al. 2011a, b; Wong et al. 2012).

What has been demonstrated in premutation carriers is that, although not showing large-scale cognitive deficits, a number of studies identified spatial and temporal attentional deficits in female and male premutation carriers (Goodrich-Hunsaker et al. 2011a, b; Hashimoto et al. 2011; Hocking et al. 2012; Wong et al. 2012). These deficits are present despite the lack of large-scale executive function deficits and gross memory disorders (Allen et al. 2011; Grigsby et al. 2008).

The analysis of behavioral deficits in the CGG KI mouse model of the fragile X premutation will emphasize the behavioral tasks included in Table 15.4. To evaluate whether the CGG KI mouse showed similar spatial and temporal attention problems as the premutation carriers reported by Goodrich-Hunsaker et al. (2011a, b), Hocking et al. (2012), and Wong et al. (2012), we applied the rationale diagrammed in Fig. 15.2. In other words, the cognitive, pathological, and neuroendocrine phenotypes of the premutation were considered in the task design for the CGG KI mouse model.

So far as neurobehavioral deficits in the CGG KI mouse are concerned, the CGG KI mouse shows a number of basic processing deficits for spatial and temporal information. The CGG KI mouse model of the fragile X premutation shows spatial memory deficits on the water maze when they are older than 52 weeks of age (van Dam et al. 2005). These deficits, however, appear to be very mild and are not as profound as the general memory deficits demonstrated in FXTAS patients (cf., Hagerman and Hagerman 2004; Jacquemont et al. 2004). Based on reports that suggest increasing CGG repeat lengths affect spatial attention (cf., Goodrich-Hunsaker et al. 2011a, b; Hocking et al. 2012; Wong et al. 2012), spatiotemporal function was specifically assayed in the CGG KI mice.

Using a pair of behavioral tasks to evaluate the resolution of spatial memory in CGG KI mice, it was demonstrated that CGG KI mice show deficits for mentally comparing the the specific distances that separate two objects in space. This was evaluated using a metric change detection task wherein mice are habituated to two objects on a tabletop separated by 45 cm for 15 min. After being removed from the table for 5 min, mice were returned to the tabletop with the objects placed at 30 cm separation. The ability of mice to notice a change in the distance between the objects required the mouse to remember the original distance and compare it with the current sensory input. Deficits for spatial attention tested by the metric task were present as early as 3 months of age, but in cross-sectional studies these deficits did not appear to become increasingly profound in mice that were 6, 9, or 12 months of age (Hunsaker et al. 2009, 2012).

However, performance of CGG KI mice on a task that required the mice to remember which side of the tabletop was occupied by an object after the objects were transposed was not impaired at early ages. In fact, CGG KI mice did not show deficits for this topological change detection task until they were 9 and 12 months of age, with their performance not differing from wildtype littermate controls at 3 and 6 months of age (Hunsaker et al. 2009, 2012). This tasks also requires spatial attention, but of a different type than the metric task. The topological task required the mouse to remember an object–place relationship that did not require the fine-scale spatial attention required by the metric task.

What can be learned from these data are twofold: (1) that the resolution of spatial attention in CGG KI mice is profoundly reduced from a very young age compared to wildtype littermates, presumably affected relatively early during development, and this resolution appears to be fixed across time, such that the resolution does not deteriorate/progressively worsen as a function of age. (2) Performance of CGG KI mice on spatial memory tasks that do not require fine spatial attention such as the topological change detection task is not impaired at an early age, but these mice do show a progressive worsening of spatial memory across age, with deficits emerging in middle life and worsening at advanced ages in CGG KI mice, a pattern similar to those seen with the water maze (Hunsaker et al. 2009, 2012; van Dam et al. 2005).

An easily overlooked element in this pattern of deficits in the CGG KI mouse model is the dissociation between the developmental course of deficits present across the metric and topological tasks. The finding that spatial attention deficits evaluated by the metric processing task are present at a young age and do not worsen across the lifespan suggests a fundamental developmental alteration that renders the CGG KI mouse unable to overcome the interference between the distances among spatial locations before and after the metric change (i.e., the metric change was not profound enough for the mouse to discriminate the new distance between objects from the remembered object distance). CGG KI mouse performance on the topological task, however, showed a somewhat degenerative pattern. Performance of the topological task was intact in young animals, suggesting that spatial memory per se was not disrupted in the CGG KI mice. As the mice age, however, deficits for this task emerged, suggesting some effect of age and the premutation compounding to result in general spatial memory impairments. This dissociation in task performance is important because it suggests that the premutation results in reduced resolution of spatial attention, not general spatial memory deficits (i.e., the mice can identify changes in object–place associations, but lack the ability to perform comparisons between an observed distance between objects and one retrieved from memory).

To evaluate temporal memory in CGG KI mice, a temporal ordering for visual objects task was used. In this task, in a clear box mice were presented with two copies of an object for 5 min, and then removed from the box for 5 min. The mice were then presented with two copes of a second object in the box for 5 min. After another 5 min break, they were exposed to two copies of a third object for 5 min. After the mice were removed after this third object exposure for a 5 min break, they received one of two tests. The first test is a temporal ordering test during which the mouse is presented with a copy of the first and a copy of the third object and allowed to explore. Typically, mice will preferentially explore the first over the third object. It has been suggested that this paradigm requires sequential learning and fine-scale temporal attention for the mouse to remember the order the stimuli were experienced and to later compare the relative order of these memories to guide behavioral choice.

On another day after a difference set of object presentation, the mice receive a second test, a novelty detection task. In this task, the first object they were presented that day as well as a never before seen novel object were presented. In general, mice will preferentially explore the novel object over the familiar one. Intact performance during this novelty task suggest that any deficits on the temporal ordering task are not due to an inability to discriminate the stimuli, general memory deficits, or forgetting the first object before the test session, as they can discriminate a familiar object from a novel object, suggesting intact visual object memory.

On these tasks, the CGG KI mice showed intact performance for the novelty detection task, but profound impairments for temporal attention as assessed by the temporal ordering task (Hunsaker et al. 2010, 2012). Again, these data suggest that the CGG KI mouse has intact sensory/perceptual processing and intact overall memory but impaired temporal attention that results in temporal ordering deficits.

As a follow up to these experiments, an explicit spatiotemporal processing task was performed based on spatiotemporal working memory tasks used in human populations (Borthwell et al. 2012; cf., Kesner et al. 1994). In this task, mice were presented with a large object in a first spatial location for 5 min in a large box with prominent visual cues present. After a 5 min break, the mice explored the same object in a second location. This was repeated for a third location. In this way, the mouse explored the same object in three locations, which we will call exploration of a location. After these presentations, one of three tests was given (over three days with new object–location pairings each day).

The first is a temporal ordering for spatial locations test wherein the first and third locations were marked with identical objects identical to that used to present the locations. Importantly, these locations were always 180° from the mouse's starting location, thus minimizing spatial interference between the remembered spatial locations to allow for an analysis of temporal attention for spatial location information. Preferential exploration the first over the third location was used to index spatiotemporal attention.

The second test was a pure spatial memory control during which the first location and a novel fourth location were marked by identical objects, which were 180° from the mouse's starting posits. This again minimized any crowding between the remembered spatial location and a novel spatial location. Preferential exploration do the novel location suggests intact general memory processing.

The final test was a spatial resolution test during which the first and novel object were only separated by 45–90° from the mouse's starting position, increasing the spatial interference to isolate the ability of the CGG KI mice to overcome the interference between the remembered location and a novel spatial location.

On these tasks, the CGG KI mice showed no deficit for spatiotemporal novelty detection when the locations were separated by 180° and thus spatial interference was minimized because there was no temporal or spatial interference among remembered spatiotemporal memory and a novel location. However, the CGG KI mice did show impairments when the spatial interference was maximized by placing the novel spatial location very close to a remembered spatial location. The CGG KI mice also showed temporal attention deficits for spatial information during the temporal ordering test—strongly suggesting an inability to overcome spatial and temporal memory interference and providing clear evidence for impaired spatiotemporal attention (Borthwell et al. 2012).

An important element to these behavioral results is that both male and female CGG KI mice showed deficits. This is not a minor point as female fXPCs show reduced disease severity due to the protective effect of a second, nonmutated *FMR1* gene on the second X chromosome, which males lack (cf., Jacquemont et al. 2004; Schluter et al. 2012; Tassone et al. 2012). Finding these deficits in both male and female CGG KI mice suggests cognitive deficits within the domain of spatiotemporal attention are fundamental consequences of the premutation, since these deficits are present and identifiable even in the least affected subgroup within the fXPC population (cf., logic provided by Goodrich-Hunsaker 2011a, b).

As stated above, FXTAS patients often present in the clinic with an intention tremor and/or a cerebellar gait ataxia. Importantly, the tremor and ataxia seem to present with an oscillatory component, such that the gait ataxia becomes more profound as the individual walks until they lose balance. They appear relatively normal for the first few steps, but then a postural away emerges that grows in amplitude with each step until the patient either braces against a wall or falls over. For the intention tremor, FXTAS patients appear to show normal motor function at first, but as the trial continues (i.e., spiral or Archimedes or drawing a third line in the space separating two lines), a minute oscillation emerges which increases amplitude until the patient stops. Gait ataxia shows a similar tendency with the amplitude of the postural sway increasing with each step until the patient braces themselves with a cane or against a wall, after which the pattern of increasing instability repeats (Hagerman and Hall, unpublished observations). These data suggest there may be some sort of abnormal feedback among cortical and cerebellar systems that prevents the fine online correction of movements so errors accumulate and exacerbate out of control. In other words, it is possible the cerebellum never receives the vestibular/kinesthetic feedback that signals the accumulating error present during each movement, so the amplitude of the error term sums exponentially with each subsequent movement until the patient loses control and has to completely stop the movement to reset. The implication for these data is that tasks requiring temporally extended performance of motor movements (e.g., long trials) and/or be sufficiently difficult to induce stress may be required to induce an intention tremor or ataxic gait in any mouse model for the fragile X premutation and FXTAS.

Although this hypothesis remains untested in fXPCs, the CGG KI mouse model does in fact show the visuospatial/visuomotor deficits predicted by the above model. To specifically evaluate visuomotor functioning CGG KI mice, a skilled forelimb reaching task was developed. In this task, the mouse was required to reach through a narrow window to obtain a reward pellet just out of reach of the tongue at a 30° angle from the edge of the window (this required the mouse to reach with the nonpreferred paw). The number of pellets the mouse was able to obtain without dropping or knocking away the pellet was recorded, as was the number of errors. The CGG KI mice showed a different learning curve than wildtype mice, with CGG KI mice learning the task on average 1–2 days later than wildtype litter mates and never quite learning the task to the same level of asymptotic performance (Diep et al. 2012). Importantly, these deficits were subtle, only becoming apparent when the mice were forced to perform a rather difficult task. These data suggest there is a fundamental impairment in one of two neural systems: (1) the parietal cortex and its interactions with the superior colliculus and cerebellum were unable to provide adequate spatiotemporal updating to allow the CGG KI mice to reach the same level of success as the wildtype mice. (2) The pontocerebellar system shows disruptions (as has been suggested in FXTAS) in a way we could not identify histologically and the deficits arise from an inability of the cerebellum to control the fine motor skills required to skillfully reach, grasp, and consume the reward.

The qualitative data suggests that the CGG KI mice reached with more of a circular or radial motion rather than a directed vector toward the reward pellet, and that mice with longer CGG repeat lengths showed less directed/more radial trajectories than wildtype littermate mice. This resulted in the CGG KI mice knocking the reward away or reaching through the reward, rather than showing difficulty in the grasping of the reward. Once the CGG KI mice grasped the reward pellet, however, they were able to consume it, not showing any difference in the ability to hold onto the pellet and consume it.

To evaluate potential subclinical gait ataxia or general clumsiness in the CGG KI mice, a skilled ladder walking task was employed. The apparatus developed to perform these experiments was a manual ladder rung task (Hunsaker et al. 2011b). The apparatus consisted of clear plexiglass walls separated by approximately 5 cm with 2 mm diameter steel rungs making up the floor of the apparatus. For this initial study, the mice were placed at one end of the apparatus and were allowed to walk back and forth for 2 min. The number of foot slips we recorded for the duration of the 2 min, except for when the mouse was turning around. The number of times the mouse went from one end of the apparatus to the other was also recorded as a general locomotor measure. On this task, mice as young as 2 months of age already showed an increased number of foot slips than wildtype littermate controls. Importantly, the mice showed both forelimb and hindlimb slips, something that suggests concurrent visuospatial and basic motor deficits. These data indicate the presence of visuospatial processing deficits in that there were a high number of forelimb slips in the CGG KI mice, suggesting a difficulty in planning where in space to place the forepaw as well as a difficulty for updating the movement as the step progressed (i.e., as the mouse moved forward the initial planned step has to be modified as the visual space and intended movement interact to guide correct foot placement and an inability to do so results in a foot slip). Hind foot slips however, do not have a visuospatial planning component, but rather reflect a dysfunction in motor function, albeit a subtle one. An inability, or at least increased difficulty with the hind limb placement may reflect some form of ataxia that has not been picked up using other apparatus and methods. Additionally, as a model of FXTAS, during performance of this task, the CGG KI mice showed a high frequency, low amplitude shaking behavior that was visually similar to the description of intention tremors in human FXTAS. This is important as this task was rather difficult, and may have required a high degree of effort from the CGG KI mice that was not required from the wildtype mice, that in no cases did wildtype mice ever present similar tremoring or shaking behaviors.

Now it can be clearly seen that by using a series of more specific tasks than those commonly used to behaviorally phenotype mice, a clear behavioral phenotype emerges in the CGG KI mouse (Borthwell et al. 2012; Diep et al. 2012; Hunsaker et al. 2009, 2010, 2011b, 2012; van Dam et al. 2005). More importantly, however, was the fact that the mouse behavioral phenotype phenocopies results in human premutation carriers without FXTAS (cf., Wong et al. 2012). The initial cognitive deficits were followed up by an analysis of subclinical motor deficits that may correlate with subclinical apraxia mentioned on numerous occasions by a collaborator working with the clinical FXTAS population (RJ Hagerman, personal communication).

These data suggest the CGG KI mouse is an appropriate model for the cognitive deficits present in the fragile X premutation—at least so far as a mouse can serve as a proxy for human cognitive function. Importantly, the CGG KI mouse shows the same neuropathological features as well as systemic organ pathology present in carriers of the fragile X premutation, proving the CGG KI mouse is a valid model for the neurologic and pathological consequences of the premutation (i.e., the mouse itself shows construct validity with the fragile X premutation; Greco et al. 2006; Hunsaker et al. 2011a; Schluter et al. 2012; Tassone et al. 2012; Wenzel et al. 2010).

Although it may at first glance seem unimportant that the mouse model phenocopies the human disorder, it is critical to understand that only by directly phenocopying the human disorder can a mouse be truly used for translational research (i.e., demonstrate predictive validity; Hunsaker 2012a). Once a mouse model is shown to recapitulate the pattern of deficits observed in the clinical population, the behavioral results in the mouse can be used as biomarkers or targets for treatment studies or experimental risk prodrome for studies into gene x environment interactions as related to the incomplete penetrance of FXTAS among premutation carriers.

Conclusions

In recent years, there has been impetus placed on developing behavioral biomarkers that can be used to predict not only later disease onset or progression, but perhaps disease severity. These collections of intermediate or behavioral endophenotypes serve as outcome measures for pharmacological interventions (Cannon and Keller 2006; Gottesman and Gould 2003; Gould and Gottesman 2006; Gur et al. 2007; Hunsaker 2012a, b). This search for behavioral biomarkers, however, has not consistently been extended into the mouse models of genetic disorders. To date, the closest research into mouse disease models comes to developing behavioral biomarkers is to thoroughly parameterize a single task and apply the biomarker as a single screen for various mouse models to choose candidates for drug studies (e.g., attenuated PPI response or audiogenic seizures for the *Fmr1* KO and 22q11.2 deletion syndrome mouse models; cf., Long et al. 2006; Paylor and Lindsay 2006). The strength of the standard approach is the ability to define a canon against which to gauge later models; however, the limitation of this approach is that it lacks the ability to evaluate complimentary models of a given disease to get at the fundamental processes disrupted in the human mutation.

This limitation occurs because a model may fail to model one phenotype, even though the mouse may model any number of other phenotypes that are not included in the standard behavioral screen. This lack of sensitivity is a major limitation as studies into the therapeutic effects of pharmacological agents will be incomplete in the absence of predefined behavioral biomarkers as outcome measures. The tools available in the NIH Toolbox will hopefully alleviate a number of these issues by expanding the number of commonly used, clinically reliable tests that can be modeled in the mouse. This not only increases the amount of clinical data that can be reliably accumulated in the human disease populations, but also expands the number of potential behavioral phenotypes to be tested in the mouse model. This reciprocal dialogue among levels of research should facilitate the usefulness of mouse disease models as has never been previously possible (cf., Hunsaker 2012a, b; Figs. 15.1, 15.2).

If the recent advances in the cognitive neuroscience of neurodevelopmental disorders are extended to their respective mouse models, perhaps the associated behavioral biomarkers of such disorders may not only be complimented by, but extended though use of mouse models studying the component processes underlying disease states. These well-defined behavioral biomarkers can be used as correlates or covariates with molecular studies of underlying disease mechanisms in mice that cannot be directly studied in human patient populations.

References

Allen, E. G., Hunter, J. E., Rusin, M., Juncos, J., Novak, G., Hamilton, D., et al. (2011). Neuropsychological findings from older premutation carrier males and their noncarrier siblings from families with fragile X syndrome. *Neuropsychology, 25*(3), 404–411.

Amann, L. C., Gandal, M. J., Halene, T. B., Ehrlichman, R. S., White, S. L., McCarren, H. S., & Siegel, S. J. (2010). Mouse behavioral endophenotypes for schizophrenia. *Brain Research Bulletin, 83*(3–4), 147–161.

Baker, K. B., Wray, S. P., Ritter, R., Mason, S., Lanthorn, T. H., & Savelieva, K. V. (2010). Male and female *Fmr1* knockout mice on C57 albino background exhibit spatial learning and memory impairments. *Genes, Brain, and Behavior, 9*(6), 562–574.

Banik, A., & Anand, A. (2011). Loss of learning in mice when exposed to rat odor: A water maze study. *Behavioural Brain Research, 216*(1), 466–471.

Barkus, C., McHugh, S. B., Sprengel, R., Seeburg, P. H., Rawlins, J. N., & Bannerman, D. M. (2010). Hippocampal NMDA receptors and anxiety: At the interface between cognition and emotion. *European Journal of Pharmacology, 626*(1), 49–56.

Berge, O. G. (2011). Predictive validity of behavioral animal models for chronic pain. *British Journal of Pharmacology, 164*(4), 1195–1206.

Bohlen, M., Cameron, A., Metten, P., Crabbe, J. C., & Wahlsten, D. (2009). Calibration of rotational acceleration for the rotarod test of rodent motor coordination. *Journal of Neuroscience Methods, 178,* 10–14.

Borthwell, R. M., Hunsaker, M. R., Willemsen, R., & Berman, R. F. (2012). Spatiotemporal processing deficits in female CGG KI mice modeling the fragile X premutation. *Behavioural Brain Research, 233*(1), 29–34.

Cannon, T. D., & Keller, M. C. (2006). Endophenotypes in the genetic analyses of mental disorders. *Annual Review of Clinical Psychology, 2,* 267–290.

Chonchaiya, W., Schneider, A., & Hagerman, R. J. (2009a). Fragile X: A family of disorders. *Advances in Pediatrics, 56,* 165–186.

Chonchaiya, W., Utari, A., Pereira, G. M., Tassone, F., Hessl, D., & Hagerman, R. J. (2009b). Broad clinical involvement in a family affected by the fragile X premutation. *Journal of Developmental and Behavioral Pediatrics, 30,* 544–551.

Crabbe, J. C., & Wahlsten, D. (2003). Of mice and their environments. *Science, 299,* 1313–1314.

Crabbe, J. C., Wahlsten, D., & Dudek, B. C. (1999). Genetics of mouse behavior: Interactions with laboratory environment. *Science, 284,* 1670–1672.

Crawley, J. N. (2004). Designing mouse behavioral tasks relevant to autistic-like behaviors. *Mental Retardation and Developmental Disabilities Research Reviews, 10,* 248–258.

Devanand, D. P., Michaels-Marston, K. S., Liu, X., Pelton, G. H., Padilla, M., Marder, K., et al. (2000). Olfactory deficits in patients with mild cognitive impairment predict Alzheimer's disease at follow-up. *The American Journal of Psychiatry, 157,* 1399–1405.

Diep, A. A., Hunsaker, M. R., Kwock, R., Kim, K. M., Willemsen, R., & Berman, R. F. (2012). Female CGG knock-in mice modeling the fragile X premutation are impaired on a skilled forelimb reaching task. *Neurobiology of Learning and Memory, 97,* 229–234.

Frick, K. M., Stillner, E. T., & Berger-Sweeney, J. (2000). Mice are not little rats: Species differences in a one-day water maze task. *Neuroreport, 11*(16), 3461–3465.

Gershon, R. C. (2007). NIH Toolbox: Assessment of neurological and behavioral function. NIH (contract HHS-N-260-2006 00007-c) http://www.nihtoolbox.org.

Gershon, R. C., Cella, D., Fox, N. A., Havlik, R. J., Hendrie, H. C., & Wagster, M. V. (2010). Assessment of neurological and behavioural function: The NIH Toolbox. *Lancet Neurology, 9*(2), 138–139.

Gilbert, P. E., & Murphy, C. (2004a). Differences between recognition memory and remote memory for olfactory and visual stimuli in nondemented elderly individuals genetically at risk for Alzheimer's disease. *Experimental Gerontology, 39*, 433–441.

Gilbert, P. E., & Murphy, C. (2004b). The effect of the ApoE epsilon4 allele on recognition memory for olfactory and visual stimuli in patients with pathologically confirmed Alzheimer's disease, probable Alzheimer's disease & healthy elderly controls. *Journal of Clinical and Experimental Neuropsychology, 26*, 779–794.

Goodrich-Hunsaker, N. J., Wong, L. M., McLennan, Y., Srivastava, S., Tassone, F., Harvey, D., et al. (2011a). Young adult female fragile X premutation carriers show age- and genetically-modulated cognitive impairments. *Brain and Cognition, 75*, 255–260.

Goodrich-Hunsaker, N. J., Wong, L. M., McLennan, Y., Tassone, F., Harvey, D., Rivera, S. M., & Simon, T. J. (2011b). Adult female fragile X premutation carriers exhibit age- and CGG repeat length-related impairments on an attentionally based enumeration task. *Frontiers in Human Neuroscience, 5*, 63.

Gottesman, I. I., & Gould, T. D. (2003). The endophenotype concept in psychiatry: Etymology and strategic intentions. *The American Journal of Psychiatry, 160*, 636–645.

Gould, T. D., & Einat, H. (2007). Animal models of bipolar disorder and mood stabilizer efficacy: A critical need for improvement. *Neuroscience and Biobehavioral Reviews, 31*, 825–831.

Gould, T. D., & Gottesman, I. I. (2006). Psychiatric endophenotypes and the development of valid animal models. *Genes, Brain, and Behavior, 5*, 113–119.

Greco, C. M., Berman, R. F., Martin, R. M., Tassone, F., Schwartz, P. H., Chang, A., et al. (2006). Neuropathology of fragile X-associated tremor/ataxia syndrome (FXTAS). *Brain: A Journal of Neurology, 129*(1), 243–255.

Greene-Schloesser, D. M., Van der Zee, E. A., Sheppard, D. K., Castillo, M. R., Gregg, K. A., et al. (2011). Predictive validity of a non-induced mouse model of compulsive-like behavior. *Behavioural Brain Research, 221*, 55–62.

Grigsby, J., Brega, A. G., Engle, K., Leehey, M. A., Hagerman, R. J., Tassone, F., et al. (2008). Cognitive profile of fragile X premutation carriers with and without fragile X-associated tremor/ataxia syndrome. *Neuropsychology, 22*(1), 48–60.

Guion, R. M. (1977). Content validity-the source of my discontent. *Applied Psychological Measurement, 1*, 1–10.

Gur, R. E., Calkins, M. E., Gur, R. C., Horan, W. P., Nuechterlein, K. H., Seidman, L. J., & Stone, W. S. (2007). The consortium on the genetics of schizophrenia: Neurocognitive endophenotypes. *Schizophrenia Bulletin, 33*, 49–68.

Gurkoff, G. G., Gahan, J. D., Ghiasvand, R. T., Hunsaker, M. R., Feng, J. F., Berman, R. F., et al. (2012). Moderate lateral fluid percussion injury results in the metric and temporal ordering but not topological working memory tasks. *Journal of Neurotrauma*.

Hagerman, P. J., & Hagerman, R. J. (2004). The fragile-X premutation: A maturing perspective. *American Journal of Human Genetics, 74*(5), 805–816.

Hasler, G., Drevets, W. C., Gould, T. D., Gottesman, I. I., & Manji, H. K. (2006). Toward constructing an endophenotype strategy for bipolar disorders. *Biological Psychiatry, 60*, 93–105.

Hatcher, J. P., Jones, D. N., Rogers, D. C., Hatcher, P. D., Reavill, C., Hagan, J. J., & Hunter, A. J. (2001). Development of SHIRPA to characterise the phenotype of gene-targeted mice. *Behavioural Brain Research, 125*, 43–47.

Hocking, D. R., Kogan, C. S., & Cornish, K. M. (2012). Selective spatial processing deficits in an at-risk subgroup of the fragile X premutation. *Brain and Cognition, 79*(1), 39–44.

Hoffman, H. J., Cruickshanks, K. J., & Davis, B. (2009). Perspectives on population-based epidemiological studies of olfactory and taste impairment. *Annals of the New York Academy of Sciences, 1170,* 514–530.
Hunsaker, M. R. (2012a). Comprehensive neurocognitive endophenotyping strategies for mouse models of genetic disorders. *Progress in Neurobiology, 96*(2), 220–241.
Hunsaker, M. R. (2012b). The importance of considering all attributes of memory in behavioral endophenotyping of mouse models of genetic disease. *Behavioral Neuroscience, 126*(3), 371–380.
Hunsaker, M. R., Fieldsted, P. M., Rosenberg, J. S., & Kesner, R. P. (2008a). Dissociating the roles of dorsal and ventral CA1 for the temporal processing of spatial locations, visual objects, and odors. *Behavioral Neuroscience, 122*(3), 643–650.
Hunsaker, M. R., Tran, G. T., & Kesner, R. P. (2008b). A double dissociation of subcortical hippocampal efferents for encoding and consolidation/retrieval of spatial information. *Hippocampus, 18,* 699–709.
Hunsaker, M. R., Wenzel, H. J., Willemsen, R., & Berman, R. F. (2009). Progressive spatial processing deficits in a mouse model of the fragile X premutation. *Behavioral Neuroscience, 123*(6), 1315–1324.
Hunsaker, M. R., Goodrich-Hunsaker, N. J., Willemsen, R., & Berman, R. F. (2010). Temporal ordering deficits in female CGG KI mice heterozygous for the fragile X premutation. *Behavioral Brain Research, 213*(2), 263–268.
Hunsaker, M. R., Greco, C. M., Spath, M. A., Smits, A. P. T., Navarro, C. S., Tassone, F., et al. (2011a). Widespread non-central nervous system organ pathology in fragile X premutation carriers with fragile X-associated tremor/ataxia syndrome and CGG knock-in mice. *Acta Neuropathologica, 122*(4), 467–479.
Hunsaker, M. R., von Leden, R. E., Ta, B. T., Goodrich-Hunsaker, N. J., Arque, G., Kim, K. M., et al. (2011b). Motor deficits on a ladder rung task in male and female adolescent CGG knock-in mice. *Behavioural Brain Research, 222,* 117–121.
Hunsaker, M. R., Kim, K. M., Willemsen, R., & Berman, R. F. (2012). CGG Trinucleotide repeat length modulates neural plasticity and spatiotemporal processing in a mouse model of the fragile X premutation. *Hippocampus, 22,* 2260–2275.
Jacquemont, S., Hagerman, R. J., Leehey, M. A., Hall, D. A., Levine, R. A., Brunberg, J. A., et al. (2004). Penetrance of the fragile X-associated tremor/ataxia syndrome in a premutation carrier population. *JAMA, 291*(4), 460–469.
Jerman, T. S., Kesner, R. P., & Hunsaker, M. R. (2006). Disconnection analysis of CA3 and DG in mediating encoding but not retrieval in a spatial maze learning task. *Learning and Memory (Hove, England), 13*(4), 458–464.
Karayiorgou, M., Simon, T. J., & Gogos, J. A. (2010). 22q11.2 microdeletions: Linking DNA structural variation to brain dysfunction and schizophrenia. *Nature Reviews Neuroscience, 11,* 402–416.
Kendler, K. S., & Neale, M. C. (2010). Endophenotype: A conceptual analysis. *Molecular Psychiatry, 15,* 789–797.
Kesner, R. P., & Hunsaker, M. R. (2010). The temporal attributes of episodic memory. *Behavioural Brain Research, 215*(2), 299–309.
Kesner, R. P., & Rogers, J. R. (2004). An analysis of independence and interactions of brain substrates that subserve multiple attributes, memory systems & underlying processes. *Neurobiology of Learning and Memory, 82*(3), 199–215.
Kesner, R. P., Farnsworth, G., & DiMattia, B. V. (1989). Double dissociation of egocentric and allocentric space following medial prefrontal and parietal cortex lesions in the rat. *Behavioral Neuroscience, 103*(5), 956–961.
Kesner, R. P., Hopkins, R. O., & Fineman, B. (1994). Item and order dissociation in humans with prefrontal cortex damage. *Neuropsychologia, 32*(8), 881–891.
Long, J. M., LaPorte, P., Merscher, S., Funke, B., Saint-Jore, B., Puech, A., et al. (2006). Behavior of mice with mutations in the conserved region deleted in velocardiofacial/DiGeorge syndrome. *Neurogenetics, 7,* 247–257.

Llano Lopez, L., Hauser, J., Feldon, J., Gargiulo, P. A., & Yee, B. K. (2010). Evaluating spatial memory function in mice: A within-subjects comparison between the water maze test and its adaptation to dry land. *Behavioural Brain Research, 209,* 85–92.

Manji, H. K., Gottesman, I. I., & Gould, T. D. (2003). Signal transduction and genes-to-behaviors pathways in psychiatric diseases. *Science STKE, 2003,* pe49.

McClelland, M. M., & Cameron, C. E. (2011). Self-regulation and academic achievement in elementary school children. In R. M. Lerner, J. V. Lerner, E. P. Bowers, S. Lewin-Bizan, S. Gestsdottir, & J. B. Urban (Eds.), Thriving in childhood and adolescence: The role of self-regulation processes. New Directions for Child and Adolescent Development, 133, 29–44.

Nakazawa, K., McHugh, T. J., Wilson, M. A., & Tonegawa, S. (2004). NMDA receptors, place cells and hippocampal spatial memory. *Nature Reviews Neuroscience, 5,* 361–372.

Olton, D. S., Becker, J. T., & Handelmann, G. E. (1979). Hippocampus, space & memory. *Brain and Behavioral Science, 2,* 313–365.

Paylor, R., & Lindsay, E. (2006). Mouse models of 22q11 deletion syndrome. *Biological Psychiatry, 59,* 1172–1179.

Pilkonis, P. A., Choi, S. W., Salsman, J. M., Butt, Z., Moore, T. L., Lawrence, S. M., et al. (2012). Assessment of self-reported negative affect in the NIH Toolbox. *Psychiatry Research.*

Pirogovsky, E., Goldstein, J., Peavy, G., Jacobson, M. W., Corey-Bloom, J., & Gilbert, P. E. (2009). Temporal order memory deficits prior to clinical diagnosis in Huntington's disease. *Journal of the International Neuropsychological Society, 15,* 662–670.

Quatrano, L. A., & Cruz, T. H. (2011). Future of outcomes measurement: Impact on research in medical rehabilitation and neurologic populations. *Archives of Physical Medicine and Rehabilitation, 92*(10 Suppl), S7–S11.

Rogers, D. C., Fisher, E. M., Brown, S. D., Peters, J., Hunter, A. J., & Martin, J. E. (1997). Behavioral and functional analysis of mouse phenotype: SHIRPA, a proposed protocol for comprehensive phenotype assessment. *Mammalian Genome, 8,* 711–713.

Rogers, D. C., Jones, D. N., Nelson, P. R., Jones, C. M., Quilter, C. A., Robinson, T. L., & Hagan, J. J. (1999). Use of SHIRPA and discriminant analysis to characterise marked differences in the behavioural phenotype of six inbred mouse strains. *Behavioural Brain Research, 105,* 207–217.

Rogers, D. C., Peters, J., Martin, J. E., Ball, S., Nicholson, S. J., Witherden, A. S., et al. (2001). SHIRPA, a protocol for behavioral assessment: Validation for longitudinal study of neurological dysfunction in mice. *Neuroscience Letters, 306,* 89–92.

Rondi-Reig, L., Petit, G. H., Tobin, C., Tonegawa, S., Mariani, J., & Berthoz, A. (2006). Impaired sequential egocentric and allocentric memories in forebrain-specific NMDA receptor knockout mice during a new task dissociating strategies of navigation. *Journal of Neuroscience, 26*(15), 4071–4081.

Rupp, J., Blekher, T., Jackson, J., Beristain, X., Marshall, J., Hui, S., et al. (2009). Progression in prediagnostic Huntington disease. *Journal of Neurology, Neurosurgery, and Psychiatry, 81*(4), 379–384.

Rustay, N. R., Wahlsten, D., & Crabbe, J. C. (2003). Influence of task parameters on rotarod performance and sensitivity to ethanol in mice. *Behavioural Brain Research, 141,* 237–249.

Salomonczyk, D., Panzera, R., Pirogovosky, E., Goldstein, J., Corey-Bloom, J., Simmons, R., & Gilbert, P. E. (2010). Impaired postural stability as a marker of premanifest Huntington's disease. *Movement Disorders, 25*(14), 2428–2433.

Schluter, E. W., Hunsaker, M. R., Greco, C. M., Willemsen, R., & Berman, R. F. (2012). Distribution and frequency of intranuclear inclusions in female CGG KI mice modeling the fragile X premutation. *Brain Research, 1472,* 124–137.

Simon, T. J. (2007). Cognitive characteristics of children with genetic syndromes. *Child and Adolescent Psychiatric Clinics of North America, 16,* 599–616.

Simon, T. J. (2008). A new account of the neurocognitive foundations of impairments in space, time and number processing in children with chromosome 22q11.2 deletion syndrome. *Developmental Disabilities Research Reviews, 14,* 52–58.

Simon, T. J. (2011). Clues to the foundation of numerical cognitive impairments: Evidence from genetic disorders. *Developmental Neuropsychology, 36*(6), 788–805.

Spencer, C. M., Alekseyenko, O., Hamilton, S. M., Thomas, A. M., Serysheva, E., Yuva-Paylor, L. A., & Paylor, R. (2011). Modifying behavioral phenotypes in Fmr1 KO mice: Genetic background differences reveal autistic-like responses. *Autism Research, 4,* 40–56.

Tassone, F., Greco, C. M., Hunsaker, M. R., Berman, R. F., Seritan, A. L., Gane, L. W., et al. (2011). Neuropathological, clinical, and molecular pathology in female fragile X premutation carriers with and without FXTAS. *Genes, Brain and Behavior, 11*(5), 577–585.

Taylor, T. N., Greene, J. G., & Miller, G. W. (2010). Behavioral phenotyping of mouse models of Parkinson's disease. *Behavioural Brain Research, 211*(1), 1–10.

Van Dam, D., Errijgers, V., Kooy, R. F., Willemsen, R., Mientjes, E., Oostra, B. A., & De Deyn, P. P. (2005). Cognitive decline, neuromotor and behavioural disturbances in a mouse model for fragile-X-associated tremor/ataxia syndrome (FXTAS). *Behavioural Brain Research, 162*(2), 233–239.

Wahlsten, D. (1972). Genetic experiments with animal learning: A critical review. *Behavioral Biology, 7,* 143–182.

Wahlsten, D. (2001). Standardizing tests of mouse behavior: Reasons, recommendations & reality. *Physiology and Behavior, 73,* 695–704.

Wahlsten, D., Metten, P., Phillips, T. J., Boehm, S. L., Burkhart-Kasch, S., Dorow, J., et al. (2003a). Different data from different labs: Lessons from studies of gene-environment interaction. *Journal of Neurobiology, 54,* 283–311.

Wahlsten, D., Rustay, N. R., Metten, P., & Crabbe, J. C. (2003b). In search of a better mouse test. *Trends in Neurosciences, 26,* 132–136.

Wahlsten, D., Metten, P., & Crabbe, J. C. (2003c). Survey of 21 inbred mouse strains in two laboratories reveals that BTBR T/+ tf/tf has severely reduced hippocampal commissure and absent corpus callosum. *Brain Research, 971,* 47–54.

Wahlsten, D., Bachmanov, A., Finn, D. A., & Crabbe, J. C. (2006). Stability of inbred mouse strain differences in behavior and brain size between laboratories and across decades. *Proceeding of the National Academy of Sciences of the United States of America, 103,* 16364–16369.

Wang, Y. C., Magasi, S. R., Bohannon, R. W., Reuben, D. B., McCreath, H. E., Bubela, D. J., et al. (2011). Assessing dexterity function: A comparison of two alternatives for the NIH Toolbox. *Journal of Hand Therapy, 24*(4), 313–320.

Wenzel, H. J., Hunsaker, M. R., Greco, C. M., Willemsen, R., & Berman, R. F. (2010). Ubiquitin positive intranuclear inclusions in neuronal and glial cells in a mouse model of the fragile X premutation. *Brain Research, 1318,* 155–166.

Wesson, D. W., Nixon, R. A., Levy, E., & Wilson, D. A. (2011). Mechanisms of neural and behavioral dysfunction in Alzheimer's disease. *Molecular Neurobiology, 43*(3), 163–179.

Weiser, M., Van Os, J., & Davidson, M. (2005). Time for a shift in focus in schizophrenia: From narrow phenotypes to broad endophenotypes. *The British Journal of Psychiatry, 187,* 203–205.

Whishaw, I. Q., & Tomie, J. (1996). Of mice and mazes: Similarities between mice and rats on dry land but not water mazes. *Physiology and Behavior, 60*(5), 1191–1197.

White, N. M., & McDonald, R. J. (2002). Multiple parallel memory systems in the brain of the rat. *Neurobiology of Learning and Memory, 77*(2), 125–184.

Wong, L. M., Goodrich-Hunsaker, N. J., McLennan, Y. A., Tassone, F., Harvey, D., Rivera, S. M., & Simon, T. J. (2012). Young adult male carriers of the Fragile X premutation exhibit genetically modulated impairments in visuospatial tasks controlled for psychomotor speed. *Journal of Neurodevelopmental Disorders, 4*(1), 26.

Xu, B., Karayiorgou, M., & Gogos, J. A. (2010). microRNAs in psychiatric and neurodevelopmental disorders. *Brain Research, 1338,* 78–88.

Yan, Q. J., Asafo-Adjei, P. K., Arnold, H. M., Brown, R. E., & Bauchwitz, R. P. (2004). A phenotypic and molecular characterization of the *Fmr1-tm1Cgr* fragile X mouse. *Genes, Brain, and Behavior, 3,* 337–359.

Yong-Kee, C. J., Salomonczyk, D., & Nash, J. E. (2010). Development and validation of a screening assay for the evaluation of putative neuroprotective agents in the treatment of Parkinson's disease. *Neurotoxicity Research, 19*(4), 519–526.

Part IV
Personal

Chapter 16
The Life and Science of Raymond P. Kesner

Pamela A. Jackson

The chapter aims to provide a brief biography of Ray Kesner. I would like to direct the reader to his excellent autobiography "Tapestry of Memory" (Kesner 2009), which covers the more scientific aspects of Ray's career till that point in time. The current approach will be much more personal in nature. It is based on interviews with Ray as well as personal communication provided at the Festschrift conference held in January 2013. A photograph of the presenters at the conference is provided in Fig. 16.1.

Ray's parents lived in Belgium until 1939, when they were forced to flee in order to avoid the German invasion. They moved to southern France but were again forced by the German army's occupation to move on. Ray was born during this time in Oran, Algeria, in 1940. The family continued to flee the German armies, traveling to many different cities in North Africa (e.g., Casablanca, Morocco, Tangiers) in order to avoid capture. His father finally decided to make a stand, so he joined the underground resistance movement and began to fight while they were in Tangiers. Due to a strategically placed bomb which made him a bit too notorious, and his discovery of a Dutch spy in Gibraltar, Ray's father was able to attain help relocating the family to London. He then became a bombardier in the Dutch part of the British Royal Airforce where the return rate from bombing missions was an abysmal 50%. Although Ray was only about four years old at this point, he has isolated memories of that time: the black-out curtains, the v-bombs, and hiding under a piano during raids. At the end of the War his family moved back to Belgium for a short period but ultimately spent the next 10 years in Rotterdam, Holland. Ray attended school there until he was 15. According to Ray: "one of the things that always blew me away was we had to take four languages … English, and German, and French, and Dutch!" (Fig. 16.2).

P. A. Jackson (✉)
Department of Psychology, Radford University, Radford, VA 24142-6946, USA
e-mail: pjackson@radford.edu

P. A. Jackson et al. (eds.), *The Neurobiological Basis of Memory*,
DOI 10.1007/978-3-319-15759-7_16

Fig. 16.1 The presenters at the Kesner Festschrift Conference held in Salt Lake City, Utah, January 2013

The family moved to Detroit, Michigan, in 1955, choosing Detroit because their sponsor lived there. Ray and his sister were not happy about immigrating to America and leaving their friends, but it turned out to be a very good decision. Ray completed 10th, 11th, and 12th grades over the next two years in the USA (Fig. 16.3).

Ray applied to college because that was what was expected of Jewish youth in America at that time. He attended Wayne State University and worked at a variety of different jobs to cover tuition, including delivering drugs for a pharmacy. Originally he thought he would major in business accounting, but he "didn't like it at all." Serendipitously, he joined a psychotherapy session with his father, and was very impressed with the psychiatrist, who described Ray as effervescent (a very accurate assessment, then and now). Ray was captivated and thought he would like to be a clinical psychologist. Unfortunately, he had been enjoying the social side of college a bit too much, had joined a fraternity, and had not been all that interested in studying (Fig. 16.4). Two things happened to alter his trajectory: his advisor steered him toward a physiological psychology course, and he met his future wife, Laya (Fig. 16.5).

After excelling on the first exam in the physiological psychology class, the professor offered Ray the job of taking care of a colony of rats. He was excited at the opportunity. Ray was surprised and taken aback to discover that the rats were not being used at all, merely maintained. This was the beginning of Ray's very prolific research career, because, of course, he had to conduct an experiment! The study

Fig. 16.2 Sixth grade (1952) in Rotterdam, the Netherlands

involved comparing dextrose and saccharin solution preference in rats that were food or water deprived. He even presented the study at the Michigan Academy of Sciences in 1961 (his first scientific presentation). His interest in research solidified when he took a full-time job, while still attending college, with the Lafayette Clinic doing teratology research on rats. One of the publications stemming from this opportunity, with Werboff, examined learning deficits in the offspring of rat mothers that were exposed to tranquilizing drugs (Werboff and Kesner 1963). These early forays into research provide a glimpse into the origins of his abiding interest in the underlying mechanisms of learning and memory.

In the meantime, he met Laya (Fig. 16.5). She was a chemistry major at Wayne State University, had been a valedictorian in high school, and was very invested in her studies. Laya transferred to the University of Michigan after a semester but that did not slow the relationship down. Ray traveled to see her on weekends. They

Fig. 16.3 Central High School Graduation, Detroit, Michigan

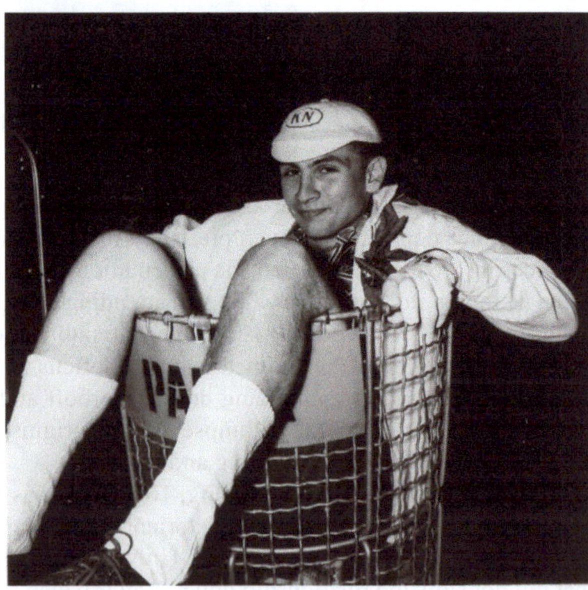

Fig. 16.4 Kappa Nu Fraternity Pledging Night (1960)

Fig. 16.5 Ray and Laya in 1990

would go out on Saturday nights and study all day on Sundays. Needless to say, with the extra Laya-induced study time, Ray became a straight-A student. When he graduated he was offered graduate fellowships in the Ph.D. programs at both the University of Illinois at Urbana-Champaign and Wayne State University. He chose the University of Illinois and majored in physiological psychology with a minor in neurophysiology.

The University of Illinois was a large school and there were many famous psychologists there during Ray's tenure. One of his first graduate classes was a proseminar in which professors from each area had one week to convey everything they thought was important in their field. It meant great exposure to many of the faculty but, in order to be prepared, there was a lot of reading! Ray felt first impressions were critical at this point so he really studied hard, and it paid off. He earned the second highest grade out of 65 graduate students in that class, which kept him in good graces throughout his graduate career, even though he says he rarely scored that high again.

Ray worked on his Master's thesis with Lawrence O'Kelly in 1964. Earlier, while working with Dr. Werboff at the Lafayette Clinic in Detroit, Michigan, Ray had also examined the effect of administering phenobarbital or metrazol to pregnant female rats on the susceptibility of the offspring to audiogenic seizures. Stemming from this experience, Ray's Master's thesis examined the effect of cortical spreading depression due to application of potassium chloride to the cortex on audiogenic seizure susceptibility. He completed his Master's that summer with O'Kelly and then he switched mentors and began working with Jerry Hirsch developing a learning paradigm in fruit flies (*Drosophila melanogaster*). They assessed preference in fruit flies for peppermint odor in a T-maze. Then they bred the peppermint-preferring flies with each other, and the non-peppermint-preferring fruit flies with each other. Ray was among the earliest epigenetic researchers of our time! However, Ray found the fruit fly model less than attractive, and, in addition, he continued to be very interested in epilepsy. So, when Garth Thomas offered him a job in his lab to carry out brain system lesions, Ray took it. And he continued with Thomas for his dissertation. Working under Garth Thomas, Ray studied the subcortical mechanisms of audiogenic seizures (Kesner 1966). They examined alterations in seizure responsiveness in seven different subcortical areas in rats that were susceptible, and in rats that were not susceptible, to sound-induced seizures.

For Ray and the other students in his lab, Garth Thomas' influence extended way beyond the science arena; he was an excellent mentor and role model. According to Ray "Garth Thomas was a fantastic person…every day we'd go to lunch with him…" and he would cleverly tweak their interest in a particular topic. Thomas would begin conversations during lunch and the graduate students would read up on that topic so they could converse intelligently the next time they met. Before long one of the famous people in that specific area would visit campus and/or the lab, and the students would be prepared. Ray thought that approach was not only sneaky, but very effective. That early training probably accounts for Ray's own approach with his students and postdoctoral fellows. Social opportunities were excellent occasions for discussing science, and Ray ensured there were many during my time in his lab: we had lunch as a group fairly frequently, we played racquetball with him one-on-one, he took us skiing in groups, we attended conferences together, and the list goes on. Sitting on a ski lift with your mentor, who lives for science, was a surefire way of having a new study to conduct by the time you unloaded!

While Ray was working on his Ph.D. degree at Illinois, Laya was finishing her B.S. degree at Michigan. They continued to date but were waiting until they obtained their respective degrees to get married. Laya graduated, they were married, and Ray had one last summer of writing before he finished his dissertation in 1965. He applied for a postdoctoral fellowship with Robert Doty at the Brain Research Institute at the University of Rochester, which was one of the two main brain research institutes at the time (the other being at UCLA). They moved to Rochester where Laya taught math for a year, and then she worked for a year at Xerox. She was an analytical chemist, and had much more fun in her year at Xerox. Interestingly, the Brain Research Institute also hired Garth Thomas around the same time! Ray worked closely with Robert Doty and continued to associate with Garth Thomas as well. He said "it was marvelous…your post-doctoral years are really your best years."

So how did Ray segue into memory? Doty had been doing research in which he stimulated the ventral tegmental area (VTA) in cats and measured evoked potentials in the olfactory tubercle. When asked, Ray was eager to work on this line of research in order to learn surgery techniques and electrophysiology in cats. As he worked with the cats, Ray discovered some anomalies, and when they pursued the issue they found that the evoked potential disappeared if the animal had been given curare, which was surprising. They then stimulated the VTA and observed eye muscle movements. It turned out that the cats' oculomotor nerve was activated whenever they stimulated the VTA, causing eye movements, which resulted in evoked potentials in the olfactory tubercle. This invalidated Doty's earlier work and they had to publish a retraction (Kesner and Doty 1966). Because Doty felt some guilt over this, he compensated by allowing Ray to do anything he wanted to do for research. And Ray was interested in memory!

According to Ray, the choices were pretty limited at that time: eyelid conditioning, electroconvulsive shock, or self-stimulation. Ray liked the McGaugh paradigm for studying retrograde amnesia but he did not really like using electroconvulsive shock (ECS), although he conducted several studies with it (e.g., Kesner and Doty 1967; Kesner et al. 1970; Kesner and D'Andrea 1971). Ray's solution was to implant electrodes in the amygdala or the hippocampus in cats, and use low-level electrical stimulation. This level of stimulation to either the amygdala or the hippocampus produced a mild seizure, which was applied immediately following exposure to a shock in an inhibitory avoidance task. They obtained a beautiful retrograde amnesia effect 24 h later (Kesner and Doty 1968), but more importantly, they also observed amnesia when there was no seizure activity at all (McDonough and Kesner 1971; Berman and Kesner 1976). It was the beginning of a beautiful career studying the neurobiology of memory! Ray was the first to do this and show that ECS, or seizure activity in general, was not necessary to produce amnesia. This was just one of the many "firsts" for Ray. He continued to track down the underlying cause of retrograde amnesia and was also the first (in the 1970s) to demonstrate amnesia after the injection of cholinergic agonists and antagonists, as well as protein synthesis inhibitors, directly into the amygdala of rats (Berman et al. 1978; Todd and Kesner 1978).

At the end of his two-year postdoc at the Brain Research Institute in Rochester in 1967, Ray took up a job in the psychology department at the University of Utah. And he stayed there until his retirement in 2014, over 47 years later. He and Laya drove cross-country from New York to Salt Lake City their first time. Laya had never been to Utah and Ray had been only once, for his interview. They carved out a life for themselves and clearly thrived. They became active in the Jewish community. Laya went to graduate school at the University of Utah and obtained her Ph.D. She worked under her Ph.D. mentor for a while because there were not any industry jobs in chemistry available in the Salt Lake area. She taught at an upscale private high school as well, and then she took over teaching the undergraduate chemistry labs at the University of Utah. She loved it but was very busy much of the time. Ray and Laya have two children, a girl named Debbie and a boy named Benjamin, and now Debbie has her own two boys. Ray was a wonderful father, and loves being a grandfather!

When Ray began at the University of Utah in 1967, the psychology department was housed in the old Fort Douglas army barracks. His primary complaint was the need to kill cockroaches every morning, probably because the operant conditioning faculty used sugar reinforcers. But the department moved to the new social and behavioral sciences building fairly soon after he arrived, so he was not forced to deal with the cockroaches for long. Ray continued to study amnesia using the electrical stimulation paradigm (Fig. 16.6). He and Jim McGaugh became friends and have engaged in many a heated discussion about the role of the amygdala and the hippocampus on the inhibitory avoidance task, among other things. Ray considers McGaugh to be one of the more important influences on his thinking.

Two more major events had far-ranging importance in shaping Ray's career. Early on, the University of Utah was asked to send two scientists to be interviewed for a very prestigious fellowship at the Stanford Center for Advanced Study in the Behavioral Sciences. The fellowships were funded via grants and they chose approximately 8 people out of 150 interviewees to come to the Center and study together for a year. Due to several factors, including that they were young and productive, Ray and Charlie Shimp (also in the Department of Psychology) were the two faculty members out of everybody in the behavioral sciences at Utah to be chosen to go for the interviews. And, both were accepted into the program! Ray and Laya decided to attend in 1971. Not only did the fellowship affect his career, but Laya became pregnant with their first child, Debbie, while they were in Stanford.

The fellowship at the Stanford Center had a huge impact on his trajectory because, for the first time ever, Ray was exposed to cognitive psychology in the con-

Dr. Kesner

Fig. 16.6 Early electrical stimulation experiments (1968)

text of memory. According to Ray "from that point on learning theory wasn't *it*, it was cognitive psychology… so, I became a behavioral neuropsychologist …with an interest in memory and cognition." Ray is justly famous for his ability to design animal studies to answer complex questions. A quote from Jim McGaugh at Ray's Festschrift: "he's been the most innovative person in designing tests for laboratory animals that I know. I don't know anyone who's been more ingenious. I remember when I used to visit him in the laboratory. I would just go away puzzled because I wouldn't know how on earth he got the idea to do something, and then the fact that it worked was even more important than that." A great deal of this creativity is due to Ray's ability to read the human cognitive literature and come away from it thinking how could I test this in a rat? This also led to him falling in love with Underwood's approach to memory and creating his own theory, the Attribute Model of Memory. Another result of his exposure to the cognitive literature was his study of serial position curves in rats (e.g., DiMattia and Kesner 1984), which led to testing his ideas on both humans and animals in comparable paradigms (e.g., Kesner and Hopkins 2001). And the list goes on.

Ray thinks of his career in phases, and developing and testing his Attribute Model of Memory was phase two (early renditions of the full model include Kesner and DiMattia 1987, and Kesner 1991). Phase one was the research on learning, memory, and amnesia which began during his postdoc years and continued through the first decade or more at Utah (e.g., Berman and Kesner 1976; D'Andrea and Kesner 1973). Once Ray decided that it was critical to look at memory in terms of different attributes, the realm of possibilities exploded. He developed paradigms to study different regions subsuming different attributes (e.g., Kesner et al. 1989, 1993; Chiba et al. 2002). He developed paradigms to address the different attributes themselves (e.g., Chiba et al. 1994; DeCoteau et al. 1997; Adams et al. 2001). The Attribute Model was rich in terms of new experiments to conduct, and according to Ray, about 80 % of the studies that were conducted in his lab supported the model! The other 20 % probably account for some of the changes to the model over the years.

Ray considers himself in phase three of his career now, which is essentially an expansion of the Attribute Model to a computational approach, heavily inspired by his work and friendship with Edmund Rolls (e.g., Kesner and Rolls 2001; Rolls and Kesner 2006). The studies involved in this phase include the research on processing of mnemonic information, exemplified by his involvement in studying pattern separation and completion (e.g., Gilbert and Kesner 2002) as well as the analysis of subregions of the brain, including the hippocampus, and their involvement in memory processes (e.g., Gilbert et al. 2001; Lee and Kesner 2003). Several chapters in the current volume provide excellent descriptions of these concepts and projects, starting in Ray's lab and going into the next generation (e.g., Kesner 2016; Rolls 2016; Gilbert et al. 2016; Kirwan and Nash 2016; Lee and Lee 2016).

The other major event that impacted Ray's career was the Winter Conference. He, Jim McGaugh, Larry Squire, Aryeh Routtenberg, and Stuart Zola-Morgan (aka the "Founding Fathers") put together the first Winter Conference on the Neurobiology of Learning and Memory (NBLM) in Park City, Utah in 1977. The executive committee has changed somewhat over the years, but the 2013 conference (during which Ray was honored) was the 37th consecutive meeting. The impetus for Ray to develop and pour so much energy into the NBLM Winter Conference every year stemmed a great deal from his feeling of geographical isolation. The University of Utah, located in Salt Lake City, was fairly distant from other neuroscientists studying memory. In addition, the founding fathers of the NBLM Conference all felt the need to have a smaller venue where discussion would be given equal importance to presentation of scientific results. This is the aspect that I have always found to be most attractive, besides the skiing, because you learn so much more from the ensuing question and answer session. Ray, and many of the attendees, love skiing, so the conference sessions begin at three or four in the afternoon. It is the ideal setting for discussion, argument, and general camaraderie. Invitees could apply to bring one or two students to the meeting, and it meant the world to many of those that came. I attended the conference for the first time in order to interview for a postdoctoral fellowship position with Ray. I fell in love with the conference structure, with Ray's effervescence, and with skiing (another first for me). And I am not sure which love was the strongest; they have all endured the test of time! One of the things that impresses me the most every time I attend the NBLM Winter Conference is watching the great memory researchers of the time argue the fine points of interpretation. Not to mention Dave Olton using a timer to limit the presentation time so that half was devoted to discussion (back in the day). It was liberating and it was inspirational.

Describing the NBLM Winter Conference provides the perfect transition into Ray's career as a mentor. The editors of this volume decided to include a personal chapter with essays about Ray, and letters written to him from former students and colleagues (Recollections 2016). Almost everyone that encounters Ray loves his infectious enthusiasm for science and his kindness. It is a common theme throughout those letters. What is even more remarkable is that those attributes do not wax and wane; they were present every day in his lab. He treated his undergraduate students in the same manner as the graduate students and the postdoctoral fellows. He not only discussed his ideas with everyone equally, he listened to their ideas, and incorporated them into his own thinking. This equal exchange created a wonderful dynamic that empowered the students, and spilled over into successful research endeavors. Ray has published over 250 peer-reviewed journal articles. Over 70 of his co-authors were undergraduate students. In fact, three of the authors of chapters in this volume worked with Ray as undergraduates before obtaining their Ph.D. elsewhere (Ryan Hunsaker, Naomi Goodrich-Hunsaker, and Brock Kirwan). Table 16.1 provides a list of Ray's Ph.D. students and postdoctoral fellows that continued in neuroscience-related careers, along with their affiliations. Many of them conduct research that stems from their time with Ray, and several have contributed chapters to this volume (Berman, Chiba, Cho, Gilbert, Hopkins, Jackson, Ragozzino, Lee, Morris, and Weeden; see Table 16.1). The very characteristics that make Ray such an excellent mentor (imparting and expecting high-quality ideas, as well as effort,

Table 16.1 Current affiliation of Kesner students and postdoctoral fellows that continued in academics and/or a related field

Name	Training	~Year	Most recent affiliation
Margaret W. Wilburn	Ph.D.	1973	Retired as program director of The Hope Unit at Utah State Hospital, Provo, UT. (Deceased, 2011)
[a]Robert F. Berman	Ph.D.	1976	Department of Neurological Surgery, University of California, Davis, CA
John A. D'Andrea	Ph.D.	1976	Retired as science director at Naval Medical Research Unit, San Antonio, TX
Linda J. Baker	Ph.D.	1982	Began in Department of Medicine, University of Wisconsin, Madison, WI; switched to private practice; retired to become an artist
Rex A. Bierley	Ph.D.	1982	Permanente Medical Group, Inc., Dept of Psychiatry, Redwood City, CA
Maureen E. Ellis	Ph.D.	1982	Cross-disciplinary life science consultant, Seattle, WA
Hideki Kametani	Postdoc	1987	Department of Psychology, Saitama Institute of Technology, Saitama, Japan
Mary E. Hunt	Ph.D.	1989	Department of Psychology, Lake Sumter State College, Clermont, FL
[a]Pamela A. Jackson	Postdoc	1990	Department of Psychology, Radford University, Radford, VA
[a]Yoon H. Cho	Postdoc	1992	Institut de Neurosciences Cognitives et Intégratives d'Aquitaine, University of Bordeaux, Talence Cedex, France
Debra Johnson	Ph.D.	1992	Department of Psychology, University of Iowa, Iowa City, IA
David R. Beers	Ph.D.	1993	Department of Neurology, Houston Methodist Research Institute, Houston Methodist Hospital, TX
[a]Andrea A. Chiba	Ph.D.	1993	Department of Cognitive Science and Program in Neuroscience; Temporal Dynamics of Learning Center; University of California, San Diego, La Jolla, CA
[a]Ramona O. Hopkins	Ph.D.	1996	Psychology Department and Neuroscience Center, Brigham Young University Provo, UT
Jeffrey M. Long	Ph.D.	1996	National Institute on Aging; National Institutes of Health, Bethesda, MD

Table 16.1 (continued)

Name	Training	~Year	Most recent affiliation
[a]Michael E. Ragozzino	Postdoc	1995	Department of Psychology, University of Illinois at Chicago, Chicago, IL
W. (Bill) E. DeCoteau	Ph.D.	1999	Psychology and Neuroscience; St. Lawrence University, Canton, NY
[a]Paul E. Gilbert	Ph.D.	2002	San Diego State University—University of California San Diego Joint Doctoral Program in Clinical Psychology, San Diego, CA
[a]Inah Lee	Ph.D.	2002	Department of Brain and Cognitive Sciences; Laboratory for Behavioral Neurophysiology; Seoul National University, Seoul, Korea
Jason L. Rogers	Ph.D.	2005	Noldus Information Technology, Leesburg, VA
David R. Vago	Ph.D.	2005	Harvard Medical School; Brigham & Women's Hospital; Functional Neuroimaging Lab; Department of Psychiatry, Boston, MA
David P. Daberkow	Ph.D.	2006	Department of Biology, Eastern Washington University, Cheney, WA
John C. Churchwell	Ph.D.	2008	Instructional designer at Pearson; Los Angeles, CA
[a]Andrea M. Morris	Ph.D.	2011	Los Angeles Fielding School of Public Health; Department of Health Policy and Management; University of California, Los Angeles, CA
[a]Christy S. S. Weeden	Ph.D.	2012	National Institute on Aging; National Institutes of Health, Bethesda, MD

[a] Chapter authors in the current volume

from all who come in contact with him, his extreme enthusiasm, and his respect, kindness, and generosity toward others) also served to enrich his thought process, and therefore his research. To me, this constitutes the final major influence on Ray's career.

Ray, I absolutely agree with you, my postdoc years were the very best! The positive impact you have had on all areas of my life is enormous. I cannot thank you enough for giving me the opportunity to work with you and for being such a marvelous mentor ever since (Fig. 16.7).

Fig. 16.7 At the Neurobiology of Learning and Memory Conference in Park City, Utah (1998)

References

Adams, S., Kesner, R. P., & Ragozzino, M. E. (2001). Role of the medial and lateral caudate-putamen in mediating an auditory conditional response association. *Neurobiology of Learning and Memory, 76,* 106–116.

Berman, R. F., & Kesner, R. P. (1976). Posttrial hippocampal, amygdaloid and lateral hypothalamic electrical stimulation: Effects upon memory of an appetitive experience. *Journal of Comparative and Physiological Psychology, 90,* 260–267.

Berman, R. F., Kesner, R. P., & Partlow, L. M. (1978). Passive avoidance impairment in rats following cycloheximide injection into the amygdala. *Brain Research, 158,* 171–188.

Chiba, A. A., Kesner, R. P., & Reynolds, A. (1994). Memory for spatial location as a function of temporal lag in rats: Role of hippocampus and medial prefrontal cortex. *Behavioral and Neural Biology, 61,* 123–131.

Chiba, A. A., Kesner, R. P., & Jackson, P. (2002). Two forms of spatial memory: A double dissociation between the parietal cortex and the hippocampus in the rat. *Behavioral Neuroscience, 116,* 874–883.

D'Andrea, J. A., & Kesner, R. P. (1973). The effects of ECS and hypoxia on information retrieval. *Physiology and Behavior, 11,* 747–752.

DeCoteau, W. E., Kesner, R. P., & Williams, J. M. (1997). Short-term memory for food reward magnitude: The role of the prefrontal cortex. *Behavioural Brain Research, 88,* 239–249.

DiMattia, B. V., & Kesner, R. P. (1984). Serial position curves in rats: Automatic vs effortful information processing. *Journal of Experimental Psychology: Animal Behavior Processes, 10,* 557–563.

Gilbert, P. E., & Kesner, R. P. (2002). The amygdala but not the hippocampus is involved in pattern separation based on reward value. *Neurobiology of Learning and Memory, 77,* 338–353.

Gilbert, P. E., Kesner, R. P., & Lee, I. (2001). Dissociating hippocampal subregions: A double dissociation between dentate gyrus and CA1. *Hippocampus, 11,* 626–636.

Gilbert, P. E., Holden, H. M., Sheppard, D. P., & Morris, A. (2016). Pattern separation: A key processing deficit associated with aging? In P. A. Jackson, A. A. Chiba, R. F. Berman, & M. E. Ragozzino (Eds.), *The neurobiological basis of memory: A system, attribute, and process analysis*. New York: Springer.

Kesner, R. P. (1966). Subcortical mechanisms of audiogenic seizure. *Experimental Neurology, 15,* 192–205.

Kesner, R. P. (1991). The role of the hippocampus within an attribute model of memory. *Hippocampus, 1,* 279–282.

Kesner, R. P. (2009). Tapestry of memory. *Behavioral Neuroscience, 123,* 1–13.

Kesner, R. P. (2016). Exploration of the neurobiological basis for a three-system, multi-attribute model of memory. In P. A. Jackson, A. A. Chiba, R. F. Berman, & M. E. Ragozzino (Eds.), *The neurobiological basis of memory: A system, attribute, and process analysis.* New York: Springer.

Kesner, R. P., & D'Andrea, J. A. (1971). Electroconvulsive shock disrupts both information storage and retrieval. *Physiology and Behavior, 7,* 73–76.

Kesner, R. P., & DiMattia, B. V. (1987). Neurobiology of an attribute model of memory. In A. R. Morrison and A. N. Epstein (Eds.), *Progress in psychobiology and physiological psychology* (pp. 207–277). New York: Academic Press.

Kesner, R. P., & Doty, R. W. (1966). A "tegmento-tubercular projection" as an artifact of third nerve stimulation. *Experimental Brain Research, 2,* 328–329.

Kesner, R. P., & Doty, R. W. (1967). Dependence of amnestic effects upon locus of electroconvulsive stimulation. *The Physiologist, 10,* 219 (Abstract).

Kesner, R. P., & Doty, R. W. (1968). Amnesia produced in cats by local seizure activity initiated from the amygdala. *Experimental Neurology, 21,* 58–68.

Kesner, R. P., & Hopkins, R. O. (2001). Short-term memory for duration and distance in humans: Role of the hippocampus. *Neuropsychology, 15,* 58–68.

Kesner, R. P., & Rolls, E. T. (2001). Role of long term synaptic modification in short term memory. *Hippocampus, 11,* 240–250.

Kesner, R. P., McDonough, J. H., & Doty, R. W. (1970). Diminishing amnestic effects of a second electroconvulsive seizure. *Experimental Neurology, 27,* 527–533.

Kesner, R. P., Walser, R. D., & Winzenried, G. (1989). Central but not basolateral amygdala mediates memory for positive affective experiences. *Behavioral Brain Research, 33,* 189–195.

Kesner, R. P., Bolland, B., & Dakis, M. (1993). Memory for spatial locations, motor responses, and objects: Triple dissociations among the hippocampus, caudate nucleus and extrastriate visual cortex. *Experimental Brain Research, 93,* 462–470.

Kirwan, C. B., & Nash, M. I. (2016). Resolving interference: The role of the human hippocampus in pattern separation. In P. A. Jackson, A. A. Chiba, R. F. Berman, & M. E. Ragozzino (Eds.), *The neurobiological basis of memory: A system, attribute, and process analysis*. New York: Springer.

Lee, I., & Kesner, R. P. (2003). Differential roles of dorsal hippocampal subregions in spatial working memory with short versus intermediate delay. *Behavioral Neuroscience, 117,* 1044–1053.

Lee, I., & Lee, C.-H. (2016). Attribute memory model and behavioral neurophysiology of memory. In P. A. Jackson, A. A. Chiba, R. F. Berman, & M. E. Ragozzino (Eds.), *The neurobiological basis of memory: A system, attribute, and process analysis*. New York: Springer.

McDonough, J. R., Jr., & Kesner, R. P. (1971). Amnesia produced by brief electrical stimulation of the amygdala or dorsal hippocampus in cats. *Journal of Comparative and Physiological Psychology, 77,* 171–178.

Recollections of, and letters to, Ray Kesner. (2016). Chapter 17 In P. A. Jackson, A. A. Chiba, R. F. Berman, & M. E. Ragozzino (Eds.), *The neurobiological basis of memory: A system, attribute, and process analysis*. New York: Springer.

Rolls, E. T. (2016). Pattern completion and pattern separation mechanisms in the hippocampus. In P. A. Jackson, A. A. Chiba, R. F. Berman, & M. E. Ragozzino (Eds.), *The neurobiological basis of memory: A system, attribute, and process analysis*. New York: Springer.

Rolls, E. T., & Kesner, R. P. (2006). A computational theory of hippocampal function, and empirical tests of the theory. *Progress in Neurobiology, 79,* 1–48.

Todd, J. W., & Kesner, R. P. (1978). Effects of posttraining injection of cholinergic agonists and antagonists into the amygdala on retention of passive avoidance training in rats. *Journal of Comparative and Physiological Psychology, 92,* 958–968.

Werboff, J., & Kesner, R. (1963). Learning deficits of offspring after administration of tranquilizing drugs to the mothers. Nature, 197, 106–107.

Chapter 17
Recollections of, and Letters to, Ray Kesner

Compiled by Michael E. Ragozzino, Andrea A. Chiba, Robert F. Berman and Pamela A. Jackson

Knowing and interacting with Raymond P. Kesner has been a privilege for many of us that has left indelible memories and impacted our own careers and personal relationships. Ray Kesner's accomplishments range far beyond the major impact he has made by advancing knowledge about the neurobiology of learning and memory, and even beyond the development and empirical support of his "attribute model of memory." In an attempt to portray the magnitude of Ray's impact, we are obliged to document at least a glimpse of his presence by providing commentary from individuals for whom Ray was an important mentor, colleague, and/or friend. All of the positive attributes we cherish in academe have been exemplified by Ray throughout his career. His boundless energy, insatiable intellectual curiosity, generosity, and warm-heartedness continue to enrich many of our lives. We hope the recollections of Ray as a mentor and scientist by Bill DeCoteau and subsequent letters from colleagues captures why Raymond P. Kesner has left us with such wonderful, enduring memories. For those of you who read this section and do not know Ray, we hope that he inspires you too.

M. E. Ragozzino (✉)
Department of Psychology, University of Illinois at Chicago, Chicago, IL 60607, USA
e-mail: mrago@uic.edu

A. A. Chiba
Department of Cognitive Science and Program in Neuroscience, Temporal Dynamics of Learning Center, University of California, San Diego, 9500 Gilman Drive, La Jolla, CA 92093-0515, USA

R. F. Berman
Department of Neurological Surgery, University of California, Davis, 1515 Newton Ct., Davis, CA 95616, USA

P. A. Jackson
Department of Psychology, Radford University, P.O. Box 6946, Radford, VA 24142, USA

P. A. Jackson et al. (eds.), *The Neurobiological Basis of Memory,*
DOI 10.1007/978-3-319-15759-7_17

The Attributes of a Model Mentor

William E. DeCoteau

Psychology Department
St. Lawrence University

I was fortunate enough to be a graduate student of Ray Kesner's at the University of Utah in the mid-1990s during what I consider the heyday of the brain and cognitive sciences division of the department. That era's distinction had little to do with me and, instead, was derived mainly from the quality of the other students present just before and after me, including those in both Ray's and Sheri Mizumori's lab. In fact, I was a small lump of coal in that, otherwise, golden age. But that puts me in a particularly good position to speak about Ray's talents as a mentor. Below I discuss, in turn, the most salient features of Ray's mentorship. Because these qualities are both numerous and complex, I have organized them in terms of a familiar heuristic (Table 17.1).

The Place Domain

Availability Ray was no "absent adviser." Most often he was located in the lab, ready to talk science. On those rare mornings that I happened to be the first student to arrive to the lab, Ray would inevitably already be there, coffee in hand, working out the intricate details of an experiment on the chalkboard with "Bobby the custodian" whose night shift had ended upward of an hour earlier.

Fairness It did not matter what your *position* was in the lab: postdoc, grad student, undergrad, visiting Russian scientist, Bobby the custodian, etc., Ray treated you like an equal, like a colleague. As a mentee, to be treated as an equal to by Ray, was empowering.

Table 17.1 Attributes of a model mentor

Attribute		
Domain	Event-Based	Knowledge-Based
Place	Availability	Fairness
Time	Persistence	Patience
Response	Guidance	Growth
Affect	Humility	Kindness
Sensory-Perceptual	Enthusiasm	Focus

The Time Domain

Persistence All students who have done experiments with Ray know his memory for the data collected in his lab is remarkably immune to temporal decay. It mattered not how long ago you had collected it, where you currently were, or what you were doing … if you collected the data, Ray made sure with gentle, repeated reminders that it still needed to be published. In the end, you and your curriculum vita (CV) were always thankful for Ray's perpetual prodding.

Patience I was personally responsible for my share of screw ups in the lab (Froot Loop spills, misplaced rats, the odd stereotaxic coordinate miscalculation). Ray was never flustered by these setbacks; he always reassured you that things could be salvaged. A misaimed lesion electrode simply provided an opportunity to explore an, as yet, uncharted brain area!

Response Domain

Guidance From the outset, Ray would actively encourage his new students to think about doing experiments. Ray tailored his mentorship to fit the capabilities of the student. The more gifted he perceived the incoming student, the more autonomy he allowed them in conceiving their own first experiment. In my case, Ray decided it best to place me on a project that was *already* well underway! The study had been initiated by his undergraduate Joe Williams who had just moved on to graduate school. I finished collecting data on the project and … if you collected the data in the lab, you wrote it up. That work became my very first publication (DeCoteau et al. 1997).

Growth Ray understands the importance of letting his mentees take ownership of their work. For the most part, Ray gave his students the freedom to do their own writing, their own presenting, mentoring, and teaching. He provided help and advice when needed, but generally was "hands-off" in these areas. He also encouraged students to branch out intellectually and explore their own ideas. In my case, Ray encouraged my own growing interest in the basal ganglia. Those ideas became the topic of my dissertation and that work eventually led to my postdoc position in Ann Graybiel's lab; an opportunity and cherished experience that was entirely owing to Ray's mentorship.

Affect Domain

Humility Everyone knows about Ray's publication record but we also know how humble he is when it comes to his amazing productivity. He always gives credit for his ideas and success to his colleagues and, in particular, his student colleagues (for a particularly detailed homage to his students see, Kesner 2009).

Kindness What is the secret weapon of Ray's mentoring success? For me, it is the incredible motivating power he wields from simply being a genuinely decent human being. The man is so kind, gentle, and loving that you simply *do not* want to disappoint the guy. And that is why you somehow find yourself continuing to publish work done in his lab from a decade or two earlier!

Sensory-Perceptual Domain

Enthusiasm Ray's positive energy is noticeable and infectious. As a teaching assistant in his Physiological Psychology course I witnessed how his students would become riveted by him … especially during the sex and hormone lecture! As a mentee, you could not help but get excited when discussing your work with Ray. You always left thinking … "yeah, you're right Ray, this study is *Nature* worthy … even if the lesion location is not quite where we intended!"

Focus St. Lawrence University holds an annual lecture in neuroscience where an esteemed neuroscientist who is known for promoting undergraduate education and research is invited to give a lecture. In my second year at the institution, I was given the task of organizing the lecture. Ray, of course, came immediately to my mind and he accepted my invitation. The lecture is a *big deal* at the school: the president of the school and his wife host a dinner for the speaker; they and people from all over the community and university attend the lecture. Given my central role in its organization, and the fact that I was not yet tenured, the lecture was a *big deal* for me too! I went "all-out" designing posters (Fig. 17.1) and personally posted them all over campus.

Fig. 17.1 A model mentor

I also requested the lecture be held in the auditorium room of a brand new student union building located at the center of campus … the room was used to show movies so it had a big seating capacity and a nice large screen … a perfect location! Unfortunately, it turned out that the stage in front of the screen lacked depth and so the lecture podium had to be located off to the side of the screen. Ray, who, as we all know, can be pretty animated when lecturing, turned out to be a little confined behind the podium. About two thirds of the way through the talk, Ray moved out from behind the podium and started ambling enthusiastically around the stage while continuing with his lecture. I recall their being a picture of a rat brain slice projected on the screen and Ray was discussing the hippocampus and pointing out the CA1 region with his laser pointer … all the while getting more and more animated as he looked up at this enormous image of the hippocampus! In his excitement, he fell backward off the stage … right in front of the university's president and his wife. Perhaps it is all the skiing he does at the Winter Conference, but, for whatever reason, Ray landed on the ground, perfectly … and then immediately hopped right back up on a stage that was at least two feet above the ground. Ray's graceful recovery did not stop the president's wife from emailing me the next morning sternly criticizing my choice of lecture locations … something about how "our esteemed guest could have been seriously injured!" Luckily for me, it was the President and not his wife who had the final say in my tenure decision! But there is more, and here is where this story connects to "Focus" and the "Sensory-Perceptual" domain: If you were a blind person simply listening to Ray's lecture, (and you ignored the gasps from the audience) you would not have known the fall occurred for, throughout the entire fall and recovery, Ray continued talking about the CA1 region, without missing a beat! If you were a deaf person, attending to the screen, you also would not have noticed the fall … for, while the tumble may have caused the dot from Ray's laser pointer to briefly skip from the CA1 to the CA3 region, I swear the dot never moved outside of the hippocampus! What truly amazing enthusiasm, focus, and motor skill for someone nearing retiring age.

On behalf of all of Ray's mentees I would like to thank him for all that he has done for us. Because of all these wonderful attributes of his we will always have strong and favorable memories of him!

Letters and Recollections

Linda J. Baker and Timothy Baker

Retired School of Medicine
Mental Health Private Practice University of Wisconsin-Madison

Dear Fans of Ray,

We both worked with Ray under his mentorship and want to take this opportunity to say, albeit briefly, what a pleasure it was to know Ray and work with him. Ray stands out amongst the many colleagues with whom we have worked for his enthusiasm, creativity, and thirst of knowledge and insight. He was consistently supportive and generous with his time and ideas. At a time when we greatly needed a receptive and enthusiastic sounding board, Ray filled the bill brilliantly. Few people can infuse excitement into a discussion of research and theory as well as Ray.

Ray, we both hope that you have a keen awareness of how many lives, including ours, you have enriched. Thank you, Ray! You should be very, very proud of your stellar record as a mentor, teacher, researcher—and mensch!

With very fond memories and great appreciation,

Tim and Linda Baker

Rebecca D. Burwell, Elisabeth A. (Betsy) Murray, and Peter R. Rapp

Brown University National Institute of Mental Health National Institute on Aging
On the Occasion of Your Festschrift

Dear Ray,

We write to convey our deep appreciation for your scientific contributions to the neurobiology of learning and memory as well as your teaching, mentoring, and service to the field.

We all began attending the Winter Conference on the Neurobiology of Learning and Memory (WCNLM) in the 1980s (Peter and Betsy circa 1985, Rebecca in 1989). In those early, formative years of our careers, it was a tremendous privilege to be part of the talks, debates, and sometimes arguments among you and the other remarkable scientists at the pinnacle of our field. For us, the participants, format, and content of the meeting provided the opportunity for what we now call networking, setting the path for our research directions, collaborations, and career choices.

With the benefit of hindsight we can now appreciate that the careful, programmatic approach you adopted throughout your career has been a key to fundamental progress in defining the structure and organization of memory in the brain. This is amply reflected in your festschrift, "The Neurobiological Basis of Memory: A System, Attribute, and Process Analysis," and it is a contribution that is likely to influence the field for years to come. It is a fitting tribute to your many experimental and theoretical contributions.

Congratulations on the publication of this remarkable volume honoring your career in neuroscience.

Warm wishes,

Rebecca, Betsy, and Peter

David G. Cook

Geriatrics Research Education and Clinical Center
Department of Veterans Affairs
Seattle, WA

Dear Ray and friends:

When you are a little kid the grown-ups will sometimes ask, "What do you want to be when you grow up?" That is when it starts. Because, even though I (and I assume nearly everyone else) had no idea what to say for many decades, it is still the kind of question that can stoke the fertile imagination of a child. My childhood was an era where new scientific endeavors were at the forefront of public consciousness—the space program, as well as many other advances in medicine and engineering. It seemed pretty cool at the time, but was very diffused and superficial; with no clear path for exploring how such things might apply to me.

It was not until I met Ray, who taught my introductory psychobiology course in college that the idea of doing science became more than a concept. It became a specific question. Would I like to work in his research laboratory? Ray has always been an inspiring professor. So, it was easy to say yes.

The experience of working in Ray's lab as an undergraduate was transformational in my life. The scientific training in behavioral neuroscience that I received there was nothing less than superb. However, as I look back it was not until much later that I began to understand a fraction of what he was saying. The value of his mentorship notwithstanding, what I will always cherish most is the memory of Ray's kindness and absolutely boundless enthusiasm.

As I reflect now on my role as a principal investigator and scientific mentor, any similarities that I or others might see in myself that remind me of Ray, I take as something to be proud of and something one can clearly attribute to the things I learned from him.

Ray, even many years after I left your lab and moved into research areas more remote from your scientific interests, you have remained a steadfast supporter, friend, and mentor. This means everything to me. Some of your relentless optimism that rubbed off makes the challenges of running and funding labs easier to deal with. Similarly, the successes are all the more sweet when framed by the sort of enthusiasm for science that I learned from you.

With sincere gratitude,
David Cook

John Dalrymple-Alford

Brain Research Institute and Department of Psychology
University of Canterbury, New Zealand

It is an honor to know Professor Ray Kesner and to write a few comments. Ray is clearly among the most influential scientists to have studied the neurobiology of memory. What makes his work stand out, in an often crowded field, is that he addresses both structure and function, in ways that allow him to address many different structures and many different functions. Whether you agree or disagree with any of his views, he is never dogmatic and always gracious and thoughtful when discussing the field. Where many seem to follow technologies and apparatus, asking how those may be blended to help answer a research question, Ray's approach is the exact opposite. He finds questions and then seeks answers by whatever technology or new-fangled apparatus will help him test his questions. His attitude toward others, in my experience, has been exemplary. Ray has been a relatively frequent visitor to New Zealand for about 15 years or so. It was a delight to host Ray and his wife, Laya, recently in Christchurch, during 2014. Due to the whims of nature, they experienced a somewhat different place to their previous visits. And despite some of the wildest weather on record for a century, Ray was always in good spirits and keen as ever to debate the latest research. I suspect that his inquisitive mind will continue for many years ahead, irrespective of any formal retirement. I wish him and Laya all the very best and look forward to more discussion in the near future.

Michael Davis

Yale University, 1965–1998
Emory University, 1998–2011

Professor Ray Kesner has been a pioneer in delineating multiple memory systems, showing that memories have different attributes that depend on different, but interconnecting and interacting, brain areas. His brilliant dissection of these memory systems has been based on wonderfully creative, ingenious experiments in rodents, and now humans. His results have been enormously influential theoretically, as well as practically, with major clinical implications. Ray and his many superb students led the way in these endeavors and served as an inspiration to those of us who watched this unfold over time. We are indebted to his insights, his leadership and his great kindness.

John F. Disterhoft

Department of Physiology
Feinberg School of Medicine
Northwestern University

Dear Ray,

I want to join the chorus who are congratulating you on your retirement. I think I should qualify this, because I know your retirement is only from your position at the University of Utah. I am quite certain that you will continue with your involvement in the Winter Conference on the Neurobiology of Learning, finishing manuscripts, lecturing, and sharing what you have learned about learning and cognition.

I am trying to pin a date on when we first met. But it must have been about 1980, almost 35 years ago, most likely at the Park City meeting. You were always a generous and positive friend, even to someone from the Heart Land with no knowledge of skiing but attempting to learn as an adult. I certainly enjoyed many days of skiing at Park City and Deer Valley, although my technique never did improve that much!

Not only did I not ski but I persisted in studying associative learning, with its limitations in testing cognition, rather than expanding the behaviors used in an attempt to probe the cognitive capacity of the animals I studied. This may have been a good strategy, since I studied only learning in rabbits for many years and many folks feel that rabbits are at best "fuzzy doorstops" from the cognitive perspective. You should be congratulated on the progress you have made in devising creative and insightful behavioral tasks to help us better understand the functional capacities of various portions of the rodent neocortex, hippocampus, and temporal lobe. You have done many definitive studies and helped develop theories of how the brain handles learning and memory processes. The type of systems neuroscience that you have been doing throughout your career is coming back into fashion, now that we can do lesion/behavior studies with even more precision and specificity. I will be intrigued to find out if these new approaches really give us more information than those you have been using with such great insight. We shall see!

I look forward to seeing you often and catching up, as I have all these years. I know you will always continue to be positive and upbeat, something I have always valued in our relationship. Enjoy your continuing scientific work—as much as you want to do. Also enjoy your expanded opportunities to travel, and to spend more time with your wife, your children, and, of course, your grandchildren.

Congratulations on this new stage in your life!

Howard Eichenbaum

Center for Memory and Brain
Boston University

Dear Ray,

Thank you, Ray, for being Ray. There have been exceedingly few researchers who are truly fluent in both the behavioral and biological aspects of neuroscience, and even fewer who have been able to combine these as well as you. I find this somewhat amazing because, as I often have to remind my molecular-cellular colleagues, the phenotype of the nervous system is, after all, behavior! Very few indeed have the capacity to operationalize cognition in ways truly amenable to biological exploration. You are one of those few who are so creative, and you have inspired me and many others to follow your model, and we can only aspire to do it as well as you have.

Aside from the science, one other thing: Just after the dinner at your Festschrift I regretted not standing up during the audience speeches and making an observation on the festivities of the day. So, I am really happy to have the chance to say this now and to the larger audience who will read this volume. I know most of your former students, and so was not surprised to see so many of them and take count of the expanse of your influence on the field through these scientific progeny. What struck me in addition on this occasion was the level of gratitude and affection expressed by each of them for your mentoring and friendship throughout their careers. I have been to several of these kinds of events, and never seen as much love from students for their mentors. You are blessed indeed by those rewards.

Best to you, and I look forward to our next vigorous discussion of the latest data!

–Howard

Vince Filoteo

School of Medicine
University of California, San Diego (UCSD)

Dear Ray,

There are so many reasons I feel I need to acknowledge you and your work at this, the time of your retirement. First, I have to thank you for your generosity and mentorship when I first arrived at the University of Utah as an assistant professor in 1994. It was clear from our initial interactions that we had a number of overlapping research interests that were solidified by us co-mentoring one of my first graduate students during her doctoral training. Her success, as is the case for all of your students, is a direct reflection on your commitment to teaching and mentoring. Second, I have to thank you for your enthusiasm in both your teaching and research. I recall having the same conversation with several of our colleagues about how we all hoped we would be as enthusiastic as you throughout our careers. Your love for your work was truly contagious among your fellow faculty, but especially among your students. Third, I have to thank you for your approach to research and your emphasis on trying to understand how the “brain” works, regardless of the species you studied. You truly are one of the rare individuals in our field who can “bridge the gap” between human and nonhuman research and showed me the importance of trying to understand brain–behavior relationships at several different levels of analysis. Finally, I would like to say that even though we have not had much contact since I returned to UCSD, I can honestly say that our interactions have left an indelible mark on my career and I still feel your influence today. I cannot say that about many colleagues.

I wish you the best of luck in your retirement.

Thank you!

Joaquin Fuster

Cognitive Neuroscience
UCLA Semel Institute for Neuroscience and Human Behavior
University of California, Los Angeles

Dear Ray,

Why, I am not going to let these guys close this Festschrift without my letter! (As usual I am late, "*mañana*," you know…) I owe you much, much more than a simple letter when you call it quits. But are you? Hope not completely. For today, these few words must do (I still have deadlines). Anyway, it does not take much print to say thanks. Thanks for the fun in Park City, over snow and over beer, and in Salt Lake, surrounded by your loyal students. And thanks for so much I learned from you. You taught me that rats are like little monkeys and have a frontal lobe too. And that space is big in it, in so little space! Oh yes, in their glorious hippocampus too! In your proverbial intellectual pose, looking at my eyes over those tiny glasses, that disarming smile with a tinge of irony, you could make any point and win any argument. Who could resist your power of conviction? I never saw you in a bad mood. Even when things did not go entirely your way, your contagious enthusiasm carried the day, especially among your students, who followed you, Pied Piper walking on water (salty of course!), with their clever experiment inspired by you. Do not quit.

I love you, and everybody else I know does too, Joaquín.

Debra Johnson

Department of Psychology
University of Iowa

Dear Ray,

I want to take this opportunity to thank you for touching my life in so many ways. As I reflect back on the years as a graduate student in your lab I am able to see a number of life lessons I learned from you:

- *Love what you do*—Your enthusiasm for your research was evident from the first day I met you. You always seemed happy and excited to come to the lab and to get to work on the next big question. You had a real sense of wonder and curiosity that was very fun to be around.
- *Be creative*—You have a wonderful way of identifying the right questions and designing an experimental protocol to address that question. Over the years I heard many of your colleagues comment that if anyone could figure out a way to do X– it was Ray Kesner.
- *Never stop learning*—Even though you were successful and at the top of your field, you were eager to learn new methodology and new technology.
- *Stay balanced*—Despite being incredibly productive and committed at work, you clearly had a private life. It was clear that you had a family and a community outside of work and that those things were very important to you.
- Even though I have not remained in a research field I have tried to apply these lessons in my day-to-day life. I will always look back fondly at my time in your lab.

I wish you a long and happy retirement filled with many adventures and new discoveries.

All the best,
Deb

Kristen A. Keefe

Department of Pharmacology and Toxicology
University of Utah (U of U)

Dear Ray,

It has been my pleasure to work with you over the past 15 years or so, since I arrived on the U of U campus. I consider myself lucky to have arrived on campus at a time when you and Bill DeCoteau were conducting your seminal work showing the double dissociation of the effects of hippocampal and striatal lesions on spatial and nonspatial sequential motor learning tasks. Your work allowed me as a young investigator to move into a new area examining the impact of methamphetamine-induced neurotoxicity on basal ganglia-mediated learning and memory function. I benefitted so significantly from your work, your willingness to have students work in your space to do those studies, and your thoughtful intellectual contributions to the ideas being pursued. I have been extremely lucky to have had you as a collaborator on these projects for so many years.

Of course, I have always tried to drag you into the basal ganglia and drug abuse and addiction work, while you have tirelessly indoctrinated me into the field of pattern completion and pattern separation, as well as subregional differences in hippocampal function. I am sure that I have been a frustrating colleague with respect to this, but I think now after 20 years, I have got it! Especially, it has been nice seeing our interactions over the years lead to a true melding of our research interests with the R21 grant using pharmacology and Arc expression to assess pattern completion processes in cocaine abuse. Seeing this project come to fruition has been rewarding and exciting for me as a capstone to our interactions over the years.

While the research projects on which we have interacted have been significant in my time at Utah and the development of my own research program, what I am most grateful for is having had the opportunity to experience your positive attitude and your passion for your science, both of which are unwavering. I have been shaped in each and every interaction I have had with you by how positive you always are. On the one hand, there has been your enthusiastic "yes" and smile when your hypothesis is proven correct as the data come in. More so, it is been the smile and enthusiasm despite experimental difficulties or unexpected outcomes occurring that have been formative for me as a scientist. That is, you have been an exceptional mentor in teaching me to remain optimistic and positive in the face of the common scientific difficulties. So, thank you for teaching me to be more resilient! Perhaps your positivity arises in large part from your infectious enthusiasm for your science in particular, and science in general. What has always struck me is how engaged you are at every seminar, every symposium, and how much you seem to just simply enjoy talking about your research and the research of others that you have just heard.

Ray, I consider myself privileged to have been your collaborator and colleague all these years at the U of U. I thank you for the time that you have invested in me as I have developed as a scientist, your continual enthusiasm for collaboration, and your unyielding positivity. I look forward to continued opportunities to collaborate and talk about science even though you are "retired," because we all know that word does not mean much and that you will never retire because of your passion for science!

Fondly,
Kristen

Donna L. Korol

Department of Biology
Syracuse University

Dear Ray,

I write to congratulate you on your long and illustrious career and to share my warm memories of you. Though we have never actually collaborated, you have contributed significantly to my scientific, professional, and personal growth through your insightful investigations of multiple memory systems within memory systems, your development of the most clever tasks that tap these multiple memory subsystems, and your warm, gentle, yet intellectually probing manner.

Your decades-long investment in the WCNLM (Neurobiology of Learning and Memory Winter Conference) alone has made a deep and wide mark on the behavioral neuroscience community. Though of course many people contribute to the success of the conference, you and your open style have created a safe space for the exchange of ideas—some wise, others rash—but all acceptable food for thought at the conference. So many junior and senior scientists, myself included, have benefited from the informal, incidental, and always intellectually stimulating learning that happens at the conference. And we have you to thank.

You probably never realized that I first started to attend, or better, crash, the WCNLM during my early years in graduate school. It all started in 1985 en route to my new graduate student home at UVA (University of Virginia) from Jim's lab at UC-Irvine. Paul convinced me to accompany him to his annual "ski meeting," knowing that I loved the moguls and thinking that, as a newly minted graduate student, I would probably enjoy the luxury of an après ski soak in the hot tub.

I knew attendance at the meeting was by invitation only and that graduate students were not typically invited, but could not hold back after my first first-night pizza party when I sat around a table talking with conference participants about the inner workings of the brain. I did not know then that these were world-renown experts; all I knew was that they were engaging me in discourse about learning, memory, and plasticity, and that I was smitten. Never being one to follow rules to a

T, during the second session talks, I decided to stand in the back of the conference room to avoid taking a chair from a registered participant (and to allow an inconspicuous getaway in case I was discovered). Perhaps thinking I was one of your diligent students, Stuart Zola asked me to work the lights, which I did. The next year you asked me to help with the slide projector, which I did. In my later graduate school years, you always welcomed me to the meeting as a colleague, providing me with the confidence and the community that helped me to grow as an independent thinker. I thank you for that.

While my warmest memories have you welcoming us to the conference, wishing us luck as we wait at the slalom gate, or afterward handing out the highly revered trophies, your influence on my work extends well beyond the conference highlights. We have used your many creative tasks in our own work, including, several years ago, your computerized X-test for humans and now your MCOL (Metric Change in Object Location) task that my postdoc has incorporated into our models of estrogenic shifts in learning strategies and bioenergetics across multiple memory systems. So, once again, thank you.

I hope you find that retirement from University of Utah gives you new-found time for developing new theories of brain function and for honing your skills on the ski slopes, as if those need honing. I look forward to hear about this exciting next phase of your life.

Best wishes always,
Donna

Inah Lee

Department of Brain and Cognitive Sciences
Seoul National University
Korea

Dear Ray,

People know it very well that you have contributed significantly to the field of memory research with your theory and all the works you have done in the past years. People may not know (but I know it very well) that, besides those official academic contributions, you have also influenced many people's lives by being such a positive mentor. I now realize, after being a mentor myself for almost 10 years, that an academic mentor for a student (particularly for a graduate student) is an important influence to the student and perhaps, the student's life is shaped mostly during that period of time by the mentor. Having said that, I think I was one of those people who were lucky to meet you at that critical moment of life and learned a lot from you about how to become a good mentor in the future. You were patient when things did not go as we expected. You were generous when I needed your generosity. You looked like the happiest person on earth when I had achieved something. You were encouraging when I was down. I now realize that these do not come easy to every mentor and certainly I am not sure if I will ever be a good mentor like you both academically and personally. Although it is very sad that we need to let you go for your retirement, as a memory researcher, you have engraved the most cherishable memories and good values to so many people in the field. So I believe that the Kesner tradition, the most joyful way of doing science while playing with those small creatures in various mazes with Froot Loops in hand, will never die in the field of memory research. And I thank you very much for that!

Best wishes,
Inah

Jeffrey M. Long

National Institute on Aging
National Institutes of Health

Dear Ray,

This Festschrift is a well-deserved recognition of your important and lasting contributions toward understanding the neurobiology of learning and memory, and on this occasion I wish to convey what a privilege it was to be a graduate student of yours in the early 1990s. As impressive as your publications and professional achievements are, they do not capture your infectious excitement for life and for science in particular. Your passion for research made the lab a vibrant, intellectually challenging, productive, and fun place to work. Your enthusiasm continues to energize your laboratory, the psychology department, and your interactions with colleagues worldwide. The successful research careers of your ex-students and postdocs, along with the continuous high quality of the Annual Winter Conference on Learning and Memory are but two examples.

I had a wonderful time in graduate school. Enjoying nature and developing lifelong friendships played a part, but the largest component was the environment you, Charlie, and Sheri created. Although the term is a bit old fashioned, the program provided a fantastic broad-based education in psychobiology. Scientific rigor and rational thought was emphasized and critical but cordial discussion of data and theories were the norm. You were a true mentor, passing along knowledge, skills, and opportunities. I recall the large pink chalkboard outside the elevators where you always seemed to be diagramming your latest experiment to anyone that happened by (I swear I saw you explaining the attribute model to a janitor!). Our own spontaneous and stimulating discussions in front of this chalkboard are some of my most treasured memories. Your enthusiasm for behavioral neuroscience cemented my own interest and I have been fortunate to spend the past 15 plus years investigating the neurobiology of learning and memory. Your mentorship made this possible and I am truly grateful.

Congratulations on your research legacy and for molding a generation of scientists proud to have worked in your laboratory. I look forward to following your "post-festschrift" research career, as I harbor no illusions that you will fully retire anytime soon.

Sincerely,
Jeff

Jim McGaugh

Center for the Neurobiology of Learning & Memory
University of California, Irvine

Dear Ray,

From one "old timer" to another somewhat less of an "old timer" I want to tell you how much I have appreciated interacting with you over these many years—and learning from your work and that of your many talented and productive students. You know more about how to find out what an animal knows than anyone else I know—or whose works I have read. And in your many decades of research you have made great inroads into understanding the brain processes that underlie the animals' "knowing." And, as I summarized in my presentation at your Festschrift symposium, my research on the amygdala involvement in learning and memory was significantly influenced by your early findings of amygdala regulation of affect.

And, of course, it was great to work with you in the creation of the Winter Conference on Learning and Memory over these past several decades. The conference has had, as you know, a major impact on the development of young behavioral neuroscientists and enabled us to keep in personal contact. You deserve enormous credit for your sustained and effective role in organizing and managing the conference. Finally, and most importantly I have appreciated our personal friendship. I wish you the best as you segue to other stages of your academic and personal life.

With warm regards,
Jim McGaugh

Andrea Morris and John Churchwell

University of California, Los Angeles

I have met few people in my life as charismatic as Ray Kesner. Ray's enthusiasm for life, science, and his students is unparalleled and is also quite contagious. During my time as one of Ray Kesner's graduate students (2007–2011) I learned a great deal about the neurobiology of learning and memory and hippocampal function. In addition, and perhaps more importantly, Ray taught me how to be a good researcher, mentor, and teacher. While in the Kesner lab, I also made some great friends (Brian Curtis, Dave Maasberg, and Christy Weeden) and met my wonderful husband, John Churchwell (another Kesner graduate student).

John and I thank you for your contribution to neuroscience, mentorship, and above all, we thank you for your lasting friendship.

All the best,

Andrea Morris and John Churchwell

Lynn Nadel

Regents Professor of Psychology and Cognitive Science
University of Arizona

Ray Kesner is a cherubic ball of energy, whose first words whenever I meet him invariably are "I have some data you will love." Sometimes I do and sometimes I do not but every time the results are interesting, and reflect a superb experimentalist at work. But Ray is more than an experimentalist—he has also tried over the years to embed his research within a theoretical framework that, in my view, has not received the attention and credit it deserves. Perhaps this book will change that. Science, like life, can be cruel, but an important thing about Ray is that he is way bigger than that—his infectious enthusiasm for the daily doing of research has persisted to this day, proper recognition or not.

I shared a room with Ray at a meeting a few years ago at which I got to see, once again, just how clever he is. We talked for hours about a group of his studies looking at spatial and memory functions in a mouse model of Down syndrome. Science at its most enjoyable.

I have also been part of summer schools and the like with Ray where another facet of his nature is obvious: students love him. And why not—he is always enthusiastic and he will talk to students on their own terms. He also listens.

Bottom line: Ray Kesner is a complete mensch—in science and in life.

Neil McNaughton

Department of Psychology
University of Otago
New Zealand

Dear Ray,

Thanks for all the great times: talking about memory while skiing in Park City; talking about learning while buying cheese in Salt Lake City; talking about behavioral inhibition while visiting the Catlins; and finely slicing and dicing the functions of subregions of the hippocampus while cooking together in Dunedin. Above all, thanks for the many, many times and places when over a beer (or two) my thinking was challenged with "Hey Neil, I got a task for you," and yes, I am still working on getting satisfactory answers for you. All the best for the future,

Neil

Jason Rogers

Noldus Information Technology
Leesburg, VA

I first met Ray Kesner in 1999 at the Society for Neuroscience Meeting in Miami. Even then it was obvious that Ray was a beacon. But what really struck me was his passion and excitement for all things research. Four years into my doctoral training, Ray would be the one waiting at my office door, ready to discuss today's readings and talk about the experiments we could do. Yet despite his extensive knowledge base, he always listened to my opinion, no matter how insightful or foolhardy. I remember when Ray told me that he doubted my dissertation experiments would pan out, but he was pleasantly surprised when they did. That was the type of mentor he was; gifted and kind, driven and patient. His work has inspired a generation of scientists. The span of his influence is longer than that of my own life. I personally owe so much to Ray. He took a chance on a 21 year old kid from Indiana State University and raised me into the scientist I am now. In this, I am not alone. In my travels, both domestic and abroad, I have learned two things about Ray from others: his academic influence is profound, and the depth of respect shown to him is even greater. I am lucky to share in that academic halo. People know me because I am a Kesnerian.

As Ray moves into the next phase of his life, the torch passes on to us to continue his work. Although no longer an academic, per se, I consider my work to be very much an extension of the 9th Floor of the Behavioral Science Building. My obsession no less tempered over the years, I get "Ray-level excited" when discussing spatial memory, especially around those who have no idea why they should care about it. No longer creating lesions or Plexiglas mazes, I now spend my time thinking about human mazes and how to apply my training in learning and memory to improve upon the consumer experience. I talk extramaze cues with retail shops. I discuss landmark-based navigation with theme parks. Although this work will scarcely be read outside a corporate white paper, nor will it ever reach the clout of my true academic colleagues, that is okay: I will always be a Kesnerian. It was at a different Society for Neuroscience Meeting when Ray taught me a simple truth: there are two types of scientists. Those who like to see their name in press, and those who just really like science. I hope, like my mentor, I am the latter.

I am honored to know Ray. I think of him often and miss his astute advice. I am blessed to be one of his academic kids. But like all things, the time has come for Ray to become greater than just Professor, or Professor Emeritus. Ray has his kids' kids to enjoy. Along with his Zumba. And his Community. And, of course, his lovely Laya. The VW bus may be gone, but the miles continue. In the words of Harvey Mackay, "A great accomplishment shouldn't be the end of the road, just the starting point for the next leap forward."

Ray, may your next leap be the best yet.

Jeffrey Rosen

Department of Psychology
University of Delaware

I first met Ray in 1985 when I started in grad school as a student under Rob Berman. The lab all went to the Society for Neuroscience meeting, and since no one had much money, we stayed with Ray and his lab in a single hotel room. There must have been 6–8 people in the room, doubling up in beds and sleeping on the floor. In the morning we went to breakfast. This is when I heard Ray talk very fast and excitedly about win–stay and win–shift maze tasks and how one experimental group will demonstrate this and another group will demonstrate that, and will learn something important about hippocampal function. And then I was asked what did I think about the experiment? My head was spinning and I have no recollection of what I said. It probably was not anything very intelligent.

I continued to go to the Society for Neuroscience meetings and stay with Rob and Ray throughout grad school. My dissertation work focused on the effects of adenosine drugs on kindled seizures, so my work deviated from Ray's for a while. However, I still learned about Ray's research and his attribution model in courses on the neurobiology of learning and memory.

I also attended and participated in the Park City Neurobiology of Learning and Memory meeting for many, many years. I cut my scientific teeth there—it gave me a forum to present my work and develop relationships with colleagues that I have maintained for many years. I also learned to ski by trying to keep up with Ray. No way I could ski that well, but it was a great fun going up the chairlift with Ray and trying to keep up going down.

The attribution model became very important to me once I started studying the amygdala and fear conditioning as a postdoc with Mike Davis. I studied more of Ray's work and particularly the attribution memory model, but the role of the amygdala in memory, and not the hippocampus, was central. Also, Ray's experiments on electrical stimulation of the amygdala on learning and memory were precursors to my own work. The attribution model of memory was my framework for understanding my own research on the role of the amygdala in fear conditioning.

Throughout my career I have continued to have dinner with Ray and Rob at the Society for Neuroscience meetings. We have had dinner more often than not every year, and I love hearing about Ray's latest ideas and experiments. My head still spins, but it makes me excited about science, especially when I lose my way in my own research.

I thank Ray for being a colleague, a supporter of my career, an inspiration, and a friend.

Aryeh Routtenberg

Departments of Psychology and Neurobiology
Northwestern University
Banana Man

Ray, may you enjoy many more sunrises and sunsets … and the wit … pattern separation … to appreciate the difference.

Paul Solomon

Department of Psychology
Williams College, MA

One of the best things about an academic career is the people you meet along the way. Some are mentors—they teach, inspire, cajole, and criticize. Others are colleagues who can agree or disagree with our views but always (well almost always) in a collegial manner that makes us better at what we do. Some are friends, and a rare few good friends. These are the people who genuinely make our lives better. Once in a long while we meet people along the way who fulfill all of these roles. For me, Ray Kesner is one of these rare people.

I first got to know Ray when I was beginning my career as an assistant professor and Ray was generous enough to invite me to one of the first meetings of the WCN-LM in Park City. A conference that 38 years later continues, due in large part to Ray's dedication to this wonderful meeting. Ray arranged (or plotted) for me to go skiing on my first day with Rob Berman, one of Ray's former students who was also attending the meeting. Rob suggested (I suspect with encouragement from Ray) that it would be a good idea for my first run ever skiing in deep powder and in near blizzard conditions to ski "Regulator Johnson" off the top of the tram at Snowbird. Ray somehow had the notion that I was in the same league as a skier as him and Rob. Ray then joined us for skiing over the next few days and I quickly realized that he was really not trying to kill me, but just to show me a bit of adventure. And skiing with Ray over the next 36 years has been an adventure. Ray is the first one out in the morning, the first one down the run, and the last one to stop for lunch. For as many years as I have skied with him, for the last run of the day he has quietly stopped on the side of the "Pay Day" run at Park City, started his chronometer, pointed his "Cheeseburger Deluxe" K2 skis straight downhill and skied full tilt down the mountain (mothers pulled their children off the slopes as his skis that genuinely looked like a cheeseburger—a brown hamburger color inner core covered by an orange/yellow Velveeta cheese looking layer, with streaks of red that were supposed to be Ketchup—streaked by them). It was Ray's goal to break his land speed record every year. As I later learned, he approached most things in life with the same vigor.

A few years after my first foray with Ray in the deep powder, he came to visit me while I was a visiting professor in Dick Thompson's lab at Irvine. Ray was interested in using multiple-unit recording in his behaving rats, and where better to learn this technique than in the laboratory of Richard Thompson. So Ray and I worked together during the day and hung out at night (he stayed with us in a beach house we were renting on Balboa Island in Newport Beach). I showed him what I knew about electrophysiology and he introduced me to Moo Shu Pork (still one of my favorite things) and his views about how brains must be organized to process memories. I clearly got the better of the deal. Over the years we always got to spend time together at the winter conference, skiing and talking about memory and memories. Our paths also crossed at many other meetings and during my visits to Salt Lake and Ray's visits to Massachusetts. I often sought his advice about research and his company to share Moo Shu pork. He helped me both as a mentor and a colleague. While I greatly value this relationship, what I most admire (and try to emulate) about Ray is that although he is a dedicated, talented, creative, and productive scientist (as the chapters in this book by his students and colleagues clearly attest), he has always kept a perspective on his work and a balance in his life. It is this quality that has enabled him to complete the trifecta of mentor, colleague, and good friend.

Paul Solomon, 8/11/14

Wendy Suzuki

Center for Neural Science
New York University

Dear Ray,

I want to acknowledge the many ways you have contributed to the field of learning and memory over your long and very productive career. Certainly your experimental and theoretical work has helped elucidate our current understanding of the way the hippocampus and its specific subdivisions contribute to memory. But for me the contribution that stands out most in my mind has been your unwavering and enthusiastic leadership of the Winter Conference for Learning and Memory that is now in its 38th year and still going strong. Along with your own body of research, this meeting that you have nurtured with grace and a wonderful sense of humor over the years has mentored and educated generations of students, myself included. Its continued vibrancy is a testament to your scientific leadership.

More than any other meeting I attended as a young graduate student I felt I grew up scientifically at this meeting. All those evenings spent listening to the leaders in our field discussing their latest findings, theories, and ideas were filled with more hippocampal excitement than you could hope for in one place. I remember workshops on the latest anatomical findings in the hippocampus or controversial new hippocampal theories being debated and discussed by night and those discussions continuing the next day on those amazing ski slopes full of powder snow (most of the time) in Park City. I have many found memories of skiing with you, Ray and the many other ski-loving neuroscientists chatting about the hippocampus on our way up the mountain. This meeting provides a wonderful venue for young and senior scientists alike to meet and interact in a truly beautiful part of the world.

One of the most wonderful things about this meeting has been the loyalty and consistency of the participants that come back year after year. I think the main reason for this loyalty is because of the warm, welcoming, and interactive sprit that you have infused into the meeting ever since it found its "home" in Park City. The meeting has boasted an executive board of world-class learning and memory researchers (including you), but it was the joy and enthusiasm that you put into each year's banquet, including live music, prizes, raffles, and the odd magic show thrown in for fun along with the traditional awarding of the coveted ski trophies that really defined the sense of community that is at the heart and soul of this meeting.

Yes, this meeting will continue even after your official retirement, but the fact that several of us have taken over the job that you did on your own for so many years speaks of the energy you brought to this task. I only hope that we can keep up the high standard that you set and help the next generation of students get the up close and personal view of the learning and memory field that I benefited from and continue to benefit from each year at this beloved meeting. Ray, I thank you for all your contributions to the field and especially for all the effort you have put into keeping this meeting active and exciting for all of us.

With Gratitude,
Wendy

Christy S. S. Weeden

National Institute of Mental Health

Dear Ray,

I arrived for my interview to work as your "last" graduate student and was both delighted at the opportunity to draw on your many years of experience but also a bit concerned that as such a distinguished scientist, you may not be as accessible as a junior professor. But I was happily surprised at just how much time you still dedicated to those around you. From the moment you greeted me personally at the airport to our recent conversations about manuscripts, you have radiated a passion for teaching and research that I, and many others, admire. I especially appreciate the energy with which you delivered lectures and your excitement to share in each student's journey toward those "aha!" moments when higher concepts are realized. I watched you give your time generously, even to novice lab assistants, as you engaged them about their particular tasks, often with a suggestion for further reading and offering a follow-up conversation. Throughout 4 years of graduate mentoring, I am sure I suggested my share of less-than-stellar ideas. You steered me in the right direction with the same energy you had for my better plans. Thank you, and I hope to one day emulate the enthusiastic guidance you have provided throughout the years to so many students, colleagues, and friends.

Thank You,
Christy S.S. Weeden

Norman M. White

Department of Psychology
McGill University
Canada

Dear Ray,

The occasion of your retirement brings several things to mind. First and foremost is the scope, volume, and value of your many contributions to the literature on learning and memory. Your ability to design tasks that test specific memory functions is unmatched, and your theoretical framework for organizing and interpreting your findings and those of others has provided a valuable catalog of knowledge about how different kinds of information are organized in the brain. I think your continuous refinement of these tasks is unmatched by anyone else in the field. I have used information from the catalog on many occasions in my own work—on more than one occasion; in fact reference to your findings has set me on a more productive path than I was taking on my own.

Another thing that your retirement brings to mind is, of course, the Winter Conference on the Neurobiology of Learning and Memory. The first one I attended was in the mid 1970s; it was the first small conference I experienced, and it was a revelation. Almost all the major researchers in the field were there, one got to know them and to learn from their work and informative conversation. I have no doubt that my more-or-less regular attendance at the Conference benefitted both my research and my career. I have you (and the other organizers) to thank for that, and I am grateful for this opportunity to do so.

Finally, I remember what a good companion you were on a few occasions when we found ourselves together at a conference somewhere. I particularly remember a meal shared in Barcelona: good food, drink, and excellent company.

Thanks, Ray, for all those things.
Sincerely,
Norm

Joe Williams

Department of Psychology
Illinois Wesleyan University

I first met Ray as a skinny, 17-year-old, first-semester freshman at the University of Utah. I looked like I was 12 (as gleefully and unhelpfully pointed out to me by a Psychology faculty member every time she saw me in the elevator) and I did not weigh enough to even be able to donate blood, but I did fortunately have the attributes to work as a research assistant in Ray's lab. I had the allocentric spatial and temporal skills to show up at the right place at the right time, the linguistic skills to communicate my thoughts properly, a positive enough affect not to set off Ray's fear response system and enough egocentric response and sensory-perceptual skills to be able to carry out the responsibilities required of a research assistant. It was only much later that I developed the ability to take a scientific theory such as the attribute model and make an awkward analogy out of it.

Perhaps one of the first things that always comes up when people reminisce about Ray is the pure enthusiasm that takes over Ray when he talks about his latest idea or newest task that popped in his head as he was walking down the hall. The old lab joke was that if you rode on a ski lift with Ray, by the time you managed to get off the lift, you had three new projects you had to run when you got back to the lab. But that enthusiasm is one of the things I loved most about Ray. As a 17-year-old new to the field, to have someone you deeply respect take a random half hour out of his day to stand at a chalkboard (and yes, it was actually a chalkboard … this was before the days when dry erase markers took over the academic world) and just explain to you the workings of his mind was an inspiring event. Ray could take a blank chalkboard and fill it up in no time with new ideas. Even decades later, Ray still maintains this level of enthusiasm. When I walk down the learning and memory poster aisle at the annual SFN (Society for Neuroscience) conference, I always see Ray talking excitedly to people, both established researchers in the field and young graduate students alike, about his ideas. I knew after my first semester in Ray's lab that I wanted to switch from a premedical path to a behavioral neuroscience path. As a researcher who now has his own lab, I try to impart that same level of enthusiasm to all the students who come through my lab. If I am lucky, 20 years from now, hopefully I will have had as much an influence on my students as Ray had on the students who passed through his lab.

What also strikes me about the Ray Kesner lab experience (perhaps not as cool as the Jimi Hendrix Experience, but certainly more hippocampally awesome), is the sense of community Ray develops in his research group. The people I met in his lab 20 years ago are people I still consider as my colleagues and friends. It was a community where people were excited to share data with each other, bounce new ideas off of each other and quite often, just hang out and enjoy each other's company. Ray created an atmosphere that did not just make you enjoy coming to work every day, but one that used that sense of community to lead to incredibly rich, intellectually productive times. Just last weekend, I was asked to deliver a keynote address on successful leadership at a National Society of Leadership and Success student induction ceremony and I found myself just randomly incorporating anecdotes from my time 20 years ago in Ray's lab.

In closing, the lessons I learned in Ray's lab are still with me to this day. From the sense of community Ray built up in his lab to the pure enthusiasm Ray exhibited when talking about his latest ideas to his constant striving to develop new behavioral tasks to tackle unanswered questions—these are all things I try to impart to current students working in my research lab. Ray, I hope you realize the impact you have had not just on my career, but the careers of all the students who came through your lab. What may seem like small things to you (a brief talk in the hallway, sharing a new idea and asking undergraduates for their input, and just expressing joy and enthusiasm for scientific research) have had immeasurable influence on us all. For that, I thank you with much deeper gratitude than the words of a one-page letter can truly express.

Michael A. Yassa

Department of Neurobiology and Behavior
University of California, Irvine

Dear Ray,

I am grateful for the opportunity to thank you in this forum for your innumerable contributions over the years. Everyone thanking you in this book will undoubtedly speak of your many accomplishments in the learning and memory field over the last 40 years. They will speak of how your mentorship and supportive nature have contributed greatly to their personal and academic success. They will speak of how you organized the single most influential meeting in our field and kept it at the forefront for 36 years. The truth is, Ray, they will probably highlight these contributions far better than I ever could.

I want to thank you for something a little different. When I first came to the winter meeting in Park City as a young graduate student I was blown away. I had never been to anything like this. The intimate setting, the collegiate atmosphere, and the presence of so many pioneers in the field in the same room were nothing short of breathtaking. I was hooked. I remember speaking with you about rules for bringing graduate students and postdoctoral fellows, as I felt quite privileged to have attended so early in my career. I recall vividly your answer to this day. You said that anyone who would like to come to the meeting should be able to come. This inclusivity made me feel secure in asking to keep coming back. Since then I have never missed a meeting. I felt at home. Those words still echo in my mind as I bring my graduate students and postdocs to the meeting today to learn how science is done. In short, Ray, I want to thank you for allowing several generations of young scientists over the years to learn how science is done.

It was not just about being in the room with my idols. It was about the discussions I had with them about their work and mine. My most memorable conversations happened with you, Ray. Your work and your advice have had a profound impact on my research trajectory. Your behavioral designs targeting hippocampal and medial temporal lobe computations have inspired and continue to inspire how my lab studies memory in humans. You have been a supportive mentor to me over the years, whether it is by providing feedback on papers and grant proposals or planning an experiment with me on a napkin at a Thai restaurant in Park City. I feel privileged and honored to have had your support despite not being your student directly. I think that everyone who knows you must feel like your student in some way. To know you, Ray, is truly to learn from you, and I have learned a great deal from you and hope to continue to learn more.

Finally, I wish to congratulate you on your illustrious career and myriad accomplishments, which have shaped the learning and memory field in powerful ways. Here is to a happy "retirement." I wish you and your family all the best.

Mike Yassa

References

DeCoteau, W. E., Kesner, R. P., & Williams, J. M. (1997). Short-term memory for food reward magnitude: The role of the prefrontal cortex. *Behavioural Brain Research, 88,* 239–249.

Kesner R. P. (2009). Tapestry of memory. *Behavioral Neuroscience, 123,* 1–13.

Chapter 18
Epilogue

Pamela A. Jackson, Andrea A. Chiba, Robert F. Berman and Michael E. Ragozzino

Understanding how our individual experiences are remembered has been an enduring interest and mystery to our species. This lasting intrigue in trying to understand the nature of memory may result from our ability to learn and remember a plethora of information that fundamentally shapes and defines who we are as individuals. The emergence of neuroscience focused attention on how the brain represents memory. One contemporary view is that memory is a fundamental property that emerges from the operations of the brain and that this fundamental property of memory is distributed throughout the brain in various neural systems (Fuster 1999; Eichenbaum and Cohen 2001). This conceptualization has led to the idea that memory as a basic component of various brain operations leads to multiple forms of memory represented in numerous, but distinct neural systems. In the neurobiology of learning and memory field, this is simply referred to as multiple memory systems.

Ray Kesner was truly a pioneer in developing one of the earliest and most comprehensive neurobiological models of multiple memory systems. From its original conceptualization, the Kesner attribute model viewed the neurobiology of memory beyond a single brain system; a single type of memory and beyond a single temporal domain, e.g. working memory. Instead, the model proposed that memory can be represented as various types of attribute information in both event-based and

P. A. Jackson (✉)
Department of Psychology, Radford University, P.O. Box 6946, Radford, VA 24142, USA
e-mail: pjackson@radford.edu

A. A. Chiba
Department of Cognitive Science and Program in Neuroscience, Temporal Dynamics of Learning Center, University of California, San Diego, 9500 Gilman Drive, La Jolla, CA 92093-0515, USA
e-mail: achiba@ucsd.edu

R. F. Berman
Department of Neurological Surgery, University of California, Davis, 1515 Newton Ct., Davis, CA 95616, USA

M. E. Ragozzino
Department of Psychology, University of Illinois at Chicago, Chicago, IL 60607, USA

P. A. Jackson et al. (eds.), *The Neurobiological Basis of Memory*,
DOI 10.1007/978-3-319-15759-7_18

knowledge-based systems that are supported by different brain systems. A major strength of this model was the wealth of knowledge gathered about how various cortical and subcortical systems process information underlying memory. Arguably, the Kesner attribute model has led to the broadest exploration of how memory is represented in the brain compared to other neurobiological models of memory.

Another major strength of the attribute model of memory is its evolution overtime, integrating empirical results based on tests of the model and findings from the broader neuroscience field. While early studies based on the attribute model often focused on a brain region as a whole, subsequent experiments investigated how various subregions or systems within a brain area support memory. Applying the attribute model in this manner has led to a deeper understanding of how neural systems underlie memory, as well as how different subregions in a brain area act in a complementary manner to support various mnemonic functions. In more recent versions of the attribute model of memory, different processes were added to explain how specific neural systems process different types of attribute information. This included concepts like pattern separation and pattern completion to explain how hippocampal circuitry supports learning and memory in the spatial domain. The latest version of the attribute model includes not only event-based and knowledge-based systems, but also a rule-based system. This represents a core memory system that uses attribute information based on specific cognitive operations to be performed. Thus, the Kesner attribute model of memory has expanded over time to build a more comprehensive framework that allows a rich path to view and explore the neurobiology of learning and memory.

The breadth and depth of the Kesner attribute model of memory provides a structure to investigate the neurobiology of learning and memory resulting in many novel and impactful findings about how memory is represented in the brain. Still there is much to be learned about the neurobiology of learning and memory that can be accomplished by further testing of the Kesner attribute model. Early findings largely emerged from studies employing lesions. Neuroscience continues to expand the number of techniques used to study the brain. Significant advances in understanding the neurobiology of learning and memory can be obtained by testing the Kesner attribute model of memory with the use of cutting edge neuroscience techniques. Furthermore, there is an increasing use of animal models of various neurological and psychiatric disorders. Many of these disorders are characterized by deficits in mnemonic functioning. Here is another area of neuroscience in which the Kesner model can be exploited to better understand the cognitive deficits in various disorders, as well as to understand whether certain treatments alleviate the cognitive deficits in animal models of neurological and psychiatric disorders. Finally, the quest to study the neurobiology of memory has been motivated to comprehend how the brain represents memory in humans. The Kesner attribute model has been largely tested in nonhuman subjects, although there have been some studies done in humans. As more neuroscience tools become available for studying human brain functions, utilizing these tools to test the Kesner attribute model offers the possibility of shining new light on one of our enduring mysteries.

References

Eichenbaum, H., & Cohen, N. J. (2001). *From conditioning to conscious recollection: Memory systems of the brain*. Oxford University Press.

Fuster, J. M. (1999). *Memory in the cerebral cortex: An empirical approach to neural networks in the human and nonhuman primate*. Cambridge: MIT press.

Index

P. A. Jackson et al. (eds.), *The Neurobiological Basis of Memory*,
DOI 10.1007/978-3-319-15759-7

MIX
Papier aus verantwortungsvollen Quellen
Paper from responsible sources
FSC® C105338

If you have any concerns about our products,
you can contact us on
ProductSafety@springernature.com

In case Publisher is established outside the EU,
the EU authorized representative is:
Springer Nature Customer Service Center GmbH
Europaplatz 3, 69115 Heidelberg, Germany

Printed by Libri Plureos GmbH
in Hamburg, Germany